CONTINUOUS ADVANCES IN QCD 2008

CONTINUOUS ADVANCES IN QCD 2008

Proceedings of the Conference

William I. Fine Theoretical Physics Institute
University of Minnesota, USA 15 – 18 May 2008

Editor

M. Peloso

University of Minnesota, USA

World Scientific

NEW JERSEY • LONDON • SINGAPORE • BEIJING • SHANGHAI • HONG KONG • TAIPEI • CHENNAI

Published by

World Scientific Publishing Co. Pte. Ltd.
5 Toh Tuck Link, Singapore 596224
USA office: 27 Warren Street, Suite 401-402, Hackensack, NJ 07601
UK office: 57 Shelton Street, Covent Garden, London WC2H 9HE

British Library Cataloguing-in-Publication Data
A catalogue record for this book is available from the British Library.

CONTINUOUS ADVANCES IN QCD 2008

ISBN-13 978-981-283-865-0
ISBN-10 981-283-865-1

Printed in Singapore by World Scientific Printers

FOREWORD

The Eighth Workshop "Continuous Advances in QCD 2008," was held at William I. Fine Theoretical Physics Institute, University of Minnesota, on May 15-18, 2008. It was the latest in a series, begun in 1994, which has grown into a major international event. The Workshop has attracted the most active researchers working at the cutting edge of QCD and more generally on the dynamics of gauge theories.

The main themes of the workshop were the latest high-order calculations in QCD and in more general gauge theories, the physics of strongly interacting particles containing heavy quarks, the behavior of quarks and gluons under the extreme conditions of high temperature and/or density, and the novel methods of studying gauge filed dynamics using its correspondence with a higher-dimensional gravity theory. These areas of research and the problems at their intersection are currently the most rapidly developing, and the Workshop has attracted the leading experts in these areas.

The new fascinating results of complex high-order calculations presented at the Workshop find their applications, in addition to QCD proper, in a broad variety of subjects ranging from positronium molecules and ions to supergravity. The heavy quark physics, including that of charmed and beauty hadrons as well as heavy quarkonia attracts an unprecedented interest due to the abundance of new experimental data, some of those being quite unexpected. The talks by the experts in this field presented new ideas on such topics as understanding the new types of heavy hadrons, oscillations of charm and beauty, and others.

The behavior of matter made of quarks and gluons at high temperature and/or density opens the ways of understanding phenomena in the early Universe in super dense stars and in the collisions of heavy ions experimentally studied at the collider RHIC in Brookhaven. The leaders in these studies have presented new ideas and perspectives in this field.

The newly developed method of study based on holographic correspondence were presented by the pioneers of this method in its general aspects as well as in its application to the high density quark-gluon matter and to description of the hadronic spectra in QCD. Traditionally, a lively interest was attracted by the talks on general aspects of gauge dynamics, including the supersymmetric versions and topological configurations.

By and large, it is fair to say that the Workshop has further developed "interdisciplinary" studies in the most vibrant areas of high-energy theory and made a significant contribution towards cross-fertilization of ideas of string/brane theory, traditional gauge field theory and the studies of specific particles, processes and phenomena.

Most of the electronic slides of the talks are available online. The files as well as additional information about the Workshop can be accessed from the URL

http://www.ftpi.umn.edu/qcd08/

M. Voloshin
September 17, 2008

CONTENTS

SECTION 1.

HIGH-ORDER CALCULATIONS IN QCD AND IN GENERAL GAUGE THEORIES

NLO EVOLUTION OF COLOR DIPOLES

I. BALITSKY*

Physics Dept, Old Dominion Univ., Norfolk, VA 23529, and
Theory Group, JLab, 12000 Jefferson Ave, Newport News, VA 23606
**E-mail: balitsky@jlab.org*

The small-x deep inelastic scattering in the saturation region is governed by the non-linear evolution of Wilson-line operators. In the leading logarithmic approximation it is given by the BK equation for the evolution of color dipoles. I discuss recent calculation of the next-to-leading order evolution of color dipoles in QCD and $\mathcal{N} = 4$ SYM.

Keywords: Small-x evolution, Wilson lines.

1. Small-x evolution of color dipoles

A general feature of high-energy scattering is that a fast particle moves along its straight-line classical trajectory and the only quantum effect is the eikonal phase factor acquired along this propagation path. In QCD, for the fast quark or gluon scattering off some target, this eikonal phase factor is a Wilson line - the infinite gauge link ordered along the straight line collinear to particle's velocity n^μ:

$$U^\eta(x_\perp) = \mathrm{Pexp}\Big\{ig\int_{-\infty}^{\infty} du\; n_\mu\; A^\mu(un + x_\perp)\Big\}, \tag{1}$$

Here A_μ is the gluon field of the target, $x_\perp$ is the transverse position of the particle which remains unchanged throughout the collision, and the index η labels the rapidity of the particle. Repeating the above argument for the target (moving fast in the spectator's frame) we see that particles with very different rapidities perceive each other as Wilson lines and therefore these Wilson-line operators form the convenient effective degrees of freedom in high-energy QCD (for a review, see ref.[1]).

Let us consider the deep inelastic scattering from a hadron at small $x_B = Q^2/(2p{\cdot}q)$. The virtual photon decomposes into a pair of fast quarks moving along straight lines separated by some transverse distance. The propagation of this quark-antiquark pair reduces to the "propagator of the color dipole" $U(x_\perp)U^\dagger(y_\perp)$ - two Wilson lines ordered along the direction collinear to quarks' velocity. The structure function of a hadron is proportional to a matrix element of this color dipole operator

$$\hat{\mathcal{U}}^\eta(x_\perp, y_\perp) = 1 - \frac{1}{N_c}\mathrm{Tr}\{\hat{U}^\eta(x_\perp)\hat{U}^{\dagger\eta}(y_\perp)\} \tag{2}$$

switched between the target's states ($N_c = 3$ for QCD). (As usual, we denote operators by "hat"). The energy dependence of the structure function is translated then into the dependence of the color dipole on the slope of the Wilson lines determined by the rapidity η.

To find the rapidity evolution of the color dipole (2) we consider the matrix element of this color dipole between (arbitrary) target states and integrate over the gluons with rapidities $\eta_1 > \eta > \eta_2 = \eta_1 - \Delta\eta$ leaving the gluons with $\eta < \eta_2$ as a background field (to be integrated over later). In the frame with $\eta \sim \eta_1$ the fields with $\eta < \eta_2$ shrink to a pancake and we obtain the four diagrams shown in Fig. 1. Technically, to find the kernel in the leading-ordrer approximation we write down the general form of the operator equation for the evolution of the color dipole

$$\frac{d}{d\eta}\mathrm{Tr}\{\hat{U}_x\hat{U}_y^\dagger\} = K_{\mathrm{LO}}\mathrm{Tr}\{\hat{U}_x\hat{U}_y^\dagger\} + ... \tag{3}$$

(where dots stand for the higher orders of the expansion) and calculate the l.h.s. of Eq. (3) in the shock-wave background

$$\frac{d}{d\eta}\langle\mathrm{Tr}\{\hat{U}_x\hat{U}_y^\dagger\}\rangle_{\mathrm{shockwave}} = \langle K_{\mathrm{LO}}\mathrm{Tr}\{\hat{U}_x\hat{U}_y^\dagger\}\rangle_{\mathrm{shockwave}} \tag{4}$$

In what follows we replace $\langle ...\rangle_{\mathrm{shockwave}}$ by $\langle ...\rangle$ for brevity.

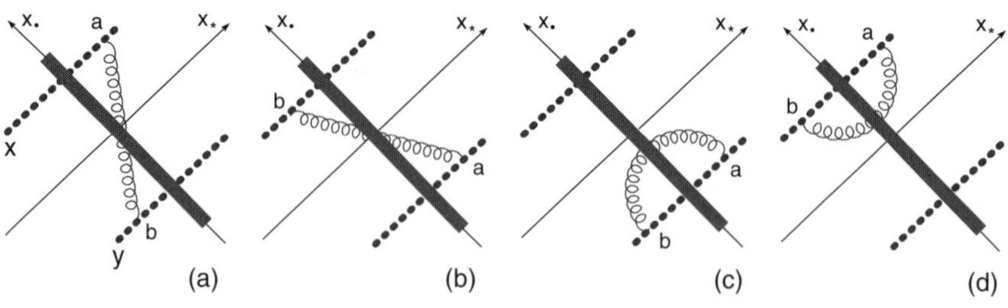

Fig. 1. Leading-order diagrams for the small-x evolution of color dipole. Gauge links are denoted by dotted lines.

Throughout the paper, we use the Sudakov variables $p = \alpha p_1 + \beta p_2 + p_\perp$ with the momenta p_1 and p_2 being the light-like vectors such that $q = p_1 - x_B p_2$ and $p = p_2 + \frac{m^2}{s} p_1$ where p is the momentum of the target and m is the mass. The Sudakov variables for coordinates are $x_\bullet \equiv x_\mu p_1^\mu$ and $x_* \equiv x_\mu p_2^\mu$ related to the light-cone coordinates: $x_* = x^+ \sqrt{s/2}$, $x_\bullet = x^- \sqrt{s/2}$.

The gluon propagator in a shock-wave external field has the form[2–4]

$$\langle \hat{A}_\mu^a(x) \hat{A}_\nu^b(y) \rangle \stackrel{x_*>0>y_*}{=} - \int_0^\infty đ\alpha \, \frac{e^{-i\alpha(x-y)_\bullet}}{2\alpha} (x_\perp| e^{-i\frac{p_\perp^2}{\alpha s} x_*} \tag{5}$$
$$\times \Big[g_{\mu\xi}^\perp - \frac{2}{\alpha s} (p_\mu^\perp p_{2\xi} + p_{2\mu} p_\xi^\perp) \Big] U^{ab} \Big[g_\nu^{\perp\xi} - \frac{2}{\alpha s} (p_2^\xi p_\nu^\perp + p_{2\nu} p_\perp^\xi) \Big] e^{i\frac{p_\perp^2}{\alpha s} y_*} |y_\perp)$$

where we use the light-like $p_2^\mu A_\mu = 0$ gauge (the propagator at $x_*, y_* > 0$ is a bare propagator). Hereafter we use Schwinger's notations $(x_\perp|F(p_\perp)|y_\perp) \equiv \int đp \, e^{i(p,x-y)_\perp} F(p_\perp)$ (the scalar product of the four-dimensional vectors in our notations is $x \cdot y = \frac{2}{s}(x_* y_\bullet + x_* y_\bullet) - (x,y)_\perp$). Note that the interaction with the shock wave does not change the α-component of the gluon momentum.

To restrict the rapidity of the color dipole we impose the cutoff on the maximal α emitted by any gluon from the Wilson lines so

$$\begin{aligned} U_x^\eta &= \text{Pexp}\Big[ig \int_{-\infty}^\infty du p_1^\mu A_\mu^\eta(up_1 + x_\perp) \Big] \\ A_\mu^\eta(x) &= \int đ^4 k \theta(e^\eta - |\alpha_k|) e^{-ik\cdot x} A_\mu(k) \end{aligned} \tag{6}$$

We obtain

$$\begin{aligned} \langle \hat{U}_x^\eta \hat{U}_y^{\dagger\eta} \rangle_{\text{Fig.1a}} &= g^2 \int_0^\infty du \int_{-\infty}^0 dv \, \langle \hat{A}_\bullet^a(un + x_\perp) \hat{A}_\bullet^b(vn + y_\perp) \rangle_{\text{Fig.1a}} \\ = \, - 4\alpha_s \int_0^{e^{-\eta_1}} \frac{d\alpha}{\alpha} (x_\perp| \frac{p_i}{p_\perp^2} U^{ab} \frac{p_i}{p_\perp^2} |y_\perp) &= \, - 4\alpha_s \Delta\eta (x_\perp| \frac{p_i}{p_\perp^2} U^{ab} \frac{p_i}{p_\perp^2} |y_\perp) \end{aligned}$$

(Formally, the integral over α diverges at the lower limit, but since we integrate over the rapidities $\eta > \eta_2$ we get $\Delta\eta$ in the LLA) and therefore

$$\langle \hat{U}_x \otimes \hat{U}_y^\dagger \rangle_{\text{Fig.1a}}^{\eta_1} = - \frac{\alpha_s}{\pi^2} \Delta\eta \, (t^a U_x \otimes t^b U_y^\dagger) \int d^2 z_\perp \frac{(x-z, y-z)_\perp}{(x-z)_\perp^2 (y-z)_\perp^2} U_z^{ab} \tag{7}$$

The contribution of the diagram in Fig. 1b is obtained from Eq. (7) by the replacement $t^a U_x \otimes t^b U_y^\dagger \to U_x t^b \otimes U_y^\dagger t^a$, $x \leftrightarrow y$ and the two remaining diagrams are obtained from Eq. 7 by taking $y = x$ (Fig. 1c) and $x = y$ (Fig.

1d). After some algebra one obtains

$$\langle \mathrm{Tr}\{\hat{U}_x\hat{U}_y^\dagger\}\rangle_{\mathrm{Fig.1}}^{\eta_1} \tag{8}$$
$$= \frac{\alpha_s\Delta\eta}{2\pi^2}\int d^2z_\perp \frac{(x-y)_\perp^2}{(x-z)_\perp^2(y-z)_\perp^2}[\mathrm{Tr}\{U_xU_z^\dagger\}\mathrm{Tr}\{U_zU_y^\dagger\} - \frac{1}{N_c}\mathrm{Tr}\{U_xU_y^\dagger\}]$$

There are also contributions coming from the diagrams shown in Fig. 2 (plus graphs obtained by reflection with respect to the shock wave).

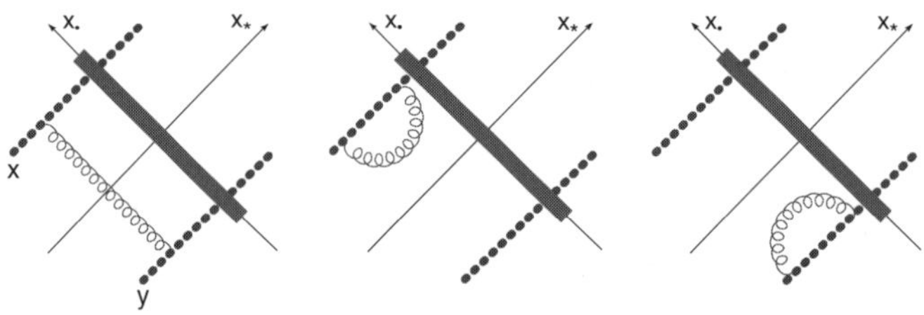

Fig. 2. Leading-order diagrams proportional to the original dipole.

These diagrams are proportional to the original dipole $\mathrm{Tr}\{U_xU_y^\dagger\}$ and therefore the corresponding term can be derived from the contribution of Fig. 1 graphs using the requirement that the r.h.s. of the evolution equation should vanish for $x = y$ since $\lim_{x\to y}\frac{d}{d\eta}\mathrm{Tr}\{U_xU_y^\dagger\} = 0$). It is easy to see that this requirement leads to

$$\langle \mathrm{Tr}\{\hat{U}_x\hat{U}_y^\dagger\}\rangle^{\eta_1} \tag{9}$$
$$= \frac{\alpha_s\Delta\eta}{2\pi^2}\int d^2z_\perp \frac{(x-y)_\perp^2}{(x-z)_\perp^2(y-z)_\perp^2}[\mathrm{Tr}\{U_xU_z^\dagger\}\mathrm{Tr}\{U_zU_y^\dagger\} - N_c\mathrm{Tr}\{U_xU_y^\dagger\}]$$

which is equivalent to the BK equation for the evolution of the color dipoles:[2,5]

$$\frac{d}{d\eta}\hat{\mathcal{U}}(x,y) = \frac{\alpha_sN_c}{2\pi^2}\int d^2z\,\frac{(x-y)^2}{(x-z)^2(z-y)^2}$$
$$\times\ [\hat{\mathcal{U}}(x,z)+\hat{\mathcal{U}}(y,z)-\hat{\mathcal{U}}(x,y)-\hat{\mathcal{U}}(x,z)\hat{\mathcal{U}}(z,y)] \tag{10}$$

The first three terms correspond to the linear BFKL evolution[6] and describe the parton emission while the last term is responsible for the parton annihilation. For sufficiently high x_B the parton emission balances the parton annihilation so the partons reach the state of saturation[7] with the characteristic transverse momentum Q_s growing with energy $1/x_B$ (for a review, see[8]).

2. NLO evolution of color dipoles in QCD

As usual, to get the region of application of the leading-order evolution equation one needs to find the next-to-leading order (NLO) corrections. In the case of the small-x evolution equation (10) there is another reason why NLO corrections are important. Unlike the DGLAP evolution, the argument of the coupling constant in Eq. (10) is left undetermined in the LLA, and usually it is set by hand to be Q_s. Careful analysis of this argument is very important from both theoretical and experimental points of view. From the theoretical viewpoint, we need to know whether the coupling constant is determined by the size of the original dipole $|x-y|$ or of the size of the produced dipoles $|x-z|$ and/or $|z-y|$ since we may get a very different behavior of the solutions of the equation (10). On the experimental side, the cross section is proportional to some power of the coupling constant so the argument determines how big (or how small) is the cross section. The typical argument of α_s is the characteristic transverse momenta of the process. For high enough energies, they are of order of the saturation scale Q_s which is $\sim 2\div 3$ GeV for the LHC collider, so even the difference between $\alpha(Q_s)$ and $\alpha(2Q_s)$ can make a substantial impact on the cross section. The precise form of the argument of α_s should come from the solution of the BK equation with the running coupling constant, and the starting point of the analysis of the argument of α_s in Eq. (10) is the calculation of the NLO evolution.

In the next-to-leading order there are three types of diagrams. Diagrams of the first type have two intersections of the emitted gluons with the shock wave, diagrams of the second type have one intersection, and finally there are diagrams of the third type without intersections. Some typical diagrams are shown in Figs. 3,4 and 5.

The calculation of these diagrams using the propagator in the shock-wave background (5) is straightforward except for two subtle points.

First, the typical NLO contribution (comong from the diagram in Fig. 3I) has the form

$$\frac{d}{d\eta}\langle \mathrm{Tr}\{\hat U_x \hat U^\dagger_y\}\rangle \tag{11}$$

$$= \frac{\alpha_s^2}{2\pi^4}\int_0^\sigma \frac{d\alpha_2}{\alpha_2}\int\frac{d^2z d^2z'}{(z-z')^2}\,U_z^{bb'}U_{z'}^{cc'}\,\frac{\sigma\alpha_1 f^{abc}f^{a'b'c'}(X,Y)\mathrm{Tr}\{t^aU_xt^{a'}U^\dagger_y\}}{(\alpha_1X^2+\alpha_2X'^2)(\alpha_1Y^2+\alpha_2Y'^2)}$$

$$= \frac{\alpha_s^2}{4\pi^4}\int_0^1\frac{du}{\bar u}\int d^2zd^2z'\;U_z^{bb'}U_{z'}^{cc'}\,\frac{uf^{abc}f^{a'b'c'}\mathrm{Tr}\{t^aU_xt^{a'}U^\dagger_y\}(X,Y)}{(z-z')^2(uX^2+\bar uX'^2)(uY^2+\bar uY'^2)}$$

where $z=z_\perp$ and $z'=z'_\perp$ are the points of intersection of the gluon

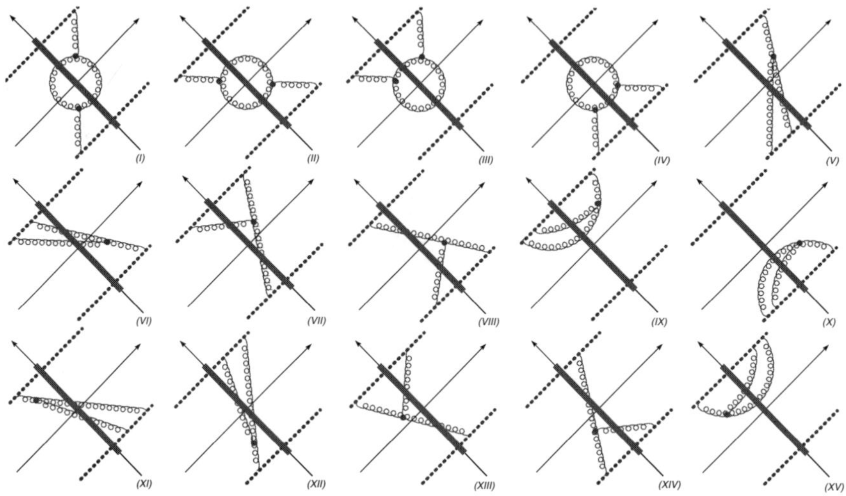

Fig. 3. Typical diagrams with two gluon-shockwave interactions.

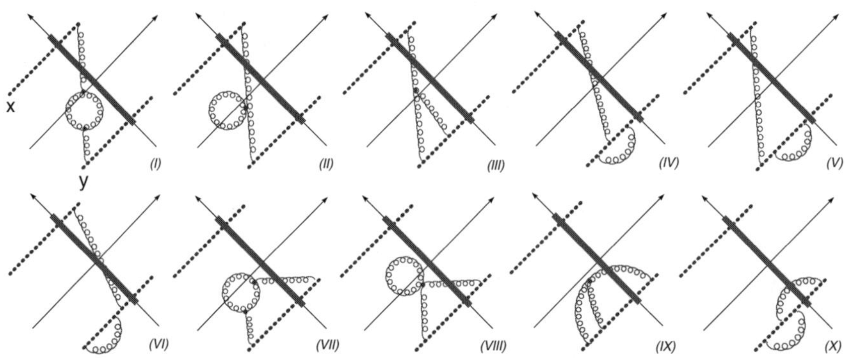

Fig. 4. Typical diagrams with one gluon-shockwave interaction.

propagators with the shock wave (hereafter we use notations $X \equiv x - z$, $X' \equiv x - z'$, $Y \equiv y - z$, and $Y' \equiv y - z'$). This integral diverges as $\bar{u} \to 0$ which corresponds to $\alpha_1 \gg \alpha_2$. This region, however, is taken into account in the LLA approximation and should be subtracted while constructing the NLO kernel.

Indeed, if we put a lower cutoff $\alpha > \sigma'$ on the α integrals we would get a contribution $\sim \ln^2 \frac{\sigma}{\sigma'}$ coming from the region $\alpha_2 \gg \alpha_1 > \sigma'$ (or $\alpha_1 \gg \alpha_2 > \sigma'$) which corresponds to the the square of the leading-order

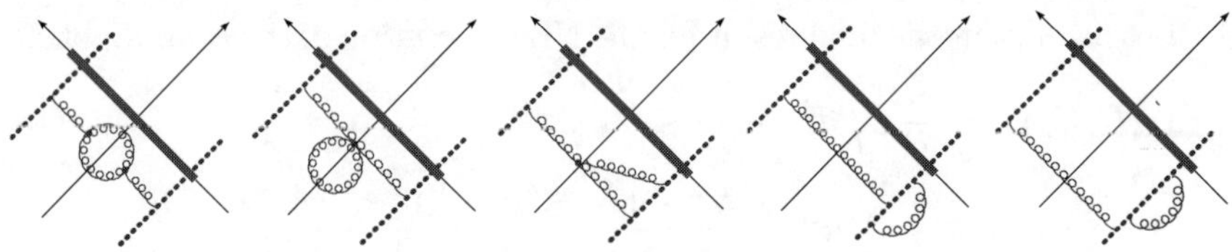

Fig. 5. Typical diagrams without gluon-shockwave interactions.

BK kernel rather than to the NLO kernel. To get the NLO kernel we need to subtract this $(LO)^2$ contribution. Indeed, the operator form of the evolution equation for the color dipole up to the next-to-leading order looks like

$$\frac{d}{d\eta}\mathrm{Tr}\{\hat{U}_x\hat{U}^\dagger_y\} = K_{\mathrm{LO}}\mathrm{Tr}\{\hat{U}_x\hat{U}^\dagger_y\} + K_{\mathrm{NLO}}\mathrm{Tr}\{\hat{U}_x\hat{U}^\dagger_y\} \tag{12}$$

where $\eta = \ln\sigma$. Our goal is to find K_{NLO} by considering the l.h.s. of this equation in the external shock-wave background $\bar{A}$ so

$$\langle K_{\mathrm{NLO}}\mathrm{Tr}\{\hat{U}_x\hat{U}^\dagger_y\}\rangle_{\bar{A}} = \frac{d}{d\eta}\langle\mathrm{Tr}\{\hat{U}_x\hat{U}^\dagger_y\}\rangle_{\bar{A}} - \langle K_{\mathrm{LO}}\mathrm{Tr}\{\hat{U}_x\hat{U}^\dagger_y\}\rangle_{\bar{A}} \tag{13}$$

The subtraction (13) leads to the $\left[\frac{1}{u}\right]_+$ prescription for the terms divergent as $\frac{1}{u}$ (and similarly $\frac{1}{\bar{u}} \to \left[\frac{1}{\bar{u}}\right]_+$ for the contribution divergent as $u \to 1$).

Another subtle point is related to the UV divergence of the r.h.s. of Eq. (11) at $z \to z'$. It is convenient to separate the divergent term by subtracting and adding the contribution at $z = z'$:

$$U^{bb'}_z U^{\dagger cc'}_{z'} = (U^{bb'}_z U^{\dagger cc'}_{z'} - U^{bb'}_z U^{\dagger cc'}_z) + U^{bb'}_z U^{\dagger cc'}_z \tag{14}$$

The first term is now UV finite while the second term looks like the contribution of the diagrams with one intersecton. It is combined with similar contributions coming from Fig. 4 to form the renormalization of coupling constant.

Let us present the final result for the NLO evolution of the color dipole[14]

$$\frac{d}{d\eta}\mathrm{Tr}\{\hat U_x\hat U^\dagger_y\} = \frac{\alpha_s}{2\pi^2}\int d^2z\,\frac{(x-y)^2}{X^2Y^2} \tag{15}$$
$$\times\Big\{1+\frac{\alpha_s}{4\pi}\Big[b\ln(x-y)^2\mu^2 - b\frac{X^2-Y^2}{(x-y)^2}\ln\frac{X^2}{Y^2}+(\frac{67}{9}-\frac{\pi^2}{3})N_c-\frac{10}{9}n_f$$
$$-\,2N_c\ln\frac{X^2}{(x-y)^2}\ln\frac{Y^2}{(x-y)^2}\Big]\Big\}\,[\mathrm{Tr}\{\hat U_x\hat U^\dagger_z\}\mathrm{Tr}\{\hat U_z\hat U^\dagger_y\}-N_c\mathrm{Tr}\{\hat U_x\hat U^\dagger_y\}]$$
$$+\,\frac{\alpha_s^2}{16\pi^4}\int d^2z d^2z'\Big[\Big(-\frac{4}{(z-z')^4}+\Big\{2\frac{X^2Y'^2+X'^2Y^2-4(x-y)^2(z-z')^2}{(z-z')^4[X^2Y'^2-X'^2Y^2]}$$
$$+\,\frac{(x-y)^4}{X^2Y'^2-X'^2Y^2}\Big[\frac{1}{X^2Y'^2}+\frac{1}{Y^2X'^2}\Big]$$
$$+\,\frac{(x-y)^2}{(z-z')^2}\Big[\frac{1}{X^2Y'^2}-\frac{1}{X'^2Y^2}\Big]\Big\}\ln\frac{X^2Y'^2}{X'^2Y^2}\Big)$$
$$\times\,[\mathrm{Tr}\{\hat U_x\hat U^\dagger_z\}\mathrm{Tr}\{\hat U_z\hat U^\dagger_{z'}\}\mathrm{Tr}\{\hat U_{z'}\hat U^\dagger_y\}-\mathrm{Tr}\{\hat U_x\hat U^\dagger_z\hat U_{z'}U^\dagger_y\hat U_z\hat U^\dagger_{z'}\}-(z'\to z)]$$
$$+\,\Big\{\frac{(x-y)^2}{(z-z')^2}\Big[\frac{1}{X^2Y'^2}+\frac{1}{Y^2X'^2}\Big]-\frac{(x-y)^4}{X^2Y'^2X'^2Y^2}\Big\}\ln\frac{X^2Y'^2}{X'^2Y^2}$$
$$\mathrm{Tr}\{\hat U_x\hat U^\dagger_z\}\mathrm{Tr}\{\hat U_z\hat U^\dagger_{z'}\}\mathrm{Tr}\{\hat U_{z'}\hat U^\dagger_y\}$$
$$+\,4n_f\Big\{\frac{4}{(z-z')^4}-2\frac{X'^2Y^2+Y'^2X^2-(x-y)^2(z-z')^2}{(z-z')^4(X^2Y'^2-X'^2Y^2)}\ln\frac{X^2Y'^2}{X'^2Y^2}\Big\}$$
$$\times\,\mathrm{Tr}\{t^a\hat U_x t^b\hat U^\dagger_y\}[\mathrm{Tr}\{t^a\hat U_z t^b\hat U^\dagger_{z'}\}-(z'\to z)]\Big]$$

Here μ is the normalization point in the $\overline{MS}$ scheme and $b=\frac{11}{3}N_c-\frac{2}{3}n_f$ is the first coefficient of the β-function (the quark part of Eq. (15) proportional to n_f was found earlier[10,11]). The NLO kernel is a sum of the running-coupling part (proportional to b), the non-conformal double-log term $\sim\ln\frac{(x-y)^2}{(x-z)^2}\ln\frac{(x-y)^2}{(x-z)^2}$ and the three conformal terms which depend on the two four-point conformal ratios $\frac{X^2Y'^2}{X'^2Y^2}$ and $\frac{(x-y)^2(z-z')^2}{X^2Y'^2}$. Note that the logarithm of the second conformal ratio $\ln\frac{(x-y)^2(z-z')^2}{X^2Y'^2}$ is absent.

It should be emphasized that the NLO result itself does not lead automatically to the argument of coupling constant α_s in Eq. 10. In order to get this argument one can use the renormalon-based approach:[12] first get the quark part of the running coupling constant coming from the bubble chain of quark loops and then make a conjecture that the gluon part of the β-function will follow that pattern. The Eq. (15) proves this conjecture in the first nontrivial order: the quark part of the β - function $\frac{2}{3}n_f$

calculated earlier gets promoted to full b. The analysis of the argument of the coupling constant was performed in Refs.[10,11] andr, oughly speaking, the argument of α_s is determined by the size of the smallest dipole $min(|x-y|,|x-z|,|y-z|)$.

It is possible to compare linearized NLO BK equation (15) with NLO BFKL in the case of forward scattering. The result (15) is in agreement with NLO BFKL equation[13] up to up to a term proportional $\alpha_s^2\zeta(3)$ times the original dipole. We think that the difference could be due to different definitions of the cutoff in the longitudinal momenta.

3. NLO evolution of color dipoles in $\mathcal{N}=4$ SYM

In $\mathcal{N}=4$ SYM theory we have two additional types of diagrams: with scalar loops and with gluino loops. Both of them look like diagrams with two and one intersections shown in Fig. 6 and Fig. 7, respectively.

Fig. 6. Diagrams with scalar/gluino loops bisected by the shock wave.

The calculation of these diagrams is similar to (but much simpler than) the QCD case and we will present here only the final result for the NLO

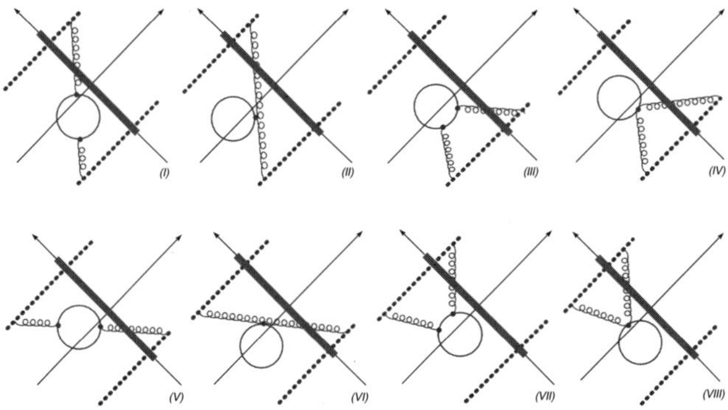

Fig. 7. Diagrams with one gluon-shockwave intsection.

evolution of the color dipole in $\mathcal{N}=4$ SYM:

$$\frac{d}{d\eta}\mathrm{Tr}\{\hat U_x\hat U^\dagger_y\} \tag{16}$$
$$= \frac{\alpha_s}{2\pi^2}\int d^2z\,\frac{(x-y)^2}{X^2Y^2}\Big\{1+\frac{\alpha_sN_c}{4\pi}\Big[\frac{1-\pi^2}{3} - 2\ln\frac{X^2}{(x-y)^2}\ln\frac{Y^2}{(x-y)^2}\Big]\Big\}$$
$$\times\,[\mathrm{Tr}\{\hat U_x\hat U^\dagger_z\}\mathrm{Tr}\{\hat U_z\hat U^\dagger_y\}-N_c\mathrm{Tr}\{\hat U_x\hat U^\dagger_y\}]$$
$$+\frac{\alpha_s^2}{16\pi^4}\int d^2zd^2z'\Big\{\frac{(x-y)^4}{[X^2Y'^2-X'^2Y^2]X^2Y'^2}+\frac{(x-y)^2}{(z-z')^2}\frac{1}{X^2Y'^2}\Big\}\ln\frac{X^2Y'^2}{X'^2Y^2}$$
$$\times\,[\mathrm{Tr}\{\hat U_x\hat U^\dagger_z\}\mathrm{Tr}\{\hat U_z\hat U^\dagger_{z'}\}\mathrm{Tr}\{\hat U_{z'}\hat U^\dagger_y\}-\mathrm{Tr}\{\hat U_x\hat U^\dagger_z\hat U_{z'}U^\dagger_y\hat U_z\hat U^\dagger_{z'}\}$$
$$-\,\mathrm{Tr}\{\hat U_x\hat U^\dagger_{z'}\hat U_zU^\dagger_y\hat U_{z'}\hat U^\dagger_z\}-(z'\to z)]$$

The scalar and gluin contributions cancel two of the four terms in the QCD kernel so the $\mathcal{N}=4$ NLO kernel is a sum of the non-conformal double-log term $\sim\ln\frac{(x-y)^2}{(x-z)^2}\ln\frac{(x-y)^2}{(x-z)^2}$ multiplied by the LO $1\to2$ dipoles color structure and the conformal term describing the $1\to3$ dipoles transition which depend on the two four-point conformal ratios $\frac{X^2Y'^2}{X'^2Y^2}$ and $\frac{(x-y)^2(z-z')^2}{X^2Y'^2}$.

In conclusion I would like to discuss the conformal invariance of the evolution equation (16). Formally, he light-like Wilson line is invariant inder SL(2,C) group of conformal transformations of the transverse plane. For example, it is easy to see that the inversion $x_\mu\to x_\mu/x^2$ with respect $(x^+,x_\perp)^2=-x_\perp^2$ so after the inversion $x_\perp\to x_\perp/x_\perp^2$ and $x^+\to x^+/x_\perp^2$

and therefore

$$[\infty p_1 + x_\perp, -\infty p_1 + x_\perp] \qquad (17)$$
$$\rightarrow \text{Pexp}\left\{ig\int_{-\infty}^{\infty} d\frac{x^+}{x_\perp^2}\, A_+(\frac{x^+}{x_\perp^2}, x_\perp)\right\} = [\infty p_1 + x_\perp, -\infty p_1 + x_\perp]$$

It looks like one should expect that the corresponding evolution kernel is conformal. It should be emphasized, however, that the matrix elements of the light-like Wilson-line operators diverge in the longitudinal direction, and when we impose cutoff in rapidity we destroy the conformal invariance. The conformal invariance of the $\mathcal{N} = 4$ amplitudes should berestored after multiplication of the evolution kernel by the coefficient functions of the high-energy operator expansion - the so-called "impact factors". The study is in progess.

Acknowledgments

This work was supported by contract DE-AC05-84ER40150 under which the Southeastern Universities Research Association (SURA) operates the Thomas Jefferson National Accelerator Facility.

References

1. I. Balitsky, *"High-Energy QCD and Wilson Lines"*, In *Shifman, M. (ed.): At the frontier of particle physics, vol. 2*, p. 1237-1342 (World Scientific, Singapore,2001) [hep-ph/0101042]
2. I. Balitsky, *Nucl. Phys.* **B463**, 99 (1996); *"Operator expansion for diffractive high-energy scattering"*, [hep-ph/9706411];
3. I. Balitsky, *Phys. Rev.* **D60**, 014020 (1999).
4. I. Balitsky and A.V. Belitsky, *Nucl. Phys.* **B629**, 290 (2002).
5. Yu.V. Kovchegov, *Phys. Rev.* **D60**, 034008 (1999); *Phys. Rev.* **D61**,074018 (2000).
6. V.S. Fadin, E.A. Kuraev, and L.N. Lipatov, *Phys. Lett.* **B 60**, 50 (1975); I. Balitsky and L.N. Lipatov, *Sov. Journ. Nucl. Phys.* **28**, 822 (1978).
7. L.V. Gribov, E.M. Levin, and M.G. Ryskin, *Phys. Rept.* **100**, 1 (1983), A.H. Mueller and J.W. Qiu, *Nucl. Phys.* **B268**, 427 (1986); A.H. Mueller, *Nucl. Phys.* **B335**, 115 (1990).
8. E. Iancu and R. Venugopalan , In *Hwa, R.C. (ed.) et al.: Quark gluon plasma* 249-3363, [e-Print: hep-ph/0303204];
   H. Weigert , *Prog.Part.Nucl.Phys.***55**, 461(2005);
   J. Jalilian-Marian and Yu.V. Kovchegov, *Prog.Part.Nucl.Phys.***56**, 104(2006).
9. I. Balitsky and G.A. Chirilli, *Phys.Rev.***D77**, 014019(2008)
10. I. Balitsky, *Phys.Rev.***D75**,014001(2007).
11. Yu. V. Kovchegov and H. Weigert, *Nucl. Phys.* **A784**, 188 (2007),

12. M. Beneke, *Phys.Rept.* **317**,1(1999); M. Beneke and V.M. Braun, *"Renormalons and power corrections."*, In *Shifman, M. (ed.): At the frontier of particle physics, vol. 3*, p. 1719-1773 (World Scientific, Singapore,2001) [hep-ph/0010208]
13. V.S. Fadin and L.N. Lipatov, *Phys. Lett.* **B429**, 127 (1998); G. Camici and M. Ciafaloni, *Phys. Lett.* **B430**, 349 (1998).
14. I. Balitsky and G.A. Chirilli, in preparation

Recent Perturbative Results on Heavy Quark Decays

J. H. Piclum*, M. Dowling and A. Pak

Department of Physics, University of Alberta,
Edmonton, AB T6G 2G7, Canada
** E-mail: jpiclum@phys.ualberta.ca*

We present new results for heavy quark decays at next-to-next-to-leading order. In particular, we discuss the calculation of the charm-quark mass effects and the expansion in the intermediate recoil limit in semi-leptonic bottom-quark decays and the evaluation of the helicity fractions of the W boson from top-quark decays.

1. Introduction

Heavy fermion decays play an important role in precision tests of the Standard Model (SM) and the search for physics beyond the SM. A good example is the muon decay, which is used to determine the Fermi constant, G_F, one of the fundamental constants in the SM. The present experimental accuracy, exceeding 1 ppm,[1] makes it necessary to evaluate effects due to the tiny mass of the electron at next-to-next-to-leading order (NNLO). We find that they reduce the rate by -0.43 ppm,[2] which is larger than a previous estimate.[3]

In the quark sector, the decays of the bottom and top quarks are very interesting due to their large masses. The semi-leptonic $b \to c\ell\bar{\nu}$ decay allows the extraction of V_{cb}, the element of the Cabibbo-Kobayashi-Maskawa matrix governing b to c transitions. This decay is closely related to the muon decay, however, the effects due to the mass of the final state fermion are much larger.

In the SM the top quark decays almost always to a W^+ boson and a bottom quark. Due to the large mass of the top quark, its coupling to the W boson is very sensitive to physics beyond the SM. The polarization of the W boson allows for a particularly detailed study of this process.

Experimentally, the longitudinal ($\mathcal{F}_L$), transverse-plus ($\mathcal{F}_+$), and transverse-minus ($\mathcal{F}_-$) helicity fractions of the W boson from top-quark

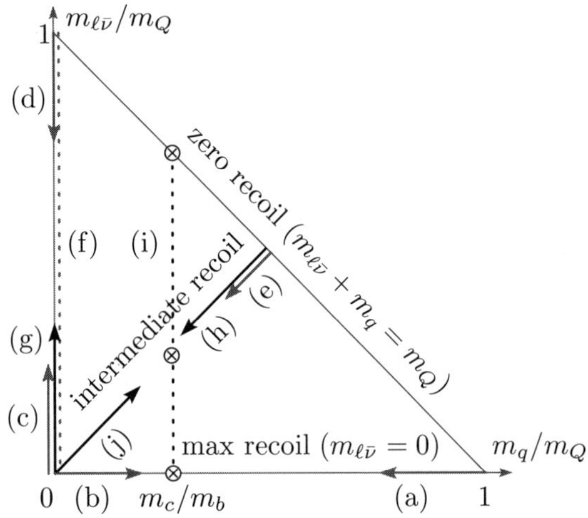

Fig. 1. NNLO corrections to decays $Q \to q\ell\bar{\nu}$. Outside the triangle, the decay is prohibited. Gray arrows indicate previously known expansions; the gray dotted line represents the result of Ref. 20. Black arrows and the black dotted line denote the results reported in these proceedings.

decays have been determined by the CDF[4–8] and the D0[9–12] collaborations at the Tevatron at Fermilab. While their results are in agreement with the SM prediction, the experimental errors are still quite large. However, they are dominated by the statistical errors and will be reduced when larger data samples become available from the Tevatron and in particular from the LHC. A recent Monte Carlo study based on 10 fb^{-1} at the LHC finds measurement uncertainties at the percent level.[13] Here, we report on the recent NNLO evaluation of the helicity fractions.[14,15]

A convenient way to represent the different kinematic limits of weak decays is the triangle diagram in Fig. 1. The expansions (a), (b), (c), (d), (e) have been known previously;[16–19] (f) corresponds to the result integrated over the phase space of leptons with a zero final quark mass.[20,21] The limit of large β_0, the so-called BLM corrections,[22] was considered in Ref. 23.

The new results presented in this contribution are shown in Fig. 1 as black arrows and the black dotted line. The arrow labeled as (h) extends and crosschecks the existing expansion from the zero recoil line,[24] matching against the complementary expansion along the intermediate recoil line (j).

The dotted line (i) represents the incorporation of mass effects due to the final state fermion.[2] This result confirms the numerical calculation of Ref. 25 and provides the electron-mass corrections to the muon decay.

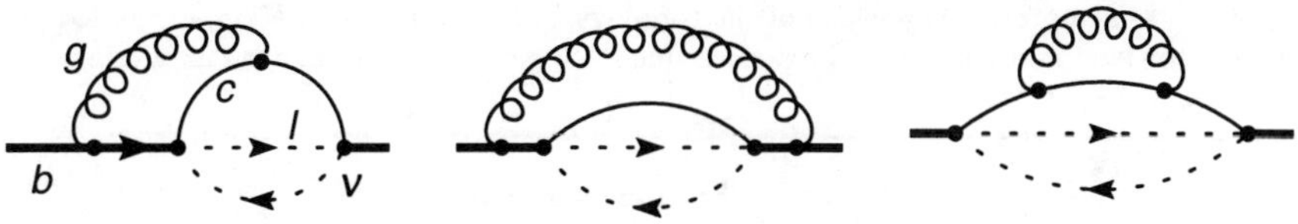

Fig. 2. Diagrams corresponding to the decay rate $\Gamma(b \to c\ell\bar{\nu})$ at NLO.

Finally, arrow (g) stands for the calculation of the helicity fractions of W bosons from top-quark decays.[14,15] The corresponding next-to-leading order (NLO) corrections have been calculated in Refs. 26–28 in closed analytical form. Electroweak and finite width corrections of the same order have been obtained in Ref. 29. NNLO corrections to the total rate were calculated in Refs. 30–32 by exploiting the fact that $x^2 = m_W^2/m_t^2$ is a small number.

2. Calculation of NNLO corrections

We will illustrate the main techniques with an example of one-loop corrections to the semi-leptonic b-quark decay $b \to c\ell\bar{\nu}$ (Fig. 1(i)).

With $\rho \equiv m_c/m_b \ll 1$, the decay rate is

$$\Gamma(b \to c\ell\bar{\nu}) = \Gamma_0 \left[X_0 + C_F \frac{\alpha_s}{\pi} X_1 + C_F \left(\frac{\alpha_s}{\pi}\right)^2 X_2 + \dots \right], \qquad (1)$$

$$\begin{aligned} \Gamma_0 &= \frac{G_F^2 |V_{cb}|^2 m_b^5}{192\pi^3}, \qquad X_0 = 1 - 8\rho^2 - 24\rho^4 \ln\rho + 8\rho^6 - \rho^8, \\ X_1 &= \frac{25}{8} - \frac{\pi^2}{2} - (34 + 24\ln\rho)\,\rho^2 + 16\pi^2\rho^3 \\ &\quad - \left(\frac{273}{2} - 36\ln\rho + 72\ln^2\rho + 8\pi^2\right)\rho^4 + \dots . \end{aligned} \qquad (2)$$

Here α_s is normalized in the $\overline{\text{MS}}$ scheme at m_b, $C_F = 4/3$ in QCD, and X_2 is evaluated with $N_L = 3$ massless quarks.

The calculation in the presence of two masses can be explained on the example of X_1. We consider the self-energy diagrams in Fig. 2, whose imaginary part reproduces the total decay rate due to the optical theorem. After taking the traces, contracting and Wick rotating, each diagram is written in terms of scalar integrals such as in Fig. 3, with various integer exponents of denominator factors $D_1, \dots, D_6$. To regularize the integrals, we use dimensional regularization with $D = 4 - 2\epsilon$.

According to the asymptotic expansion technique,[33–35] each of the three loop momenta should be assigned either a scale m_b or m_c, and the integral

Table 1. Asymptotic expansions of the topology in Fig. 3. In each region, expanded denominator factors are shown. The double line in Region 4 represents a static propagator.

	Region 1: $\mathbf{|k_1|, |k_2|, |k_3| \sim m_b}$: $D_2 \to k_2^2$, $D_3 \to k_1^2$
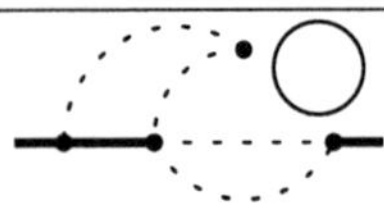	Region 2: $\mathbf{|k_1| \sim m_c}$, $\mathbf{|k_2|, |k_3| \sim m_b}$: $D_2 \to k_2^2$, $D_5 \to k_2^2$, $D_6 \to (p-k_2)^2 + m_b^2$
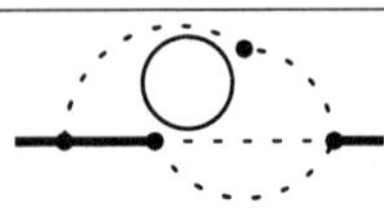	Region 3: $\mathbf{|k_2| \sim m_c}$, $\mathbf{|k_1|, |k_3| \sim m_b}$: $D_3 \to k_1^2$, $D_1 \to (p+k_3)^2$, $D_5 \to k_1^2$, $D_6 \to (p+k_1)^2 + m_b^2$
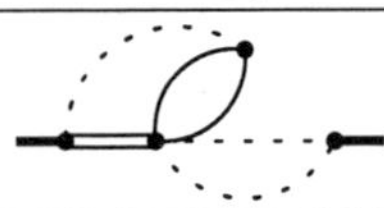	Region 4: $\mathbf{|k_1|, |k_2| \sim m_c}$, $\mathbf{|k_3| \sim m_b}$: $D_1 \to (p+k_3)^2$, $D_6 \to 2p(k_1 - k_2) + i0$

expanded in a Taylor series to the required order in ρ. After that, in each region the integration is performed without limitations, and the results are added together. Four such assignments corresponding to non-zero contributions are sketched in Tab. 1 (in the two-gluon case there are up to eleven such regions for some topologies). This procedure produces a large number of integrals that have to be evaluated. To simplify the problem, we use recurrence relations originating from integration-by-parts identities.[36,37] In our example, we may consider some D-dimensional divergence, which vanishes when integrated over the whole space due to Stokes' theorem,

$$\int \left(\frac{d}{dk_j^\mu} k_i^\mu \right) \frac{d^D k_1 d^D k_2 d^D k_3}{D_1^{a_1} D_2^{a_2} D_3^{a_3} D_4^{a_4} D_5^{a_5} D_6^{a_6}} = 0. \tag{3}$$

Using the chain differentiation rule, the same condition leads to a linear equation on integrals of Fig. 3 with coefficients depending on the dimension, D, and, in this example, the ratio m_c/m_b. Solving[38,39] the systems of such

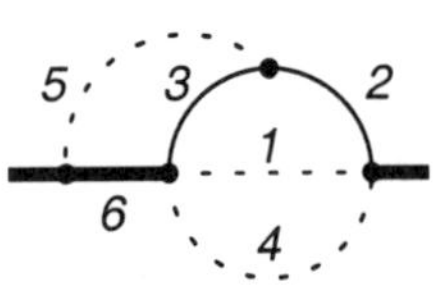

$$\mathrm{I}(a_1, a_2, a_3, a_4, a_5, a_6) = \int \frac{d^D k_1 d^D k_2 d^D k_3}{D_1^{a_1} D_2^{a_2} D_3^{a_3} D_4^{a_4} D_5^{a_5} D_6^{a_6}},$$
$$D_1 = (p + k_3 - k_2)^2,\ D_2 = k_2^2 + m_c^2,\ D_3 = k_1^2 + m_c^2,$$
$$D_4 = k_3^2,\ D_5 = (k_1 - k_2)^2,\ D_6 = (p + k_1 - k_2)^2 + m_b^2,$$
$$p^2 = -m_b^2.$$

Fig. 3. General double-scale topology and its denominator factors. Thin lines represent mass m_c, thick – mass m_b, dashed lines are massless.

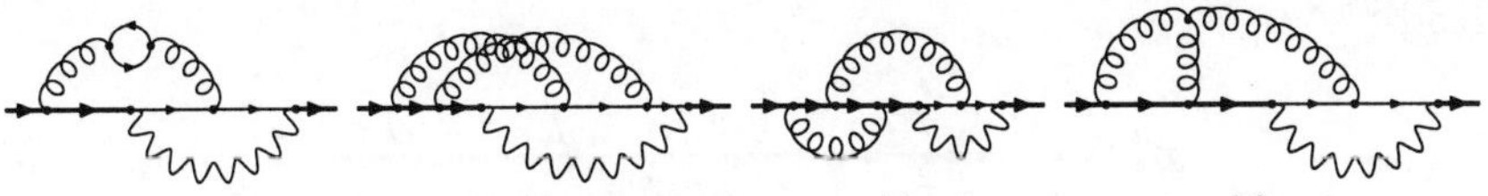

Fig. 4. Sample diagrams corresponding to the decay rate $\Gamma(t \to bW)$ ($\Gamma(b \to cW)$) at NNLO. Wavy and curly lines denote W bosons and gluons, respectively. Thick lines denote top (bottom) quarks and thin lines denote bottom (charm) quarks.

equations obtained for various momenta k_i and k_j, we find an expression of every integral of that topology in terms of a few so-called master integrals. This is the most CPU-intensive part in calculation of X_2, requiring weeks for the reduction of all necessary integrals, with intermediate expressions growing to several gigabytes.

Finally, unknown master integrals need to be evaluated. For that purpose, we consider derivatives of the double-scale master integrals, e.g.

$$\begin{aligned}\frac{\partial}{\partial m_c^2} I(a_1,a_2,a_3,a_4,a_5,a_6) = & -\, a_2 I(a_1,a_2+1,a_3,a_4,a_5,a_6) \\ & -\, a_3 I(a_1,a_2,a_3+1,a_4,a_5,a_6).\end{aligned} \qquad (4)$$

These expressions with "shifted" indices are then reduced back to master integrals using recurrence relations. Thus, a system of coupled differential equations is built, which can be solved[40,41] analytically for boundary conditions obtained in the simpler limit $m_c \to \infty$. In practice, though, it is not convenient to use the "natural" second parameter m_c, and we used artificial mass assignments chosen to simplify the differential equations.

Treating polarization

The results of Fig. 1(h,i,j) were obtained as described above. However, the polarized result (g) involved a modification of the procedure, which will be discussed in the following. Sample diagrams are shown in Fig. 4.

Compared to the NNLO calculation of the total decay width, one has to replace the total rate projector [a]

$$\mathbb{P}_0^{\mu\nu} = -g^{\mu\nu} + \frac{q^\mu q^\nu}{m_W^2}\,, \qquad (5)$$

where q is the momentum of the W boson, by the longitudinal projector $\mathbb{P}_L^{\mu\nu}$ and the transverse-plus/minus projectors $\mathbb{P}_\pm^{\mu\nu}$. The longitudinal projector

[a] In unitary gauge, the W boson propagator reads $i\,\mathbb{P}_0^{\mu\nu}/(q^2 - m_W^2)$.

reads[27]

$$\mathbb{P}_L^{\mu\nu} = \frac{(m_W^2\, p^\mu - p\cdot q\, q^\mu)(m_W^2\, p^\nu - p\cdot q\, q^\nu)}{m_W^2 m_t^2\, |\vec{q}\,|^2}\,, \tag{6}$$

where p is the external momentum with $p^2 = m_t^2$. The transverse projectors can be obtained with the help of the forward-backward projector[27]

$$\mathbb{P}_F^{\mu\nu} = -\frac{1}{m_t|\vec{q}\,|}\, i\varepsilon^{\mu\nu\sigma\rho}\, p_\sigma\, q_\rho\,, \tag{7}$$

$$\mathbb{P}_\pm^{\mu\nu} = \frac{1}{2}\,(\mathbb{P}_0^{\mu\nu} - \mathbb{P}_L^{\mu\nu} \pm \mathbb{P}_F^{\mu\nu})\,. \tag{8}$$

Thus, we have to deal with the additional propagator-like structure $1/|\vec{q}\,|^n$, $n \in \{1, 2\}$.

Since we set $m_b = 0$ in this calculation, we only have to consider two regions. In the region where $q \sim m_t$, we have to construct an expansion of $1/|\vec{q}\,|$ in m_W/m_t. This can be done by rewriting $|\vec{q}\,|^2$ as follows:

$$|\vec{q}\,|^2 = q_0^2 - m_W^2 = \frac{(2p\cdot q)^2}{4m_t^2} - m_W^2 \tag{9}$$

$$= \frac{1}{4m_t^2}\left(q^2 + 2p\cdot q - m_W^2\right)^2 - m_W^2 \tag{10}$$

$$= \frac{1}{4m_t^2}\left(S^2 - 2m_W^2 S + m_W^4 - 4m_t^2 m_W^2\right)\,, \tag{11}$$

where $S = q^2 + 2p\cdot q$ is the denominator of a top-quark propagator with momentum $q + p$. In Eq. (10) we used the fact that we are only interested in the imaginary part and that $q^2 = m_W^2$ on the cut. Now we can construct the desired expansion as

$$\frac{1}{|\vec{q}\,|^2} = \frac{4m_t^2}{S^2}\sum_{i=0}^{\infty}\left(\frac{2m_W^2 S - m_W^4 + 4m_t^2 m_W^2}{S^2}\right)^i\,, \tag{12}$$

which we truncate at some order. Thus, the additional propagator-like structure from the projectors is transformed into a scalar on-shell propagator with momentum $p + q$ and mass m_t, raised to integer powers.

In the region where $q \sim m_W$, we cannot perform an expansion of $|\vec{q}\,|$, since $|\vec{q}\,|^2 = q_0^2 - m_W^2$. However, in this region the W-boson loop factorizes. Therefore, we only have to replace the usual one-loop vacuum integrals with integrals including the additional propagator-like structure.

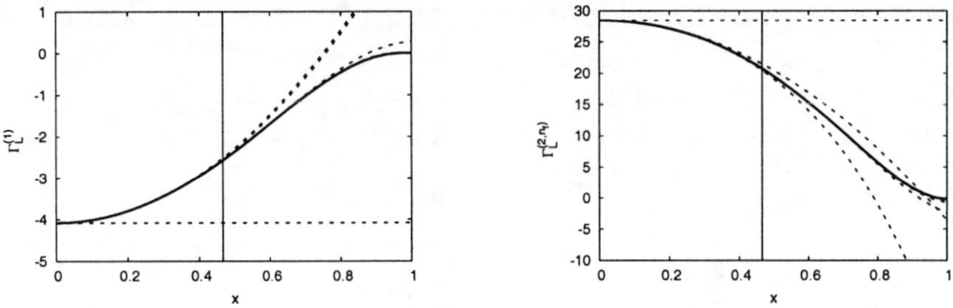

(a) (b)

Fig. 5. The $\mathcal{O}(\alpha_s)$ and $\mathcal{O}(\alpha_s^2)$ corrections to the decay rate into longitudinally polarized W bosons as functions of x. The dashed lines denote expansions truncated at different orders in x. The solid line denotes the full correction in (a) and the expansion up to $\mathcal{O}(x^{10})$ in (b). The vertical lines denotes the physical value of x.

3. Helicity Fractions

In order to check our calculation we have computed the NLO corrections up to terms of order $\mathcal{O}(x^{16})$. We find perfect agreement with the expansion of the closed form results of Refs. 26–28.

As an example of our NNLO results, we give the corrections due to diagrams with closed fermion loops (cf. Fig. 4(*a*)) in numerical form. Using $m_W = 80.403$ GeV,[42] $m_t = 172.5$ GeV and $\alpha_s(m_t) = 0.1073$ we find

$$\mathcal{F}_L = 0.6971\left(1 - 0.0108 + 0.0018\big|_{n_f}\right), \tag{13}$$

$$\mathcal{F}_+ = 0 + 0.00103 - 0.00017\big|_{n_f}, \tag{14}$$

$$\mathcal{F}_- = 0.3029\left(1 + 0.0214 - 0.0037\big|_{n_f}\right), \tag{15}$$

where the results have been split-up into the contributions from LO, NLO, and NNLO, respectively, and we have included terms up to $\mathcal{O}(x^{10})$ in the NNLO result.

It is interesting to look at the convergence behavior of the series in x. For this purpose, we write the decay rate for the top-quark decay into longitudinally polarized W bosons as

$$\Gamma_L = \frac{G_F\, m_t^3\, |V_{tb}|^2}{8\sqrt{2}\,\pi}\left[\Gamma_L^{(0)} + \frac{\alpha_s}{2\pi} C_F \Gamma_L^{(1)} + \left(\frac{\alpha_s}{2\pi}\right)^2 C_F \left(\Gamma_L^{(2,n_f)} + \dots\right)\right], \tag{16}$$

where the ellipsis denotes non-fermionic contributions.

Fig. 5 shows $\Gamma_L^{(1)}$ and $\Gamma_L^{(2,n_f)}$ as functions of x. It shows several expansions which have been truncated at different orders in x. For $\Gamma_L^{(1)}$ the full result is shown as well. The convergence behavior of the expansion is very

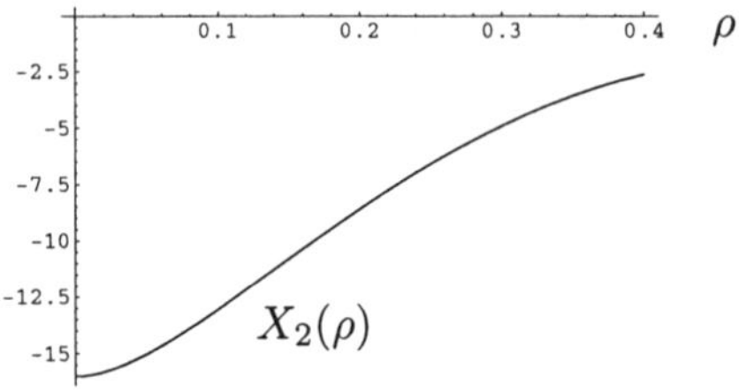

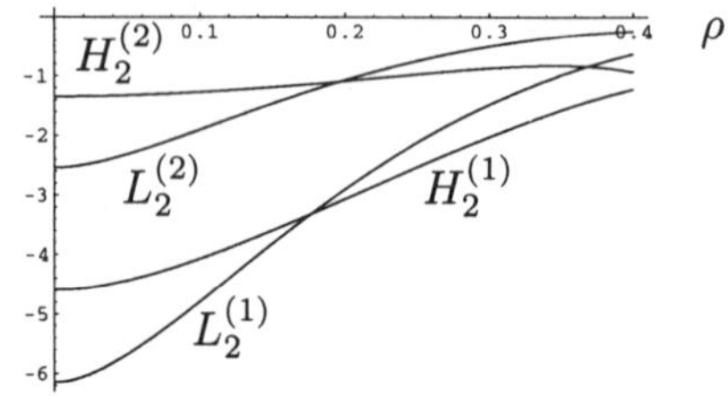

Fig. 6. Mass dependence of X_2 and moments of distributions of lepton and hadron system energies

good. In fact, it is sufficient to include terms of order $\mathcal{O}\left(x^{10}\right)$ to reproduce the full one-loop correction even at $x = 1$. The behavior of the two-loop term is similar. This means that for all practical purposes, our expansion is equivalent to the (unknown) full result. Analogous plots for Γ_+ and Γ_- show an equally good convergence behavior.

4. Semi-leptonic b decays

4.1. *Charm-mass effects*

Using the techniques above, we have been able to evaluate terms through $\mathcal{O}(\rho^7)$ in X_2, as well as in the corrections to the first two moments in the lepton energy, E_l, and the hadronic system energy, E_h. The former are defined as

$$\int \left(\frac{E_l}{m_b}\right)^n d\Gamma = \Gamma_0 \left(L_0^{(n)} + C_F \left(\frac{\alpha_s}{\pi}\right) L_1^{(n)} + C_F \left(\frac{\alpha_s}{\pi}\right)^2 L_2^{(n)} + \dots\right), \tag{17}$$

and by analogy the moments of E_h with coefficients $H_j^{(n)}$. The energies are evaluated in the rest frame of the b quark, and the integral is taken over the phase space of the decay products. The dependence of all those quantities on ρ is shown in Fig. 6.

An important feature of the result is the presence of a term linear in ρ (in contrast to the estimate of Ref. 3), which appears due to the chosen pole mass definition. This scheme is not appropriate in QCD calculations since effects at scales smaller than $\Lambda_{\rm QCD}$ cannot be treated perturbatively, and has to be changed to a more appropriate short-distance mass scheme before applying the results to phenomenology. However, QED traditionally uses pole masses, and the linear term together with large logarithms then results in the aforementioned half ppm correction to the decay rate.

The exact analytic results being known, we may check the previous logarithmic resummation attempts.[43] Of three terms predicted in that reference, we confirm two. Comparing our results with another prediction made by polynomial extrapolation from three kinematic limits,[18] we support the statement[25] that this extrapolation contains errors. This is discussed in the following.

4.2. *Intermediate recoil limit*

In Ref. 18 an estimate of the second order corrections was made by fitting a polynomial to three known points along the line describing $d\Gamma(b \to c\ell\nu_\ell)/dq^2$ (the three circles along the dotted line (i) in Fig. 1). By integrating along this polynomial, the authors were able to provide an estimate of the difference between the BLM contribution and the full second order correction: the non-BLM contribution. They found that the non-BLM contribution is $\Delta_{fit}^{nBLM} = 0.9(3)$. The calculation of the NNLO corrections to the full decay,[2,25] however, provide a different value: $\Delta_{exact}^{nBLM} = 1.73(4)$. This disagreement has prompted a check of the original intermediate recoil expansion, (e) in Fig. 1, by completing a complementary expansion, (j).

The methods used to calculate this complementary expansion are the same as described earlier with no new master integrals needed. Sample diagrams are shown in Fig. 4. The asymptotic expansion produced as many as eleven regions for a single topology, much like the $b \to c\ell\nu_\ell$ expansion discussed in Sec. 2. This provides a strong crosscheck of the previous calculation because of the different methods used in both expansions. (In Ref. 18 the expansion was calculated by considering all of the contributing amplitudes separately and integrating over the phase space to obtain the corrections to the decay width.)

When completing this complementary expansion, (j), we found that the expansion from zero recoil, (e), did not include enough terms to provide an accurate value for the intermediate recoil point used in the polynomial fit. To resolve this issue, we have extended the expansion so that the value of the result at the physical point of interest, $m_c/m_b = 0.25$, gives consistent results with the complementary expansion presented here.

Furthermore, we found a problem with the charge renormalization. In the original calculation, a normalization point of $\sqrt{m_b m_c}$ was used and the authors included a charge renormalization term scaling the b-quark loop contribution, N_H,

$$Z_{\alpha_s} \propto (m_b m_c)^{-2\epsilon}(T_R N_H). \tag{18}$$

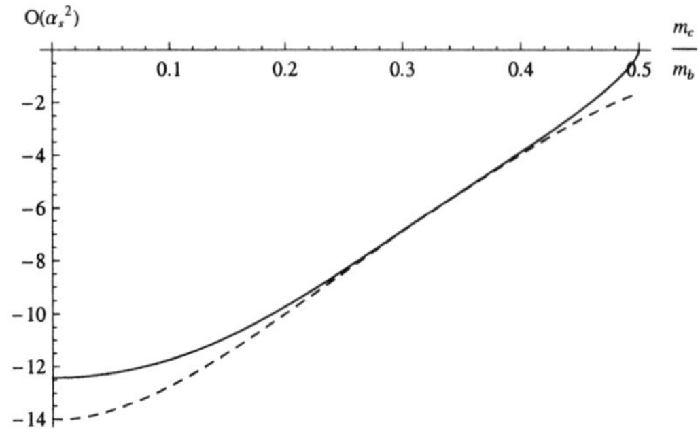

Fig. 7. Matching between the two intermediate recoil expansions. The solid line corresponds to the zero recoil expansion and the dashed line is the expansion in the opposite limit.

At this normalization scale, however, this term is incorrect because the b-quark mass is larger than the normalization scale. Instead it should be,

$$Z_{\alpha_s} \propto m_b^{-2\epsilon}(T_R N_H). \tag{19}$$

The combination of these two corrections to the expansion from zero recoil results in good matching with the opposite expansion that we have done. Our result is shown in Fig 7.

With these new results we have carried out the polynomial fit (following Ref. 18) to obtain a new estimate for the non-BLM contribution,[24] consistent with Refs. 2,25,

$$\Delta_{fit}^{nBLM} = 1.5(2). \tag{20}$$

To investigate the dependence of the result on the details of the fitting procedure, we considered another normalization of the fitted points. This resulted in the value

$$\Delta_{fit(2)}^{nBLM} = 1.76(4), \tag{21}$$

which is consistent with the above results.

Acknowledgments

We thank Andrzej Czarnecki for his collaboration. JHP also thanks Jürgen Körner. This work was supported by the Natural Sciences and Engineering Research Council of Canada. JHP is grateful to the Institut für Physik at the Universität Mainz for hospitality, to the Graduiertenkolleg "Eichtheorien" for travel support, and to the organizers of "Continuous Advances in QCD 2008" for an interesting conference.

Some of our algebraic computations were done using `GiNaC`[44] and `FORM`.[45] The Feynman diagrams were drawn with `JaxoDraw`.[46]

References

1. D. B. Chitwood *et al.* [MuLan Collaboration], *Phys. Rev. Lett.* **99**, 032001 (2007) [arXiv:0704.1981 [hep-ex]].
2. A. Pak and A. Czarnecki, *Phys. Rev. Lett.* **100**, 241807 (2008) [arXiv:0803.0960 [hep-ph]].
3. T. van Ritbergen and R. G. Stuart, *Nucl. Phys. B* **564**, 343 (2000) [arXiv:hep-ph/9904240].
4. A. A. Affolder *et al.* [CDF Collaboration], *Phys. Rev. Lett.* **84**, 216 (2000) [arXiv:hep-ex/9909042].
5. D. E. Acosta *et al.* [CDF Collaboration], *Phys. Rev. D* **71**, 031101 (2005) [Erratum-ibid. *D* **71**, 059901 (2005)] [arXiv:hep-ex/0411070].
6. A. Abulencia *et al.* [CDF-Run II Collaboration], *Phys. Rev. D* **73**, 111103 (2006) [arXiv:hep-ex/0511023].
7. A. Abulencia *et al.* [CDF Collaboration], *Phys. Rev. Lett.* **98**, 072001 (2007) [arXiv:hep-ex/0608062].
8. A. Abulencia *et al.* [CDF II Collaboration], *Phys. Rev. D* **75**, 052001 (2007) [arXiv:hep-ex/0612011].
9. V. M. Abazov *et al.* [D0 Collaboration], *Phys. Rev. D* **72**, 011104 (2005) [arXiv:hep-ex/0505031].
10. V. M. Abazov *et al.* [D0 Collaboration], *Phys. Lett. B* **617**, 1 (2005) [arXiv:hep-ex/0404040].
11. V. M. Abazov *et al.* [D0 Collaboration], *Phys. Rev. D* **75**, 031102 (2007) [arXiv:hep-ex/0609045].
12. V. M. Abazov *et al.* [D0 Collaboration], *Phys. Rev. Lett.* **100**, 062004 (2008) [arXiv:0711.0032 [hep-ex]].
13. J. A. Aguilar-Saavedra, J. Carvalho, N. Castro, A. Onofre and F. Veloso, *Eur. Phys. J. C* **53**, 689 (2008) [arXiv:0705.3041 [hep-ph]].
14. J. H. Piclum, A. Czarnecki and J. G. Körner, *Helicity fractions of W bosons from top quark decays*, to appear in *Nucl. Phys. B (Proc. Suppl.)*.
15. J. H. Piclum, A. Czarnecki and J. G. Körner, in preparation.
16. A. Czarnecki, *Phys. Rev. Lett.* **76**, 4124 (1996) [arXiv:hep-ph/9603261].
17. A. Czarnecki and K. Melnikov, *Phys. Rev. Lett.* **78**, 3630 (1997) [arXiv:hep-ph/9703291].
18. A. Czarnecki and K. Melnikov, *Phys. Rev. D* **59**, 014036 (1999) [arXiv:hep-ph/9804215].
19. A. Pak, I. Blokland and A. Czarnecki, *Phys. Rev. D* **73**, 114009 (2006) [arXiv:hep-ph/0604233].
20. T. van Ritbergen and R. G. Stuart, *Phys. Rev. Lett.* **82**, 488 (1999) [arXiv:hep-ph/9808283].
21. M. Steinhauser and T. Seidensticker, *Phys. Lett. B* **467**, 271 (1999) [arXiv:hep-ph/9909436].
22. S. J. Brodsky, G. P. Lepage and P. B. Mackenzie, *Phys. Rev. D* **28**, 228

(1983).
23. M. E. Luke, M. J. Savage and M. B. Wise, *Phys. Lett. B* **343**, 329 (1995) [arXiv:hep-ph/9409287].
24. M. Dowling, A. Pak, A .Czarnecki, Alberta-Thy 09-08, to be published.
25. K. Melnikov, arXiv:0803.0951 [hep-ph].
26. M. Fischer, S. Groote, J. G. Körner, M. C. Mauser and B. Lampe, *Phys. Lett. B* **451**, 406 (1999) [arXiv:hep-ph/9811482].
27. M. Fischer, S. Groote, J. G. Körner and M. C. Mauser, *Phys. Rev. D* **63**, 031501 (2001) [arXiv:hep-ph/0011075].
28. M. Fischer, S. Groote, J. G. Körner and M. C. Mauser, *Phys. Rev. D* **65**, 054036 (2002) [arXiv:hep-ph/0101322].
29. H. S. Do, S. Groote, J. G. Körner and M. C. Mauser, *Phys. Rev. D* **67**, 091501 (2003) [arXiv:hep-ph/0209185].
30. K. G. Chetyrkin, R. Harlander, T. Seidensticker and M. Steinhauser, *Phys. Rev. D* **60**, 114015 (1999) [arXiv:hep-ph/9906273].
31. I. Blokland, A. Czarnecki, M. Ślusarczyk and F. Tkachov, *Phys. Rev. Lett.* **93**, 062001 (2004) [arXiv:hep-ph/0403221].
32. I. Blokland, A. Czarnecki, M. Ślusarczyk and F. Tkachov, *Phys. Rev. D* **71**, 054004 (2005) [arXiv:hep-ph/0503039].
33. F. V. Tkachov, *Sov. J. Part. Nucl.* **25**, 649 (1994) [arXiv:hep-ph/9701272].
34. A. Czarnecki and V. A. Smirnov, *Phys. Lett. B* **394**, 211 (1997) [arXiv:hep-ph/9608407].
35. V. A. Smirnov, *Springer Tracts Mod. Phys.* **177**, 1 (2002).
36. F. V. Tkachov, *Phys. Lett. B* **100**, 65 (1981).
37. K. G. Chetyrkin and F. V. Tkachov, *Nucl. Phys. B* **192**, 159 (1981).
38. S. Laporta and E. Remiddi, *Phys. Lett. B* **379**, 283 (1996) [arXiv:hep-ph/9602417].
39. S. Laporta, *Int. J. Mod. Phys. A* **15**, 5087 (2000) [arXiv:hep-ph/0102033].
40. A. V. Kotikov, *Phys. Lett. B* **254**, 158 (1991).
41. E. Remiddi, *Nuovo Cim. A* **110**, 1435 (1997) [arXiv:hep-th/9711188].
42. W. M. Yao *et al.* [Particle Data Group], *J. Phys. G* **33** 1, (2006).
43. C. W. Bauer, A. F. Falk and M. E. Luke, *Phys. Rev. D* **54**, 2097 (1996) [arXiv:hep-ph/9604290].
44. C. W. Bauer, A. Frink and R. Kreckel, arXiv:cs/0004015.
45. J. A. M. Vermaseren, arXiv:math-ph/0010025.
46. D. Binosi and L. Theussl, *Comput. Phys. Commun.* **161**, 76 (2004) [arXiv:hep-ph/0309015].

LEADING AND NON-LEADING SINGULARITIES IN GAUGE THEORY HARD SCATTERING

GEORGE STERMAN

C.N. Yang Institute for Theoretical Physics, Stony Brook University
Stony Brook, NY 11794-3840 USA
E-mail: george.sterman@stonybrook.edu

I review some classic results and recent progress in the resummation of leading and nonleading singularities in QCD cross sections and dimensionally-regularized hard-scattering amplitudes.

Keywords: Factorization, resummation, infrared safety

1. Introduction

Interest in perturbative methods for QCD and related gauge theories arises from the phenomenology of high energy scattering, and also from the study of weak-strong duality, as inspired by string theory. In the following, I'll review some methods and techniques that have a long history but remain of continuing interest, along with a few recent advances. The talk starts with a perspective on the place of perturbation theory in an asymptotically free theory, goes on to recall ideas of factorization and resummation in pQCD, which leads to a review of one of its classic successes, so-called Q_T resummation. It concludes with applications of these same ideas to dimensionally-regulated amplitudes for the scattering of massless partons, which have been the subject of much recent work.

2. How We Use Perturbative QCD

It's worth recalling that despite the early successes of asymptotic freedom as a qualitative explanation of scaling, the applicability of perturbative methods beyond the parton model was met with a fair amount of skepticism. The underlying problems, of course, remain with us. First, confinement ensures that the quantities we would most naturally compute in perturbative

QCD (pQCD), time-ordered products of fields,

$$\int d^4x \, e^{-iq\cdot x} \langle 0 | \, T[\phi_a(x) \ldots] \, |0\rangle \,, \tag{1}$$

have no $q^2 = m^2$ poles for any field (particle) ϕ_a that transforms nontrivially under color, while the "physical" poles at $q^2 = m_\pi^2$, for example, in

$$\int d^4x \, e^{-iq\cdot x} \langle 0 | \, T[\pi(x) \ldots] \, |0\rangle \,, \tag{2}$$

are not accessible to perturbation theory directly. And yet we use asymptotic freedom, up to power-suppressed corrections,

$$\begin{aligned} Q^2 \, \hat{\sigma}_{\rm SD}(Q^2, \mu^2, \alpha_s(\mu)) &= \sum_n c_n(Q^2/\mu^2) \, {\alpha_s}^n(\mu) + \mathcal{O}\left(1/Q^p\right) \\ &= \sum_n c_n(1) \, {\alpha_s}^n(Q) + \mathcal{O}\left(1/Q^p\right) \,, \end{aligned} \tag{3}$$

for single-scale cross sections $\sigma(Q)$, so long as they are finite in the zero-mass limit in perturbation theory, a property known as "infrared safety". Various total and jet cross sections as well as predictions based on evolution are of this type, and their phenomenological successes are well-known.

So, what are we really calculating? In many cases, we are computing matrix elements for color singlet currents, of the general form $\int e^{-iq\cdot x} \langle 0 | \, T[J(x)J(0) \ldots] \, |0\rangle$, related to observables by the optical theorem. Of course, the optical theorem requires a complete sum over final states. But, in fact, there is another class of infrared (IR) safe color singlet matrix elements, related to jets and event shapes, that have received attention of late. These matrix elements accompany currents with the energy-momentum tensor, $T_{\mu\nu}$, schematically,[1]

$$\lim_{R\to\infty} R^2 \int dx_0 \int d\hat{n} \, f(\hat{n}) \, e^{-iq\cdot y} \langle 0 | \, J(0) T[\hat{n}_i \, T_{0i}(x_0, R\hat{n}) J(y)] \, |0\rangle \,, \tag{4}$$

with $f(\hat{n})$ a "weight" that controls the contributions of particles flowing to infinity in different directions, $\hat{n}$. With the operator T_{0i} placed at infinity, these matrix elements rather directly represent the action of a calorimeter. If the weight is a smooth function of angles, then even though the matrix elements for individual final states have IR divergences in general, they cancel in sums over collinear splitting/merging and soft parton emission, precisely because these rearrangements respect energy flow. We regularize these divergences dimensionally (typically) and "pretend" to calculate the long-distance enhancements only to cancel them in infrared safe quantities.

3. Factorization and Resummation

Beyond the relatively limited class of cross sections that are directly IR safe, the predictive power of pQCD depends on factorization.[2,3] From factorization we can derive the evolution familiar from deep-inelastic scattering and other single-scale problems, and generalizing this viewpoint, we can motivate resummations of enhancements in multiscale problems. A factorized cross section takes the general form

$$Q^2\sigma_{\rm phys}(Q,m) = \omega_{\rm SD}(Q/\mu,\alpha_s(\mu)) \otimes f_{\rm LD}(\mu,m) + \mathcal{O}\left(1/Q^p\right) , \tag{5}$$

where μ is a factorization scale, m represents IR scales, perturbative or nonperturbative, and where $\otimes$ represents a convolution, typically in parton fraction or transverse momentum, often accurate to power corrections as shown. Speculations on new physics are contained $\omega_{\rm SD}$, as perturbative (as in SUSY) or nonperturbative (as in technicolor) extensions of the Standard Model; $f_{\rm LD}$ represents parton distributions of various sorts, universal among cross sections sharing the same factorization.

The familiar "DGLAP" evolution equations[4] can be derived from factorization, just by observing that physical cross sections cannot depend on the choice of factorization scale

$$0 = \mu\frac{d}{d\mu}\ln\sigma_{\rm phys}(Q,m) , \tag{6}$$

which, combined with (5) leads to a separation of variables,

$$\mu\frac{d\ln f}{d\mu} = -P(\alpha_s(\mu)) = -\mu\frac{d\ln\omega}{d\mu} , \tag{7}$$

where the "separation constant" P can depend only on the variables held in common between the short- and long-distance functions in the factorized expression, α_s and the convolution variable(s).

The solutions to evolution equations like Eq. (7) are examples of resummation, in this case summarizing leading (and nonleading) logarithms of Q,

$$\ln\sigma_{\rm phys}(Q,m) \sim \exp\left\{\int^Q \frac{d\mu'}{\mu'}P\left(\alpha_s(\mu')\right)\right\} . \tag{8}$$

This result is most familiar in the form of DGLAP evolution; as we shall see, however, its applications are even more wide-ranging.

This sequence of methods and results: factorization → evolution → resummation, varies between observables, and must be verified for each case.

Such verifications, or "factorization proofs",[2,3] rely in general on four features of gauge theory: (1) The operator product expansion, from which short-distance dynamics in $\omega_{\rm SD}$ is incoherent with long-distance dynamics; (2) Jet-jet factorization, or the mutual incoherence of the dynamics of particles with $v_{\rm rel} = c$; (3) Jet-soft factorization, by which wide angle soft radiation depends only on the overall color flow in jets;[5,6] (4) Dimensionless couplings and renormalizability, which ensure that infrared singularities are no worse than logarithmic.[7]

4. The Classic Case: Q_T Resummation

What makes factorization necessary, and evolution and resummation so rewarding, is that every final state from a hard scattering carries the imprint of QCD dynamics from all distance scales. We will illustrate how these ideas play out in *the* classic application of resummed pQCD, the transverse momentum distribution for Dell-Yan pairs.[6,8] We start with the transverse momentum distribution at order α_s for the purely partonic process

$$q(p_1) + \bar{q}(p_2) \rightarrow \gamma^*(Q) + g(k)\,. \tag{9}$$

At lowest order (LO), $\mathbf{k} = -\mathbf{Q}_T$, and the partonic cross section is free of infrared divergences. The corresponding factorized expression for the LO *hadronic* cross section is

$$\frac{d\sigma_{NN\rightarrow\mu^+\mu^-+X}(Q,p_1,p_2)}{dQ^2 d^2\mathbf{Q}_T} = \int_{\xi_1,\xi_2} \sum_{a=q\bar{q}} \frac{d\hat{\sigma}_{a\bar{a}\rightarrow\mu^+\mu^-(Q)+X}(Q,\mu,\xi_i p_i,\mathbf{Q}_T)}{dQ^2 d^2\mathbf{Q}_T} \times f_{a/N}(\xi_1,\mu)\, f_{\bar{a}/N}(\xi_2,\mu)\,. \tag{10}$$

The LO diagrams for the measured-Q_T cross section are shown in Fig. 1, where the short-distance factor (the analog of $\omega_{\rm SD}$ above) is

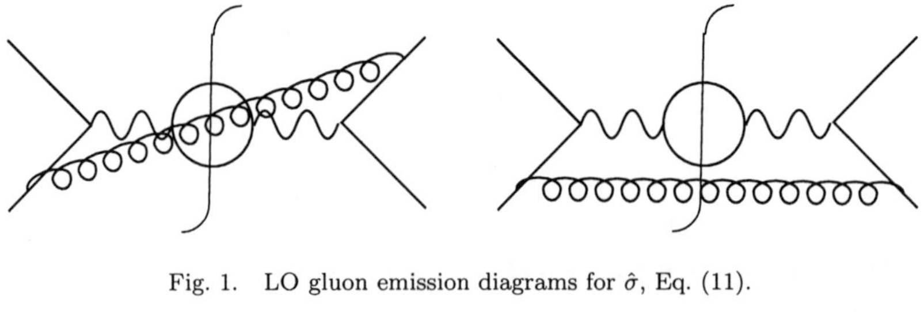

Fig. 1. LO gluon emission diagrams for $\hat{\sigma}$, Eq. (11).

$$\frac{d\hat{\sigma}^{(1)}_{q\bar{q}\to\gamma^* g}}{dQ^2\, d^2\mathbf{Q}_T} = \sigma_0 \frac{\alpha_s C_F}{\pi^2} \left(1 - \frac{4\mathbf{Q}_T^2}{(1-z)^2 \xi_1 \xi_2 S}\right)^{-1/2} \times \left[\frac{1}{\mathbf{Q}_T^2}\frac{1+z^2}{1-z} - \frac{2z}{(1-z)Q^2}\right], \qquad (11)$$

with σ_0 the LO total cross section, This expression, and the corresponding factorized cross section (10), is well-defined as long as $\mathbf{Q}_T \neq 0$ and $z = Q^2/\xi_1\xi_2 S \neq 1$.

Now the leading behavior for $\mathbf{Q}_T \ll Q$ can be found by considering the z integral in the case when Q^2/S is not too close to unity. In this case, the phase space factor in (11) and the parton distribution functions (PDFs) can be treated as nearly constant over the physical range of z, which then gives a logarithmic integral,

$$\frac{1}{\mathbf{Q}_T^2} \int_{Q^2/S}^{1-2|\mathbf{Q}_T|/Q} \frac{dz}{1-z} \sim \frac{1}{\mathbf{Q}_T^2} \ln\left[\frac{Q}{|\mathbf{Q}_T|}\right]. \qquad (12)$$

This approximation gives a neat prediction for Q_T dependence at fixed Q,

$$\frac{d\sigma_{NN\to\mu^+\mu^-+X}(Q,\mathbf{Q}_T)}{dQ^2 d^2\mathbf{Q}_T} \sim \frac{\alpha_s C_F}{\pi} \frac{1}{\mathbf{Q}_T^2} \ln\left[\frac{Q}{|\mathbf{Q}_T|}\right] \times \sum_{a=q\bar{q}} \int_{\xi_1\xi_2} \frac{d\hat{\sigma}_{a\bar{a}\to\mu^+\mu^-(Q)+X}(Q,\mu)}{dQ^2} f_{a/N}(\xi_1,\mu)\, f_{\bar{a}/N}(\xi_2,\mu), \quad (13)$$

which we can compare, for example, to the transverse momentum of the Z boson at the Tevatron. As can be seen from Fig. 2, taken from Ref.,[9] a simple $\ln Q/Q_T$-dependence works pretty well for "large" Q_T, less than but of the order of $Q = m_Z$, but at smaller Q_T the distribution reaches a maximum, then decreases near the "exclusive" limit, at $Q_T = 0$, corresponding to parton model kinematics. Indeed, most events are at "low" $Q_T \ll m_Z$, where the LO cross section diverges. To understand the distribution in this range, we turn to transverse momentum resummation, which, as we shall see, controls logarithms of Q_T to all orders in α_s. As suggested above, we can resum logarithms of Q_T by developing variant factorizations and separations of variables.

In brief, the factorization we will exhibit reflects a relatively simple physical picture. The active quark and antiquark "arrive" at the point of annihilation with nonzero transverse momenta, due to gluons radiated in the transition from the initial state. Now before the collision, the quark and antiquark radiate independently, reflecting a lack of overlap between their Coulomb fields. Similarly, after the collision, final-state radiation occurs too

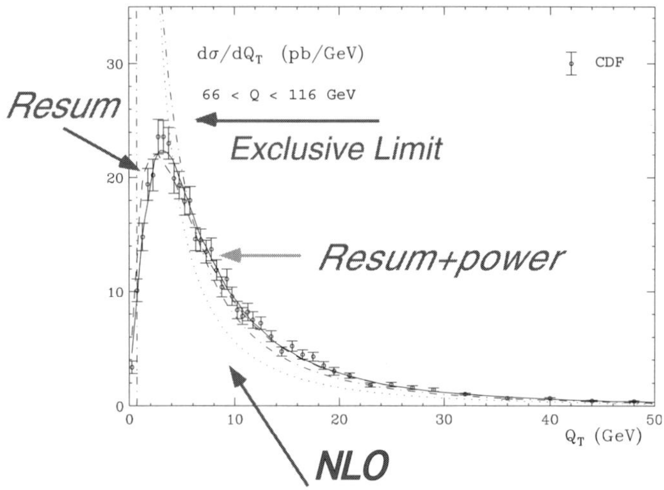

Fig. 2. Transverse momentum dependence of the Z boson as observed by the CDF collaboration.

late to affect the cross section, that is, the net probability of annihilation into an electroweak vector boson with a given Q_T. These considerations are summarized by Q_T-factorization, in the form[8]

$$\frac{d\sigma_{NN\to QX}}{dQ^2d^2Q_T} = \int d\xi_1 d\xi_2\ d^2\mathbf{k}_{1T}d^2\mathbf{k}_{2T}d^2\mathbf{k}_{sT}\,\delta^2\left(\mathbf{Q}_T - \mathbf{k}_{1T} - \mathbf{k}_{2T} - \mathbf{k}_{sT}\right)$$
$$\times\ H(\xi_1 p_1, \xi_2 p_2, Q, n)_{a\bar{a}\to Q+X}$$
$$\times\ \mathcal{P}_{a/N}(\xi_1, p_1\cdot n, \mathbf{k}_{1T})\,\mathcal{P}_{\bar{a}/N}(\xi_2, p_2\cdot n, \mathbf{k}_{2T})\ U_{a\bar{a}}(\mathbf{k}_{sT}, n)\,. \qquad (14)$$

Here the $\mathcal{P}'s$ are new *transverse momentum-dependent* PDFs, and in the general case we also need a new function labelled, U, a *soft function* that describes wide-angle radiation. Symbolically, in the spirit of the general factorization, Eq. (5), we can write

$$\frac{d\sigma_{NN\to QX}}{dQ^2d^2Q_T} = H \times \mathcal{P}_{a/N}(\xi_1, p_1\cdot n, k_{1T})\,\mathcal{P}_{\bar{a}/N}(\xi_2, p_2\cdot n, k_{2T})$$
$$\otimes_{\xi_i, k_{iT}}\ U_{a\bar{a}}(k_{sT}, n)\,. \qquad (15)$$

What we are going to do is *derive* the k_T dependence of the $\mathcal{P}$'s. For the purposes of this talk, we proceed intuitively and with broad strokes; much more careful analyses can be found in,[6,8] and.[10]

In Eq. (14) we encounter new invariants, $p_i\cdot n$, formed from a fixed vector n^μ. We can think of n^μ as being used to apportion real and virtual gluons of momentum k into the various factors in (14), according to the scheme:

$$\begin{aligned} p_a \cdot k < n \cdot k \ &\Rightarrow \ k \in \mathcal{P}_a \\ p_a \cdot k,\ p_{\bar{a}} \cdot k > n \cdot k \ &\Rightarrow \ k \in U\,. \end{aligned} \tag{16}$$

It is the variables $p_a \cdot n$ that will play the role of factorization scales. Before reviewing this analysis, we go to impact parameter space, replacing the convolution in $k_{i,T}$ by a product after the Fourier transform with $e^{i\vec{Q}_T\cdot\vec{b}}$, giving, in place of (14),

$$\begin{aligned} \frac{d\sigma_{NN\to QX}(Q,b)}{dQ^2} &= \int d\xi_1 d\xi_2\ H(\xi_1 p_1, \xi_2 p_2, Q, n)_{a\bar{a}\to Q+X} \\ &\quad \times \mathcal{P}_{a/N}(\xi_1, p_1\cdot n, b)\, \mathcal{P}_{\bar{a}/N}(\xi_2, p_2\cdot n, b)\ U_{a\bar{a}}(b,n)\,. \end{aligned} \tag{17}$$

We are now ready once again to resum by separating variables.

The physical impact parameter cross section of Eq. (17) is independent of both $\mu_{\rm ren}$ and of the vector n^μ. As a result, we have two equations that express this independence,

$$\mu_{\rm ren}\frac{d\sigma}{d\mu_{\rm ren}} = 0\,, \qquad\qquad n^\alpha \frac{d\sigma}{dn^\alpha} = 0\,. \tag{18}$$

These equations represent the scale variation and the boost invariance of the theory. The solutions to pairs of equations of this kind were developed in this context by Collins and Soper[6] and by Sen.[11]

Now variations from the jets must cancel variations from the short-distance function H and from the soft function U, which depend on different variables. This analysis gives

$$p\cdot n\,\frac{\partial}{\partial p\cdot n}\ \ln \mathcal{P}(p\cdot n/\mu, b\mu) = \frac{1}{2}G(p\cdot n/\mu, \alpha_s(\mu)) + \frac{1}{2}K(b\mu, \alpha_s(\mu)), \tag{19}$$

where G matches H, and K matches U. On the other hand, renormalization is independent of n^μ, which implies

$$\mu\,\frac{\partial}{\partial\mu}\left[\,G(p\cdot n/\mu, \alpha_s(\mu)) + K(b\mu, \alpha_s(\mu))\,\right] = 0\,, \tag{20}$$

from which we find

$$\mu\,\frac{\partial}{\partial\mu}G(p\cdot n/\mu, \alpha_s(\mu)) \ = \ \gamma_K(\alpha_s(\mu)) \ = \ -\,\mu\,\frac{\partial}{\partial\mu}K(b\mu, \alpha_s(\mu))\,. \tag{21}$$

It is the combination of Eqs. (19) and (21) that gives the basic results.

We solve Eq. (21) first,

$$G(p\cdot n/\mu,\alpha_s(\mu))+K(b\mu,\alpha_s(\mu))=G(1,\alpha_s(p\cdot n))+K(1,\alpha_s(1/b)) - \int_{1/b}^{p\cdot n}\frac{d\mu'}{\mu'}\,\gamma_K(\alpha_s(\mu'))\,. \qquad (22)$$

Inserting this result in the consistency equation (19) for the jet enables us to integrate $p\cdot n$ and get double logs in b, which, when inverted back to Q_T space, produce the leading behavior $\alpha_s^n\,\frac{\ln^{2n-1}(Q/Q_T)}{Q_T}$ at each order, along with nonleading contributions (which require an analysis of the soft function U). When carried out in detail (with attention paid to nonperturbative corrections from large b), this approach can describe the data of Fig. 2 all the way to $Q_T=0$.[9,12] The resulting expression can be summarized as

$$\frac{d\sigma_{NN\mathrm{res}}}{dQ^2\,d^2\vec{Q}_T}=\sum_a\int\frac{d^2b}{(2\pi)^2}\,e^{i\vec{Q}_T\cdot\vec{b}}\exp\left[\,E_{a\bar{a}}^{\mathrm{PT}}(b,Q,\mu)\,\right]$$
$$\times\sum_{a=q\bar{q}}\int_{\xi_1\xi_2}H(\xi_1p_1,\xi_2p_2,Q,n)_{a\bar{a}\to Q+X}\;f_{a/N}(\xi_1,1/b)\,f_{\bar{a}/N}(\xi_2,1/b)\,, \qquad (23)$$

with a "Sudakov" exponent that, as anticipated, links large and low virtuality,

$$E_{q\bar{q}}^{\mathrm{PT}}=-\int_{1/b^2}^{Q^2}\frac{d\mu^2}{\mu^2}\left[2A_q(\alpha_s(\mu))\,\ln\left(\frac{Q^2}{\mu^2}\right)+2B_q(\alpha_s(\mu))\right]\,, \qquad (24)$$

where[8] B_q is related to $(K+G)_{\mu=p\cdot n}$ and at lowest order $A_q=\gamma_K/2$, and where the lower limit $1/b$ of the integral in the exponent generates the leading logarithmic Q_T dependence.

5. Poles in Color Exchange Amplitudes

Color exchange is a feature central to the analysis of hard scattering and jet and heavy particle production at hadron colliders. As intermediate results in calculations of short-distance functions, and as a subject of interest in their own right, multiloop scattering amplitudes in dimensional regularization have received considerable attention.[13–16] We conclude with a sketch of how the general methods described above lead to important results for these ampliudes.

We consider a partonic process, denoted: f : $f_A(p_A,r_A)+f_B(p_B,r_B)\to f_1(p_1,r_1)+f_2(p_2,r_2)+\ldots$, where we restrict ourselves to

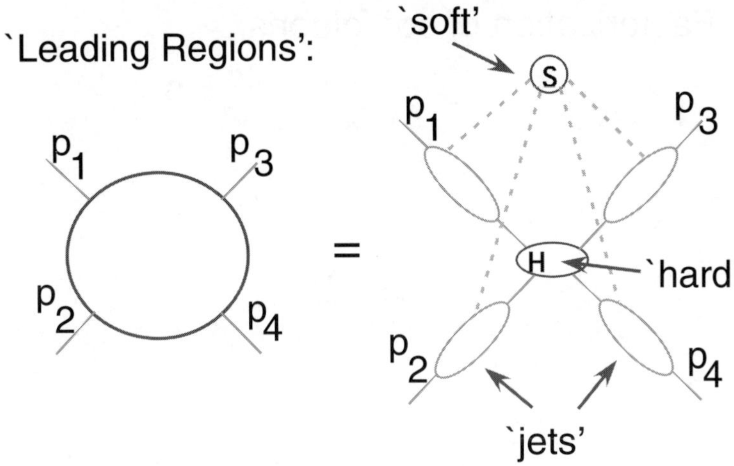

Fig. 3. Leading regions for 2 → 2 scattering.

wide-angle scattering. The amplitude for any such process can be expanded in a basis of color tensors c_L linking the external partons,

$$\mathcal{M}^{[f]}_{\{r_i\}}\left(p_j, \frac{Q^2}{\mu^2}, \alpha_s(\mu^2), \epsilon\right) = \mathcal{M}^{[f]}_{L}\left(p_j, \frac{Q^2}{\mu^2}, \alpha_s(\mu^2), \epsilon\right) (c_L)_{\{r_i\}} \,, \tag{25}$$

with infrared singularities regularized by going to $4 - 2\epsilon$ dimensions with $\epsilon < 0$, after renormalization has been performed. Examples of the c_Is are singlet and octet exchange in the s-channel of quark-antiquark scattering. We need to control poles in ϵ for factorized calculations at fixed order, and, for resummation, to all orders.

Double logs and poles in dimensional regularization are associated with "leading regions"[7,17] in the loop momentum space for arbitrary graphical contributions to the amplitude. These take the general form shown in Fig. 3. Leading regions are characterized by *jet subdiagrams*, consisting of lines parallel to the external momenta p_i, a *short-distance subdiagram* (H), with only lines off-shell by order of the momentum transfer(s), and a *soft subdiagram* (S) with lines whose momenta vanish. Historically, it was the soft subdiagram that was seen as a problem for the control of infrared behavior.[18] In Fig. 3, however, we encounter the same cast of characters as for the Q_T analysis in Drell-Yan, and the analogous factorization is shown schematically in Fig. 4, in which jet-soft factorization separates jet and soft dynamics in this more complex scattering process.[19] For the amplitude, the IR regularization variable $\epsilon = 2 - d/2$ plays the role of b in the Q_T cross

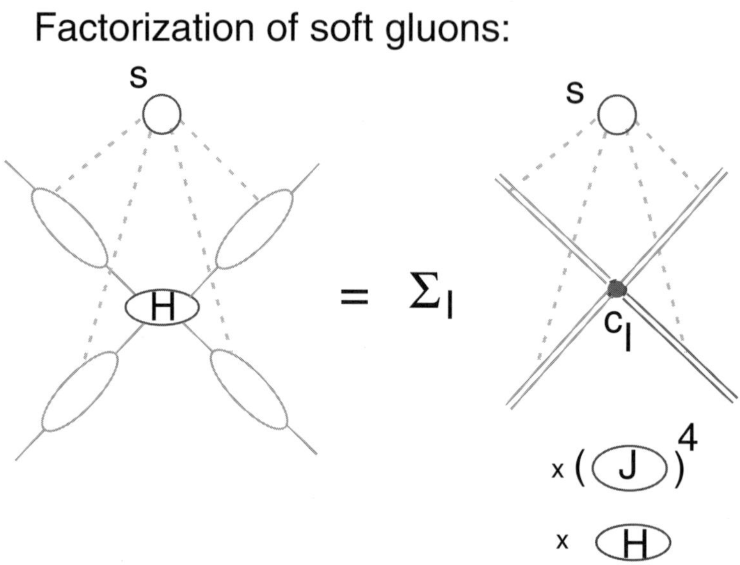

Fig. 4. Soft-jet factorization for wide-angle scattering.

section.

In summary, we can write a factorized expression for $\mathcal{M}$,

$$\mathcal{M}_L^{[f]}\left(p_i,\frac{Q^2}{\mu^2},\alpha_s(\mu^2),\epsilon\right)=\prod_{f=A,B,1,2} J_f^{[\mathrm{virt}]}\left(\frac{Q^2}{\mu^2},\alpha_s(\mu^2),\epsilon\right)$$
$$\times\mathbf{S}_{LI}^{[f]}\left(p_i,\frac{Q^2}{\mu^2},\alpha_s(\mu^2),\epsilon\right)h_I^{[f]}\left(\wp_i,\frac{Q^2}{\mu^2},\alpha_s(\mu^2)\right), \quad (26)$$

where the jet functions J_f for parton f can be identified with the square roots of the corresponding singlet form factors, $\sqrt{\Gamma^f_{\mathrm{singlet}}(Q^2)}$,[15] the soft functions are matrices labelled by color exchange (singlet, octet ...), and all factors require dimensional regularization. We return to the soft function $\mathbf{S}^{[f]}$ below.

The same analysis as for Drell-Yan Q_T described above, starting with factorization and arriving at resummation, gives the following explicit expression[13] modeled on the work of Collins and Soper[6] and of Sen:[11]

$$\Gamma\left(\frac{Q^2}{\mu^2},\alpha_s(\mu^2),\mathrm{e}\right) = \exp\left\{\frac{1}{2}\int_0^{-Q^2}\frac{d\xi^2}{\xi^2}\left[\mathcal{K}\left(\epsilon,\alpha_s(\mu^2)\right)\right.\right. \quad (27)$$

$$\left.\left. + \ \mathcal{G}\left(-1,\alpha_s\left(\xi^2,\epsilon\right),\epsilon\right) + \frac{1}{2}\int_{\xi^2}^{\mu^2}\frac{d\lambda^2}{\lambda^2}\gamma_K\left(\alpha_s\left(\lambda^2,\epsilon\right)\right)\right]\right\},$$

where the running coupling is treated as ϵ-dependent. All levels of exponentiating poles are generated by the anomalous dimensions $\mathcal{G}$, $\mathcal{K}$ and $\gamma_K = -\mu d\mathcal{K}/d\mu$. (The functions are $\mathcal{G}$ and $\mathcal{K}$ are related to, but not identical with the analogous functions above.) The relations of such QCD results to supersymmetric Yang-Mills theories were explored in several talks at this workshop (see also the recent review by Alday and Roiban[20]). Double poles are generated from γ_K, which is familiar as the so-called "cusp" anomalous dimension.[21] A complete portrait of single poles at each order requires the ϵ-dependent function $\mathcal{G}$,[22] which also generates finite coefficient functions in Γ_{singlet}.[23] To find a field-theoretic interpretation for $\mathcal{G}$, we once again turn to the factorization approach, this time for the singlet form factor itself, as illustrated in Fig. 5. In the figure, soft radiation is organized in a singlet product of light-like Wilson lines,

$$\mathcal{S}\left(\alpha_s(\mu^2),\epsilon\right) = \langle 0|\Phi_{\beta_2}(\infty,0)\,\Phi_{\beta_1}(0,-\infty)\,|0\rangle\,, \quad (28)$$

where $\Phi_\beta(\infty,0) \equiv P\exp[-ig\int_0^\infty d\lambda\beta\cdot A(\lambda\beta)]$, and where we may take $\beta_1\cdot\beta_2 = 1$. Such an expectation value obeys[24]

$$\mu\frac{d}{d\mu}\log\mathcal{S}\left(\alpha_s(\mu^2),\varepsilon\right) = G_{\text{eik}}\left(\alpha_s(\mu^2)\right) - \frac{1}{2}\int_0^{\mu^2}\frac{d\xi^2}{\xi^2}\gamma_K\left(\overline{\alpha}(\xi^2,\varepsilon)\right)\,, \quad (29)$$

in terms of the same γ_K and a new anomalous dimension G_{eik} that organizes non-collinear poles.

Following this analysis, the full $\mathcal{G}$ for the form factor in Eq. (27) can be written as[22]

$$\mathcal{G} = 2B + G_{\text{eik}} + \beta(g)\frac{\partial}{\partial g}C(\alpha_s(Q))\,, \quad (30)$$

with B the N-independent coefficient in spin-N leading-twist operators for parton i, and with C the short-distance function shown in Fig. 5. Similar combinations have been encountered in analyses of deep-inelastic scattering and Drell Yan in Refs.[25–27]

The remainder of the dimensional dependence in the general amplitude, Eq. (26) is generated by a matrix of anomalous dimensions for the soft

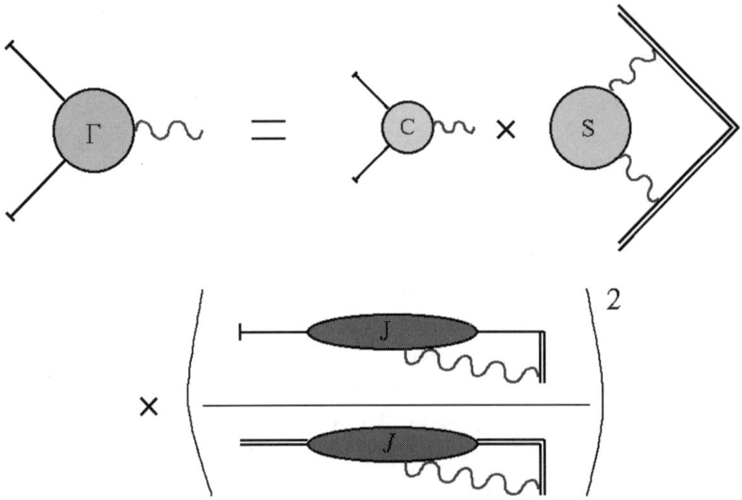

Fig. 5. Factorization of the singlet form factor. From Ref.[22].

functions[15,28]

$$\mathbf{S}^{[\mathrm{f}]}\left(\frac{Q^2}{\mu^2}, \alpha_s(\mu^2), \epsilon\right) = \mathrm{P}\ \exp\left[-\frac{1}{2}\int_0^{Q^2} \frac{d\tilde{\mu}^2}{\tilde{\mu}^2}\mathbf{\Gamma}_S^{[\mathrm{f}]}\left(\alpha_s\left(\tilde{\mu}^2, \epsilon\right)\right)\right] . \tag{31}$$

The one-loop expressions for arbitrary $\mathbf{\Gamma}_S^{[\mathrm{f}]}$ were computed in,[28] and the two-loop expressions in.[16] Remarkably, the one- and two-loop contributions are proportional,[16]

$$\mathbf{\Gamma}_S = \frac{\alpha_s}{\pi}\left(1 + \frac{\alpha_s}{2\pi}K\right)\mathbf{\Gamma}_S^{(1)} + \cdots , \tag{32}$$

with with the same constant, $K = C_A(67/18 - \zeta_2) - (5/9)n_f$, that appears in γ_K. This suggests an exact one-loop anomalous dimension, supplemented by a "CMW" scheme for α_s.[29] If this conjecture turns out to hold, there is a deep simplicity inherent in infrared vector exchange, even in QCD.

6. Summary

I have shown how the factorization properties of gauge theories serve as keys to resummation. For double-logarithmic, or "Sudakov" corrections, resummation follows from two equations, one associated with boost invariance, and another with scale variations (invariance for conformal theories). The basic factorization structure and its consequences are not limited to weak coupling. Whether at weak or strong coupling, many of the the long-distance properties of gauge theories can be organized quite explicitly in both cross sections and the perturbative S-matrix.

Acknowledgments

I thank the organizers for an invitation, and for support in the course of this stimulating workshop. I also thank my collaborators in the recent work reported on here, Mert Aybat, Lance Dixon and Lorenzo Magnea. This work was supported in part by by the National Science Foundation, grants PHY-0354776, PHY-0354822 and PHY-0653342.

References

1. N. A. Sveshnikov and F. V. Tkachov, Phys. Lett. B 382, 403 (1996) [arXiv:hep-ph/9512370];
   G. P. Korchemsky, G. Oderda and G. Sterman, in *DIS 97*, AIP Conf. Proc. No. 407, ed. J. Repond and D. Krakauer, (American Institute of Physics, Woodbury, NY, 1997), p. 988. arXiv:hep-ph/9708346;
   C. W. Bauer, S. P. Fleming, C. Lee and G. Sterman, arXiv:0801.4569 [hep-ph];
   D. M. Hofman and J. Maldacena, JHEP **0805**, 012 (2008) [arXiv:0803.1467 [hep-th]].
2. G.T. Bodwin, Phys. Rev. D31, 2616 (1985), Erratum ibid. D34, 3932 (1986);
   J. C. Collins, D. E. Soper and G. Sterman, Nucl. Phys. B261, 104 (1985); ibid. B308, 833 (1988).
3. J. C. Collins, D. E. Soper and G. Sterman, in *Perturbative quantum chromodynamics*, ed. A. H. Mueller (World Scientific, Singapore, 1989), p. 1, hep-ph/0409313.
4. G. Altarelli and G. Parisi, Nucl. Phys. B126, 298 (1977);
   V. N. Gribov and L. N. Lipatov, Sov. J. Nucl. Phys. 15, 438, 675 (1972);
   Yu. L. Dokshitzer, Sov. Phys. JETP 46, 641 (1977).
5. J. C. Collins and G. Sterman, Nucl. Phys. B **185** (1981) 172.
6. J. C. Collins and D. E. Soper, Nucl. Phys. **B 193** (1981) 381 [Erratum-ibid. **B 213** (1983) 545].
7. G. Sterman, Phys. Rev. D17, 2773; 2789 (1978); Phys. Rev. D **19**, 3135 (1979).
8. J. C. Collins, D. E. Soper and G. Sterman, Nucl. Phys. B **250**, 199 (1985);
   E. Laenen, G. Sterman and W. Vogelsang, Phys. Rev. **D 63** (2001) 114018, [arXiv:hep-ph/0010080].
9. A. Kulesza, G. Sterman and W. Vogelsang, Phys. Rev. D **66**, 014011 (2002) [arXiv:hep-ph/0202251].
10. H. Contopanagos, E. Laenen and G. Sterman, Nucl. Phys. **B 484** (1997) 303, [arXiv:hep-ph/9604313].

11. A. Sen, Phys. Rev. **D 24** (1981) 3281.
12. F. Landry, R. Brock, P. M. Nadolsky and C. P. Yuan, Phys. Rev. D **67**, 073016 (2003) [arXiv:hep-ph/0212159];
E. L. Berger, J. w. Qiu and X. f. Zhang, Phys. Rev. D **65**, 034006 (2002) [arXiv:hep-ph/0107309].
13. L. Magnea and G. Sterman, Phys. Rev. **D 42** (1990) 4222.
14. S. Catani, Phys. Lett. B **427**, 161 (1998) [arXiv:hep-ph/9802439].
15. G. Sterman and M. E. Tejeda–Yeomans, Phys. Lett. **B 552** (2003) 48, [arXiv:hep-ph/0210130].
16. S. Mert Aybat, L. J. Dixon and G. Sterman, Phys. Rev. **D 74** (2006) 074004, [arXiv:hep-ph/0607309].
17. R. Akhoury, Phys. Rev. D **19**, 1250 (1979).
18. C. G. Callan and D. J. Gross, Phys. Rev. **D 11** (1975) 2905.
19. A. Sen, Phys. Rev. **D 28** (1983) 860.
20. L. F. Alday and R. Roiban, arXiv:0807.1889 [hep-th].
21. G. P. Korchemsky and A. V. Radyushkin, Nucl. Phys. B **283**, 342 (1987).
22. L. J. Dixon, L. Magnea and G. Sterman, JHEP **0808**, 022 (2008) [arXiv:0805.3515 [hep-ph]].
23. S. Moch, J. A. M. Vermaseren and A. Vogt, Phys. Lett. **B 625** (2005) 245, [arXiv:hep-ph/0508055].
24. G. P. Korchemsky and G. Marchesini, Nucl. Phys. **B 406** (1993) 225, hep-ph;
A. V. Belitsky, Phys. Lett. **B 442** (1998) 307, hep-ph.
25. A. Idilbi, X. d. Ji and F. Yuan, Nucl. Phys. **B 753** (2006) 42, [arXiv: 0605068 [hep-ph]] .
26. T. Becher, M. Neubert and B. D. Pecjak, JHEP **0701** (2007) 076, [arXiv: 0607228 [hep-ph]].
27. T. Becher, M. Neubert and G. Xu, JHEP **0807** (2008) 030 [arXiv:0710.0680 [hep-ph]].
28. N. Kidonakis, G. Oderda and G. Sterman, Nucl. Phys. B 531, 365 (1998) [arXiv:hep-ph/9803241].
29. S. Catani, B. R. Webber and G. Marchesini, Nucl. Phys. B **349**, 635 (1991).

The Space-Cone Gauge, Lorentz Invariance and On-Shell Recursion for One-Loop Yang-Mills amplitudes

Diana Vaman[1], York-Peng Yao[2]

[1] *Department of Physics, The University of Virginia*
Charlottesville, VA, 22904
and
KITP, University of California
Santa Barbara, CA, 93106 E-mail: dv3h@virginia.edu

[2] *Randall Laboratory of Physics, The University of Michigan*
Ann Arbor, MI 48109-1120
E-mail: yyao@umich.edu

Recursion relations are succinctly obtained for $(++\cdots+)$ and $(-++\cdots+)$ amplitudes in the context of the space-cone gauge in QCD. We rely on the helicity symmetry of the problems to dictate our choices of reference twistors and the momentum shifts used to analytically continue the amplitudes. Of great importance is the power of Lorentz invariance, which is enough to determine the soft factors in the latter cases.

1. Introduction

An extremely efficient way to calculate multi-particle amplitudes is to use recurrence relations, if in fact they can be devised. Then, one can recycle the results obtained earlier for amplitudes with smaller number of particles to generate those with larger numbers. In other words, the aim is to factorize the amplitudes into sums of two groups through the exchange of some states, where both groups correspond to physical on-shell amplitudes which have already been constructed. The feasibility to accomplish this relies on analytic continuation of the amplitudes into complex momenta, so as to be able to set the sub-amplitudes on-shell. Obviously, one has to have sufficient understanding and some handles on the analytic properties of the amplitudes to push this program through. For tree amplitudes in QCD, because they are made of rational functions of holomorphic $\langle , \rangle$ and anti-holomorphic $[,]$ products, the analyticity is simple. Britto, Cachazo, Feng and Witten (BCFW)[1] have exploited it to obtain their celebrated recurrence relations, in which the intermediate exchanges are one-particle states. At the one-loop level, there are other amplitudes, such as those in which all or all but one of the gluons share the same helicity, which are also rational functions of $\langle , \rangle$ and $[\ ,]$ products. Their evaluation requires extra physical input, such as collinearity of massless particles.These considerations have been discussed by.[2] The case when all gluons but one have the same helicity was solved by Mahlon,[3] who obtained a somewhat complicated off-shell recursion relation, extending the off-shell tree level recursion relations of Berends and Giele.[4]

In,[5] among other results, we re-derived BCFW relations by using QCD in the space-cone gauge. The motivation for us to seek guidance from Lagrangian field theory is definitely not to regress to the cumbersome perturbative diagrammatic computation, but to better understand the make-up of the Feynman rules and therefore to explore various ways to continue into the complex plane.

One distinct feature in the space-cone gauge of Chalmers and Siegel[6] is that QCD is like a field theory with two scalars, corresponding to the two polarizations of a gluon. The redundant degrees of freedom have been eliminated to produce non-local vertices. The structure for a process is better organized, because of fewer terms in each vertex, and because the number of diagrams is highly reduced. It behooves to preserve these virtues when we analytically continue. In other words, let the continuation be carried out by shifting some of the external momenta parameterized by a variable z, then we shall restrict to those shifts such that the vertices and the polarizations are not affected. The dependence on z will come only from the scalar propagators.[5] If we call the continued amplitude $A(z)$, then $A(0)$ is the original physical amplitude. A recurrence relation follows after we complexify z and evaluate an integral

$$I = \oint dz \frac{A(z)}{z}, \tag{1}$$

through a closed contour at infinity. Because of the way we introduce z dependence, as long as there are enough numbers of scalar propagators, we shall never pick up any surface term due to $A(z \to \infty) \neq 0$ (see also[7]).

One of our objectives in this present article is to continue investigating this aspect and to apply it to derive on-shell recurrence relations for one-loop gluon amplitudes which are rational functions of external momenta. One of our main points is that we can offer a guiding principle for the essential step of performing the analytic continuation of the loop amplitudes. In addition, we are able to derive the soft factors which were conjectured by Bern, Dixon and Kosower (BDK),[8] and which are a crucial ingredient in writing the recurrence relation, each time there is a double z-pole in the analytically continued amplitude. Our derivation of the soft factors is based on making a certain analytic continuation, with an arbitrary twistor-dependence, using the space-cone gauge to select this analytic continuation, and requiring that the end result is independent on the choice made for that arbitrary twistor. To some extent this is tantamount to requiring that the amplitude is Lorentz invariant. However, there is more to it than just Lorentz invariance, as the analytic continuation we perform is intimately related to the gauge choice. We shall refer to this as reference-twistor invariance. The present considerations may be extended to the other one-loop amplitudes, but this falls outside the scope of this article. For recent results on on-shell methods in perturbative QCD see,[9],[10],[11] and for an extension of on-shell recursions to higher-dimensions and of the space-cone gauge for gravity see.[12]

2. Possible Analytic Continuations

One way to continue an amplitude into complex external momenta is to shift some of them with a parameter z, about which we would like to make a comment. In Ref.,[5] the on-shell recurrence relations were obtained as a consequence of the largest time equation, in which some time-ordered (or null-ordered) step functions appear. z is the variable of integration along the <u>*real*</u> axis to represent such step functions:

$$\theta(\eta \cdot (x_1 - x_2)) = \frac{1}{2\pi i} \int_{-\infty}^{\infty} dz \frac{e^{iz\eta\cdot(x_1-x_2)}}{z - i\epsilon}. \tag{2}$$

The ensuing complexification of z is used as a means to evaluate subsequent integrals by using the residue theorem in (1). We shall abide by this understanding and the parameter z that we shall introduce in the following shifts is assumed to be real *ab initio*. A crucial ingredient in our analysis is the space-cone gauge which was introduced in Ref.[6] This gauge is specified by a null vector, which is given in terms of some reference twistors

$$\eta = |+\rangle[-|, \tag{3}$$

such that for the gauge field $A_\mu = a^+|+\rangle[+| + a^-|-\rangle[-| + a|-\rangle[+| + \bar{a}|+\rangle[-|$ a condition is imposed:

$$\eta \cdot A = 0 \Rightarrow a = 0. \tag{4}$$

After re-expressing the dependent component $\bar{a}$ by $a^{\pm}$, we have the Yang-Mills Lagrangian

$$L = a_a^+ \partial_\mu \partial^\mu a_a^- - g f_{abc} (\frac{\partial^-}{\partial} a_a^+) a_b^+ \partial a_c^- - g f_{abc} (\frac{\partial^+}{\partial} a_a^-) a_b^- \partial a_c^+ \\ + \frac{g^2}{2} f_{abc} a_a^+ \partial a_b^- \frac{1}{\partial^2} f_{a'b'c} a_{a'}^- \partial a_{b'}^+ . \tag{5}$$

f_{abc} are the structure constants of the gauge group, which will be chosen as $SU(N)$. Note that the derivative couplings are independent of $\bar{\partial}$. After Fourier-transforming to momentum space, the vertices will be independent of the four-momentum component $\bar{p}$. This will be relevant in all our future considerations.

In (2), η need not be identified with the null space-cone gauge fixing vector, but this can be a winning strategy once the analytic continuation in z has been performed. Before proceeding with the analytic continuation, there are several restrictions we need to observe. Let the set of momenta to be shifted among the n external momenta be $\{P_{i_1}, P_{i_2}, \ldots P_{i_m}\}$, $n > m \geq 2$, which we shall relabel as $\{Q_1, Q_2, \ldots Q_m\}$ for ease of notation. First of all, we want the shifted momenta $\hat{Q}_i$ to remain massless. This means that we write for each one

$$\hat{Q}_i = Q_i + \delta Q_i = |q_i\rangle[q_i| + z(|q_i\rangle[\xi_i| + |\zeta_i\rangle[q_i|) \tag{6}$$

where $|\zeta_i\rangle$ and $[\xi_i|$ are to be determined. Now the overall energy-momenta must be conserved, which gives

$$\sum_{i=1}^{m} (|Q_i\rangle[\xi_i| + |\zeta_i\rangle[Q_i|) = 0. \tag{7}$$

Then we must take into account that when we evaluate the integral in (2) we are looking for singularities in $A(z)$. It will make the task easier if a pole in a multi-particle channel with non-vanishing residues corresponds to a pole in z. Such a pole $(\sum P_k)^2 = 0$, where P_k are some of the n external momenta, contains products of some of the shifted momenta

$$-2\hat{Q}_i \cdot \hat{Q}_j = \langle \hat{Q}_i \hat{Q}_j \rangle [\hat{Q}_j \hat{Q}_i]. \tag{8}$$

Clearly, we have terms $\sim z^{0,1,2}$. To have just a pole in z when we solve for it from $(\sum P_j)^2 = 0$, we must demand that the coefficients to z^2 vanish. Then some of the $|\zeta_i\rangle$ vanish and that those which do not must be aligned. The same is said for $[\xi_i|$.Hence, we can write

$$|\zeta_i\rangle = c_i |\zeta\rangle, \qquad |\xi_i] = d_i |\xi], \tag{9}$$

for those Q_i and Q_j which can be grouped together into at least one of the possible channels that form single particle poles.

At this point, we may have some Q_i's and Q_j's which cannot be grouped together onto the same side of any channel $\sum P_k$. As a consequence, we may have several pairs of $|\zeta\rangle$ and $[\xi|$, but in the space-cone gauge, the most natural thing to do is to collapse them into just one pair:

$$|\zeta\rangle = |+\rangle, \qquad |\xi] = |-]. \tag{10}$$

There are other conditions on c_i, d_i, $|\zeta\rangle$ and $|\xi]$ which follow from the requirement that the relevant polarization vectors and vertices should not have any z dependence. We want to enforce these conditions in order for the analytically continued amplitude $A(z)$ to be well behaved at infinity.

For those Q_j with positive helicity, the external line factors are

$$\epsilon^+(Q_j) = \frac{[-Q_j]}{\langle +Q_j\rangle}, \tag{11}$$

and those Q_k with negative helicity

$$\epsilon^-(Q_k) = \frac{\langle +Q_k\rangle}{[-Q_k]}. \tag{12}$$

We note that because of the identification in (10), we have

$$\epsilon^+(\hat{Q}_j) = \epsilon^+(Q_j), \quad \epsilon^-(\hat{Q}_k) = \epsilon^-(Q_k), \quad \hat{q}_i = q_i. \tag{13}$$

The next important observation is that the vertices in this gauge depend only on $(k^\pm, k)$ and $(p_i^\pm, p_i)$ but not on $\bar{k}$ or $\bar{p}_i$, where K_μ is the loop momentum. After performing the loop integration, we shall have only tensors made of $(p_i^\pm, p_i)$. Therefore, our choice of c_i and d_i is to require

$$\hat{q}_i^\pm = q_i^\pm, \tag{14}$$

whichever appear in the amplitude. Whence, all the z dependence is confined in the denominators of some scalar integrals of Feynman parameters.

To summarize, we find that the shifts of (6) have to satisfy (9, 14) and (7). The latter is written as

$$\left(\sum_{i=1}^{m} d_i|Q_i\rangle\right)[-| + |+\rangle\left(\sum_{i=1}^{m} c_i[Q_i|\right) = 0. \tag{15}$$

We shall consider $m = 2, 3$ in the following.

For $m = 2$, we first look at Q_1 and Q_2 both of which are external momenta of positive helicity gluons. Let us first consider the consequence of $\hat{q}_1^- = q_1^-$ and $\hat{q}_2^- = q_2^-$, which yield $d_1\langle +Q_1\rangle = d_2\langle +Q_2\rangle = 0$, or

$$d_1 = d_2 = 0, \tag{16}$$

because otherwise we shall need $\langle +Q_1\rangle = 0$, and/or $\langle +Q_2\rangle = 0$. These conditions will give rise to very badly behaved polarizations $\epsilon^+(Q_1)$ and/or $\epsilon^+(Q_2)$.

Enforcing (16), (15) is now $c_1|Q_1] + c_2|Q_2] = 0$, which in turn gives

$$c_1 = c_2 = 0, \tag{17}$$

meaning no shift. We are then led to the conclusion that we cannot analytically continue into complex momenta by shifting only two of the momenta, if both of their gluons have the same helicity. It then follows trivially that for amplitudes with all positive or negative helicity, one has to shift more than two momenta for continuation.

Now we turn to the case when Q_1 comes with positive helicity but Q_2 with negative helicity. Our requirement $\hat{q}^{\pm}_{1,2} = q^{\pm}_{1,2}$ yields

$$c_1[Q_1-] = 0, \qquad c_2[Q_2-] = 0, \tag{18}$$

and

$$d_1\langle +Q_1\rangle = 0, \qquad d_2\langle +Q_2\rangle = 0. \tag{19}$$

We pick

$$d_1 = 0 \quad \text{and} \quad c_2 = 0 \tag{20}$$

so that $\epsilon^+(Q_1)$ and $\epsilon^-(Q_2)$ behave properly. In order not to have a trivial shift, we must have also $[Q_1-] = 0$ and $\langle Q_2+\rangle = 0$, or

$$|Q_1] = |-], \quad \text{and} \quad |Q_2\rangle = |+\rangle. \tag{21}$$

An additional condition is from (15), which is $c_1 + d_2 = 0$. Because z is arbitrary, we can just scale them to

$$c_1 = -d_2 = 1. \tag{22}$$

This is the standard shift used by BCFW.[1]

Next we look at $m = 3$. Since one-loop rational gluon amplitudes are our focus, we first look at the case when all external gluons have the same helicity. To be specific, we consider a one-loop amplitude with all external gluons having positive helicities. At the one-loop level, the space-cone gauge will not produce vertices which depend on p_j^+ or q_i^+. This leads to

$$c_1 = [Q_2Q_3], \quad c_2 = [Q_3Q_1], \quad c_3 = [Q_1Q_2]. \tag{23}$$

These shifts were used for a different purpose by Risager in Ref.[13]

For a one-loop amplitude of the type $(- + \cdots +)$, the space-cone gauge Feynman diagrams are constructed from trivalent vertices only. We are also making the identification $|+\rangle \equiv |Q_1\rangle$, where $Q_1 = |Q_1\rangle[Q_1|$ is the momentum of the external negative helicity gluon. Then the only $(--+)$ vertex is the one where one of the legs is Q_1^-, and all other vertices are of the type $(++-)$. Moreover, the vertex times external line factor for a vertex (Q_1^-, Q_2^+, K^-) is $\epsilon^+(Q_2)k$. This means that the vertices depend only on q_i^+, q_i momentum components. Therefore, a possible triple momentum shift which leaves the vertex structure and polarizations invariant is

$$\begin{aligned} &|+\rangle \equiv |Q_1\rangle, |-] = \text{arbitrary}, |\zeta\rangle = |+\rangle, |\xi] = |-] \\ &d_1 = [Q_2 Q_3], \qquad c_2 = [Q_3 -], \qquad c_3 = [-Q_2]. \end{aligned} \tag{24}$$

3. On-shell recurrence relations for one-loop rational gluon amplitudes

3.1. *One-loop same helicity gluon amplitudes*

We begin by making the observation that in the space-cone gauge a one loop same helicity gluon amplitude is built only out of 3-point vertices. To be specific, let us consider 1-loop amplitudes with positive helicity external gluons. Then all the vertices will be trivalent: $(++-)$.

As discussed in the previous section, we can select three external gluons and shift their momenta according to

$$\begin{aligned} P_1 = |1\rangle[1| \to \hat{P}_1 = |\hat{1}\rangle[1| = (|1\rangle + z[23]|+\rangle)[1| \\ P_2 = |2\rangle[2| \to \hat{P}_2 = |\hat{2}\rangle[2| = (|2\rangle + z[31]|+\rangle)[2| \\ P_3 = |3\rangle[3| \to \hat{P}_3 = |\hat{3}\rangle[3| = (|3\rangle + z[12]|+\rangle)[3|. \end{aligned} \tag{25}$$

This leads to a recurrence relation of the type

$$\begin{aligned} A_n^{(1)}(P_1, P_2, P_3, P_4, \dots P_n) &= A_3^{(0)}(\widehat{P}_1, \widehat{P}_2, K)\frac{1}{2P_1\cdot P_2}A_{n-1}^{(1)}(K, \widehat{P}_3, P_4 \dots P_n)\Big|_{z_{12}} \\ &+ A_3^{(0)}(\widehat{P}_2, \widehat{P}_3, K)\frac{1}{2P_2\cdot P_3}A_{n-1}^{(1)}(K, P_4, \dots P_n \widehat{P}_1)\Big|_{z_{23}} \\ &+ A_3^{(0)}(\widehat{P}_3, P_4, K)\frac{1}{2P_3\cdot P_4}A_{n-1}^{(1)}(K, \dots P_n, \widehat{P}_1, \widehat{P}_2)\Big|_{z_3} \\ &+ A_3^{(0)}(P_n, \widehat{P}_1, K)\frac{1}{2P_n\cdot P_1}A_{n-1}^{(1)}(K, \widehat{P}_2, \dots P_{n-1})\Big|_{z_n} \end{aligned}$$

where the superscripts $(0,1)$ indicate whether the on-shell amplitude is tree or one-loop level and the hats denote the shifts made such that the line cut is put on-shell. The z shifts corresponding to the four terms which appear in the previous recurrence relation are given by

$$z_{12} = -\frac{\langle 12\rangle}{[23]\langle +2\rangle + [31]\langle 1+\rangle}, \quad z_{23} = -\frac{\langle 23\rangle}{[31]\langle +3\rangle + [21]\langle +2\rangle}$$
$$z_{34} = -\frac{\langle 43\rangle}{[12]\langle 4+\rangle}, \qquad z_{n1} = -\frac{\langle n1\rangle}{[23]\langle n+\rangle}. \tag{26}$$

3.2. *Reference-Twistor Independence and One-Loop $(- + + \cdots +)$ Amplitudes*

As argued previously, the analytic continuation is performed by shifting the momenta P_1, P_2, P_n by:

$$\begin{aligned} P_1 &\to \hat{P}_1 = |1\rangle[1| + z[2n]|1\rangle[-| \\ P_2 &\to \hat{P}_2 = |2\rangle[2| + z[n-]|1\rangle[2| \\ P_n &\to \hat{P}_n = |n\rangle[n| + z[-2]|1\rangle[n|. \end{aligned} \tag{27}$$

In addition we shall make the choice $|+\rangle \equiv |1\rangle$, while keeping $|-]$ arbitrary (recall that the space-cone gauge fixing vector is $\eta = |+\rangle[-|$). With this analytic continuation prescription, $A(z)$ vanishes as $z \to \infty$. The physical amplitude $A(z=0)$ can be recovered from a closed contour integral (1):

$$A(z=0) = -\sum_i (z - z_i)\frac{A(z)^{s.p.}}{z}\bigg|_{z=z_i} + \sum_j (z - z_j)^2 \frac{A(z)^{d.p.}}{z^2}\bigg|_{z=z_j},$$

where we have assumed that $A(z)$ has a set of simple poles at z_i in $A(z)^{s.p.}$ and a set of double poles at z_j in $A(z)^{d.p.}$. This is an extended recursion relation.

Now z_i and z_j all depend on the reference twistors $|+\rangle$ and $|-]$. The gauge-fixing vector $\eta = |+\rangle[-|$, if present in a physical amplitude, would even destroy Lorentz invariance. However, if we focus on the right hand side, we actually avail ourselves of an opportunity to study the interplay of reference twistors dependence of individual terms. The dependence must be quite special, because the sum being the physical amplitude should not allow it. In other words, the pieces which depend on $|+\rangle$ and/or $[-|$ must cancel. For a tree amplitude, all the terms on the right hand side are made of known tree amplitudes of smaller number of gluons and therefore the independence is just a check on a calculation. For a loop amplitude, the situation is more interesting, because there are one particle irreducible diagrams. They give rise to contributions to partial amplitudes with simple poles which are buried under those with double poles with $z_i = z_j$. Thus, the right hand side must contain some factors which have not been encountered before. We shall show that, for $(-++\cdots+)$ amplitudes in QCD, the requirement that the sum should be independent of the gauge twistors is enough to determine these unknown 'soft factors' uniquely. That is to say the constraints due to Lorentz invariance are so stringent that the contributions from the irreducible diagrams can be obtained without having to calculate them explicitly.

3.2.1. *The Five-Point (-++++) Amplitude*

We want to further support our contention that the reference twistor (η)-invariance requirement can be stringent enough to be used to determine contributions from irreducible graphs in some cases, by giving another example: $(1^-2^+3^+4^+5^+)$. The analytic continuation is performed as in (27), shifting the external momenta (P_1, P_2, P_5).

We shall make our presentation as much as possible in the z-plane, consistent with what is to be expected of a complexified amplitude $A_5(z)^{(1)}$. Then, the poles in the z-plane come from the vanishing of the invariants

$$\hat{s}_{12} = \langle 12\rangle[2\hat{1}], \quad \hat{s}_{15} = \langle 15\rangle[5\hat{1}], \quad \hat{s}_{23} = \langle\hat{2}3\rangle[32], \quad \hat{s}_{45} = \langle 4\hat{5}\rangle[54],$$

or the vanishing of $[2\hat{1}], [5\hat{1}], \langle\hat{2}3\rangle, \langle 4\hat{5}\rangle$ respectively. They yield

$$[\hat{1}2] = [25][-2](z - z_{12}), \quad [5\hat{1}] = [25][5-](z - z_{15}),$$

$$\langle\hat{2}3\rangle = [5-]\langle 13\rangle(z - z_{23}), \quad \langle 4\hat{5}\rangle = [-2]\langle 41\rangle(z - z_{45}),$$

where

$$z_{12} = -\frac{[12]}{[25][-2]}, \quad z_{15} = -\frac{[51]}{[25][5-]},$$
$$z_{23} = -\frac{\langle 23\rangle}{[5-]\langle 13\rangle}, \quad z_{45} = -\frac{\langle 45\rangle}{[-2]\langle 41\rangle}. \tag{28}$$

To understand the residue at z_{12}, one constructs the vector which corresponds to the intermediate cut channel $\hat{P}_1 + \hat{P}_2 \equiv \hat{K}_{12}$ at that position

$$\hat{K}_{12} \equiv (|1\rangle[\hat{1}| + |\hat{2}\rangle[2|)_{z=z_{12}} = (\frac{[15]}{[25]}|1\rangle + |2\rangle)[2|, \tag{29}$$

after using

$$|\hat{1}]_{z=z_{12}} = \frac{[1-]}{[2-]}|2], \tag{30}$$

and

$$|\hat{2}\rangle_{z=z_{12}} = |2\rangle + \frac{[12][5-]}{[25][2-]}|1\rangle. \tag{31}$$

Thus, its twistors are

$$|\hat{K}_{12}] = |2], \quad |\hat{K}_{12}\rangle = \frac{[15]}{[25]}|1\rangle + |2\rangle, \tag{32}$$

which are η-independent. With these, we calculate

$$A_4^{(1)}(\hat{K}_{12}{}^+, 3^+, 4^+, \hat{5}^+)|_{z=z_{12}} = N\frac{[\hat{K}_{12}3][4\hat{5}]}{\langle\hat{K}_{12}3\rangle\langle 4\hat{5}\rangle}|_{z=z_{12}} = N\frac{[25]^2}{\langle 34\rangle^2}, \tag{33}$$

and

$$A_3^{(0)}(\hat{1}^-, \hat{2}^+, -\hat{K}_{12}{}^-)|_{z=z_{12}} = \frac{\langle 12\rangle[2-]}{[1-]},$$

and form the partial amplitude

$$\begin{aligned} A_{12} &\equiv A_3^{(0)}(\hat{1}^-, \hat{2}^+, -\hat{K}_{12}{}^-)|_{z=z_{12}} \frac{1}{K_{12}^2} A_4^{(1)}(\hat{K}_{12}{}^+, 3^+, 4^+, \hat{5}^+)|_{z=z_{12}} \\ &= N\frac{[25]^2[2-]}{\langle 34\rangle^2[1-][12]}, \end{aligned} \tag{34}$$

where we substituted

$$K_{12} = (P_1 + P_2)^2 = \langle 12\rangle[12].$$

In a similar fashion, we evaluate the residue of $A_5^{(1)}$ at $z = z_{15}$

$$\begin{aligned} A_{15} &\equiv A_3^{(0)}(\hat{1}^-, \hat{5}^+, -\hat{K}_{15}{}^-)|_{z=z_{15}} \frac{1}{K_{15}^2} A_4^{(1)}(\hat{K}_{15}{}^+, \hat{2}^+, 3^+, 4^+)|_{z=z_{15}} \\ &= N\frac{[25]^2[5-]}{\langle 34\rangle^2[1-][15]}. \end{aligned} \tag{35}$$

We note that the reference-twistor dependent part of A_{12} cancels that of A_{15}.

In Figure 1 we have depicted the cut sub-amplitudes A_{12} and A_{15}. The line which is cut (with the cut depicted by a small vertical red line) is placed on-shell by the shifting the momenta P_1, P_2, P_5. The shifts of the *on-shell* momenta, which are always those of the lone negative helicity gluon 1^- and of the two positive helicity gluons adjacent to it, are represented by blue lines. The letter L stands for a one-loop amplitude. The helicities of the on-shell gluons are specified as well:+/− for a positive/negative helicity gluon.

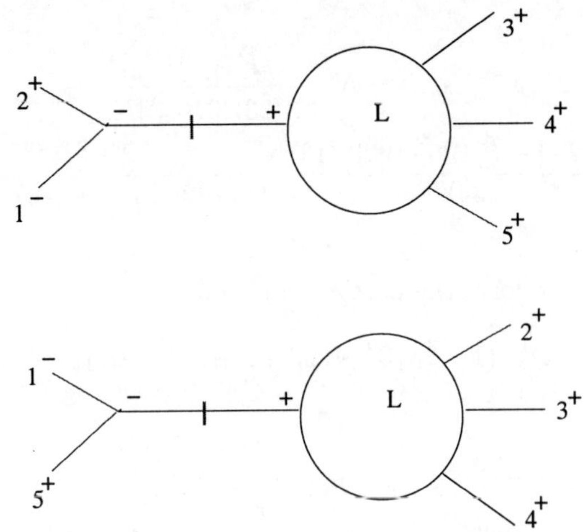

Fig. 1. A_{12} and A_{15}

We turn to the residues at z_{45}. The relevant twistors for the cut line are

$$|\hat{K}_{45}\rangle = |4\rangle, \quad |\hat{K}_{45}] = |4] + \frac{\langle 51\rangle}{\langle 41\rangle}|5],$$

which form

$$\hat{K}_{45} \equiv (|4\rangle[4| + |\hat{5}\rangle[5|)_{z=z_{45}}.$$

With these, after a fair amount of algebra, we have

$$\begin{aligned} A_3^{(0)}(-\hat{K}_{45}{}^-, 4^+, \hat{5}^+) &= \frac{[4\hat{5}]^3}{[4\hat{K}_{45}][\hat{K}_{45}\hat{5}]} \\ &= \frac{\langle 14\rangle[45]}{\langle 15\rangle}, \end{aligned} \tag{36}$$

$$\begin{aligned} A_3^{(1)}(-\hat{K}_{45}{}^+, 4^+, \hat{5}^+) &= -N\frac{[4\hat{5}][\hat{5}\hat{K}_{45}][\hat{K}_{45}4]}{K_{45}^2} \\ &= N\frac{\langle 15\rangle[45]^3}{\langle 14\rangle}\frac{1}{K_{45}^2}, \end{aligned} \tag{37}$$

$$\begin{aligned} A_4^{(0)}(\hat{1}^-, \hat{2}+, 3^+, \hat{K}_{45}{}^-) &= \frac{\langle \hat{K}_{45}\hat{1}\rangle^3}{\langle \hat{1}\hat{2}\rangle\langle \hat{2}3\rangle\langle 3\hat{K}_{45}\rangle} \\ &= \frac{\langle 41\rangle^3}{\langle 12\rangle\langle 34\rangle}\cdot\frac{[-2]\langle 41\rangle}{\langle 23\rangle[-2]\langle 41\rangle + \langle 45\rangle[5-]\langle 13\rangle}, \end{aligned} \tag{38}$$

and

$$A_4^{(1)}(\hat{1}^-,\hat{2}^+,3^+,\hat{K}_{45}{}^+) = N\frac{\langle\hat{2}\hat{K}_{45}\rangle[\hat{2}\hat{K}_{45}]^3}{[\hat{1}\hat{2}]\langle\hat{2}3\rangle\langle 3\hat{K}_{45}\rangle[\hat{K}_{45}\hat{1}]}$$
$$= N\frac{(\langle 24\rangle[2-] + \langle 54\rangle[5-])[23]\langle 13\rangle^3}{\langle 34\rangle^2}\frac{[-2]\langle 14\rangle}{(\langle 23\rangle[2-]\langle 41\rangle + \langle 45\rangle[5-]\langle 13\rangle)^2}, \tag{39}$$

From these, we form the partial amplitudes

$$A_{45}^{(a)} \equiv A_4^{(0)}(\hat{1}^-,\hat{2}^+,3^+,\hat{K}_{45}{}^-)\frac{1}{K_{45}^2}A_3^{(1)}(-\hat{K}_{45}{}^+,4^+,\hat{5}^+) \tag{40}$$

and

$$A_{45}^{(b)} \equiv A_4^{(1)}(\hat{1}^-,\hat{2}^+,3^+,\hat{K}_{45}{}^+)\frac{1}{K_{45}^2}A_3^{(0)}(-\hat{K}_{45}{}^-,4^+,\hat{5}^+). \tag{41}$$

In the like manner, we construct the residues at $z = z_{23}$, where the twistors are

$$|\hat{K}_{23}\rangle = |3\rangle, \quad |\hat{K}_{23}] = \frac{\langle 12\rangle}{\langle 13\rangle}|2] + |3],$$

for

$$\hat{K}_{23} \equiv (|\hat{2}\rangle[2| + |3\rangle[3|)_{z=z_{23}},$$

from which we obtain

$$A_3^{(0)}(-\hat{K}_{23}{}^-,\hat{2}^+,3^+) = \frac{\langle 13\rangle[23]}{\langle 12\rangle}, \tag{42}$$

$$A_3^{(1)}(-\hat{K}_{23}{}^+,\hat{2}^+,3^+) = -N\frac{\langle 12\rangle[23]^3}{\langle 13\rangle}\frac{1}{K_{23}^2}, \tag{43}$$

$$A_4^{(0)}(4^+,\hat{5}^+,\hat{1}^-,\hat{K}_{23}{}^-) = \frac{\langle 1\hat{K}_{23}\rangle^3}{\langle\hat{K}_{23}4\rangle\langle 4\hat{5}\rangle\langle\hat{5}1\rangle},$$
$$= \frac{\langle 13\rangle^3}{\langle 34\rangle\langle 51\rangle}\cdot\frac{[-5]\langle 12\rangle}{\langle 23\rangle[-2]\langle 41\rangle + \langle 45\rangle[5-]\langle 13\rangle} \tag{44}$$

and

$$A_4^{(1)}(\hat{1}^-,\hat{K}_{23}{}^+,4^+,\hat{5}^+) = N\frac{\langle\hat{K}_{23}\hat{5}\rangle[\hat{5}\hat{K}_{23}]^3}{[\hat{1}\hat{K}_{23}]\langle\hat{K}_{23}4\rangle\langle 4\hat{5}\rangle[\hat{5}\hat{1}]}$$
$$= N\frac{(\langle 32\rangle[2-] + \langle 35\rangle[5-])[45]\langle 14\rangle^3}{\langle 34\rangle^2}\frac{[5-]\langle 13\rangle}{(\langle 23\rangle[2-]\langle 41\rangle + \langle 45\rangle[5-]\langle 13\rangle)^2}. \tag{45}$$

As before, we form partial amplitudes

$$A_{23}^{(a)} \equiv A_4^{(0)}(4^+, \hat{5}^+, \hat{1}^-, \hat{K}_{23}{}^-)\frac{1}{K_{23}^2}A_3^{(1)}(-\hat{K}_{23}{}^+, \hat{2}^+, 3^+)$$
$$A_{23}^{(b)} \equiv A_4^{(1)}(4^+, \hat{5}^+, \hat{1}^-, \hat{K}_{23}{}^+)\frac{1}{K_{23}^2}A_3^{(0)}(-\hat{K}_{23}{}^-, \hat{2}^+, 3^+).$$

In Figure 2 we depicted the remaining cut sub-amplitudes, $A_{23}^{a,b}$ and $A_{45}^{a,b}$. The conventions are the same as before. In addition, the letter T now denotes a tree-level amplitude.

If we just add together $A_{12} + A_{15} + A_{23}^{(a)} + A_{23}^{(b)} + A_{45}^{(a)} + A_{45}^{(b)}$, we shall include only the reducible contributions, which do not constitute the complete amplitude $A_5^{(1)}$, as indicated by the fact that this sum is dependent on the reference twistor $|-]$. We need to add some extra terms. The central issue we posed for ourselves is whether the demand of reference-twistor invariance of $A_5^{(1)}$ will be sufficient to determine these extras. We are going to show the affirmative. As we saw $A_{12} + A_{15}$ is η-independent. Next, we should pair $A_{45}^{(a)}$ with $A_{23}^{(b)}$ and $A_{45}^{(b)}$ with $A_{23}^{(a)}$, based on the appearance of factors $\langle 13\rangle^3[23]/\langle 15\rangle$ and $\langle 14\rangle^3[45]/\langle 12\rangle$ and others. We propose to add the extra terms separately to these pairs and we further parameterize them as $A_{45}^{(a)} f_{45}$ and $A_{23}^{(a)} f_{23}$. We will show that the expression of the 'soft factors' f_{23} and f_{45} can be secured by requiring that

$$A_{23}^{(b)} + A_{45}^{(a)}(1 + f_{45}) \text{ and } A_{45}^{(b)} + A_{23}^{(a)}(1 + f_{23}) \tag{46}$$

should be made η-independent individually.

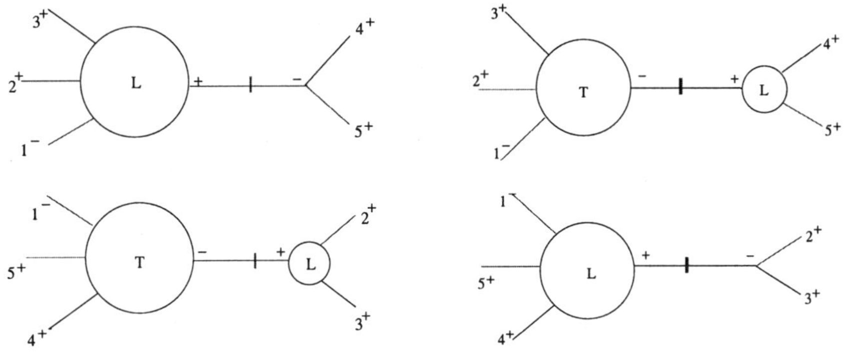

Fig. 2. A_{45}^b and A_{23}^a on the left side figure, and A_{45}^a and A_{23}^b to the right

By inspection, the soft factor f_{45} must be of the type

$$f_{45} = \frac{A[2-]+B[5-]}{C[2-]+D[5-]}$$

where A, B, C, D depend only on the external momenta. Similar requirements hold for f_{23}. The solution for f_{45} and f_{23} is unique,

$$f_{45} = -\frac{\langle 13\rangle\langle 14\rangle\langle 45\rangle(\langle 32\rangle[2-]+\langle 35\rangle[5-])}{\langle 15\rangle\langle 34\rangle(\langle 13\rangle\langle 45\rangle[-5]+\langle 14\rangle\langle 23\rangle[2-])},$$

$$f_{23} = -\frac{\langle 13\rangle\langle 14\rangle\langle 23\rangle(\langle 24\rangle[2-]+\langle 45\rangle[-5])}{\langle 12\rangle\langle 34\rangle(\langle 13\rangle\langle 45\rangle[-5]+\langle 14\rangle\langle 23\rangle[2-])}. \tag{47}$$

In summary, we have shown that the one loop complete amplitude for $(1^-2^+3^+4^+5^+)$ computed recursively yields, as expected

$$A_5^{(1)}(1^-,2^+,3^+,4^+,5^+) = \frac{N}{\langle 34\rangle^2}\left[-\frac{[25]^3}{[12][51]}+\frac{\langle 14\rangle^3[45]\langle 35\rangle}{\langle 12\rangle\langle 23\rangle\langle 45\rangle^2}-\frac{\langle 13\rangle^3[32]\langle 42\rangle}{\langle 15\rangle\langle 54\rangle\langle 32\rangle^2}\right].$$

Since now we are guaranteed to arrive at $|-]$-independent expressions, we are free to choose the reference twistor $|-]$ such that the recursive relation is as simple as possible. There are two possible choices that lead to simplifications:$|-] = |2]$, or $|-] = |5]$. Then our triple shifts reduce to BCFW double shifts, and we recover the soft factors of BDK corresponding to a BCFW shift.

3.2.2. *A generic (-++... +) Amplitude*

For a general amplitude $(1^-2^+\dots n^+)$, the soft factors are

$$f_{23} = \left(\frac{\langle 13\rangle\langle 23\rangle}{\langle 12\rangle\langle 34\rangle}\right)\left(\frac{[n\,-]\langle 14\rangle\langle n-1\;n\rangle-[-2]\langle 24\rangle\langle n-1\;1\rangle}{[n\,-]\langle 13\rangle\langle n-1\;n\rangle-[-2]\langle 23\rangle\langle n-1\;1\rangle}\right), \tag{48}$$

and

$$f_{n-1\;n} = \left(\frac{\langle 1\;n-1\rangle\langle n\;n-1\rangle}{\langle 1\;n\rangle\langle n-1\;n-2\rangle}\right)$$
$$\times\left(\frac{[n-]\langle 13\rangle\langle n-2\;n\rangle-[-2]\langle 23\rangle\langle n-2\;1\rangle}{[n\,-]\langle 13\rangle\langle n-1\;n\rangle-[-2]\langle 23\rangle\langle n-1\;1\rangle}\right), \tag{49}$$

where we have made the shifts given in (27).

References

1. R. Britto, F. Cachazo, B. Feng and E. Witten, Phys. Rev. Lett. **94**, 181602 (2005) [arXiv:hep-th/0501052].
2. Z. Bern, G. Chalmers, L. J. Dixon and D. A. Kosower, Phys. Rev. Lett. **72**, 2134 (1994) [arXiv:hep-ph/9312333].

3. G. Mahlon, Phys. Rev. D **49**, 4438 (1994) [arXiv:hep-ph/9312276].
4. F. A. Berends and W. T. Giele, Nucl. Phys. B **306**, 759 (1988).
5. D. Vaman and Y. P. Yao, JHEP **0604**, 030 (2006) [arXiv:hep-th/0512031].
6. G. Chalmers and W. Siegel, Phys. Rev. D **59**, 045013 (1999) [arXiv:hep-ph/9801220].
7. D. Vaman and Y. P. Yao, Mod. Phys. Lett. A **23**, 1 (2008)
8. Z. Bern, L. J. Dixon and D. A. Kosower, Phys. Rev. D **71**, 105013 (2005) [arXiv:hep-th/0501240].
9. C. Schwinn and S. Weinzierl, JHEP **0704**, 072 (2007) [arXiv:hep-ph/0703021].
10. Z. Bern, L. J. Dixon and D. A. Kosower, Annals Phys. **322**, 1587 (2007) [arXiv:0704.2798 [hep-ph]].
    Z. Bern *et al.* [NLO Multileg Working Group], arXiv:0803.0494 [hep-ph].
11. C. F. Berger *et al.*, arXiv:0803.4180 [hep-ph].
12. N. Arkani-Hamed and J. Kaplan, JHEP **0804**, 076 (2008) [arXiv:0801.2385 [hep-th]].
13. K. Risager, JHEP **0512**, 003 (2005) [arXiv:hep-th/0508206].
14. Z. Bern, L. J. Dixon and D. A. Kosower, Phys. Rev. Lett. **70**, 2677 (1993) [arXiv:hep-ph/9302280].
15. Z. Bern and D. A. Kosower, Nucl. Phys. B **379**, 451 (1992).

SECTION 2.
HEAVY FLAVOR PHYSICS

EXOTIC $c\bar{c}$ MESONS

ERIC BRAATEN
Department of Physics, The Ohio State University, Columbus, OH 43210, USA
** E-mail: braaten@mps.ohio-state.edu*

A surprising number of new $c\bar{c}$ mesons with masses above the $D\bar{D}$ threshold have been discovered at the B factories. Some of them are ordinary charmonium states, but others are definitely exotic mesons. The current theoretical status of the new $c\bar{c}$ mesons is summarized.

Keywords: Charmonium; QCD; exotic mesons.

1. New $c\bar{c}$ mesons

The discovery of charmonium in November 1974 marked the beginning of the modern era of particle physics. Within a few years, 10 charmonium states had been discovered. These states are shown in Fig. 1. The charmonium states below the $D\bar{D}$ threshold at 3729 MeV were the 0^{-+} ground state η_c, the 1^{--} states J/ψ and $\psi(2S)$, and the 0^{++}, 1^{++}, and 2^{++} states $\chi_{c0}(1P)$, $\chi_{c1}(1P)$, and $\chi_{c2}(1P)$. The only complete charmonium multiplet was the $1S$ multiplet consisting of η_c and J/ψ. The $c\bar{c}$ mesons above the $D\bar{D}$ threshold were four 1^{--} states: $\psi(3770)$, $\psi(4035)$, $\psi(4253)$, and $\psi(4421)$, which can be identified as members of the $1D$, $3S$, $2D$, and $4S$ charmonium multiplets. There were no definitive discoveries of any new $c\bar{c}$ mesons for more than 20 years.

The B factories, which began operating near the beginning of the century, have turned out to be surprisingly effective machines for studying $c\bar{c}$ mesons. The new $c\bar{c}$ mesons discovered since 2002 are shown in Fig. 1. There are now three complete multiplets below the $D\bar{D}$ threshold: $1S$, $1P$, and $2S$. Above the $D\bar{D}$ threshold there are four new 1^{--} charmonium states, a 1^{++} state called the $X(3872)$, the 2^{++} state $\chi_{c2}(2P)$, and three states whose J^{PC} quantum numbers are not yet known. In addition, there is a charged $c\bar{c}$ meson called $Z^{\pm}(4430)$. These new states provide a challenge to our understanding of QCD. There are several recent reviews of the new $c\bar{c}$

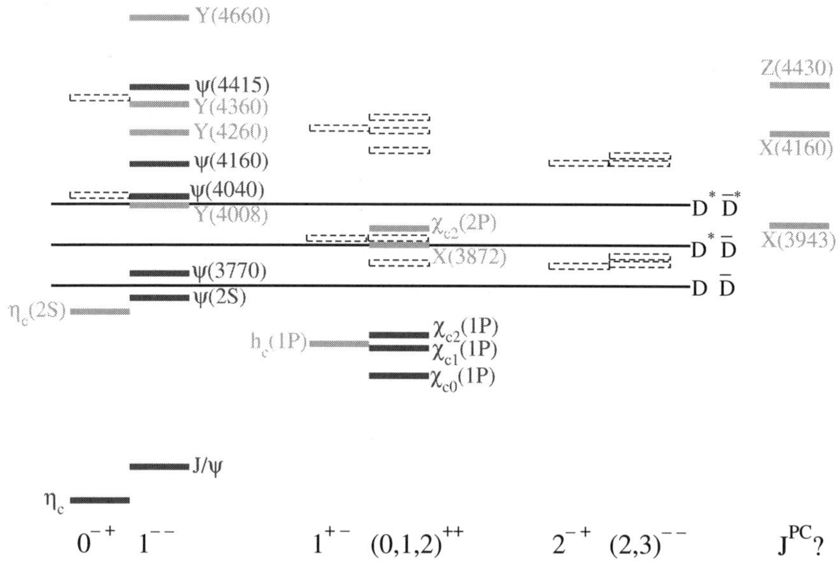

Fig. 1. The spectrum of $c\bar{c}$ mesons separated into J^{PC} channels. The darker (red) states were discovered before the B factories. The lighter (orange) states were discovered since 2002. The open boxes represent missing charmonium states predicted by potential models. The horizontal lines are thresholds for pairs of charm mesons.

mesons.[1–4] I will describe three of the discoveries in more detail.

1.1. $X(3872)$

The $X(3872)$ was discovered by the Belle Collaboration in November 2003 through the decay $B^+ \to K^+ + X$ followed by $X \to J/\psi\,\pi^+\pi^-$.[5] The discovery has been confirmed by the CDF, BaBar, and D0 collaborations.[6–8] The mass of the $X(3872)$ is 3871.4±0.6 MeV. Its width is less than 2.3 MeV at the 90% confidence level.[5] Its quantum numbers are strongly preferred to be 1^{++}.[9,10] There are two puzzling properties of the $X(3872)$ that indicate that it is not an ordinary charmonium state:

(1) Its decays violate isospin symmetry. The discovery decay mode $J/\psi\,\pi^+\pi^-$ is dominated by J/ψ and a virtual ρ meson, which is a final state with $I = 1$. However the decay of X into $J/\psi\,\pi^+\pi^-\pi^0$ has also been observed with a comparable branching fraction.[11] This decay is dominated by J/ψ and a virtual ω meson, which is a final state with $I = 0$.

(2) Its measured mass depends on the decay channel. The measured mass in the $D^0\bar{D}^0\pi^0$ decay channel is higher than in the $J/\psi\,\pi^+\pi^-$ channel by 3.8 ± 1.1 MeV.[12,13]

1.2. *New 1^{--} $c\bar{c}$ mesons*

Four new 1^{--} $c\bar{c}$ mesons have been discovered by the Belle and Babar collaborations through the initial state radiation process $e^+e^- \to \gamma + e^+e^- \to \gamma + Y$. The $Y(4008)$ and $Y(4260)$ were discovered through their decays into $J/\psi\,\pi^+\pi^-$.[14,15] The $Y(4360)$ and $Y(4660)$ were discovered through their decays into $\psi(2S)\,\pi^+\pi^-$.[16,17] They do not seem to be ordinary 1^{--} charmonium states, because no resonant peaks are observed in the inclusive e^+e^- cross section at these energies. There are also no resonant peaks in the exclusive e^+e^- cross sections into $D\bar{D}$, $D^*\bar{D}$, or $D^*\bar{D}^*$, in spite of the masses being well above the thresholds for these decays.

1.3. $Z^\pm(4430)$

The $Z^\pm(4430)$ was discovered by the Belle Collaboration in August 2007 through the decay $B^+ \to K^0 + Z^+$ followed by $Z^+ \to \psi(2S)\,\pi^+$.[18] Its mass and width are 4433 ± 5 MeV and 45^{+35}_{-18} MeV, respectively. Its J^P quantum numbers are not known, but it has $I^G = 1^+$. Since the decay products $\psi(2S)$ and π^+ have constituents $c\bar{c}$ and $u\bar{d}$, respectively, the Z^+ must have constituents $c\bar{c}u\bar{d}$. Thus $Z^\pm(4430)$ is a manifestly exotic meson.

2. What are they?

What are the new $c\bar{c}$ mesons above the $D\bar{D}$ threshold? Some of them could be ordinary charmonium ($c\bar{c}$), but some of them are definitely exotic mesons. The exotic possibilities include charmonium hybrids ($c\bar{c}g$), tetraquark mesons ($c\bar{c}q\bar{q}$), and sexaquark mesons ($c\bar{c}qq\bar{q}\bar{q}$). A tetraquark meson could be a compact color-singlet state, but it could also have substructure. The possible substructures consisting of two clusters include charm meson molecules, diquark–antidiquark states, and hadrocharmonium. I will discuss several of these possibilities in more detail.

2.1. *Charmonium*

Potential models are a phenomenological framework for charmonium that is well developed in most respects. The exception is the effect of couplings to charm meson pairs, which are essential for accurate predictions above the $D\bar{D}$ threshold. Updated predictions for the charmonium spectrum from potential models have been given by Barnes, Godfrey, and Swanson[19] and by Eichten, Lane, and Quigg.[20] The missing charmonium states (and their quantum numbers) include

- one member of the $3S$ multiplet: $\eta_c(3S)$ (0^{-+}),
- three members of the $2P$ multiplet: $\chi_{c0}(2P)$, $\chi_{c1}(2P)$, and $h_c(2P)$ $(0^{++}$, 1^{++}, and $1^{+-})$,
- three members of the $1D$ multiplet: $\psi_2(1D)$, $\psi_3(1D)$, and $\eta_{c2}(1D)$ $(2^{--}$, 3^{--}, and $2^{-+})$.

There are also entire missing multiplets: $4S$, $3P$, $2D$, ... The missing charmonium states are shown as open boxes in Fig. 1. While some of the new $c\bar{c}$ mesons lie close to the predicted mass of a missing charmonium state, most of them do not.

2.2. *Charm meson molecules*

Charm meson molecules are tetraquark $c\bar{c}$ mesons that consist of a pair of charm mesons ($c\bar{q}$ and $\bar{c}q$). The interactions of the charm mesons are constrained by experimental data. The first quantitative predictions for the spectrum of charm meson molecules were made by Tornqvist in 1991.[21,22] He used a meson potential model with one-pion-exchange interactions and an ultraviolet cutoff on the $1/r^3$ potential that was tuned to reproduce the binding energy of the deuteron. Tornqvist predicted charm meson molecules near threshold in several J^{PC} channels with $I = 0$:

- For $D\bar{D}$, whose threshold is at 3729 MeV, there should be none.
- For $D^*\bar{D}/D\bar{D}^*$, there should be molecules near the threshold at 3872 MeV in the 0^{-+} and 1^{++} channels.
- For $D^*\bar{D}^*$, there should be molecules near the threshold at 4014 MeV in the 0^{++}, 0^{-+}, 1^{+-}, and 2^{++} channels.

Tornqvist essentially predicted the $X(3872)$ in 1994, although he did not anticipate the importance of the isospin splittings between D^+ and D^0 and between D^{*+} and D^{*0}.

Updated predictions of a meson potential model have been given by Swanson.[1] He tuned the ultraviolet cutoff on the $1/r^3$ potential to obtain the observed mass of the $X(3872)$, and he also included quark-exchange interactions. Besides the $X(3872)$, whose binding energy was used as an input, the only other molecular state that is predicted to be bound is a 0^{++}

$D^*\bar{D}^*$ state at 4013 MeV. The meson potential model can also be applied to bottom meson molecules. There should be two $B^*\bar{B}/B\bar{B}^*$ molecules (0^{-+} and 1^{++}) and four $B^*\bar{B}^*$ molecules (0^{++}, 0^{-+}, 1^{+-}, and 2^{++}). Swanson predicts their binding energies to range from 40 to 70 MeV.

2.3. *Tetraquark mesons*

One of the possible substructures for a tetraquark $c\bar{c}$ meson is a diquark (cq) and an antidiquark ($\bar{c}\bar{q}$). The problem can be dramatically simplified by treating the constituent diquarks as point particles. This approach has been followed by Maiani, Piccinini, Polosa, and Riquer,[23–25] by Ishida, Ishida, and Maeda,[26] by Ebert, Faustov, and Galkin,[27] and by Karliner and Lipkin.[28,29] The most attractive color channel for a constituent diquark cq is 3^*. If the diquark has no internal orbital angular momentum, it has two possible states S and A corresponding to total spin 0 and 1, respectively. The possible S-wave tetraquarks (and their quantum numbers) are $S\bar{S}$ (0^{++}), $A\bar{S}/S\bar{A}$ (1^{++}, 1^{+-}), and $A\bar{A}$ (0^{++}, 1^{+-}, 2^{++}). For each of these 6 states, there is a flavor multiplet. If we consider the three light quarks $q = u, d, s$, the flavor multiplet consists of 9 states. If we only consider two light quarks $q = u, d$, the flavor multiplet consists of 4 states: an isospin triplet (X^-, X^0, X^+) and an isospin singlet $X^{0'}$.

One problem with the interpretation of new $c\bar{c}$ mesons as diquark–antidiquark bound states is that too many other such states are predicted. If we only consider diquarks with color 3^* and take the diquark–antidiquark system to be in its lowest state, $6 \times 9 = 54$ states are predicted. There are many additional states if we also consider diquarks in the 6 color state or if we allow orbital-angular-momentum or radial excitations of the diquark–antidiquark system. With all these possible states, it is easy to fit the mass of any new $c\bar{c}$ meson above the $D\bar{D}$ threshold. The challenge is then to explain why all the other predicted states have not been observed.

One way to avoid predicting too many tetraquark states is to start from the 4-body problem for constituent quarks. Vijande, Valcarce, et al.[30,31] and Hiyama, Suganama, and Kamimura[32] have solved this 4-body problem numerically. They have shown that there are no stable $c\bar{c}q\bar{q}$ states with only 2-body color-dependent forces.

A more fundamental approach to the problem of predicting the spectrum of tetraquark mesons is to use QCD sum rules to determine the most attractive channels. Navarra, Nielsen, et al.[33,34] have shown that the QCD sum rules are consistent with tetraquark states in channels associated with several of the new $c\bar{c}$ mesons, including $X(3872)$ and $Z^{\pm}(4430)$.

The spectrum of tetraquark $c\bar{c}$ mesons should eventually be calculable using lattice QCD. One problem is that dynamical light quarks are essential for calculating the spectrum of $c\bar{c}$ mesons above the $D\bar{D}$ threshold and this makes the calculations computationally demanding. The easiest masses to calculate are those for states with exotic quantum numbers. The calculations for states with the same quantum numbers as excited charmonium states are much more difficult.

2.4. *Charmonium hybrids*

Charmonium hybrids are $c\bar{c}$ mesons in which the gluon field is in an excited state. If the excitation of the gluon field is interpreted as a constituent gluon, the constituents of the charmonium hybrid are $c\bar{c}g$. However there is no reason to expect the excitation of the gluon field to have particle-like behavior.

The spectrum of charmonium hybrids has been calculated using lattice gauge theory. The masses obtained thus far are not definitive, because they have been calculated without dynamical light quarks. The spectrum was calculated using a Born-Oppenheimer approximation by Juge, Kuti, Morningstar.[35] In this approximation, the lowest multiplet consists of degenerate states with the quantum numbers 0^{-+}, 0^{+-}, 1^{--}, 1^{++}, 1^{+-}, 1^{-+}, 2^{-+}, and 2^{+-}. Their result for the mass is approximately 4200 MeV. The energies of charmonium hybrids with exotic quantum numbers have been calculated using conventional lattice gauge theory by Liao and Manke[36] and by Liu and Luo.[37] Their results for the masses of the lowest states with the exotic quantum numbers 1^{-+}, 0^{+-}, and 2^{+-} are 4400, 4700, and 4900 MeV, respectively. Since the lowest Born-Oppenheimer multiplet includes a 1^{--} state, there should also be a 1^{--} state in this mass range. The masses of the new 1^{--} states $Y(4260)$, $Y(4360)$, and $Y(4660)$ are compatible with this mass range.

An important selection rule for the decays of charmonium hybrids has been derived by Isgur, Kokoski, and Paton,[38] by Close and Page,[39] and by Kou and Pene.[40] Decays of a charmonium hybrid into two S-wave charm mesons, namely $D\bar{D}$, $D^*\bar{D}$, $D\bar{D}^*$, or $D^*\bar{D}^*$, are suppressed. Thus its dominant decay modes are expected to be into a P-wave charm meson and an S-wave charm meson, such as $D_1\bar{D}^{(*)}$ and $D_2\bar{D}^{(*)}$. The new 1^{--} $c\bar{c}$ mesons $Y(4260)$, $Y(4360)$, and $Y(4660)$ have only been observed in either the $J/\psi\,\pi\pi$ or $\psi(2S)\,\pi\pi$ decay channels. Despite the large phase space available, these states have not been seen in the decay modes $D\bar{D}$, $D^*\bar{D}$, or $D^*\bar{D}^*$. This makes them prime condidates for charmonium hybrids.

2.5. *Hadro-charmonium*

Dubynskiy and Voloshin have recently proposed a new possibility for the substructure of tetraquark $c\bar{c}$ mesons: hadro-charmonium, which consists of a charmonium ψ $(c\bar{c})$ and a light meson h $(q\bar{q})$.[41] This possibility is motivated by the multipole expansion for the long-wavelength gluon fields that dominate the interaction between ψ and h at long distances. They pointed out that this interaction is attractive and they suggested that it might be strong enough to form a bound state.

A hadro-charmonium ψh is expected to have substantial decay rates into final states consisting of ψ and decay modes of h. Several of the new $c\bar{c}$ mesons have been observed to decay into final states that include J/ψ but not $\psi(2S)$ or vice versa. This pattern is easily explained if these states are hadro-charmonia. The $Y(4008)$ and $Y(4260)$, which decay into $J/\psi\,\pi\pi$, might be hadro-charmonium states containing a J/ψ, while $Y(4360)$ and $Y(4660)$, which decay via $\psi(2S)\,\pi\pi$, might be hadro-charmonium states containing a $\psi(2S)$. The $Z^{\pm}(4430)$, which decays into $\psi(2S)\,\pi^{\pm}$, might be a hadro-charmonium state containing $\psi(2S)$.

3. What is the $X(3872)$?

The nature of the $X(3872)$ can be determined unambiguously from the combination of two crucial experimental inputs:

(1) Its mass is extremely close to the $D^{*0}\bar{D}^0$ threshold:

$$M_X - (M_{D^{*0}} + M_{D^0}) = -0.4 \pm 0.7 \text{ MeV}.$$

(2) Its quantum numbers are $J^{PC} = 1^{++}$.

The determination is based on the remarkable universal properties of non-relativistic particles with short-range interactions that have an S-wave resonance near the threshold.[42] The universal properties are determined by the large S-wave scattering length or, equivalently, by the small binding energy. The first fact above implies that the $X(3872)$ has an S-wave coupling to $D^{*0}\bar{D}^0$. The second fact above implies that it is a resonant coupling. We can conclude that $X(3872)$ is a charm meson molecule (bound state or virtual state) whose meson structure is

$$X = \frac{1}{\sqrt{2}}\left(D^{*0}\bar{D}^0 - D^0\bar{D}^{*0}\right). \tag{1}$$

An example of the universal properties is a simple relation between the mean separation of the constituents $\langle r\rangle_X$ and the binding energy E_X:

$\langle r \rangle_X = 1/(8\mu E_X)^{1/2}$, where μ is the reduced mass. This implies that if $E_X < 0.4$ MeV, $\langle r \rangle_X > 3.5$ fm. Thus the constituent mesons in the $X(3872)$ are usually well separated.

This determination of the nature of the $X(3872)$ provides simple explanations for the two puzzling features described in Section 1.1:

(1) The violation of isospin symmetry in decays of the $X(3872)$ is explained by its mass being much closer to the $D^{*0}\bar{D}^0$ threshold than to the $D^{*+}D^-$ threshold, which is higher by about 8 MeV. As a consequence, the meson state of the $X(3872)$, which is given in Eq. (1), is an equal superposition of $I = 0$ and $I = 1$.

(2) The larger mass of X measured in the $D^0\bar{D}^0\pi^0$ decay channel compared to the $J/\psi\,\pi^+\pi^-$ channel is explained by the difference between the line shapes in those two channels. The line shape from $B \to K + D^0\bar{D}^0\pi^0$ is the combination of a resonance below the $D^{*0}\bar{D}^0$ threshold from $B \to K + X$ and a threshold enhancement above the $D^{*0}\bar{D}^0$ threshold from $B \to K + D^{*0}\bar{D}^0$.[43,44] The line shape from $B \to K + J/\psi\,\pi^+\pi^-$ has only the resonance below the $D^{*0}\bar{D}^0$ threshold. Thus its peak represents the true mass of the $X(3872)$ resonance.

An analysis of the Belle data on $B \to K + J/\psi\,\pi^+\pi^-$ and $B \to K + D^0\bar{D}^0\pi^0$ favors the $X(3872)$ being a bound state whose mass is below the $D^{*0}\bar{D}^0$ threshold,[44] although a virtual state whose mass is above the $D^{*0}\bar{D}^0$ threshold is not excluded.[45]

Other properties of the $X(372)$ can be deduced by taking into account charged charm meson pairs, whose threshold is only about 8 MeV higher than that of the neutral charm meson pair $D^{*0}\bar{D}^0$.[46,47] The ratio of the production rates of $X(372)$ in B^0 and B^+ decays can be expressed as[47]

$$\frac{\Gamma[B^0 \to K^0 + X]}{\Gamma[B^+ \to K^+ + X]} = \left| \frac{\gamma_1}{\gamma_1 - \kappa_1} \right|^2,$$

where $\kappa_1 = 125$ MeV and γ_1 is the inverse scattering length for $D^*\bar{D}$ in the $I = 1$ channel. This result supercedes an incorrect prediction by Braaten and Kusunoki that this ratio should be much less than 1.[48] The line shape from $B^+ \to K^+ + J/\psi\,\pi^+\pi^-$ should have a zero about 6 MeV above the $D^{*0}\bar{D}^0$ threshold, while that for $B^0 \to K^0 + J/\psi\,\pi^+\pi^-$ should have a zero about 2 MeV below the $D^{*0}\bar{D}^0$ threshold.[47] This result supercedes an incorrect prediction by Voloshin.[46] In contrast, the line shapes from $B \to K + J/\psi\,\pi^+\pi^-\pi^0$ and $B \to K + D^0\bar{D}^0\,\pi^0$ should have no zeroes in the $D^*\bar{D}$ threshold region.

4. Conclusions

My main conclusions on the new $c\bar{c}$ mesons are as follows:

- The $X(3872)$ is a weakly-bound charm meson molecule whose meson content is given in Eq. (1). It has universal properties that are determined by its small binding energy and are otherwise insensitive to details of QCD, including the mechanism for the binding of the charm mesons.
- The new 1^{--} mesons $Y(4260)$, $Y(4360)$, and $Y(4660)$ are good candidates for charmonium hybrids. This interpretation could be confirmed by observing their decays into $D_1\bar{D}^{(*)}$ and $D_2\bar{D}^{(*)}$.
- The $Z^{\pm}(4430)$ is an exotic tetraquark meson with quark content $c\bar{c}u\bar{d}$. They are the charged members of an isospin multiplet (Z^-, Z^0, Z^+). The implications of this state for other $c\bar{c}$ mesons depends on its J^P quantum numbers, which have not yet been determined.

The $X(3872)$ and $Z^{\pm}(4430)$ provide existence proofs for exotic $c\bar{c}$ mesons. By heavy quark symmetry, replacing the charm quarks by bottom quarks decreases the kinetic energy without significantly changing the potential energy. This implies that exotic $b\bar{b}$ mesons should also exist and have larger binding energies. The challenge for theory is to predict their properties before they are discovered in experiments.

Acknowledgments

This research was supported in part by the U.S. Department of Energy under grant DE-FG02-91-ER40690.

References

1. E. S. Swanson, Phys. Rept. **429**, 243 (2006) [arXiv:hep-ph/0601110].
2. E. Eichten, S. Godfrey, H. Mahlke and J. L. Rosner, arXiv:hep-ph/0701208.
3. M. B. Voloshin, arXiv:0711.4556 [hep-ph].
4. S. Godfrey and S. L. Olsen, arXiv:0801.3867 [hep-ph].
5. S. K. Choi *et al.* [Belle Collaboration], Phys. Rev. Lett. **91**, 262001 (2003) [arXiv:hep-ex/0309032].
6. D. E. Acosta *et al.* [CDF II Collaboration], Phys. Rev. Lett. **93**, 072001 (2004) [arXiv:hep-ex/0312021].
7. B. Aubert *et al.* [BABAR Collaboration], Phys. Rev. D **71**, 071103 (2005) [arXiv:hep-ex/0406022].
8. V. M. Abazov *et al.* [D0 Collaboration], Phys. Rev. Lett. **93**, 162002 (2004) [arXiv:hep-ex/0405004].
9. K. Abe *et al.*, arXiv:hep-ex/0505038.

10. A. Abulencia *et al.* [CDF Collaboration], Phys. Rev. Lett. **98**, 132002 (2007) [arXiv:hep-ex/0612053].
11. K. Abe *et al.*, arXiv:hep-ex/0505037.
12. G. Gokhroo *et al.*, Phys. Rev. Lett. **97**, 162002 (2006) [arXiv:hep-ex/0606055].
13. B. Aubert *et al.* [BABAR Collaboration], Phys. Rev. D **77**, 011102 (2008) [arXiv:0708.1565 [hep-ex]].
14. B. Aubert *et al.* [BABAR Collaboration], Phys. Rev. Lett. **95**, 142001 (2005) [arXiv:hep-ex/0506081].
15. C. Z. Yuan *et al.* [Belle Collaboration], Phys. Rev. Lett. **99**, 182004 (2007) [arXiv:0707.2541 [hep-ex]].
16. B. Aubert *et al.* [BABAR Collaboration], Phys. Rev. Lett. **98**, 212001 (2007) [arXiv:hep-ex/0610057].
17. X. L. Wang *et al.* [Belle Collaboration], Phys. Rev. Lett. **99**, 142002 (2007) [arXiv:0707.3699 [hep-ex]].
18. K. Abe *et al.* [Belle Collaboration], Phys. Rev. Lett. **100**, 142001 (2008) [arXiv:0708.1790 [hep-ex]].
19. T. Barnes, S. Godfrey and E. S. Swanson, Phys. Rev. D **72**, 054026 (2005) [arXiv:hep-ph/0505002].
20. E. J. Eichten, K. Lane and C. Quigg, Phys. Rev. D **73**, 014014 (2006) [Erratum-ibid. D **73**, 079903 (2006)] [arXiv:hep-ph/0511179].
21. N. A. Tornqvist, Phys. Rev. Lett. **67**, 556 (1991).
22. N. A. Tornqvist, Z. Phys. **C61**, 525 (1994).
23. L. Maiani, F. Piccinini, A. D. Polosa and V. Riquer, Phys. Rev. D **71**, 014028 (2005) [arXiv:hep-ph/0412098].
24. L. Maiani, V. Riquer, F. Piccinini and A. D. Polosa, Phys. Rev. D **72**, 031502 (2005) [arXiv:hep-ph/0507062].
25. L. Maiani, A. D. Polosa and V. Riquer, Phys. Rev. Lett. **99**, 182003 (2007) [arXiv:0707.3354 [hep-ph]].
26. M. Ishida, S. Ishida and T. Maeda, arXiv:hep-ph/0509212.
27. D. Ebert, R. N. Faustov and V. O. Galkin, Phys. Lett. B **634**, 214 (2006) [arXiv:hep-ph/0512230].
28. M. Karliner and H. J. Lipkin, Phys. Lett. B **638**, 221 (2006) [arXiv:hep-ph/0601193].
29. M. Karliner and H. J. Lipkin, arXiv:0710.4055 [hep-ph].
30. J. Vijande, E. Weissman, N. Barnea and A. Valcarce, Phys. Rev. D **76**, 094022 (2007) [arXiv:0708.3285 [hep-ph]].
31. J. Vijande, E. Weissman, A. Valcarce and N. Barnea, Phys. Rev. D **76**, 094027 (2007) [arXiv:0710.2516 [hep-ph]].
32. E. Hiyama, H. Suganuma and M. Kamimura, Prog. Theor. Phys. Suppl. **168**, 101 (2007) [arXiv:0707.4044 [hep-ph]].
33. R. D. Matheus, F. S. Navarra, M. Nielsen and R. Rodrigues da Silva, Phys. Rev. D **76**, 056005 (2007) [arXiv:0705.1357 [hep-ph]].
34. S. H. Lee, A. Mihara, F. S. Navarra and M. Nielsen, Phys. Lett. B **661**, 28 (2008) [arXiv:0710.1029 [hep-ph]].
35. K. J. Juge, J. Kuti and C. J. Morningstar, Phys. Rev. Lett. **82**, 4400 (1999)

[arXiv:hep-ph/9902336].
36. X. Liao and T. Manke, arXiv:hep-lat/0210030.
37. X. Q. Luo and Y. Liu, Phys. Rev. D **74**, 034502 (2006) [Erratum-ibid. D **74**, 039902 (2006)] [arXiv:hep-lat/0512044].
38. N. Isgur, R. Kokoski and J. E. Paton, Phys. Rev. Lett. **54**, 869 (1985).
39. F. E. Close and P. R. Page, Nucl. Phys. B **443**, 233 (1995) [arXiv:hep-ph/9411301].
40. E. Kou and O. Pene, Phys. Lett. B **631**, 164 (2005) [arXiv:hep-ph/0507119].
41. S. Dubynskiy and M. B. Voloshin, arXiv:0803.2224 [hep-ph].
42. E. Braaten and H. W. Hammer, Phys. Rept. **428**, 259 (2006) [arXiv:cond-mat/0410417].
43. E. Braaten, M. Kusunoki and S. Nussinov, Phys. Rev. Lett. **93**, 162001 (2004) [arXiv:hep-ph/0404161].
44. E. Braaten and M. Lu, Phys. Rev. D **76**, 094028 (2007) [arXiv:0709.2697 [hep-ph]].
45. C. Hanhart, Yu. S. Kalashnikova, A. E. Kudryavtsev and A. V. Nefediev, Phys. Rev. D **76**, 034007 (2007) [arXiv:0704.0605 [hep-ph]].
46. M. B. Voloshin, Phys. Rev. D **76**, 014007 (2007) [arXiv:0704.3029 [hep-ph]].
47. E. Braaten and M. Lu, Phys. Rev. D **77**, 014029 (2008) [arXiv:0710.5482 [hep-ph]].
48. E. Braaten and M. Kusunoki, Phys. Rev. D **71**, 074005 (2005) [arXiv:hep-ph/0412268].

Search for New Physics in B_s-mixing

Alexander J. Lenz*

Faculty of Physics, University of Regensburg,
D-93040 Regensburg, Germany
**E-mail: alexander.lenz@physik.uni-regensburg.de*

We present the current status of the search for new physics effects in the mixing quantities ΔM_s, $\Delta\Gamma_s$ and ϕ_s of the neutral B_s-system.

Keywords: B_s-mixing, New Physics; Proceedings; World Scientific Publishing.

1. Introduction

Despite the enormous success of the standard model there is still room for new physics to be detected at currently running experiments. Huge efforts have been made in recent years in the precision measurement and precision calculation of flavor physics observables at the B-factories and at TeVatron, see e.g. Ref. [1] for a review and references therein. The system of the neutral B_s mesons seems to be particular promising to find hints for new physics (for a recent review of B-mixing see Ref. [2]): the standard model contribution is suppressed strongly, so even small new physics contributions might be of comparable size and the hadronic uncertainties are under good control.

In the standard model the mixing of neutral B-meson ist described, by the box diagrams, see e.g. [3–8] for more details. The absorptive part Γ_{12} of the box diagrams is sensitive light internal particles and the dispersive part M_{12} is sensitive to heavy internal particles. The two complex quantities M_{12} and Γ_{12} can be related to the following physical quantities:

- The mass difference ΔM_s between the heavy and the light mass eigenstates of the neutral B mesons:

$$\Delta M_s = 2|M_{12}|\,. \tag{1}$$

- The decay rate difference $\Delta\Gamma_s$ between the heavy and the light mass eigenstates of the neutral B mesons:

$$\Delta\Gamma_s = 2|\Gamma_{12}|\cos(\phi_s) \tag{2}$$

with the weak phase $\phi_s := \mathrm{Arg}(-M_{12}/\Gamma_{12})$.

- The tiny CP asymmetries in semileptonic B-decays a_{sl}^s

$$a_{sl}^s = \frac{|\Gamma_{12}|}{|M_{12}|}\sin(\phi_s)\ . \tag{3}$$

For the weak phase ϕ_s different notations are used in the literature, which led already to some confusion. For more details on the definitions see the *Note added* in [7].

Recently there were several claims of possible new physics effects in the B_s-mixing system in the literature:

(1) End of 2006 a 2 σ-deviation was found,[9] if all mixing quantities in the B_s-system were combined.
(2) This was more or less confirmed in july 2007 by UT-Fit.[10]
(3) With new data available UT-fit[11] claimed in march 2008 a 3.7 σ-deviation from the standard model. Since from the experiments (D0 and CDF) the full information about the likelihoods was not available at that time, the combination of the data in Ref. [11] had to rely on some assumptions.
(4) This analysis is currently redone - with the missing experimental information - by CKM Fitter in collaboration with the authors of Ref. [9],[12] preliminary results[13] show a deviation of less than 3 σ.

The above claims are based on the following experimental data for the B_s mixing system, mostly from D0 and CDF:

- The mass difference ΔM_s was measuered at CDF[14] and at D0[15] and the numbers were combined from HFAG[16] to

$$\Delta M_s = 17.78 \pm 0.12\,\mathrm{ps}^{-1}\,. \tag{4}$$

- D0[17] and CDF[18] performed a tagged analysis of the decay $B_s \to J/\Psi\phi$ to determine the decay rate difference $\Delta\Gamma_s$ and the weak mixing angle ϕ_s. HFAG[16] combines the values to, see Fig. (1)

$$\Delta\Gamma_s = 0.154^{+0.054}_{-0.070}\,\mathrm{ps}^{-1}\,, \tag{5}$$

$$\phi_s = -0.77^{+0.29}_{0.37}\,. \tag{6}$$

The result from CDF[18] is now superseeded by Ref. [19].

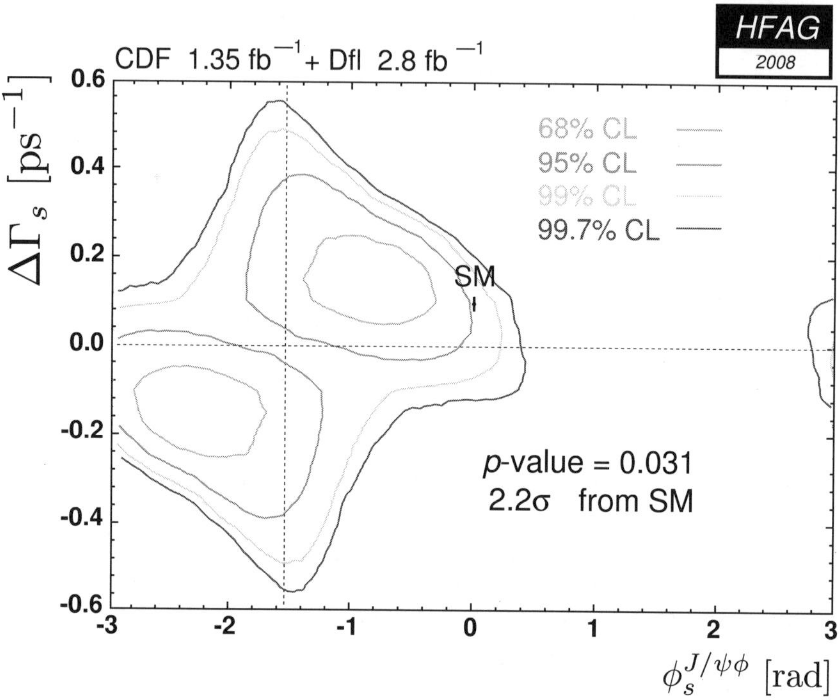

Fig. 1. The combined experimental values for $\Delta\Gamma_s$ and ϕ_s from the tagged analysis of the decay $B_s \to J/\Psi\phi$ from D0 and CDF.

- The semileptonic CP asymmetry can be obtained from the dimuon asymmetry (CDF,[20] D0[21]) or it can be measure directly (D0[22]). These numbers were combined from HFAG[16] to

$$a^s_{sl} = +0.0016 \pm 0.0085 \,. \tag{7}$$

The untagged result from Ref. [[22]] is now superseeded by the new tagged result[23]

$$a^s_{sl} = -0.0024 \pm 0.0117^{+0.0015}_{-0.0024} \,. \tag{8}$$

There are numerous applications of new physics models to the B_s mixing sector, for some recent examples, see e.g. Refs. [24–40]. TeVatron is continuing to take data and we will get more precise data from the upcomming experiments at LHC[41] or possibly at a SuperB-factory[42] running also at the $\Upsilon(5s)$-resonance.

2. Strategy to search for new physics

In [9] we worked out a model independent analysis of new physics effects in B-mixing. Γ_{12} is due to real intermediate states, i.e. particles which are lighter than m_B. Any new physics contributions to Γ_{12} affects also tree-level B-decays. Since no evidence for sizeable new physics effects in tree-level B-decays has been found so far, it reasonable to assume that Γ_{12} is described by the standard model contributions alone. Deviations from that assumption are expected to be smaller than the hadronic unvertainties in the standard model prediction for Γ_{12}. M_{12}, however, might be affected by large new physics effects. We write therefore

$$M_{12}^s = M_{12}^{SM,s} \cdot \Delta = M_{12}^{SM,s} \cdot |\Delta| \cdot e^{i\phi_s^\Delta} \,, \tag{9}$$

$$\Gamma_{12}^s = \Gamma_{12}^{SM,s} \,, \tag{10}$$

where all new physics effects are parameterized by the complex number Δ. Now we can relate the experimental observables in the mixing system with the standard model predictions[9] and with Δ.

$$\begin{aligned} \Delta M_s &= \Delta M_s^{\rm SM} \, |\Delta_s| \\ &= (19.30 \pm 6.74)\, \mathrm{ps}^{-1} \cdot |\Delta_s| \,, \end{aligned} \tag{11}$$

$$\begin{aligned} \Delta \Gamma_s &= 2|\Gamma_{12}^s| \, \cos\left(\phi_s^{\rm SM} + \phi_s^\Delta\right) \\ &= (0.096 \pm 0.039)\, \mathrm{ps}^{-1} \cdot \cos\left(\phi_s^{\rm SM} + \phi_s^\Delta\right) \,, \end{aligned} \tag{12}$$

$$\begin{aligned} \frac{\Delta \Gamma_s}{\Delta M_s} &= \frac{|\Gamma_{12}^s|}{|M_{12}^{\rm SM,s}|} \cdot \frac{\cos\left(\phi_s^{\rm SM} + \phi_s^\Delta\right)}{|\Delta_s|} \\ &= (4.97 \pm 0.94) \cdot 10^{-3} \cdot \frac{\cos\left(\phi_s^{\rm SM} + \phi_s^\Delta\right)}{|\Delta_s|} \,, \end{aligned} \tag{13}$$

$$\begin{aligned} a_{\rm fs}^s &= \frac{|\Gamma_{12}^s|}{|M_{12}^{\rm SM,s}|} \cdot \frac{\sin\left(\phi_s^{\rm SM} + \phi_s^\Delta\right)}{|\Delta_s|} \\ &= (4.97 \pm 0.94) \cdot 10^{-3} \cdot \frac{\sin\left(\phi_s^{\rm SM} + \phi_s^\Delta\right)}{|\Delta_s|} \,, \end{aligned} \tag{14}$$

$$\text{with} \qquad \phi_s^{\rm SM} = (4.2 \pm 1.4) \cdot 10^{-3} \,. \tag{15}$$

By comparing experiment and theory, we can give bounds in the complex Δ-plane [a]. If nature would be such, that Δ has the values:

$$|\Delta| = 0.9 \,, \qquad \phi_s^\Delta = \frac{\pi}{4} \,, \tag{16}$$

---

[a]The bounds in the complex Δ-plane are much more descriptive than in the $|\Delta|$-ϕ_s^Δ-plane, which is used also in the literature.

one would get the bounds shown in Fig. (2).

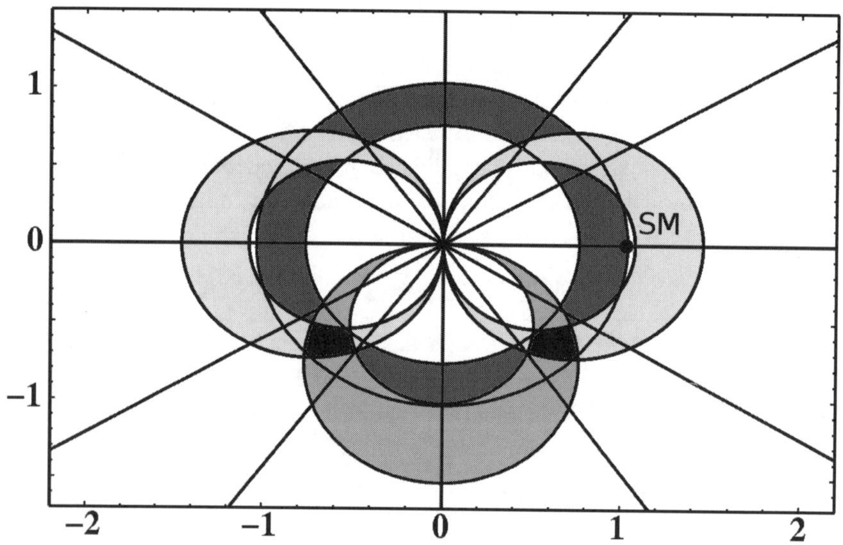

Fig. 2. The bounds in the complex Δ-plane, obtained by comparing experiment and theory for the mixing quantities. The red circle comes from ΔM_s, the yellow band from $\Delta\Gamma_s/\Delta M_s$, the light blue range from the semileptonic CP asymmetries and the rays through the origin from a direct determination of ϕ_s.

- ΔM_s gives a bound on the absolute value of Δ (c.f. Eq. (11)), which is represented by the red band in Fig. (2).
- ϕ_s^Δ can be obtained directly from the angular analysis of the decay $B_s \to J/\Psi\phi$. With a considerably worse accuracy this phase can also be obtained from $\Delta\Gamma_s$ (c.f. Eq. (12)).
- $\Delta\Gamma_s/\Delta M_s$ gives a bound on $\cos\left(\phi_s^{\rm SM} + \phi_s^\Delta\right)/|\Delta|$ (c.f. Eq. (13)), which is represented by the yellow band in Fig. (2).
- $a_{\rm fs}^s$ gives a bound on $\sin\left(\phi_s^{\rm SM} + \phi_s^\Delta\right)/|\Delta|$ (c.f. Eq. (14)), which is represented by the light blue band in Fig. (2).

The overlap of all these bounds gives the values for Re(Δ) and Im(Δ). Within the standard model one has Re(Δ)=1 and Im(Δ)=0.

3. Theoretical framework and uncertainties

In order to fulfill the above described program it is mandatory to have sufficient control over the theoretical uncertainities in the standard model predictions.

Inclusive decays can be described by the Heavy Quark Expansion (HQE),[43–50] for some recent examples see [51–53]. According to the HQE an inclusive decay rate can be expanded in inverse powers of the heavy b-quark mass

$$\Gamma = \Gamma_0 + \left(\frac{\Lambda_{QCD}}{m_b}\right)^2 \Gamma_2 + \left(\frac{\Lambda_{QCD}}{m_b}\right)^3 \Gamma_3 + \left(\frac{\Lambda_{QCD}}{m_b}\right)^4 \Gamma_4 + \dots \qquad (17)$$

In order to estimate the theoretical accuracy for the mixing quantities Γ_{12} and M_{12}, one first has to investigate the general validity of the expansion in Eq.(17). This was done many times in the literature under the name of *violations of quark-hadron duality,* see e.g. [54] and references therein. We follow a pragmatic strategy, as described in more detail in [5]: the calculation of the mixing quantities Γ_{12} is identical to the ones of the lifetimes, which are also known to NLO-QCD.[55,56] Since experiment and the HQE predicition agree very well,[5] we see no room for sizeable violations of quark-hadron duality.

All Γ_is in Eq.(17) are products of perturbatively calculable Wilson coefficients and of non-perturbative matrix elements. To be sure to achieve a reasonable theoretical accuracy we have to calculate up to a sufficient order in the HQE and in QCD (each Γ_i can be expanded as $\Gamma_i^{(0)} + \frac{\alpha_s}{\pi}\Gamma_i^{(1)} + \dots$). In addition to the leading term $\Gamma_3^{(0)}$ the following corrections we done in the literature for Γ_{12}:

- 1996: Power corrections ($\Gamma_4^{(0)}$)[57] turned out to be sizable.
- 1998: NLO-QCD corrections ($\Gamma_3^{(1)}$)[58] to the leading term are also sizeable and of conceptual importance.
- 2000: In 1998 no lattice data for all arising matrix elements of four quark operators were available, the numerical update of [58] with lattice values was given in [59].
- 2003: NLO-QCD corrections ($\Gamma_3^{(1)}$) to all CKM structures were calculated in [60] and [61]. This was a relativeley small correction for $\Delta\Gamma$, but the dominant contribution to the semileptonic CP-asymmetries.
- 2004: At that time all corrections to the leading term of $\Delta\Gamma$ seemed to be unnatural large, this bad behaviour was summarized in [62].
- 2006: A reanalysis[9] of the theoretical determination of Γ_{12}, showed that

the above shortcommings were due to the use of an unproper operator basis with large unphysical cancellations, the use of the pole b-quark mass and the neglect of subleading CKM structures. Taking all this into account the theoretical uncertainty in Γ_{12}/M_{12} could be reduced by a facor of almost three.
- 2007: Higher power corrections $(\Gamma_5^{(0)})$[63] were estimated to be negligible.

Despite considerable efforts in the non-perturbative determination of the matrix elements of four-quark operators entering Γ_3, see [64] for a recent review, we still have a relatively limited knowledge of the decay constants, see e.g.[5] for more details, wich results in large uncertainties in ΔM_s and $\Delta\Gamma_s$ [b]. In Ref. [9] we used the conservative estimate $f_{B_s} = 240 \pm 40$ MeV, while [64] obtains the lattice average $f_{B_s} = 245 \pm 25$ MeV, which is very close to the most recent QCD sum rule estimate[65] $f_{B_s} = 244 \pm 21$ MeV. In Γ_{12}/M_{12} the decay constants cancel, and therefore $\Delta\Gamma_s/\Delta M_s$ and the semileptonic CP-asymmetries are theoretical well under control.
Summarizing we can state for the theoretical uncertainties in the B_s mixing quantities: $\Delta\Gamma_s$ and ΔM_s are completely dominated by the uncertainty in the decay constant f_{B_s}, while for $\Delta\Gamma_s/\Delta M_s$ and the semileptonic CP-asymmetries conservative error estimates yield errors of about $\pm 20\%$.[9]

4. Conclusions

The system of the neutral B_s mesons is ideally suited for the search for new physics effects. In particular the standard model predicts an almost vanishing mixing phase ϕ_s, while we have currently some experimental 2-3σ hints for a sizeable value of this phase. If this hints will be confirmed, then we have an unambiguous proof for new physics in flavor physics. Depending on the actual size of Δ a confirmation of the hints might already be possible at TeVatron or at an extended $\Upsilon(5s)$ run of Belle. Precision data on Δ will be available from LHC and from a Super-B factory.

Acknowledgments

I would like to thank the organizers of CAQCD 2008 for the invitation and for the financial support.

References

1. M. Battaglia *et al.*, *hep-ph/0304132* (2003).

[b] ΔM_s and $\Delta\Gamma_s$ depend quadratically on f_{B_s}.

2. O. Schneider, *arXiv:0806.4634* (2008).
3. K. Anikeev *et al.*, *hep-ph/0201071* (2001).
4. I. I. Y. Bigi and A. I. Sanda, *Camb. Monogr. Part. Phys. Nucl. Phys. Cosmol.* **9**, 1 (2000).
5. A. Lenz, *AIP Conf. Proc.* **1026**, 36 (2008).
6. A. Lenz, *arXiv:0710.0940* (2007).
7. A. Lenz, *Nucl. Phys. Proc. Suppl.* **177-178**, 81 (2008).
8. A. Lenz, *hep-ph/0612176* (2006).
9. A. Lenz and U. Nierste, *JHEP* **06**, p. 072 (2007).
10. M. Bona *et al.*, *JHEP* **03**, p. 049 (2008).
11. M. Bona *et al.*, *arXiv:0803.0659* (2008).
12. A. Lenz, U. Nierste and CKMfitter Group, *to appear* (2008).
13. J. Charles for CKM fitter Group, *http://www.slac.stanford.edu/xorg/ckmfitter/ckm_talks.html* (2008).
14. A. Abulencia *et al.*, *Phys. Rev. Lett.* **97**, p. 242003 (2006).
15. *D0 note 5618-CONF* (2008).
16. E. Barberio *et al.*, *arXiv:0808.1297* (2008).
17. V. M. Abazov *et al.* (2008).
18. T. Aaltonen *et al.*, *Phys. Rev. Lett.* **100**, p. 161802 (2008).
19. *CDF note 9458* (2008).
20. *CDF note 9015* (2007).
21. V. M. Abazov *et al.*, *Phys. Rev.* **D74**, p. 092001 (2006).
22. V. M. Abazov *et al.*, *Phys. Rev. Lett.* **98**, p. 151801 (2007).
23. *D0 note 5730-CONF* (2008).
24. W. Altmannshofer, A. J. Buras and P. Paradisi, *arXiv:0808.0707* (2008).
25. C.-H. Chen, C.-Q. Geng and L. Li, *arXiv:0808.0127* (2008).
26. A. Soni, A. K. Alok, A. Giri, R. Mohanta and S. Nandi, *arXiv:0807.1971* (2008).
27. J. P. Lee and K. Young Lee, *arXiv:0806.1389* (2008).
28. F. J. Botella, G. C. Branco and M. Nebot, *arXiv:0805.3995* (2008).
29. A. J. Buras and D. Guadagnoli, *arXiv:0805.3887* (2008).
30. N. Kifune, J. Kubo and A. Lenz, *Phys. Rev.* **D77**, p. 076010 (2008).
31. J. M. Cabarcas, D. Gomez Dumm and R. Martinez, *Phys. Rev.* **D77**, p. 036002 (2008).
32. S.-L. Chen, X.-G. He, X.-Q. Li, H.-C. Tsai and Z.-T. Wei, *arXiv:0710.3663* (2007).
33. A. S. Joshipura and B. P. Kodrani, *Phys. Rev.* **D77**, p. 096003 (2008).
34. A. Lenz, *Phys. Rev.* **D76**, p. 065006 (2007).
35. A. Badin, F. Gabbiani and A. A. Petrov, *Phys. Lett.* **B653**, 230 (2007).
36. E. Lunghi and A. Soni, *JHEP* **09**, p. 053 (2007).
37. S.-L. Chen, X.-G. He, A. Hovhannisyan and H.-C. Tsai, *JHEP* **09**, p. 044 (2007).
38. A. S. Joshipura and B. P. Kodrani, *arXiv:0706.0953* (2007).
39. A. Dighe, A. Kundu and S. Nandi, *Phys. Rev.* **D76**, p. 054005 (2007).
40. S.-L. Chen, N. G. Deshpande, X.-G. He, J. Jiang and L.-H. Tsai, *Eur. Phys. J.* **C53**, 607 (2008).

41. M. Artuso *et al.*, *arXiv:0801.1833* (2008).
42. M. Bona *et al.*, *arXiv:0709.0451* (2007).
43. V. A. Khoze and M. A. Shifman, *Sov. Phys. Usp.* **26**, p. 387 (1983).
44. M. A. Shifman and M. B. Voloshin, *Sov. J. Nucl. Phys.* **41**, p. 120 (1985).
45. M. A. Shifman and M. B. Voloshin, *Sov. Phys. JETP* **64**, p. 698 (1986).
46. J. Chay, H. Georgi and B. Grinstein, *Phys. Lett.* **B247**, 399 (1990).
47. I. I. Y. Bigi, N. G. Uraltsev and A. I. Vainshtein, *Phys. Lett.* **B293**, 430 (1992).
48. I. I. Y. Bigi, M. A. Shifman, N. G. Uraltsev and A. I. Vainshtein, *Phys. Rev. Lett.* **71**, 496 (1993).
49. B. Blok, L. Koyrakh, M. A. Shifman and A. I. Vainshtein, *Phys. Rev.* **D49**, 3356 (1994).
50. A. V. Manohar and M. B. Wise, *Phys. Rev.* **D49**, 1310 (1994).
51. A. Lenz, U. Nierste and G. Ostermaier, *Phys. Rev.* **D56**, 7228 (1997).
52. A. Lenz, U. Nierste and G. Ostermaier, *Phys. Rev.* **D59**, p. 034008 (1999).
53. A. Lenz, *hep-ph/0011258* (2000).
54. I. I. Y. Bigi, M. A. Shifman, N. Uraltsev and A. I. Vainshtein, *Phys. Rev.* **D59**, p. 054011 (1999).
55. M. Beneke, G. Buchalla, C. Greub, A. Lenz and U. Nierste, *Nucl. Phys.* **B639**, 389 (2002).
56. E. Franco, V. Lubicz, F. Mescia and C. Tarantino, *Nucl. Phys.* **B633**, 212 (2002).
57. M. Beneke, G. Buchalla and I. Dunietz, *Phys. Rev.* **D54**, 4419 (1996).
58. M. Beneke, G. Buchalla, C. Greub, A. Lenz and U. Nierste, *Phys. Lett.* **B459**, 631 (1999).
59. M. Beneke and A. Lenz, *J. Phys.* **G27**, 1219 (2001).
60. M. Beneke, G. Buchalla, A. Lenz and U. Nierste, *Phys. Lett.* **B576**, 173 (2003).
61. M. Ciuchini, E. Franco, V. Lubicz, F. Mescia and C. Tarantino, *JHEP* **08**, p. 031 (2003).
62. A. Lenz, *hep-ph/0412007* (2004).
63. A. Badin, F. Gabbiani and A. A. Petrov, *Phys. Lett.* **B653**, 230 (2007).
64. V. Lubicz and C. Tarantino, *arXiv:0807.4605* (2008).
65. M. Jamin and B. O. Lange, *Phys. Rev.* **D65**, p. 056005 (2002).

IMPLICATIONS OF $D^0 - \overline{D}^0$ MIXING FOR NEW PHYSICS

ALEXEY A. PETROV

Department of Physics and Astronomy, Wayne State University
Detroit, MI 48201, USA
Michigan Center for Theoretical Physics, University of Michigan
Ann Arbor, MI 48109, USA
E-mail: apetrov@wayne.edu

I review constraints on possible New Physics interactions from $D^0 - \overline{D}^0$ mixing measurements. I consider the most general low energy effective Hamiltonian and include leading order QCD running of effective operators. I discuss constraints from an extensive list of popular New Physics models, each of which could be discovered at the LHC, that can generate these operators. In most of the scenarios, strong constraints that surpass those from other search techniques could be placed on the allowed parameter space using the existent evidence for observation of D meson mixing.

Keywords: charm quark, meson-anti-meson mixing, physics beyond the Standard Model

1. Introduction

Meson-antimeson mixing has traditionally been of importance because it is sensitive to heavy degrees of freedom that propagate in the underlying mixing amplitudes. Estimates of the charm quark and top quark mass scales were inferred from the observation of mixing in the K^0 and B_d systems, respectively, before these particles were discovered directly.

This success has motivated attempts to indirectly detect New Physics (NP) signals by comparing the observed meson mixing with predictions of the Standard Model (SM). K^0-$\overline{K}^0$ mixing has historically placed stringent constraints on the parameter space of theories beyond the SM and provides an essential hurdle that must be passed in the construction of models with NP. The large mixing signal in the B_d and B_s systems, observed at the B-factories and the Tevatron collider, can be precisely described in terms of the SM alone, which makes the parameter spaces of various NP models increasingly constrained. These facts influenced theoretical and experimental studies of D^0 flavor oscillations, where the SM mixing rate is sufficiently small that the NP component might be able to compete. There has been a flurry of recent experimental activity regarding the detection of D^0-$\bar{D}^0$ mixing, which marks the first time Flavor Changing Neutral Currents (FCNC) have been observed in the charged $+2/3$ quark sector. With the potential window to discern large NP effects in the charm sector and the anticipated improved accuracy for future mixing measurements, the motivation for a comprehensive up-to-date theoretical analysis of New Physics contributions to D meson mixing is compelling.

The phenomenon of meson-anti-meson mixing occurs in the presence of operators that change quark flavor by two units.[1] Those operators can be generated in both the Standard Model and many possible extensions of it. They produce off-diagonal terms in the meson-anti-meson mass matrix, so that the basis of flavor eigenstates no longer coincide with the basis of mass eigenstates. Those two bases, however, are related by a linear transformation,

$$|D_{\frac{1}{2}}\rangle = p|D^0\rangle \pm q|\overline{D}^0\rangle, \tag{1}$$

where the complex parameters p and q are obtained from diagonalizing the $D^0 - \overline{D}^0$ mass matrix. Neglecting CP-violation leads to $p = q = 1/\sqrt{2}$. The mass and width splittings between those mass eigenstates are given by

$$x_D = \frac{m_1 - m_2}{\Gamma_D}, \qquad y_D = \frac{\Gamma_1 - \Gamma_2}{2\Gamma_D}. \tag{2}$$

It is expected that x_D and y_D should be rather small in the Standard Model, which is usually attributed to the absence of superheavy quarks destroying Glashow-Iliopoulos-Maiani (GIM) cancellation. In Eq. (2), $\Gamma_{\rm D}$ is the average width of the two neutral D meson mass eigenstates. The quantities which are actually measured in most experimental determinations of the mass and width differences, $y_{\rm D}^{(\rm CP)}$, $x'_{\rm D}$, and $y'_{\rm D}$, are defined as

$$\begin{aligned} y_{\rm D}^{(\rm CP)} &= y_{\rm D}\cos\phi - x_{\rm D}\sin\phi\left(\frac{A_m}{2} - A_{prod}\right) \quad , \\ x'_D &= x_D\cos\delta_{K\pi} + y_D\sin\delta_{K\pi} \quad , \\ y'_D &= y_{\rm D}\cos\delta_{K\pi} - x_{\rm D}\sin\delta_{K\pi} \quad , \end{aligned} \tag{3}$$

where $A_{prod} = \left(N_{D^0} - N_{\overline{D}^0}\right) / \left(N_{D^0} + N_{\overline{D}^0}\right)$ is the so-called production asymmetry of D^0 and $\overline{D}^0$ (giving the relative weight of D^0 and $\overline{D}^0$ in the sample) and $\delta_{K\pi}$ is the strong phase difference between the Cabibbo favored and double Cabibbo suppressed amplitudes,[2] which is usually measured in

$D \to K\pi$ transitions. In what follows we shall neglect CP-violating parameters ϕ and A_m. In this limit $y_{\rm D}^{\rm (CP)} = y_D$. Please see recent reviews[1,3,4] for more complete analysis.

2. Experimental Constraints on Charm Mixing

The recent interest in D^0-$\bar{D}^0$ mixing started with the almost simultaneous observations by the BaBar[6] and Belle[7] collaborations of nonzero mixing signals at about the per cent level,

$$y'_{\rm D} = (0.97 \pm 0.44 \pm 0.31) \cdot 10^{-2} \qquad \text{(BaBar)}\ , \tag{4}$$

$$y_{\rm D}^{\rm (CP)} = (1.31 \pm 0.32 \pm 0.25) \cdot 10^{-2} \qquad \text{(Belle)}\ . \tag{5}$$

This was soon followed by the announcement by the Belle collaboration of mixing measurements from the Dalitz plot analyses of $D^0 \to K_S\pi^+\pi^-$,[8]

$$x_{\rm D} = (0.80 \pm 0.29 \pm 0.17) \cdot 10^{-2}\ , \qquad y_{\rm D} = (0.33 \pm 0.24 \pm 0.15) \cdot 10^{-2}\ . \tag{6}$$

A fit to the current database by the Heavy Flavor Averaging Group (HFAG) gives[5]

$$x_{\rm D} = 9.8^{+2.6}_{-2.7} \cdot 10^{-3}\ , \quad y_{\rm D} = (7.5 \pm 1.8) \cdot 10^{-3}\ , \tag{7}$$

which is obtained assuming no CP-violation affecting mixing. It is important to note that the combined analysis of $x_{\rm D}$ and $y_{\rm D}$ excludes the "no-mixing" point $x_{\rm D} = y_{\rm D} = 0$ by 6.7σ.[5] This fact adds confidence that charm mixing has indeed been observed. Then, a correct interpretation of the results is important. In addition, as with any rare low-energy transition, the question arises on how to use it to probe for physics beyond the Standard Model.

3. Standard Model "background" in $D^0 - \overline{D^0}$ mixing

Theoretical predictions for x_D and y_D obtained in the framework of the Standard Model historically span several orders of magnitude. I will not discuss predictions of the SM for the charm mixing rates here, instead referring the interested reader to recent reviews.[1,3,4] It might be advantageous to note that there are two approaches to describe $D^0 - \overline{D}^0$ mixing, neither of which give very reliable results because m_c is in some sense intermediate between heavy and light.

The inclusive approach[10,11] is based on the operator product expansion (OPE). In the formal limit $m_c \gg \Lambda$ limit, where Λ is a scale characteristic of the strong interactions, x_D and y_D can be expanded in terms of matrix elements of local operators. The use of the OPE relies on local quark-hadron duality, and on Λ/m_c being small enough to allow a truncation of the series after the first few terms. This, however, is not realized in charm mixing, as the leading term in $1/m_c$ is suppressed by four and six powers of the strange quark mass for x_D and y_D respectively. The parametrically-suppressed higher order terms in $1/m_c$ can have less powers of m_s, thus being more important numerically.[11] This results in reshuffling of the OPE series, making it a triple expansion in $1/m_c$, m_s, and α_s. The (numerically) leading term contains over twenty matrix elements of dimension-12, eight-quark operators, which are difficult to compute reliably. A naive power counting then yields $x_D, y_D < 10^{-3}$. The exclusive approach[12] sums over intermediate hadronic states. Since there are cancellations between states within a given $SU(3)$ multiplet, one needs to know the contribution of each state with high precision. However, the D is not light enough that its decays are dominated by a few final states. In the absence of sufficiently precise data, one is forced to use some assumptions. Large effects in y_D appear for decays close to D threshold, where an analytic expansion in $SU(3)_F$ violation is no longer possible. Thus, even though theoretical calculations of x_D and y_D are quite uncertain, the values $x_D \sim y_D \sim 1\%$ are quite natural in the Standard Model.[13]

It then appears that experimental results of Eq. (7) are consistent with the SM predictions. Yet, those predictions are quite uncertain to be subtracted from the experimental data to precisely constrain possible NP contributions. In this situation the following approach can be taken. One can neglect the SM contribution altogether and assume that NP saturates the result reported by experimental collaborations. This way, however, only an upper bound on the NP parameters can be placed. A subtlety of this method of constraining the NP component of the mixing amplitude is related to the fact that the SM and NP contributions can have either the same or opposite signs. While the sign of the SM contribution cannot be calculated reliably due to hadronic uncertainties, x_D computed entirely within a given NP model can be determined rather precisely. This stems from the fact that NP contributions are generated by heavy degrees of freedom making short-distance OPE reliable. This means that only the part of parameter space of NP models that generate x_D of the same sign as observed experimentally can be reliably and unambiguously constrained.

4. New Physics contributions to $D^0 - \overline{D^0}$ mixing

Any NP degree of freedom will generally be associated with a generic heavy mass scale M, at which the NP interaction will be most naturally described. At the scale m_c of the charm mass, this description will have been modified by the effects of QCD, which should be taken into account. In order to see how NP might affect the mixing amplitude, it is instructive to consider off-diagonal terms in the neutral D mass matrix,

$$\left(M - \frac{i}{2}\Gamma\right)_{12} = \frac{1}{2M_{\rm D}}\langle \overline{D}^0 | \mathcal{H}_w^{\Delta C=-2} | D^0 \rangle + \frac{1}{2M_{\rm D}} \sum_n \frac{\langle \overline{D}^0 | \mathcal{H}_w^{\Delta C=-1} | n \rangle \, \langle n | \mathcal{H}_w^{\Delta C=-1} | D^0 \rangle}{M_{\rm D} - E_n + i\epsilon} \tag{8}$$

where the first term contains $\mathcal{H}_w^{\Delta C=-2}$, which is an effective $|\Delta C| = 2$ hamiltonian, represented by a set of operators that are local at the $\mu \simeq m_D$ scale. Note that a b-quark also gives a (negligible) contribution to this term. This term only affects x_D, but not y_D.

The second term in Eq. (8) is given by a double insertion of the effective $|\Delta C| = 1$ Hamiltonian $\mathcal{H}_w^{\Delta C=-1}$. This term is believed to give dominant contribution to $D^0 - \overline{D}^0$ mixing in the Standard Model, affecting both x and y. It is generally believed that NP cannot give any sizable contribution to this term, since $\mathcal{H}_w^{\Delta C=-1}$ Hamiltonian also mediates non-leptonic D-decays, which should then also be affected by this NP contribution. I will show that there is a well-defined theoretical limit where NP contribution *dominates* lifetime difference y_D and consider implications of this limit in "real world".

4.1. *New Phyiscs in $|\Delta C| = 1$ interactions.*

Consider a non-leptonic D^0 decay amplitude, $A[D^0 \to n]$, which includes a small NP contribution, $A[D^0 \to n] = A_n^{(\rm SM)} + A_n^{(\rm NP)}$. Here, $A_n^{(\rm NP)}$ is assumed to be smaller than the current experimental uncertainties on those decay rates. This ensures that NP effects cannot be seen in the current experimental analyses of non-leptonic D-decays. One can then write $y_{\rm D}$ as

$$y_{\rm D} \simeq \sum_n \frac{\rho_n}{\Gamma_{\rm D}} A_n^{(\rm SM)} A_n^{(\rm SM)} + 2 \sum_n \frac{\rho_n}{\Gamma_{\rm D}} A_n^{(\rm NP)} A_n^{(\rm SM)} \quad . \tag{9}$$

The first term of Eq. (schematic) represents the SM contribution to $y_{\rm D}$. The SM contribution to $y_{\rm D}$ is known to vanish in the limit of exact flavor $SU(3)$. Moreover, the first order correction is also absent, so the SM contribution arises only as a *second* order effect.[13] This means that in the flavor $SU(3)$ limit the lifetime difference $y_{\rm D}$ is dominated by the second term in Eq. (9), i.e. New Physics contributions, even if their contibutions are tiny in the individual decay amplitudes[14]! A calculation reveals that NP contribution to $y_{\rm D}$ can be as large as several percent in R-parity-violating SUSY models[9] or as small as $\sim 10^{-10}$ in the models with interactions mediated by charged Higgs particles.[14]

This wide range of theoretical predictions can be explained by two observations. First, many NP affecting $|\Delta C| = 1$ transitions also affect $|\Delta B| = 1$ or $|\Delta S| = 1$ decays or kaon and B-meson mixings, which are tightly constrained. Second, a detailed look at a given NP model that can potentially affect $y_{\rm D}$ reveals that the NP contribution itself can vanish in the flavor $SU(3)$ limit. For instance, the structure of the NP interaction might simply mimic the one of the SM. Effects like that can occur in some models with extra space dimensions. Also, the chiral structure of a low-energy effective lagrangian in a particular NP model could be such that the leading, mass-independent contribution vanishes exactly, as in a left-right model (LRM). Finally, the NP coupling might explicitly depend on the quark mass, as in a model with multiple Higgs doublets. However, most of these models feature second order $SU(3)$-breaking already at leading order in the $1/m_c$ expansion. This should be contrasted with the SM, where the leading order is suppressed by six powers of m_s and term of order m_s^2 only appear as a $1/m_c^6$-order correction.

4.2. *New Phyiscs in $|\Delta C| = 2$ interactions*

Though the particles present in models with New Physics may not be produced in charm quark decays, their effects can nonetheless be seen in the form of effective operators generated by the exchanges of these new particles. Even without specifying the form of these new interactions, we know that their effect is to introduce several $|\Delta C| = 2$ effective operators built out of the SM degrees of freedom.

By integrating out new degrees of freedom associated with new interactions at a scale M, we are left with an effective hamiltonian written in the form of a series of operators of increasing dimension. Operator power counting then tells us the most important contributions are given by the operators of the lowest possible dimension, $d = 6$ in this case. This means that they must contain only quark degrees of freedom and no derivatives. Realizing this, we can write the complete basis of these effective operators, which can be done most conveniently in terms of chiral quark fields,

$$\langle f|\mathcal{H}_{NP}|i\rangle = G\sum_{i=1} \mathrm{C}_i(\mu)\ \langle f|Q_i|i\rangle(\mu)\ , \tag{10}$$

where the prefactor G has the dimension of inverse-squared mass, the C_i are dimensionless Wilson coefficients, and the Q_i are the effective operators:

$$\begin{array}{ll} Q_1 = (\overline{u}_L\gamma_\mu c_L)\,(\overline{u}_L\gamma^\mu c_L)\ , & Q_5 = (\overline{u}_R\sigma_{\mu\nu} c_L)\,(\overline{u}_R\sigma^{\mu\nu} c_L)\ , \\ Q_2 = (\overline{u}_L\gamma_\mu c_L)\,(\overline{u}_R\gamma^\mu c_R)\ , & Q_6 = (\overline{u}_R\gamma_\mu c_R)\,(\overline{u}_R\gamma^\mu c_R)\ , \\ Q_3 = (\overline{u}_L c_R)\,(\overline{u}_R c_L)\ , & Q_7 = (\overline{u}_L c_R)\,(\overline{u}_L c_R)\ , \\ Q_4 = (\overline{u}_R c_L)\,(\overline{u}_R c_L)\ , & Q_8 = (\overline{u}_L\sigma_{\mu\nu} c_R)\,(\overline{u}_L\sigma^{\mu\nu} c_R)\ . \end{array} \tag{11}$$

In total, there are eight possible operator structures that exhaust the list of possible independent contributions to $|\Delta C| = 2$ transitions. Since these operators are generated at the scale M where the New Physics is integrated out, a non-trivial operator mixing can occur when one takes into account renormalization group running of these operators between the scales M and μ, with μ being the scale where the hadronic matrix elements are computed. We shall work at the renormalization scale $\mu = m_c \simeq 1.3$ GeV. This evolution is determined by solving the RG equations obeyed by the Wilson coefficients,

$$\frac{d}{d\log\mu}\vec{C}(\mu) = \hat{\gamma}^T\vec{C}(\mu)\ , \tag{12}$$

where $\hat{\gamma}$ represents the matrix of anomalous dimensions of the operators in Eq. (11).[15] Due to the relatively simple structure of $\hat{\gamma}$, one can easily write the evolution of each Wilson coefficient in Eq. (10) from the New Physics scale M down to the hadronic scale μ, taking into account quark thresholds. Corresponding to each of the eight operators $\{Q_i\}$ $(i = 1,\ldots,8)$ is an RG factor $r_i(\mu, M)$. The first of these, $r_1(\mu, M)$, is given explicitly by

$$r_1(\mu, M) = \left(\frac{\alpha_s(M)}{\alpha_s(m_t)}\right)^{2/7}\left(\frac{\alpha_s(m_t)}{\alpha_s(m_b)}\right)^{6/23}\left(\frac{\alpha_s(m_b)}{\alpha_s(\mu)}\right)^{6/25}\ . \tag{13}$$

and the rest can be expressed in terms of $r_1(\mu, M)$ as

$$\begin{array}{ll} r_2(\mu, M) = [r_1(\mu, M)]^{1/2}\ , & \\ r_3(\mu, M) = [r_1(\mu, M)]^{-4}\ , & r_6(\mu, M) = r_1(\mu, M)\ , \\ r_4(\mu, M) = [r_1(\mu, M)]^{(1+\sqrt{241})/6}\ , & r_7(\mu, M) = r_4(\mu, M)\ , \\ r_5(\mu, M) = [r_1(\mu, M)]^{(1-\sqrt{241})/6}\ , & r_8(\mu, M) = r_5(\mu, M)\ . \end{array} \tag{14}$$

The RG factors are generally only weakly dependent on the NP scale M since it is taken to be larger than the top quark mass, m_t, and the evolution of α_s is slow at these high mass scales. In Table 1, we display numerical values for the $r_i(\mu, M)$ with $M = 1, 2$ TeV and $\mu = m_c \simeq 1.3$ GeV. Here, we compute α_s using the one-loop evolution and matching expressions for perturbative consistency with the RG evolution of the effective hamiltonian.

A contribution to $D^0 - \overline{D}^0$ mixing from a particular NP model can be obtained by calculating matching conditions for the Wilson coefficients C_i at the scale M, running their values down to μ and computing the relevant matrix elements of four-quark operators. A generic model of New Physics would then give the following contribution $x_{\rm D}$,

$$x_{\rm D}^{NP} = G \, \frac{f_D^2 B_D m_D}{\Gamma_D} \left[\frac{2}{3} [C_1(m_c) + C_6(m_c)] - \frac{5}{12} [C_4(m_c) + C_7(m_c)] \right. \\ \left. + \frac{7}{12} C_3(m_c) - \frac{5 C_2(m_c)}{6} + [C_5(m_c) + C_8(m_c)] \right] \quad (15)$$

Here we simplified the result by assuming that all non-perturbative ('bag') parameters are equal to $B_D \simeq 0.82$. The Wilson coefficients at the scale μ are related to the Wilson coefficients at the scale M by renormalization group evolution,

$$\begin{aligned}
C_1(m_c) &= r_1(m_c, M) C_1(M) \, , \\
C_2(m_c) &= r_2(m_c, M) C_2(M) \, , \\
C_3(m_c) &= \frac{2}{3} \left[r_2(m_c, M) - r_3(m_c, M) \right] C_2(M) + r_3(m_c, M) C_3(M) \, , \\
C_4(m_c) &= \frac{8}{\sqrt{241}} \left[r_5(m_c, M) - r_4(m_c, M) \right] \left[C_4(M) + \frac{15}{4} C_5(M) \right] \\
&\quad + \frac{1}{2} \left[r_4(m_c, M) + r_5(m_c, M) \right] C_4(M) \, , \\
C_5(m_c) &= \frac{1}{8\sqrt{241}} \left[r_4(m_c, M) - r_5(m_c, M) \right] \left[C_4(M) + 64 C_5(M) \right] \\
&\quad + \frac{1}{2} \left[r_4(m_c, M) + r_5(m_c, M) \right] C_5(M) \, , \\
C_6(m_c) &= r_6(m_c, M) C_6(M) \, , \qquad (16) \\
C_7(m_c) &= \frac{8}{\sqrt{241}} \left[r_8(m_c, M) - r_7(m_c, M) \right] \left[C_7(M) + \frac{15}{4} C_8(M) \right] \\
&\quad + \frac{1}{2} \left[r_7(m_c, M) + r_8(m_c, M) \right] C_7(M) \, , \\
C_8(m_c) &= \frac{1}{8\sqrt{241}} \left[r_7(m_c, M) - r_8(m_c, M) \right] \left[C_7(M) + 64 C_8(M) \right] \\
&\quad + \frac{1}{2} \left[r_7(m_c, M) + r_8(m_c, M) \right] C_8(M) \, ,
\end{aligned}$$

A contribution of each particular NP model can then be studied using

M(TeV)	$r_1(m_c, M)$	$r_2(m_c, M)$	$r_3(m_c, M)$	$r_4(m_c, M)$	$r_5(m_c, M)$
1	0.72	0.85	3.7	0.41	2.2
2	0.71	0.84	4.0	0.39	2.3

Eq. (15). Even before performing such an analysis, one can get some idea what energy scales can be probed by $D^0 - \overline{D}^0$ mixing. Setting $G = 1/M^2$ and $C_i(M) = 1$, we obtain $M \sim 10^3$ TeV. More realistic models can be probed in the region of several TeV, which is very relevant for LHC phenomenology applications.

A program described above has been recently executed[15] for 21 well-motivated NP models, which will be actively studied at LHC. The results are presented in Table 2. As can be seen, out of 21 models considered, only four received no useful constraints from $D^0 - \overline{D}^0$ mixing. More informative exclusion plots can be found in that paper[15] as well. It is interesting to note that some models *require* large signals in the charm system if mixing and FCNCs in the strange and beauty systems are to be small (as in, for example, the SUSY alignment model[16,17]).

5. Conclusions

I reviewed implications of recent measurement of $D^0 - \overline{D}^0$ mixing rates for constraining models of New Physics. A majority of considered models received competitive constraints from $D^0 - \overline{D}^0$ mixing measurements despite hadronic uncertainties that plague SM contributions. It should be noted that vast majority of predictions of NP models do not suffer from this uncertainty, and can be computed reliably, if lattice QCD community provides calculations of matrix elements of four-fermion operators Eq. (11).

Another possible manifestation of new physics interactions in the charm system is associated with the observation of (large) CP-violation.[1,4,18] This is due to the fact that all quarks that build up the hadronic states in weak decays of charm mesons belong to the first two generations. Since 2×2 Cabbibo quark mixing matrix is real, no CP-violation is possible in the dominant tree-level diagrams which describe the decay amplitudes. CP-violating amplitudes can be introduced in the Standard Model by including penguin or box operators induced by virtual b-quarks. However, their contributions are strongly suppressed by the small combination of CKM matrix elements $V_{cb}V_{ub}^*$. It is thus widely believed that the observation of (large) CP violation in charm decays or mixing would be an unambiguous sign for New

Model	Approximate Constraint
Fourth Generation	$\|V_{ub'}V_{cb'}\| \cdot m_{b'} < 0.5$ (GeV)
$Q = -1/3$ Singlet Quark	$s_2 \cdot m_S < 0.27$ (GeV)
$Q = +2/3$ Singlet Quark	$\|\lambda_{uc}\| < 2.4 \cdot 10^{-4}$
Little Higgs	Tree: see $Q = -1/3$ Singlet Quark
	Box: can reach observed $x_{\rm D}$
Generic Z'	$M_{Z'}/C > 2.2 \cdot 10^3$ TeV
Family Symmetries	$m_1/f > 1.2 \cdot 10^3$ TeV $(m_1/m_2 = 0.5)$
Left-Right Symmetric	No constraint
Alternate Left-Right Symmetric	$M_R > 1.2$ TeV $(m_{D_1} = 0.5$ TeV$)$
	$(\Delta m/m_{D_1})/M_R > 0.4$ TeV^{-1}
Vector Leptoquark Bosons	$M_{VLQ} > 55(\lambda_{PP}/0.1)$ TeV
Flavor Conserving Two-Higgs-Doublet	No constraint
Flavor Changing Neutral Higgs	$m_H/C > 2.4 \cdot 10^3$ TeV
FC Neutral Higgs (Cheng-Sher)	$m_H/\|\Delta_{uc}\| > 600$ GeV
Scalar Leptoquark Bosons	See entry for RPV SUSY
Higgsless	$M > 100$ TeV
Universal Extra Dimensions	No constraint
Split Fermion	$M/\|\Delta y\| > (6 \cdot 10^2$ GeV$)$
Warped Geometries	$M_1 > 3.5$ TeV
MSSM	$\|(\delta^u_{12})_{\rm LR,RL}\| < 3.5 \cdot 10^{-2}$ for $\tilde{m} \sim 1$ TeV
	$\|(\delta^u_{12})_{\rm LL,RR}\| < .25$ for $\tilde{m} \sim 1$ TeV
SUSY Alignment	$\tilde{m} > 2$ TeV
Supersymmetry with RPV	$\lambda'_{12k}\lambda'_{11k}/m_{\tilde{d}_{R,k}} < 1.8 \cdot 10^{-3}/100$ GeV
Split Supersymmetry	No constraint

Physics.

Acknowledgments

This work was supported in part by the U.S. National Science Foundation CAREER Award PHY–0547794, and by the U.S. Department of Energy under Contract DE-FG02-96ER41005. I would like to thank the organizers of CAQCD-08 for a great conference.

References

1. M. Artuso, B. Meadows and A. A. Petrov, Ann. Rev. Nucl. Part. Sci. **58**, 249 (2008), arXiv:0802.2934 [hep-ph]; A. A. Petrov, *In the Proceedings of Flavor*

Physics and CP Violation (FPCP 2003), Paris, France, 3-6 Jun 2003, pp MEC05 [arXiv:hep-ph/0311371].
2. S. Bergmann, *et. al*, Phys. Lett. B **486**, 418 (2000); A. F. Falk, Y. Nir and A. A. Petrov, JHEP **9912**, 019 (1999).
3. E. Golowich, arXiv:0806.1868 [hep-ph].
4. S. Bianco, F. L. Fabbri, D. Benson and I. Bigi, Riv. Nuovo Cim. **26N7**, 1 (2003).
5. A. J. Schwartz, arXiv:0803.0082 [hep-ex].
6. B. Aubert *et al.* [BABAR Collaboration], Phys. Rev. Lett. **98**, 211802 (2007).
7. M. Staric *et al.* [Belle Collaboration], Phys. Rev. Lett. **98**, 211803 (2007).
8. K. Abe *et al.* [BELLE Collaboration], Phys. Rev. Lett. **99**, 131803 (2007).
9. A. A. Petrov and G. K. Yeghiyan, Phys. Rev. D **77**, 034018 (2008).
10. A. Datta, D. Kumbhakar, Z. Phys. C**27**, 515 (1985); A. A. Petrov, Phys. Rev. D**56**, 1685 (1997); E. Golowich and A. A. Petrov, Phys. Lett. B **625**, 53 (2005).
11. H. Georgi, Phys. Lett. B297, 353 (1992); T. Ohl, G. Ricciardi and E. Simmons, Nucl. Phys. B403, 605 (1993); I. Bigi and N. Uraltsev, Nucl. Phys. B **592**, 92 (2001).
12. J. Donoghue, E. Golowich, B. Holstein and J. Trampetic, Phys. Rev. D33, 179 (1986); L. Wolfenstein, Phys. Lett. B164, 170 (1985); P. Colangelo, G. Nardulli and N. Paver, Phys. Lett. B242, 71 (1990); T.A. Kaeding, Phys. Lett. B357, 151 (1995); E. Golowich and A. A. Petrov, Phys. Lett. B **427**, 172 (1998); A. A. Anselm and Y. I. Azimov, Phys. Lett. B **85**, 72 (1979).
13. A. F. Falk, Y. Grossman, Z. Ligeti and A. A. Petrov, Phys. Rev. D **65**, 054034 (2002); A. F. Falk, *et. al*, Phys. Rev. D **69**, 114021 (2004).
14. E. Golowich, S. Pakvasa and A. A. Petrov, Phys. Rev. Lett. **98**, 181801 (2007).
15. E. Golowich, J. Hewett, S. Pakvasa and A. A. Petrov, Phys. Rev. D **76**, 095009 (2007).
16. Y. Nir and N. Seiberg, Phys. Lett. B **309**, 337 (1993).
17. M. Ciuchini et al., Phys. Lett. B **655**, 162 (2007).
18. A. A. Petrov, *In the Proceedings of International Workshop on Charm Physics (Charm 2007), Ithaca, New York, 5-8 Aug 2007, pp 11* [arXiv:0711.1564 [hep-ph]].

PRECISE DETERMINATIONS OF THE CHARM QUARK MASS

Matthias Steinhauser

Institut für Theoretische Teilchenphysik, Universität Karlsruhe (TH), 76128 Karlsruhe, Germany
E-mail: matthias.steinhauser@uka.de

In this contribution two recent analyses for the extraction of the charm quark mass are discussed. Although they rely on completely different experimental and theoretical input the two methods provide the same final results for the charm quark mass and have an uncertainty of about 1%.

Keywords: Quark masses, perturbative QCD, lattice gauge theory

1. Introduction

There has been an enormous progress in the determination of the quark masses in the recent years due to improved experimental results, many high-order calculations in perturbative QCD and precise lattice simulations.[1] In this contribution we describe two recent analyses which lead to the most precise results for the $\overline{\text{MS}}$ charm quark mass.

The first method[2–4] is based on four-loop perturbative calculations for the moments of the vector correlator which are combined with moments extracted from precise experimental input for the total hadronic cross section in electron positron collisions.

Also the second method[5] relies on four-loop calculations, however, for the pseudo-scalar rather than for the vector current correlator. It is combined with data obtained from simulations on the lattice with dynamical charm quarks. The latter are tuned such that the mass splitting between the Υ' and Υ and the meson masses m_π^2, $2m_K^2 - m_\pi^2$, m_{η_c} and m_Υ are correctly reproduced. Thus the underlying experimental data are completely different from the first approach.

2. $R(s)$ and perturbative QCD

The basic object for the first method is the total hadronic cross section in e^+e^- annihilation. Normalized to the production cross section of a muon pair it defines the quantity

$$R(s) = \frac{\sigma(e^+e^- \to \text{hadrons})}{\sigma_{\text{pt}}}, \tag{1}$$

where $\sigma_{\text{pt}} = 4\pi\alpha^2/(3s)$.

A compilation of the experimental data contributing to $R(s)$ in the charm region can be found in Fig. 1. For our analysis it is of particular importance to have precise values for the electronic widths of the narrow resonances J/Ψ and Ψ' which have been measured by various experiments.[1] Furthermore, we rely on the excellent data provided by the BES collaboration[6,7] in the region between 3.73 GeV (which is the onset of D meson production) and about 5 GeV which marks the end point of the strong variations of $R(s)$. Above 5 GeV $R(s)$ is basically flat and can be described very well within perturbative QCD taking into account charm quark effects. Thus in this region we use `rhad`,[8] a `fortran` program containing all state-of-the-art radiative corrections to $R(s)$ since between 5 GeV and 7 GeV no reliable data is available.

Since we are interested in the extraction of the charm quark mass we have to consider the part of $R(s)$ which corresponds to the production of charm quarks, usually denoted by $R_c(s)$. $R_c(s)$ is used to compute the so-called experimental moments through

$$\mathcal{M}_n^{\text{exp}} \equiv \int \frac{\mathrm{d}s}{s^{n+1}} R_c(s) \,. \tag{2}$$

It is clear that in order to perform the integration in Eq. (2) one has to subtract the contributions from the three light quarks. This has to be done in a careful manner which is described in detail in Ref.[4]

The theoretical counterpart to Eq. (2) is given by

$$\mathcal{M}_n^{\text{th}} = \left(\frac{1}{4m_c^2}\right)^n \bar{C}_n \,. \tag{3}$$

where the $\bar{C}_n$ are obtained from the Taylor coefficients of the photon polarization function for small external momentum.

Low moments are perturbative and have long been known through three-loop order [11–13] (see Ref.[14,15] for moments up to $n = 30$). More recently also the four-loop contribution for $n = 1$[16,17] and $n = 2$ could be evaluated[18] (see also Ref.[19]).

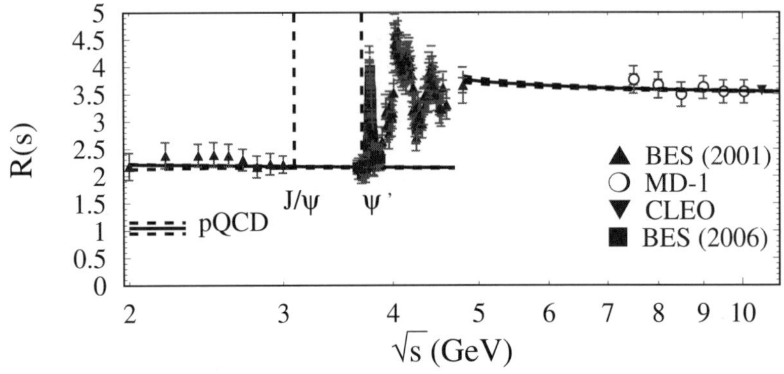

Fig. 1. $R(s)$ around the charm threshold region. The solid line corresponds to the theoretical prediction. The uncertainties, which are indicated by the dashed curves, are obtained from the variation of the input parameters and of μ. The inner and outer error bars give the statistical and systematical uncertainty, respectively. Next to the data from BES[6,7] we also show the results form MD-1[9] and CLEO.[10] The narrow resonances are indicated by dashed lines.

In the perturbative calculation we renormalize the charm quark mass in the $\overline{\mathrm{MS}}$ scheme. This enables us to extract directly the corresponding short-distance quantity avoiding the detour to the pole mass and the corresponding intrinsic uncertainty.

The results obtained for the charm quark mass from equating the experimental and theoretical moments are collected in Tab. 1. In order to obtain these numbers we set the renormalization scale to $\mu = 3$ GeV and extract as a consequence $m_c(3\text{ GeV})$. The uncertainties are due to the experimental moments, $\delta\alpha_s(M_Z) = \pm 0.002$, the variation of μ between 2 GeV and 4 GeV and the non-perturbative gluon condensate.

In contrast to the corresponding table in Ref.[4] we included in Tab. 1 the new four-loop results from Ref.[18] for $n = 2$. This leads to a shift in the central value from 0.979 GeV to 0.976 GeV. Furthermore the uncertainty of 6 MeV which was due to the absence of the four-loop result is removed.

The results in Tab. 1 show an impressive consistency when going from $n = 1$ to $n = 4$ although the relative weight form the various energy regions contributing to $\mathcal{M}^{\text{exp}}$ is completely different: whereas for $n = 1$ the region for $\sqrt{s} \geq 5$ GeV amounts to about 50% of the resonance contribution it is less than 4% for $n = 3$. Also the decomposition of the uncertainty changes

Table 1. Results for $m_c(3\text{ GeV})$ in GeV. The errors are from experiment, α_s, variation of μ and the gluon condensate. The error from the yet unknown four-loop term is kept separate.

n	$m_c(3\text{ GeV})$	exp	α_s	μ	np	total	$\delta\bar{C}_n^{(30)}$
1	0.986	0.009	0.009	0.002	0.001	0.013	—
2	0.976	0.006	0.014	0.005	0.000	0.016	—
3	0.982	0.005	0.014	0.007	0.002	0.016	0.010
4	1.012	0.003	0.008	0.030	0.007	0.032	0.016

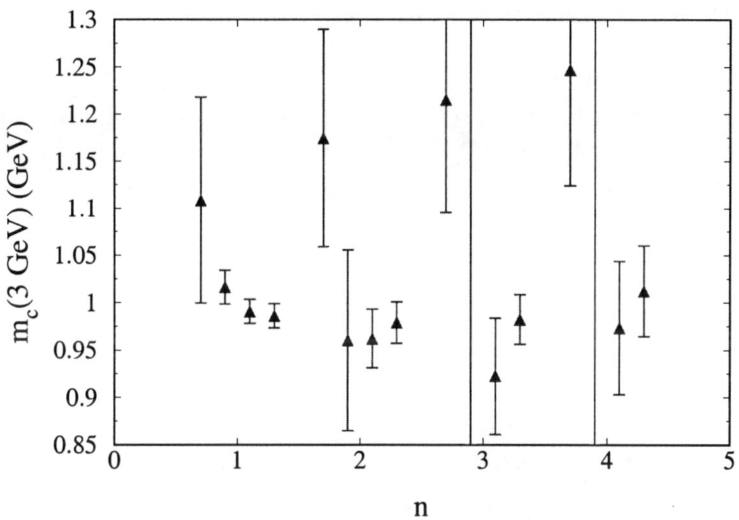

Fig. 2. $m_c(3\text{ GeV})$ for $n = 1, 2, 3$ and 4. For each value of n the results from left to right correspond the inclusion of the one-, two-, three- and four-loop terms in the theory moments.

substantially as can be seen in Tab. 1. Whereas for $n = 1$ the contribution from the μ variation is negligible it exceeds the experimental uncertainty for $n = 3$.

In Fig. 2 we show for the first four moments the result for $m_c(3\text{ GeV})$ as a function of the loop order used for $\mathcal{M}_n^{\text{th}}$. One observes a nice convergence for each n. Furthermore, the consistency among the three- and in particular the four-loop results is clearly visible from this plot.

As final result of the analysis described in this Section we quote the value given in Ref.[4] which reads

$$m_c(3\text{ GeV}) = 0.986(13)\text{ GeV}\,. \tag{4}$$

Table 2. Results for $m_c(3 \text{ GeV})$ in GeV. Both the total uncertainties are shown and the splitting into contributions from the lattice simulation, α_s, missing higher order corrections and the non-perturbative gluon condensate.

n	$m_c(3 \text{ GeV})$	lattice	α_s	h.o.	np	total
2	0.986	0.008	0.003	0.004	0.003	0.010
3	0.986	0.009	0.004	0.003	0.000	0.011

3. Lattice gauge theory and perturbative QCD

In the recent years there has been a tremendous progress in developing precise QCD simulations on the lattice. In particular, it has been possible to simulate relativistic charm quarks using the so-called Highly Improved Staggered Quark (HISQ) discretization of the quark action.[20,21] In Ref.[5] this has been used to evaluate moments of the pseudo-scalar correlator with an uncertainty below 1%. The moments from the lattice calculation are equated with the ones computed within perturbative QCD. In Ref.[19] the second non-trivial moment could be evaluated with the help the axial Ward identity from the first moment of the longitudinal part of the axial-vector current. Very recently this trick could be extended in order to arrive at the third moment for the pseudo-scalar current.[22]

Tab. 2 summarizes the results obtained for $m_c(3 \text{ GeV})$ (for $n = 2$ and 3)[a] together with the corresponding uncertainties from the lattice, α_s, missing higher order perturbative corrections and the gluon condensate.[b]

Like in the previous section we find also here an excellent agreement in the central values which leads us to the final result

$$m_c(3 \text{ GeV}) = 0.986(10) \text{ GeV}\,. \tag{5}$$

Let us mention that the dimensionless first moment can be used to extract a value for the strong coupling. We can furthermore consider ratios of moments in order to get rid of the overall dependence on m_c and again extract α_s. In Ref.[5] this has been done for the ratio of the second to the third moment which is known to four-loop order within perturbative QCD. The two determinations lead to

$$\alpha_s^{(4)}(3 \text{ GeV}) = 0.251(6)\,, \tag{6}$$

[a]Note that for $n = 1$ no charm quark mass can be determined since, in contrast to the vector correlator, the corresponding moment is dimensionless.
[b]For the presentation in this Section the notation of Ref.[5] for the numeration of the moments has been translated to the one of Ref.[4]

which corresponds to[c]

$$\alpha_s^{(5)}(M_Z) = 0.1174(12)\,. \tag{7}$$

This value agree well with the particle data group result[1] and other recent determinations (see, e.g., Refs.[24,25]).

4. Summary

In this contribution we have presented the two to date most precise determinations of the charm quark mass. Let us stress once again that, although in both cases moments of current correlators are considered, the two methods rely on completely different experimental input and on different theory calculations. Whereas in one case perturbative QCD is compared with experimental data for $R(s)$, in the second case high precision lattice simulations with dynamical charm quarks are crucial ingredients. It is quite impressive that the final results as given in Eqs. (4) and (5) coincide both in the central value and the uncertainty.

In Fig. 3 we compare the results of Section 2 and Section 3 with various other recent determinations. One observes a good agreement, however, our results are by far the most precise ones, as can be seen by the grey band.

Up to this point we have presented results for the $\overline{\text{MS}}$ charm quark mass evaluated at the scale $\mu = 3$ GeV. In general, the comparison of results from various analyses are performed for the scale-invariant mass, $m_c(m_c)$ (see, e.g., Ref.[1]). Note, however, that the scale $\mu = m_c$ is quite low and the numerical value of α_s is relatively big. Thus, it would be more appropriate to perform the comparison at a higher scale like $\mu = 3$ GeV. Let us nevertheless present the scale-invariant charm quark mass. From 0.986(10) GeV one obtains

$$m_c(m_c) = 1.268(9)\text{ GeV}\,. \tag{8}$$

The method described in Section 2 can also be used to extract the bottom quark mass. The analysis of Ref.[4] leads to

$$m_b(m_b) = 4.164(25)\text{ GeV}\,. \tag{9}$$

After including the new four-loop results from Ref.[18] the result of Eq. (9) becomes

$$m_b(m_b) = 4.162(19)\text{ GeV}\,, \tag{10}$$

which has a significantly reduced uncertainty.

---

[c]The calculation of the running and decoupling is easily done with the help of `RunDec`.[23]

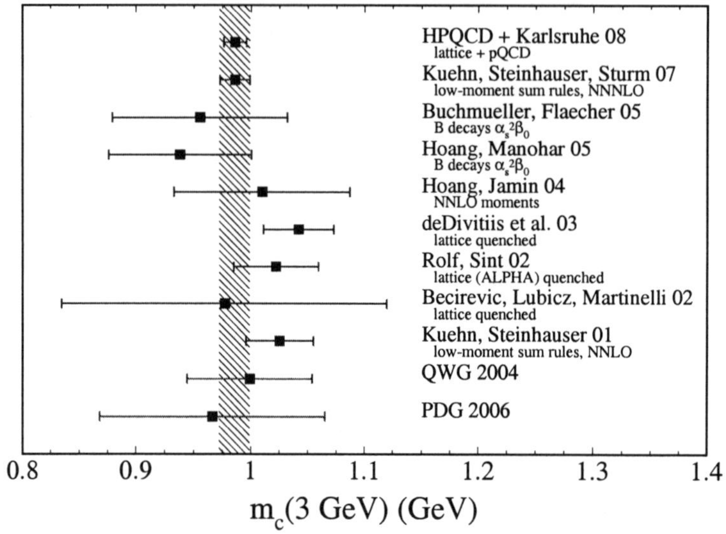

Fig. 3. Comparison of recent determinations of m_c(3 GeV).

Acknowledgments

I would like to thank Konstantin Chetyrkin, Hans Kühn, Peter Lepage, Christian Sturm and the HPQCD lattice group for a fruitful and pleasant collaboration. This work was supported by the DFG through SFB/TR 9.

References

1. W. M. Yao *et al.* [Particle Data Group], J. Phys. G **33** (2006) 1.
2. V. A. Novikov, L. B. Okun, M. A. Shifman, A. I. Vainshtein, M. B. Voloshin and V. I. Zakharov, Phys. Rept. **41** (1978) 1.
3. J. H. Kühn and M. Steinhauser, Nucl. Phys. B **619** (2001) 588 [Erratum-ibid. B **640** (2002) 415] [arXiv:hep-ph/0109084].
4. J. H. Kühn, M. Steinhauser and C. Sturm, Nucl. Phys. B **778** (2007) 192 [arXiv:hep-ph/0702103].
5. I. Allison *et al.*, in Lattice arXiv:0805.2999 [hep-lat].
6. J. Z. Bai *et al.* [BES Collaboration], Phys. Rev. Lett. **88** (2002) 101802 [arXiv:hep-ex/0102003].
7. M. Ablikim *et al.* [BES Collaboration], arXiv:hep-ex/0612054.
8. R. V. Harlander and M. Steinhauser, Comput. Phys. Commun. **153** (2003) 244 [arXiv:hep-ph/0212294].

9. A. E. Blinov *et al.* [MD-1 Collaboration], Z. Phys. C **70** (1996) 31.
10. R. Ammar *et al.* [CLEO Collaboration], Phys. Rev. D **57** (1998) 1350 [arXiv:hep-ex/9707018].
11. K. G. Chetyrkin, J. H. Kühn and M. Steinhauser, Phys. Lett. B **371** (1996) 93 [arXiv:hep-ph/9511430].
12. K. G. Chetyrkin, J. H. Kühn and M. Steinhauser, Nucl. Phys. B **482** (1996) 213 [arXiv:hep-ph/9606230].
13. K. G. Chetyrkin, J. H. Kühn and M. Steinhauser, Nucl. Phys. B **505** (1997) 40 [arXiv:hep-ph/9705254].
14. R. Boughezal, M. Czakon and T. Schutzmeier, Nucl. Phys. Proc. Suppl. **160** (2006) 160 [arXiv:hep-ph/0607141].
15. A. Maier, P. Maierhöfer and P. Marquard, Nucl. Phys. B **797** (2008) 218 [arXiv:0711.2636 [hep-ph]].
16. K. G. Chetyrkin, J. H. Kühn and C. Sturm, Eur. Phys. J. C **48** (2006) 107 [arXiv:hep-ph/0604234].
17. R. Boughezal, M. Czakon and T. Schutzmeier, Phys. Rev. D **74** (2006) 074006 [arXiv:hep-ph/0605023].
18. A. Maier, P. Maierhöfer and P. Marquard, arXiv:0806.3405 [hep-ph].
19. C. Sturm, arXiv:0805.3358 [hep-ph].
20. E. Follana *et al.* [HPQCD Collaboration and UKQCD Collaboration], Phys. Rev. D **75** (2007) 054502 [arXiv:hep-lat/0610092].
21. E. Follana, C. T. H. Davies, G. P. Lepage and J. Shigemitsu [HPQCD Collaboration and UKQCD Collaboration], Phys. Rev. Lett. **100** (2008) 062002 [arXiv:0706.1726 [hep-lat]].
22. A. Maier, P. Maierhöfer and P. Marquard, in preparation.
23. K. G. Chetyrkin, J. H. Kühn and M. Steinhauser, Comput. Phys. Commun. **133** (2000) 43 [arXiv:hep-ph/0004189].
24. P. A. Baikov, K. G. Chetyrkin and J. H. Kühn, arXiv:0801.1821 [hep-ph].
25. J.H. Kühn, TTP08-28, SFB/CPP-08-48, these proceedings.

SECTION 3.

QUARK-GLUON DYNAMICS AT HIGH DENSITY AND/OR HIGH TEMPERATURE

Crystalline Condensate in the Chiral Gross-Neveu Model*

Gerald V. Dunne and Gökçe Başar

Physics Department, University of Connecticut, Storrs CT 06269, USA
**E-mail: dunne@phys.uconn.edu; basar@phys.uconn.edu*

In this talk I review the recent discovery of a crystalline condensate solution to the gap equation for the (massless) chiral Gross-Neveu model. This work is motivated by the search for the complete phase diagram, in the temperature and chemical potential plane, of this well-known 1+1 dimensional quantum field theory that exhibits continuous chiral symmetry, asymptotic freedom and dynamical mass generation. We are also motivated by the recent solution [by M. Thies et al] of the phase diagram for the non-chiral Gross-Neveu model, which has instead a discrete chiral symmetry. The key technical advance in our work is the observation that a suitable ansatz for the Gorkov resolvent reduces the functional gap equation, for the inhomogeneous condensate, to a nonlinear Schrödinger equation, which is exactly soluble. The resulting general crystalline solution includes as special cases all previously known real and complex condensate solutions to the gap equation.

Keywords: Gross-Neveu model; gap equations; inhomogeneous condensate

1. Introduction

This work is motivated by an attempt to understand temperature/density phase diagrams of interacting fermion systems. A simple and well-studied model is the Gross-Neveu model, a model with four-fermion interactions in $1+1$ dimensions.[3] The original Gross-Neveu model, GN_2, with Lagrangian $\mathcal{L} = \bar{\psi}\, i\, \partial\!\!\!/ \psi + \frac{g^2}{2}\left(\bar{\psi}\psi\right)^2$, has a discrete chiral symmetry, while the *chiral* Gross-Neveu model, $\chi\mathrm{GN}_2$ or NJL_2, with Lagrangian $\mathcal{L} = \bar{\psi}\, i\, \partial\!\!\!/ \psi + \frac{g^2}{2}\left[\left(\bar{\psi}\psi\right)^2 + \left(\bar{\psi} i\gamma^5\psi\right)^2\right]$, has a continuous chiral symmetry. These models are renormalizable, asymptotically free, and have dynamical mass generation. With N fermion flavors there is chiral symmetry breaking in the large N limit, which is conveniently studied using powerful semiclassical techniques, pioneered by Dashen, Hasslacher and Neveu.[4] The phase

---
*Talk presented by G. Dunne

diagram of the non-chiral model, GN_2 has recently been solved analytically by Thies et al,[7] using Hartree-Fock techniques. They find that there is a crystalline phase, at low T and high μ, and for which the periodic inhomogeneous condensate can be found analytically. The crystalline phase is not seen in the mean field gap equation studies that assume a uniform condensate,[11] and an early lattice study noted this puzzle.[12] In a more recent lattice investigation, the crystalline phase has been confirmed.[10] The situation with the NJL_2 model, with continuous chiral symmetry, is less clear. Assuming a homogeneous condensate there is no difference between GN_2 and NJL_2, but the resulting phase diagram is not correct, as can be seen from its baryon content.[8] Thies and Schön have found an inhomogeneous "chiral spiral" phase[8] . Here we present the full analytic solution of the gap equation of the NJL_2 model. A Ginzburg-Landau computation[1] of the free energy suggests that the twisted chiral kink crystal is preferred over the chiral spiral. We now have the exact spectral properties of the twisted chiral kink crystal condensate,[2] and so an exact thermodynamical treatment should now be possible.

Our approach is based on the observation that a carefully motivated ansatz for the associated Gorkov resolvent reduces the gap equation to a simple ordinary differential equation, an explicitly soluble form of the nonlinear Schrödinger equation (NLSE). In general, the gap equation for an *inhomogeneous* condensate is a highly nontrivial functional differential equation, so the reduction to the NLSE represents a significant simplification.

The gap equation approach is based on the effective action [obtained by a Hubbard-Stratonovich transformation] for the condensate field. In the chiral model the condensate is most naturally written as a complex field: $\Delta \equiv \sigma - i\,\pi \equiv M\, e^{i\chi}$. The effective action for the condensate Δ is

$$S_{\rm eff} = -\frac{1}{2g^2}\int |\Delta|^2 - i\,N\,\ln\det\left[i\,\partial\!\!\!/ - \frac{1}{2}\left(1-\gamma^5\right)\Delta - \frac{1}{2}\left(1+\gamma^5\right)\Delta^*\right] \quad (1)$$

The corresponding (complex) gap equation is

$$\frac{\Delta(x)}{2Ng^2} = -i\frac{\delta}{\delta\Delta(x)^*}\ln\det\left[i\,\partial\!\!\!/ - \frac{1}{2}\left(1-\gamma^5\right)\Delta(x) - \frac{1}{2}\left(1+\gamma^5\right)\Delta^*(x)\right] \quad (2)$$

If the condensate is constant, as is commonly assumed, it is straightforward to evaluate the determinant and solve the gap equation.[11] When the condensate is inhomogeneous this is a much more difficult problem. Dashen, Hasslacher and Neveu[4] used inverse scattering to find kink-like static but spatially inhomogeneous condensates for the gap equation of the GN_2 model (where there is no pseudoscalar condensate, so Δ is real). Shei[5] extended this inverse scattering analysis to the chiral Gross-Neveu model, the NJL_2 model, and found a spatially inhomogeneous complex kink. A new approach to the inhomogeneous gap equation, based on the resolvent, was developed by Feinberg and Zee[6] and applied to the kink solutions of both the GN_2 and NJL_2 models. For the GN_2 model, Thies used a Hartree-Fock approach to find a periodic extension of the real kink solution, motivated by analogous inhomogeneous condensates in condensed matter systems.[7] In,[1,2] the present authors showed that the complex gap equation (2) can be reduced in an elementary manner to a soluble form of the nonlinear Schrödinger equation.

Another approach to finding a self-consistent condensate is to solve the relativistic Hartree-Fock problem $H\psi = E\psi$, with

$$
\begin{aligned}
H &= -i\gamma^5 \frac{d}{dx} + \gamma^0 \left[\frac{1}{2}\left(1-\gamma^5\right)\Delta(x) + \frac{1}{2}\left(1+\gamma^5\right)\Delta^*(x)\right] \\
&= \begin{pmatrix} -i\frac{d}{dx} & \Delta(x) \\ \Delta^*(x) & i\frac{d}{dx} \end{pmatrix} \quad . \qquad (3)
\end{aligned}
$$

and subject to the consistency condition $\langle\bar{\psi}\psi\rangle - i\langle\bar{\psi}i\gamma^5\psi\rangle = -\Delta/g^2$. We choose Dirac matrices $\gamma^0 = \sigma_1$, $\gamma^1 = -i\sigma_2$, $\gamma^5 = \sigma_3$, to emphasize the natural complex form of the condensate Δ. This Hamiltonian is also known as the Bogoliubov-de Gennes (BdG) Hamiltonian, and we will refer to the spectral equation $H\psi = E\psi$ as the Bogoliubov-de Gennes (BdG) equation.

The main new idea in our analysis[1,2] is the reduction of the functional gap equation (2) to the nonlinear Schrödinger equation. The key quantity is the coincident limit of Gor'kov Green's function, the "diagonal resolvent":

$$R(x;E) \equiv \langle x|\frac{1}{H-E}|x\rangle \quad . \qquad (4)$$

The resolvent (4) is clearly a 2×2 matrix, and it encodes all spectral information. Indeed, the spectral function of the single-particle spectrum of fermions in the presence of $\Delta(x)$ is $\rho(E) = \frac{1}{\pi}\mathrm{Im}\,\mathrm{Tr}_{D,x}\left[R(x;E+i\epsilon)\right]$.

The first step is to note that the form of the BdG equation places very strong constraints on the possible form of $R(x;E)$. For any static condensate $\Delta(x)$, $R(x;E)$ must satisfy the following algebraic conditions:[2,19,20]

$$R = R^\dagger \quad , \quad \mathrm{tr}_D\left(R(x;E)\sigma_3\right) = 0 \quad , \quad \det R(x;E) = -\frac{1}{4} \qquad (5)$$

Furthermore, $R(x;E)$ must satisfy the first-order differential equation

$$\frac{\partial}{\partial x} R(x;E)\,\sigma_3 = i\left[\begin{pmatrix} E & -\Delta(x) \\ \Delta^*(x) & -E \end{pmatrix}, R(x;E)\,\sigma_3\right] \qquad (6)$$

In superconductivity, (6) is known as the Eilenberger equation,[9,20] and in mathematical physics as the Dik'ii equation.[18] These conditions (5), and the Eilenberger equation (6), all follow from the simple fact that for the one-dimensional BdG equation, which involves derivatives with respect to the single variable x, the Green's function can be expressed as a product of two independent solutions : $R(x;E) = \frac{1}{2iW}\left(\psi_1\psi_2^T + \psi_2\psi_1^T\right)\sigma_1$, where W is the Wronskian of two independent solutions $\psi_{1,2}$: $W = i\psi_1^T\sigma_2\psi_2$.

The next step is to note that the gap equation provides further information about the possible form of the resolvent, and this is enough to motivate a specific ansatz form.[1,2] There are two ways of viewing the gap equation (2) in terms of the resolvent. First, for a static condensate we can write the log det term in the effective action (1) in terms of the single-particle spectral function $\rho(E)$:

$$-\frac{1}{\beta}\int_{-\infty}^{\infty} dE\,\rho(E)\ln\left(1+e^{-\beta(E-\mu)}\right) \quad . \tag{7}$$

All dependence on $\Delta(x)$ resides in the spectral function $\rho(E)$. Therefore, inserting this into the gap equation (2), this relates $\Delta(x)$ to the *diagonal* entries of $R(x;E)$. Further, from (5), these diagonal entries are equal. So, the simplest natural solution to the gap equation is for the diagonal entries of $R(x;E)$ to be linear in $|\Delta(x)|^2$. A second way to view the gap equation is to evaluate the functional derivative in (2), which for a static condensate leads to:

$$\Delta(x) = -iNg^2\mathrm{tr}_{D,E}\left[\gamma^0\left(\mathbf{1}+\gamma^5\right)R(x;E)\right] \tag{8}$$

The Dirac trace then relates the *off-diagonal* entries of $R(x;E)$ to $\Delta(x)$. Since R is hermitean, these off-diagonal entries are complex conjugates of one another. In fact, to describe *inhomogeneous* condensates we need to extend this slightly to the following ansatz form:[1,2]

$$R(x;E) = \mathcal{N}(E)\begin{pmatrix} a(E)+|\Delta(x)|^2 & b(E)\Delta(x)-i\Delta'(x) \\ b(E)\Delta^*(x)+i\Delta'^*(x) & a(E)+|\Delta(x)|^2 \end{pmatrix} \quad . \tag{9}$$

Here $a(E)$, $b(E)$ and $\mathcal{N}(E)$ are functions of E, to be determined. This ansatz (9) is consistent with the various algebraic constraints and with the Eilenberger equation (6). Indeed, substituting the ansatz (9) into the Eilenberger equation (6), we see that the diagonal entry of this equation is identically satisfied, while the off-diagonal entry implies that $\Delta(x)$ must satisfy the following nonlinear Schrödinger equation (NLSE):

$$\Delta'' - 2|\Delta|^2\,\Delta + i\,(b-2E)\,\Delta' - 2\,(a-Eb)\,\Delta = 0 \quad . \tag{10}$$

Thus, we have reduced the very difficult problem of solving the functional gap equation (2) for a self-consistent condensate $\Delta(x)$ to the much simpler problem of solving the NLSE for $\Delta(x)$. In fact, the NLSE (10) is explicitly soluble, as is discussed below.

1.1. *Real solutions of the NLSE*

We first show that the known real solutions to the gap equation of the *non-chiral* GN_2 model are indeed solutions of the NLSE (10).

(1) Homogeneous condensate: $\Delta(x) = m$. This clearly satisfies the NLSE (10), with $a(E) = 2E^2 - m^2$, $b(E) = 2E$, and $\mathcal{N}(E) = 1/(4E\sqrt{m^2 - E^2})$.
(2) Real kink condensate:[4] $\Delta(x) = m \tanh(m\,x)$. This satisfies the NLSE: $\Delta'' - 2\,\Delta^3 + 2m^2\Delta = 0$, and we deduce the exact diagonal resolvent to be of the form (9) with $a(E) = 2E^2 - m^2$, $b(E) = 2E$, and $\mathcal{N}(E) = 1/(4E\sqrt{m^2 - E^2})$.
(3) Real kink crystal condensate:[7] $\Delta(x) = \sqrt{\nu}\,\frac{2m}{1+\sqrt{\nu}}\,\mathrm{sn}\left(\frac{2m}{1+\sqrt{\nu}}\,x;\nu\right)$, where sn is the Jacobi elliptic function and $0 \leq \nu \leq 1$ is the real elliptic parameter.[21] This solution is the key to understanding the crystalline phase of the non-chiral Gross-Neveu model.[7] Interestingly, it also describes a polaron crystal in polymer physics,[13,14] a periodic pair potential in inhomogeneous superconductors.[15–17] When $\nu = 1$, it reduces to the single kink condensate. For general ν, it satisfies the NLSE: $\Delta'' - 2\,\Delta^3 + (1+\nu)\left(\frac{2m}{1+\sqrt{\nu}}\right)^2\Delta = 0$. We deduce the exact diagonal resolvent to be of the form (9) with $a(E) = 2E^2 - 2m^2\frac{1+\nu}{(1+\sqrt{\nu})^2}$, $b(E) = 2E$, $\mathcal{N}(E) = 1/(4\sqrt{m^2 - E^2}\,\sqrt{E^2 - m^2(\frac{1-\sqrt{\nu}}{1+\sqrt{\nu}})^2})$.

1.2. *Complex solutions of the NLSE*

(1) Single plane wave condensate: $\Delta = m\,e^{iqx}$ This clearly satisfies the NLSE (10), with $a(E) = 2\left(E - \frac{q}{2}\right)^2 - m^2$, $b(E) = 2E - 2q$, $\mathcal{N}(E) = 1/(4(E - q/2)\sqrt{m^2 - (E - q/2)^2})$. One can always remove such a plane wave factor from $\Delta(x)$ by a local chiral rotation: $\Delta \to e^{-iqx}\Delta$, $\psi \to e^{-iqx/2\gamma_5}\psi$, that has the effect of shifting the entire energy spectrum by $q/2$.
(2) Single complex kink condensate:[5] $\Delta(x) = m\,\frac{\cosh(m\,\sin(\theta/2)\,x - i\theta/2)}{\cosh(m\,\sin(\theta/2)\,x)}\,e^{i\theta/2}$. This complex form is plotted in Figure 1. This illustrates the role of the parameter $\theta \in [0, 2\pi]$ as the net rotation angle of the kink as x

Fig. 1. Plot of Shei's single complex kink condensate, for $\theta = 3\pi/2$, illustrating how the kink winds around zero without the amplitude vanishing. The kink is the solid line, and the surface is shown simply to illustrate that both the amplitude and the phase are changing.

goes from $-\infty$ to $+\infty$: $\Delta(x = +\infty) = e^{-i\theta}\Delta(x = -\infty)$. Observe that when $\theta = \pi$, the complex kink is in fact real, and reduces to the familiar real kink solution; this real kink changes its sign [i.e., rotates through π] in passing from $x = -\infty$ to $x = +\infty$. The complex kink condensate satisfies the NLSE: $\Delta'' - 2\,|\Delta|^2\,\Delta - 2\,i\,m\,\cos(\theta/2)\,\Delta' + 2m^2\Delta = 0$, from which we deduce the exact diagonal resolvent to be of the form (9) with $a(E) = 2E^2 - 2\,m\,\cos(\theta/2)E - m^2$, $b(E) = 2E - m\,\cos(\theta/2)$, $\mathcal{N}(E) = 1/(4(E - m\,\cos(\theta/2))\sqrt{m^2 - E^2})$. Shei[5] further showed that this condensate satisfies the gap equation provided the winding parameter $\theta/(2\pi)$ is equal to the filling fraction $\frac{n}{N}$, in the large flavor limit, of the single bound state in the gap by n flavors, with $\frac{n}{N}$ fixed as $N \to \infty$:[5,6]

$$\frac{\theta}{2\pi} = \frac{n}{N} \quad . \tag{11}$$

In our approach, precisely the same condition arises from demanding that the coefficient of the $\Delta'(x)$ term vanishes after the energy trace, a necessary requirement to satisfy the gap equation.[1,2]

(3) <u>Complex kink crystal condensate:</u> our new complex condensate is a periodic array of Shei's complex kinks.

$$\Delta(x) = -A\frac{\sigma(A\,x + i\mathbf{K}' - i\frac{\theta}{2})}{\sigma(A\,x + i\mathbf{K}')\sigma(i\frac{i\theta}{2})}\exp\left[iA\,x\left(-i\zeta(\frac{i\theta}{2}) + i\text{ns}(\frac{i\theta}{2})\right) + i\theta\frac{\eta_3}{2}\right] \tag{12}$$

The parameter A sets the scale of the condensate and its length scale: $A = A(m,\theta,\nu) \equiv -2i\,m\,\mathrm{sc}\,(i\theta/4)\,\mathrm{nd}\,(i\theta/4)$, where sc=sn/cn and nd=1/dn are Jacobi elliptic functions.[21] The functions σ and ζ are the Weierstrass sigma and zeta functions.[21] We have chosen real and imaginary half-periods: $\omega_1 = \mathbf{K}(\nu)$, and $\omega_3 = i\,\mathbf{K}' \equiv i\mathbf{K}(1-\nu)$. Both periods are therefore controlled by the single [real] elliptic parameter $0 \le \nu \le 1$. Also, $\eta_3 \equiv \zeta(i\mathbf{K}')$ is purely imaginary. The parameter $\theta \in [0, 4\mathbf{K}'(\nu)]$ is related to the angle through which the condensate rotates in one period $L = 2\mathbf{K}/A$:

$$\Delta(x+L) = e^{2i\varphi}\Delta(x) \tag{13}$$

where the angle φ is a function of θ and ν: $\varphi = \mathbf{K}\left(-i\,\zeta(i\theta/2) + i\,\mathrm{ns}(i\theta/2) - \frac{\eta\theta}{2\mathbf{K}}\right)$.

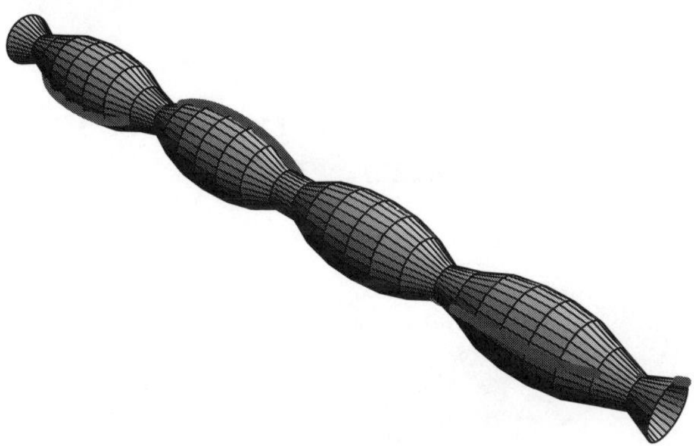

Fig. 2. Plot of the complex kink crystal condensate (12), for $\nu = 0.8$ and $\theta = 3\mathbf{K}(0.2)/2$, illustrating how the kink winds around zero each period, without the amplitude vanishing. The kink is the solid line, and the surface is shown simply to illustrate that both the amplitude and the phase are changing over each period.

The complex crystalline condensate in (12) satisfies the NLSE:

$$\Delta'' - 2\,|\Delta|^2\,\Delta - i\,\left(2A\,i\,\mathrm{ns}(i\theta/2)\right)\Delta' - A^2\left(3\,\mathcal{P}(i\theta/2) - \mathrm{ns}^2(i\theta/2)\right)\Delta = 0 \tag{14}$$

Comparing this equation with the NLSE (10) we can extract the functions $a(E)$, $b(E)$ and $\mathcal{N}(E)$ appearing in (9), thereby determining the *exact* diagonal resolvent. To express these functions in a compact form, we first define some properties of the associated fermionic spectrum for the BdG equation. This spectrum has positive and negative energy continua starting at $E = \pm m$, together with a single bound band in the gap, as depicted in Figure 3. In contrast to the case for the real kink crystal, here the bound band is not centered in the middle of the gap, but is displaced from the center. The parameter θ characterizes this asymmetry in the spectrum. The band edges are functions of both the winding angle θ and the elliptic parameter ν:

$$\begin{aligned} E_1 &= -m \\ E_2 &= m(-1 + 2\,\mathrm{nc}^2(i\theta/4;\nu)) \\ E_3 &= m(-1 + 2\,\mathrm{nd}^2(i\theta/4;\nu)) \\ E_4 &= +m \end{aligned} \tag{15}$$

In the infinite period limit $[\nu \to 1]$, the band contracts to a single bound

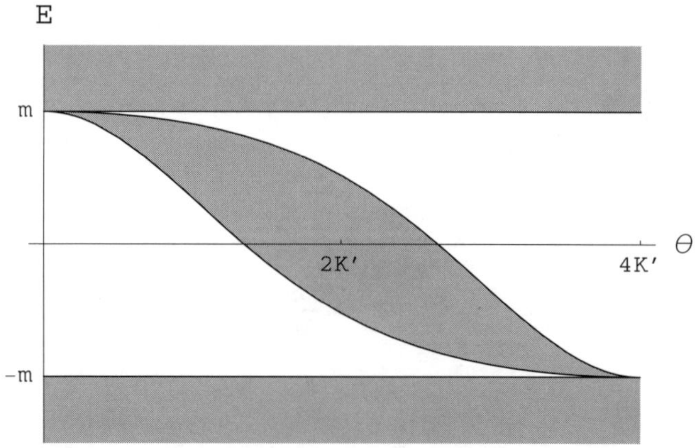

Fig. 3. Plot of the single-particle fermion spectrum for the complex kink crystal condensate (12), for $\nu = 0.1$, as a function of the winding parameter θ. Note that for $\theta = 2\mathbf{K}'$ [when the condensate is real] the band is centered symmetrically about $E = 0$, but for all other values of θ the band lies asymmetrically in the gap.

state, with $E_2 = E_3 = m\cos(\theta/2)$, and this is precisely the bound state of the single complex kink.[2,5] At finite period, but when $\theta = 2\mathbf{K}'(\nu)$, we find $E_2 = -E_3 = -\left(\frac{1-\sqrt{\nu}}{1+\sqrt{\nu}}\right) m$, and the band is centered symmetrically about 0; this is precisely the band spectrum of the real kink array.[2,7] In terms of the band edges, the resolvent functions $a(E)$, $b(E)$ and $\mathcal{N}(E)$

that appear in the resolvent (9) take the following simple form:

$$
\begin{aligned}
a(E) &= 2E^2 - (E_2 + E_3)E - \frac{(E_2 - E_3)^2}{4} - m^2 \\
b(E) &= 2E - (E_2 + E_3) \\
\mathcal{N}(E) &= \frac{1}{4\sqrt{(m^2 - E^2)(E - E_2)(E - E_3)}}
\end{aligned}
\tag{16}
$$

Here the normalization $\mathcal{N}(E)$ is fixed by the property $\det R = -\frac{1}{4}$.

1.3. *Density of states*

Remarkably, not only is it possible to solve the NLSE (10) exactly, we can also solve exactly the associated BdG equation.[22] The explicit spinor solutions can be found in.[2] Here we concentrate on the spectrum of single-particle states. The spinor solutions are parametrized by a spectral parameter α that lies on a fundamental rectangle, as shown in Figure 4. The

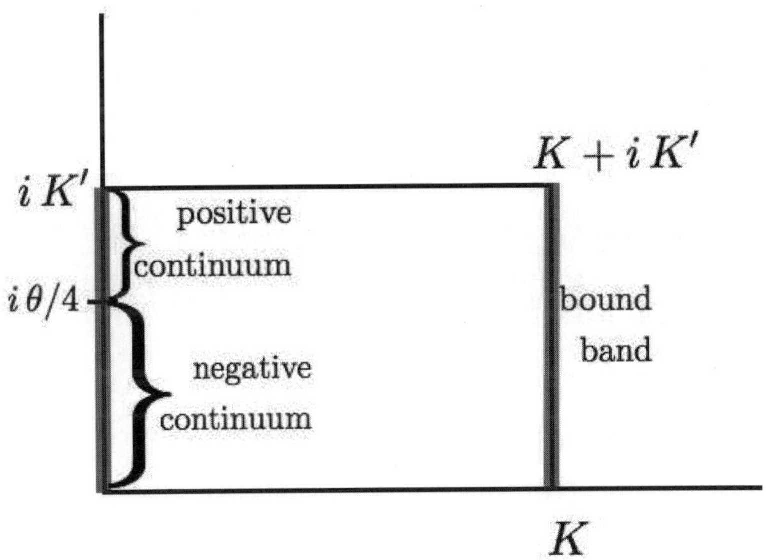

Fig. 4. The fundamental rectangle for the spectral parameter $i\alpha$ appearing in the spinor solutions to the BdG equation. The bound band is characterized by $\mathbf{K}(\nu) \leq i\alpha \leq \mathbf{K}(\nu) + i\mathbf{K}'(\nu)$, while the positive and negative energy continua are characterized by $0 \leq i\alpha \leq i\mathbf{K}'(\nu)$. The point $\alpha = \theta/4$ represents $E = \pm\infty$, depending on the side of approach.

energy E is related to α as:

$$E(\alpha) = \frac{A}{2}[i\,\zeta(i\alpha - i\theta/4) - i\,\zeta(i\alpha + i\theta/4) + i\,\zeta(i\theta/2) + i\,\mathrm{ns}(i\theta/2)] \quad (17)$$

To identify the momentum associated with these solutions, we recall that the quasiperiodic winding (13) of the condensate in (12) implies that the BdG hamiltonian (3) is invariant under a period shift, up to a global chiral rotation through the winding angle φ:

$$H\,(x+L) = e^{i\gamma_5\varphi}\,H\,(x)\;e^{-i\gamma_5\varphi} \qquad (18)$$

The spinor solutions acquire a chiral rotation and a Bloch phase:

$$\psi_\pm(x+L) = e^{\pm ikL}e^{i\varphi\gamma_5}\psi_\pm(x) \qquad (19)$$

The Bloch momentum k is expressed in terms of the spectral parameter α as:

$$k(\alpha) = -\frac{A}{2}\,[i\,\zeta(i\alpha + i\theta/4) + i\,\zeta(i\alpha - i\theta/4) + 2\eta\alpha/\mathbf{K}] \qquad (20)$$

The momentum is real for α taking values on the vertical edges of the fundamental rectangle.

Given the energy and momentum as explicit functions of the spectral parameter α, we can compute the density of states.

$$\begin{aligned}\frac{dE}{d\alpha} &= -\frac{A}{2}\,[\mathcal{P}\,(i\alpha + i\theta/4) - \mathcal{P}\,(i\alpha - i\theta/4)]\\ \frac{dk}{d\alpha} &= -\frac{A}{2}\,[\mathcal{P}\,(i\alpha + i\theta/4) + \mathcal{P}\,(i\alpha - i\theta/4) + 2\eta/\mathbf{K}]\end{aligned} \qquad (21)$$

Consequently,[2] the density of states can be expressed as

$$\begin{aligned}\frac{dk}{dE} &= \frac{dk}{d\alpha}/\frac{dE}{d\alpha}\\ &= \frac{[a(E) - A^2\mathcal{P}(i\theta/2) - A^2\eta/\mathbf{K}]}{2\sqrt{(E^2-m^2)(E-E_2)(E-E_3)}}\end{aligned} \qquad (22)$$

The density of states (22) has been derived from the energy and momentum of the spinor solutions. For consistency, we compare this with the trace of the resolvent (9) over one period:

$$\frac{1}{L}\int_L \mathrm{tr}_D\,R(x;E)dx = \frac{[a(E) - A^2\mathcal{P}(i\theta/2) - A^2\eta/\mathbf{K}]}{2\sqrt{(m^2-E^2)(E-E_2)(E-E_3)}}\;, \qquad (23)$$

2. Conclusions

To conclude, we have solved the inhomogeneous gap equation (2) in the NJL_2 model, to obtain a self-consistent condensate (12) that is crystalline in form. It is inherently complex, unlike the condensate in the discrete-chiral GN_2 model where the condensate is real. The complex crystal has a periodic amplitude, while its phase rotates though a certain angle over each period. This complex crystalline condensate contains all other previously known solutions to the gap equation as special cases. The key technical idea in our approach is to use the form of the gap equation to motivate an ansatz (9) for the Gorkov resolvent. For any condensate, the resolvent must satisfy the Eilenberger-Dikii equation (6), and so this reduces the problem to solving the nonlinear Schrödinger equation (10). Given this solution, and in particular the exact density of states (22), we are now in a position to study exactly the phase diagram of the NJL_2 model, extending the Ginzburg-Landau analysis in.[1] This will be addressed in future work.

An important possible extension of this work would be to the massive NJL_2 model, in which a bare fermion mass is included in the original Gross-Neveu Lagrangian. This mass term explicitly breaks the continuous chiral symmetry, and approximate methods have found a rich structure in the associated phase diagram.[23] Perhaps the most interesting extension would be to try to extend some of these ideas to higher dimensions, for example for higher dimensional Gross-Neveu or Nambu-Jona Lasinio models,[24] or more ambitiously to search for crystalline condensates in QCD or QCD models.[25–29]

Acknowledgments We thank the organizers for an interesting conference, the DOE for support through grant DE-FG02-92ER40716, and J. Feinberg, D. Nickel and M. Thies for helpful comments.

References

1. G. Başar and G. V. Dunne, "Self-consistent crystalline condensate in chiral Gross-Neveu and Bogoliubov-de Gennes systems," Phys. Rev. Lett. **100**, 200404 (2008), [arXiv:0803.1501 [hep-th]].
2. G. Başar and G. V. Dunne, "A Twisted Kink Crystal in the Chiral Gross-Neveu Model," [arXiv:0806.2659 [hep-th]], Phys. Rev. D in press.
3. D. J. Gross and A. Neveu, "Dynamical Symmetry Breaking In Asymptotically Free Field Theories," Phys. Rev. D **10**, 3235 (1974).
4. R. F. Dashen, B. Hasslacher and A. Neveu, "Semiclassical Bound States In An Asymptotically Free Theory," Phys. Rev. D **12**, 2443 (1975).
5. S. S. Shei, "Semiclassical Bound States In A Model With Chiral Symmetry," Phys. Rev. D **14**, 535 (1976).

6. J. Feinberg and A. Zee, "Dynamical Generation of Extended Objects in a 1+1 Dimensional Chiral Field Theory: Non-Perturbative Dirac Operator Resolvent Analysis," Phys. Rev. D **56**, 5050 (1997) [arXiv:cond-mat/9603173]; "Generalized supersymmetric quantum mechanics and reflectionless fermion bags in 1+1 dimensions," [arXiv:hep-th/0109045], in Olshanetsky, M. (ed.) et al. *Multiple facets of quantization and supersymmetry* (World Scientific, 2002).
7. M. Thies, "Analytical solution of the Gross-Neveu model at finite density," Phys. Rev. D **69**, 067703 (2004) [arXiv:hep-th/0308164]; "From relativistic quantum fields to condensed matter and back again: Updating the Gross-Neveu phase diagram," J. Phys. A **39**, 12707 (2006) [arXiv:hep-th/0601049].
8. V. Schon and M. Thies, "Emergence of Skyrme crystal in Gross-Neveu and 't Hooft models at finite density," Phys. Rev. D **62**, 096002 (2000) [arXiv:hep-th/0003195]; "2D model field theories at finite temperature and density," [arXiv:hep-th/0008175], in Shifman, M. (ed.), *At the frontier of particle physics*, Vol. 3 (World Scientific).
9. G. Eilenberger, "Transformation of Gorkov's Equation for Type II Superconductors into Transport Like Equations", Z. Phys. **214**, 195 (1968).
10. P. de Forcrand and U. Wenger, "New baryon matter in the lattice Gross-Neveu model," PoS **LAT2006**, 152 (2006) [arXiv:hep-lat/0610117].
11. U. Wolff, "The phase diagram of the infinite-N Gross-Neveu Model at finite temperature and chemical potential", Phys. Lett. **157B**, 303 (1985).
12. F. Karsch, J. B. Kogut and H. W. Wyld, "The Gross-Neveu Model At Finite Temperature And Density," Nucl. Phys. B **280**, 289 (1987).
13. B. Horovitz, "Soliton Lattice in Polyacetylene, Spin-Peierls Systems, and Two-Dimensional Sine-Gordon Systems", Phys. Rev. Lett. **46**, 742 (1981).
14. S. A. Brazovskii, S. A. Gordynin, and N. N. Kirova, "Exact solution of the Peierls model with an arbitrary number of electrons in the unit cell", Pis. Zh. Eksp. Teor. Fiz. **31**, 486 (1980) [JETP Lett. **31**, 456 (1980)]; "Excitons, polarons and bipolarons in conducting polymers", Pis. Zh. Eksp. Teor. Fiz. **33**, 6 (1981) [JETP Lett. **33**, 4 (1981)]; S. A. Brazovskii, N. N. Kirova and Matveenko, "Peierls effect in conducting polymers", Zh. Eksp. Teor. Fiz. **86**, 743 (1984) Sov. Phys. JETP **59**, 434 (1984).
15. J. Mertsching and H. J. Fischbeck, "The incommensurate Peierls phase of the quasi-dimensional Frohlich model with a nearly half-filled band", Phys. Stat. Sol. B **103**, 783 (1981).
16. J. Bar-Sagi and C. G. Kuper, "Self-Consistent Pair Potential in an Inhomogeneous Superconductor", Phys. Rev. Lett. **28** 1556 (1972).
17. K. Machida and H. Nakanishi, "Superconductivity under a ferromagnetic molecular field", Phys. Rev. B **30**, 122 (1984).
18. L. A. Dickey, *Soliton Equations and Hamiltonian Systems*, (World Scientific, Singapore, 1991).
19. D. Waxman, "The Fredholm Determinant For A Dirac Hamiltonian With A Topological Mass Term," Annals Phys. **241**, 285 (1995).
20. I. Kosztin, S. Kos, M. Stone and A. J. Leggett, "Free energy of an inhomogeneous superconductor: A wave-function approach", Phys. Rev. B **58**, 9365

(1998); S. Kos and M. Stone, "Gradient expansion for the free energy of a clean superconductor ", Phys. Rev. B **59**, 9545 (1999).

21. D. F. Lawden, *Elliptic Functions and Applications*, (Springer-Verlag, New York, 1980).
22. A. O. Smirnov, "The Dirac Operator with Elliptic Potential", Sbornik Math **186**, 1213 (1995).
23. C. Boehmer, F. Karbstein and M. Thies, "Baryons in the large N limit of the massive NJL2 model," arXiv:0803.1369 [hep-th].
24. S. Hands, A. Kocic and J. B. Kogut, "The Four Fermi Model In Three-Dimensions At Nonzero Density And Temperature," Nucl. Phys. B **390**, 355 (1993) [arXiv:hep-lat/9206024].
25. I. R. Klebanov, "Nuclear Matter In The Skyrme Model," Nucl. Phys. B **262**, 133 (1985).
26. A. S. Goldhaber and N. S. Manton, "Maximal Symmetry Of The Skyrme Crystal," Phys. Lett. B **198**, 231 (1987).
27. A. D. Jackson and J. J. M. Verbaarschot, "Phase structure of the Skyrme model," Nucl. Phys. A **484**, 419 (1988); L. Castillejo, P. S. J. Jones, A. D. Jackson, J. J. M. Verbaarschot and A. Jackson, "Dense Skyrmion Systems," Nucl. Phys. A **501**, 801 (1989).
28. N. S. Manton and P. M. Sutcliffe, "Skyrme crystal from a twisted instanton on a four torus," Phys. Lett. B **342**, 196 (1995) [arXiv:hep-th/9409182].
29. B. Bringoltz, "Chiral crystals in strong-coupling lattice QCD at nonzero chemical potential," JHEP **0703**, 016 (2007) [arXiv:hep-lat/0612010].

The Strong Coupling Constant at Low and High Energies

J. H. Kühn
Institut für Theoretische Teilchenphysik, Universität Karlsruhe, D-76128 Karlsruhe, Germany

Recent results for the cross section of electron-positron annihilation into hadrons and for the decay rates of the Z boson and the τ lepton into hadrons including corrections of order α_s^4 are reviewed. The consistency between two values of α_s measured at vastly different energies constitutes a striking test of asymptotic freedom and requires the proper matching conditions at charm and bottom thresholds. Combining the results from Z and τ decays leads to $\alpha_s(M_Z) = 0.1198 \pm 0.0015$ as one of the most precise and presently only NNNLO result for the strong coupling constant. We report a recent determination of α_s which is based on a lattice evaluation of the pseudoscalar correlator for the charm quarks combined with continuum perturbation theory.

The strong coupling constant α_s is one of the three fundamental gauge couplings constants of the Standard Model (SM) of particle physics. Its precise determination is one of the most important aims of particle physics. Experiments at different energies allow to test the predictions for its energy dependence based on the renormalization group equations, the comparison of the results obtained from different processes leads to critical tests of the theory and potentially to the discovery of physics beyond the Standard Model. Last not least, the convergence of the three gauge coupling constants related by SU(3)×SU(2)×U(1) to a common value, after evolving them to high energies, allows to draw conclusions about the possibility of embedding the SM in the framework of a Grand Unified Theory.

A characteristical example is shown in Fig.1, which demonstrates in particular the anticipated improvement in our information about coupling constant unification at the GUT scale, if the experimental uncertanty in α_s will be reduced from 0.0027 (corresponding to the value quoted by the "Electroweak Working Group"[1]) to 0.0009, a precision that might be within reach with 10^9 Z-events at a GIGA-Z.[2] In this latter case it may well be possible to deduce non-trivial constraints on threshold corrections and the structure of symmetry breaking at the GUT-scale.

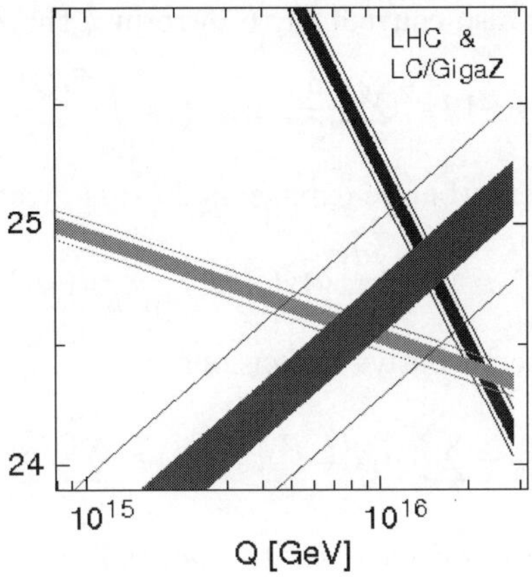

Fig. 1. Renormalization group equation of the three coupling constants α_1, α_2 and α_3[3].

One of the most precise and theoretically safe determination of α_s is based on measurements of the cross section for electron-positron annihilation into hadrons. These have been performed in the low-energy region between 2 GeV and 10 GeV and, in particular, at and around the Z resonance at 91.2 GeV. Conceptually closely related is the measurement of the semileptonic decay rate of the τ-lepton, leading to a determination of α_s at a scale below 2 GeV.

From the theoretical side, in the framework of perturbative QCD, these rates and cross sections are evaluated as inclusive rates into massless quarks and gluons.[4,5] (Power suppressed mass effects are well under control for e^+e^--annihilation, both at low energies and around the Z resonance, and for τ decays,[6–9] and the same applies to mixed QCD and electroweak corrections[10,11]).

The ratio $R(s) \equiv \sigma(e^+e^- \to \text{hadrons})/\sigma(e^+e^- \to \mu^+\mu^-)$ is expressed through the absorptive part of the correlator of the electromagnetic current j_μ:

$$R(s) = 12\pi\, \Im\Pi(-s - i\epsilon), \tag{1}$$

$$3\,Q^2\,\Pi(Q^2) = i\int \mathrm{d}^4x e^{iq\cdot x}\langle 0|\mathrm{T}j_\mu(x)j^\mu(0)|0\rangle, \tag{2}$$

with $Q^2 = -q^2$. It is also convenient to introduce the Adler function as

$$D(Q^2) = -12\,\pi^2 Q^2 \,\frac{\mathrm{d}}{\mathrm{d}Q^2}\Pi(Q^2) = \int_0^\infty \frac{Q^2\ R(s)ds}{(s+Q^2)^2},$$

Conversely, in the perturbative expansion $R(s)$ is given by

$$R(s) = D(s) - \pi^2\beta_0^2\{\frac{d_1}{3}a_s^3 + (d_2 + \frac{5}{6\,\beta_0}d_1\,\beta_1)\,a_s^4\} + \dots \,. \tag{3}$$

Here we define the perturbative expansions

$$D(Q^2) = \sum_{i=0}^{\infty}\ d_i a_s^i(Q^2),\quad R(s) = \sum_{i=0}^{\infty}\ r_i a_s^i(s), \tag{4}$$

where $a_s \equiv \alpha_s/\pi$ and we have set the normalization scale $\mu^2 = Q^2$ or to $\mu^2 = s$ for the Euclidian and Minkowskian functions respectively. The results for generic values of μ can be easily recovered with standard RG techniques.

Note that the first three terms of the perturbative series for D and R coincide. Starting from r_3, terms proportional π^2 arise which can be predicted from those of lower order. It has been speculated that these "π^2-terms", also called "kinematical terms", might constitute a major part of the full higher order corrections (see, e.g.[12]); however, the validity of this hypothesis can only be established by the full calculation. Indeed, for the scalar correlator this assumption has been shown to fail completely.[13]

For the vector correlator the terms of order a_s^2 and a_s^3 have been evaluated nearly thirty and about fifteen years ago,[14–16] respectively. The a_s^4 corrections are conveniently classified according to their power of n_f, with n_f denoting the number of light quarks. The $a_s^4 n_f^3$ term is part of the "renormalon chain", the evaluation of the next term, of order $a_s^4 n_f^2$, was a test case for the techniques used extensively in this paper and, furthermore, led to useful insights into the structure of the perturbative series already.[17]

The complete five-loop calculation has been performed in.[18] It required the evaluation of about twenty thousand diagrams generated automatically via QGRAF.[19] Using "infrared rearrangement",[20] the R^* operation[21] and the prescriptions formulated in[22] to algorithmically resolve the necessary combinatorics, it was possible to express the absorptive part of the five-loop diagrams in terms of four-loop massless propagator integrals. Using then a representation for Feynman integrals proposed in,[23,24] these integrals were reduced to a sum of 28 master integrals, with coefficients which are rational functions of the space-time dimension D. These coefficients were, in a first step, evaluated in the large-D limit, and, after evaluating sufficiently many terms of the $1/D$ expansion, fully reconstructed.[25] This direct and largely automatic procedure required enormous computing resources and was performed on a 32+8-node SGI parallel computer with SMP architecture and on the HP XC4000 supercomputer of the federal state Baden-Württemberg using a parallel version[26] of FORM.[27]

The so-called "non-singlet" diagrams, where one and the same closed quark line is connected to the external currents, and which were evaluated in,[18] are sufficient for a complete description of τ-decays. For e^+e^- annihilation, they correspond to the dominant terms proportional $\sum_i Q_i^2$. The singlet contributions proportional $(\sum_i Q_i)^2$ arise for the first time in $\mathcal{O}(\alpha_s^3)$ and are known to be small. Similar comments apply to the singlet contributions in Z decays.

The analytic result for the Adler function contains transcendental numbers up to ζ_7, with ζ_4 and ζ_6 being absent. It leads straightforwardly to R at order α_s^4. In numerical form

$$\begin{aligned} R = 1 &+ a_s + (1.9857 - 0.1152\, n_f)\, a_s^2 \\ &+ (-6.63694 - 1.20013 n_f - 0.00518 n_f^2)\, a_s^3 \\ &+ (-156.61 + 18.77\, n_f - 0.7974\, n_f^2 \\ &+ 0.0215\, n_f^3)\, a_s^4. \end{aligned} \tag{5}$$

It is also instructive to explicitly display the genuine five-loop contributions to d_4 (underlined in (6,7) below) and the "kinematical" terms originating from the analytic continuation:

$$\begin{aligned} r_3 = \underline{18.2} - 24.9 + (\underline{-4.22} + 3.02)\, n_f \\ + (\underline{-0.086} + 0.091)\, n_f^2, \end{aligned} \tag{6}$$

$$\begin{aligned} r_4 = \underline{135.8} - 292.4 + (\underline{-34.4} + 53.2)\, n_f \\ + (\underline{1.88} - 2.67)\, n_f^2 + (\underline{-0.010} + 0.032)\, n_f^3. \end{aligned} \tag{7}$$

Since it will presumably take a long time until the next term of the perturbative series will be evaluated, it is of interest to investigate the predictive power of various optimization schemes empirically. Using the principles of "Fastest Apparent Convergence" (FAC)[28] or of "Minimal Sensitivity" (PMS),[29] which happen to coincide in this order, the central values of the predictions[30,31]

$$d_4^{\rm pred}(n_f = 3, 4, 5) = 27 \pm 16, 8 \pm 18, -8 \pm 44$$

differ significantly from the exact result

$$d_4^{\rm exact}(n_f = 3, 4, 5) = 49.08, \quad 27.39, \quad 9.21. \tag{8}$$

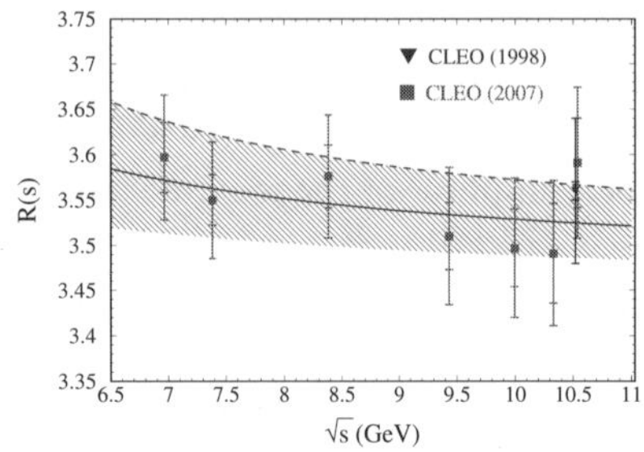

Fig. 2. $R(s)$ as measured by CLEO .[35,36] The full solid line and the hatched band correspond to the theory prediction based on $\alpha_s(M_Z^2) = 0.110^{-0.014}_{-0.017}$. The dashed line represents theory prediction where results from the CLEO analysis ,[36] $\alpha_s^{(5)}(M_Z) = 0.126$, has been used.

However, within the error estimates[31] predicted and exact values are in agreement. The picture changes, once these estimates are used to predict the coefficient r_4. Although sizable cancellations between "dynamical" and "kinematical" terms are observed for the individual n_f coefficients in (7) the predictions for the final results are significantly closer (in relative sense) to the results of the exact calculation:

$$r_4^{\text{pred}}(n_f = 3,4,5) = -129 \pm 16, \quad -112 \pm 30, -97 \pm 44,$$
$$r_4^{\text{exact}}(n_f = 3,4,5) = -106.88, \quad -92.898, -79.98.$$

(This is in striking contrast to the case of the scalar correlator, where the predictions for the dynamical terms work well, but, as a consequence of the strong cancellations between dynamical and kinematical terms fail completely in the Minkowskian region.) Using FAC and the exact result for d_4, the coefficients d_5 and r_5 can be predicted (following[31,32]) for $n_f = 3,4,5$, namely

$$d_5^{\text{pred}}(n_f = 3,4,5) = 275, \quad 152, \quad 89, \tag{9}$$
$$r_5^{\text{pred}}(n_f = 3,4,5) = -505, \quad -134, \quad 168. \tag{10}$$

These terms may become of relevance for the International Linear Collider (ILC) running in the GIGA-Z mode with an anticipated precision of $\delta\alpha_s = 0.0005 - 0.0007$[2], and already today for the analysis of τ-decays.

Let us now move to the analysis of present data for e^+e^- annihilation and τ decays. Measurements of $R(s)$ at lower energies, with their correspondingly larger values of α_s, are in principle more sensitive to $\alpha_s(M_Z)$

if the same relative precision could be obtained. At present, however, the systematic experimental error of 2% is a limiting element for a competitive measurement. Nevertheless, it is interesting, that the result of a recent analysis $\alpha_s(9^2\ \mathrm{GeV}^2) = 0.160 \pm 0.034$[33] is a clear 3σ signal for the influence of QCD corrections also in this energy range. Furthermore, as shown in,[33] proper inclusion of quark mass effects and matching at flavour threshold is required to obtain the correct result (Fig. 2). A convenient FORTRAN program, called RHAD, which predicts $R_{\mathrm{had}}(s)$ and where all quark mass effects and higher order QCD corrections are implemented, is described in.[34] Ignoring quark mass effects and proper matching conditions,[35] leads to a shift of α_s by one standard deviation, as shown in Fig. 2 (dashed line).

Evolving the result from 9 GeV to M_Z leads to $\alpha_s(M_Z^2) = 0.110^{+0.014}_{-0.017}$. It is instructive to combine this value with the one obtained in Ref.,[37] $\alpha_s^{(4)}(5^2\mathrm{GeV}^2) = 0.235^{+0.047}_{-0.047}$, which was based on earlier measurements by BES,[38] MD-1[39] and CLEO[35] in the region between 3 and 10 GeV. Adding the correlated and uncorrelated errors of the different experiments in quadrature, the final result

$$\alpha_s(9^2\mathrm{GeV}^2)^{NNLO} = 0.182^{+0.022}_{-0.025} \tag{11}$$

represents the combined information on the strong coupling from measurements in the region below the bottom threshold. The shift in α_s from the inclusion of the α_s^4 term amounts to $\delta\alpha_s(9^2\mathrm{GeV}^2) = 0.002$ and is thus irrelevant compared to the large experimental error.

In contrast to the situation below 10 GeV the extraction of α_s from Z decays is affected by the α_s^4 terms. The analysis of the electroweak working group[1] is based on eq. (5) with $n_f = 5$, including terms up to $\mathcal{O}(\alpha_s^3)$ and leads to

$$\alpha_s(M_Z)^{NNLO} = 0.1185 \pm 0.0026^{\mathrm{exp}}. \tag{12}$$

Including the α_s^4 term leads to a shift $\delta\alpha_s(M_Z) = 0.0005$

$$\alpha_s(M_Z)^{NNNLO} = 0.1190 \pm 0.0026^{\mathrm{exp}}. \tag{13}$$

The theory error may either be conservatively based on the shift produced by the last term (0.0005) or on the scale variation with $\mu/M_Z = \frac{1}{3} \div 3$, leading to ± 0.0002 and can now be neglected in both cases.

Higher orders are of particular relevance in the low-energy region, for example in τ decays. The correction from perturbative QCD to the ratio

$$\begin{aligned} R_{\tau,V+A} &= \frac{\Gamma(\tau \to \mathrm{hadrons}_{S=0} + \nu_\tau)}{\Gamma(\tau \to l + \bar{\nu}_l + \nu_\tau)} \\ &= 3|V_{ud}|^2 S_{EW}(1 + \delta_0 + \delta'_{EW} + \delta_2 + \delta_{NP}) \end{aligned} \tag{14}$$

is given by

$$1 + \delta_0 = 2 \int_0^{M_\tau^2} \frac{ds}{M_\tau^2} \left(1 - \frac{s}{M_\tau^2}\right)^2 \left(1 + \frac{2s}{M_\tau^2}\right) R(s). \tag{15}$$

The perturbative quantity δ_0 can be evaluated in Fixed Order perturbation theory or with "Contour Improvement" as proposed in[40,41]

$$\delta_0^{FO} = a_s + 5.202\, a_s^2 + 26.366\, a_s^3 + 127.079\, a_s^4, \tag{16}$$

$$\delta_0^{CI} = 1.364\, a_s + 2.54\, a_s^2 + 9.71\, a_s^3 + 64.29\, a_s^4. \tag{17}$$

From the experimental side $\delta_0^{\mathrm{exp}} = 0.1998 \pm 0.0043_{\mathrm{exp}}$ has been used.[42] More recent evaluations of the experimental input correspond to slightly larger values $\delta_0^{\mathrm{exp}} = 0.2066 \pm 0.0070$[43] and $\delta_0^{\mathrm{exp}} = 0.2042 \pm 0.0050$,[44] but are still consistent with this previous value. Starting from δ_0^{exp} and using eq. (16) and eq. (17) respectively, one easily obtains $\alpha_s(M_\tau)$. Its dependence on the choice of d_4 (previous estimate and the new exact result) is summarized in Table I.

	$\alpha_s^{FO}(M_\tau)$	$\alpha_s^{CI}(M_\tau)$
	$0.337 \pm 0.004 \pm 0.03$	$0.354 \pm 0.006 \pm 0.02$
$d_4 = 25$	$0.325 \pm 0.004 \pm 0.02$	$0.347 \pm 0.006 \pm 0.009$
$d_4 = 49.08$	$0.322 \pm 0.004 \pm 0.02$	$0.342 \pm 0.005 \pm 0.01$

As stated above the theory error for α_s from Z decays is small compared to the experimental uncertainties. The situation is more problematic for τ decays and to some extent the theory error remains to be matter of choice. As anticipated in[31] it decreases significantly, once α_s^4 terms are included. However, the difference between the two methods stabilizes (this was checked in[31] by adding an estimate for the α_s^5 term) and must be considered as irreducible uncertainty, with the final result

$$\alpha_s(M_\tau) = 0.332 \pm 0.005_{\mathrm{exp}} \pm 0.015_{\mathrm{th}}. \tag{18}$$

For the central value the mean value of FO and CI has been used, for the theory error half of the difference between two methods (0.01) plus (module of) the estimated correction from α_s^5 term (-0.005), the latter being based on $d_5 = 275$ (see eq. (9)).

Recently it has been argued,[43] that predictions based on FO are unstable and should therefore be abandoned. Using CI only and the new value for d_4, and assuming small higher order terms in the expansion of the Adler function, leads to a relatively large value $\alpha_s(m_\tau) = 0.344 \pm 0.005_{\rm exp} \pm 0.007_{\rm th}$ and an optimistic theory error. This line of reasoning is in striking contrast with a new study performed in.[44] In this paper the Adler function series is described by a model which incorporates the exactly known coefficients up to d_4 and theory constraints on the large-order behaviour and it is argued, that FO smoothly approaches the correct asymptotic result. The result $\alpha_s(m_\tau) = 0.320 \pm 0.003_{\rm exp}{}^{+0.011}_{-0.006}$ is significantly lower than the previous one. As expected the two results, taken at face value, are in agreement with the result for FO and CI respectively, as listed in Table 1. On the other hand, they are mutually exclusive. Individually, after evolving α_s to M_Z the associated error corresponds to about or less than 10^{-4} and the respective result would completely dominate all determinations of α_s. For the time being it seems prudent to stay with eq. (18) which takes the difference between FO and CI into account.

Applying four-loop running and matching[45–48] as implemented in the updated version of the MATHEMATICA-routine RunDec[49] to (18) we arrive at

$$
\begin{aligned}
\alpha_s(M_Z) &= 0.1202 \pm 0.0006_{\rm exp} \pm 0.0018_{\rm th} \pm 0.0003_{\rm evol} \\
&= 0.1202 \pm 0.0019. \qquad (19)
\end{aligned}
$$

Here the evolution error receives contributions from the uncertainties in the c-quark mass (0.00003, $m_c(m_c) = 1.286(13)$ GeV[50]) and the b-quark mass (0.00001, $m_b(m_b) = 4.164(25)$ GeV[50]), the matching scale (0.0001, μ varied between $0.7\, m_q(m_q)$ and $3.0\, m_q(m_q)$), the four-loop truncation in the matching expansion (0.0001) and the four-loop truncation in the RGE equation (0.0003). (For the last two errors the size of the shift due the highest known perturbative term was treated as systematic uncertainty.) The errors had been added in quadrature.

A recent, completely different determination of α_s is based on the combination of lattice simulation and continuum pertubation theory.[51] Derivatives of the correlator of two charm-quark currents $m_c\overline{\psi}\gamma_5\psi$ were evaluated at vanishing momentum q^2. Relating lattice and perturbative results for the same quantity then allowed for a determination of m_c and α_s. Specifically, higher moments, corresponding to higher derivatives with respect to q^2, were used for an accurate determination of m_c. From the second derivative, being dimensionless, a value $\alpha_s(M_Z) = 0.113(4)$ was obtained. Although

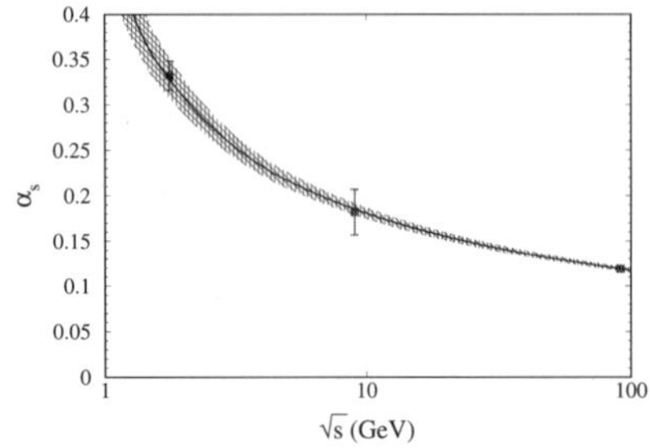

Fig. 3. Comparison of the experimental results eqs. (11), (13) and (18) with the four-loop evaluation and a starting value $\alpha_s(M_Z) = 0.1198 \pm 0.0015$

this method cannot yet compete with the determinations discussed above, the combination of lattice and continuum methods may well lead to precision determinations of α_s.

As shown in Fig. 3, it is instructive to compare the results of α_s measurements at m_τ, at intermediate energies around 9 GeV and at M_Z directly with the predictions based on four-loop running and matching, starting from a value $\alpha_s(M_Z) = 0.1198 \pm 0.0015$. Theory prediction and the three experimental results are in remarkable agreement.

Summary: The exact result for the α_s^4 term in the Adler function allows to extract the strong coupling constant from Z and τ decays with high precision. Including the exact α_s^4 leads to small shifts of the central value and to a significant reduction of the theory uncertainty. Note that the shifts in $\alpha_s(M_Z)$ from Z and τ decays, are opposite in sign and move the values in the *proper* direction, decreasing, thus, the current slight mismatch between two independent determinations of α_s. The final results

$$\alpha_s(M_Z)|_Z = 0.1190 \pm 0.0026\,, \tag{20}$$

$$\alpha_s(M_Z)|_\tau = 0.1202 \pm 0.0019\,, \tag{21}$$

from these two observables, although based on measurements of vastly different energy scales, are in remarkable agreement. This constitutes a striking test of asymptotic freedom in QCD. The two values can be combined to

$$\alpha_s(M_Z) = 0.1198 \pm 0.0015.$$

This is one of the most precise and presently only N^3LO result.

Acknowledgement

I would like to thank P. Baikov and K. Chetyrkin for a fruitful and pleasant collaberation. Work supported by SFB/TR-9 "Computational Particle Physics" and by the High Performance Computing Center Stuttgart (HLRS) under the grant "ParFORM".

References

1. J. Alcaraz et al. (LEP Collaboration and ALEPH Collaboration and DELPHI Collaboration and L3 Collaboration and OPAL Collaboration and LEP Electroweak Working Group) (2007), arXiv:0712.0929[hep-ex].
2. M. M. Winter (2001), ECFA-DESYLinearColliderNoteLC-PHSM-2001-016.
3. B. C. Allanach, G. A. Blair, S. Kraml, H. U. Martyn, G. Polesello, W. Porod and P. M. Zerwas, arXiv:hep-ph/0403133.
4. K. G. Chetyrkin, J. H. Kühn, and A. Kwiatkowski, Phys. Rep. **277**, 189 (1996), and references therein.
5. M. Davier, A. Hocker, and Z. Zhang, Rev. Mod. Phys. **78**, 1043 (2006), and references therein.
6. K. G. Chetyrkin, J. H. Kühn, and M. Steinhauser, Phys. Lett. **B371**, 93 (1996), hep-ph/9511430; Nucl. Phys. **B482**, 213 (1996), hep-ph/9606230.
7. K. G. Chetyrkin and J. H. Kühn, Phys. Lett. **B248**, 359 (1990); Nucl. Phys. **B432**, 337 (1994), hep-ph/9406299.
8. K. G. Chetyrkin, R. V. Harlander, and J. H. Kühn, Nucl. Phys. **B586**, 56 (2000), hep-ph/0005139.
9. P. A. Baikov, K. G. Chetyrkin, and J. H. Kühn, Nucl. Phys. Proc. Suppl. **135**, 243 (2004).
10. A. Czarnecki and J. H. Kühn, Phys. Rev. Lett. **77**, 3955 (1996), hep-ph/9608366.
11. R. Harlander, T. Seidensticker, and M. Steinhauser, Phys. Lett. **B426**, 125 (1998), hep-ph/9712228.
12. D. V. Shirkov (2000), hep-ph/0009106.
13. P. A. Baikov, K. G. Chetyrkin, and J. H. Kühn, Phys. Rev. Lett. **96**, 012003 (2006), hep-ph/0511063.
14. K. G. Chetyrkin, A. L. Kataev, and F. V. Tkachov, Phys. Lett. **B85**, 277 (1979).
15. S. G. Gorishnii, A. L. Kataev, and S. A. Larin, Phys. Lett. **B259**, 144 (1991).
16. K. G. Chetyrkin, Phys. Lett. B **391** (1997) 402 [arXiv:hep-ph/9608480].
17. P. A. Baikov, K. G. Chetyrkin, and J. H. Kühn, Phys. Rev. Lett. **88**, 012001 (2002), hep-ph/0108197.
18. P. A. Baikov, K. G. Chetyrkin and J. H. Kühn, Phys. Rev. Lett. **101**, 012002, arXiv:0801.1821 [hep-ph].
19. P. Nogueira, J. Comput. Phys. **105**, 279 (1993).
20. A. A. Vladimirov, Theor. Math. Phys. **43**, 417 (1980).
21. K. G. Chetyrkin and V. A. Smirnov, Phys. Lett. **B144**, 419 (1984).
22. K. G. Chetyrkin, Phys. Lett. **B391**, 402 (1997), hep-ph/9608480.
23. P. A. Baikov, Phys. Lett. **B385**, 404 (1996), hep-ph/9603267; Nucl. Instrum.

Meth. **A389**, 347 (1997), hep-ph/9611449; Phys. Lett. **B634**, 325 (2006), hep-ph/0507053.
24. P. A. Baikov, Nucl. Instrum. Meth. **A389**, 347 (1997), hep-ph/9611449.
25. P. A. Baikov, Phys. Lett. **B634**, 325 (2006), hep-ph/0507053.
26. M. Tentyukov et al. (2004), cs.sc/0407066.
27. J. A. M. Vermaseren (2000), math-ph/0010025.
28. G. Grunberg, Phys. Rev. **D29**, 2315 (1984).
29. P. M. Stevenson, Phys. Rev. **D23**, 2916 (1981).
30. A. L. Kataev and V. V. Starshenko, Mod. Phys. Lett. **A10**, 235 (1995), hep-ph/9502348.
31. P. A. Baikov, K. G. Chetyrkin, and J. H. Kühn, Phys. Rev. **D67**, 074026 (2003), hep-ph/0212299.
32. A. L. Kataev and V. V. Starshenko, Nucl. Phys. Proc. Suppl. **39BC**, 312 (1995), hep-ph/9408395.
33. J. H. Kühn, M. Steinhauser, and T. Teubner, Phys. Rev. **D76**, 074003 (2007), arXiv:0707.2589[hep-ph]
34. R. V. Harlander and M. Steinhauser, Comput. Phys. Commun. **153** (2003) 244 [arXiv:hep-ph/0212294].
35. R. Ammar *et al.* [CLEO Collaboration], Phys. Rev. D **57** (1998) 1350, [arXiv:hep-ex/9707018].
36. D. Besson *et al.* [CLEO Collaboration], Phys. Rev. D **76** (2007) 072008, arXiv:0706.2813 [hep-ex].
37. J. H. Kühn and M. Steinhauser, Nucl. Phys. B **619** (2001) 588 [Erratum-ibid. B **640** (2002) 415] [arXiv:hep-ph/0109084].
38. J. Z. Bai *et al.* [BES Collaboration], Phys. Rev. Lett. **88** (2002) 101802 [arXiv:hep-ex/0102003].
39. A. E. Blinov *et al.* [MD-1 Collaboration], Z. Phys. C **70** (1996) 31.
40. A. A. Pivovarov, Z. Phys. **C53**, 461 (1992), hep-ph/0302003.
41. F. Le Diberder and A. Pich, Phys. Lett. **B286**, 147 (1992).
42. M. Davier, A. Hocker, and Z. Zhang, Nucl. Phys. Proc. Suppl. **169**, 22 (2007), hep-ph/0701170.
43. M. Davier, S. Descotes-Genon, A. Hocker, B. Malaescu and Z. Zhang, arXiv:0803.0979 [hep-ph].
44. M. Beneke and M. Jamin, arXiv:0806.3156 [hep-ph].
45. T. van Ritbergen, J. A. M. Vermaseren, and S. A. Larin, Phys. Lett. **B400**, 379 (1997), hep-ph/9701390.
46. M. Czakon, Nucl. Phys. **B710**, 485 (2005), hep-ph/0411261.
47. K. G. Chetyrkin, J. H. Kühn, and C. Sturm, Nucl. Phys. B**744** (2006) 121, hep-ph/0512060.
48. Y. Schroder and M. Steinhauser, JHEP **01**, 051 (2006), hep-ph/0512058.
49. K. G. Chetyrkin, J. H. Kühn and M. Steinhauser, Comput. Phys. Commun. **133** (2000) 43 [arXiv:hep-ph/0004189].
50. J. H. Kühn, M. Steinhauser, and C. Sturm, Nucl. Phys. **B778**, 192 (2007), hep-ph/0702103.
51. I. Allison *et al.*, arXiv:0805.2999 [hep-lat].

Quarkyonic Matter and the Phase Diagram of QCD

Larry McLerran

Physics Department and Riken Brookhaven Research Center, Building 510A
Brookhaven National Laboratory, Upton, NY-11973, USA

Quarkyonic matter is a new phase of QCD at finite temperature and density which is distinct from the confined and de-confined phases. Its existence is unambiguously argued in the large numbers of colors limit, $N_c \to \infty$, of QCD. Hints of its existence for QCD, $N_c = 3$, are shown in lattice Monte-Carlo data and in heavy ion experiments.

1. Quarkyonic Matter and the Large N_c limit of QCD

The large N_c limit of QCD has provided numerous insights into the structure of strongly interacting matter, both in vacuum,,[1] and at finite temperature.[2] The large N_c approximation allows a correct reproduction of both qualitative and semi-quantitative features of QCD. If the number of fermions is held finite as $N_c \to \infty$, then QCD in this limit is confining in vacuum. The spectrum of the confined world consists of non-interacting confined mesons and glueballs. At finite temperature, there is a first order phase transition between a confined world of glueballs and mesons and an unconfined world of gluons. The energy density, pressure and entropy are parametrically of order one in the confined world since the confined states are colorless, but in the de-confined world are of order N_c^2, corresponding to the N_c^2 gluon degrees of freedom. The latent heat of the phase transition is of order N_c^2.

As one approaches the phase transition from lower temperature, the transition is hinted at by the existence of a Hagedorn spectrum of particles, whose density accumulates as one approaches the phase transition temperature. This accumulation resolves the paradoxical situation that at large N_c, hadrons do not interact, which seems to contradict the existence of a de-confining phase transition. If there is an accumulation of states at the Hagedorn temperature, then the the high density of states very near to the transition temperature can compensate for weak interaction strength, resulting in a strongly interacting gas of hadrons.

The results of lattice computations for QCD ($N_c = 3$) are in qualitative accord with the results at large N_c.[3] There is a rapid change at at a well defined temperature, although not a strict discontinuity when a finite number of quark flavors are included. The presence of quarks makes the order parameter for confinement not so well defined, since the order parameter is $e^{-\beta F_q}$ where inverse temperature is β and the change in the free energy of the system, F_q, is that due to the addition of a quark. In a theory with no light quarks, one can still probe the system with a heavy quark source, which corresponds to the order parameter. In such a theory, $e^{-\beta F_q} = 0$ in the confined phase because the quarks have infinite mass. In the de-confined phase, the order parameter is finite. When light quarks are included, these light quarks can form bound states with the heavy quark probes, and so the free energy need never be infinite. The presence of light quarks therefore does not allow for an order parameter for confinement, In QCD, since there is no order parameter associated with confinement, there need be no strict phase transition, and there appears not to be for realistic quark masses. There is nevertheless a quite rapid transition at a temperature of about 200 MeV, where the energy density changes by of order N_c^2, in accord with large N_c arguments.

The popular wisdom for QCD at finite temperature and density is that there is a line of cross overs, perhaps converting to a first order phase transition at high density and low temperature, that separates the confined and de-confined world. Typically plots are made as a function of temperature T and baryon chemical potential, μ. Such a hypothetical diagram is shown in Figure 1. At very high temperature and density, there may be phase transitions associated with color superconductivity, which affect transport properties of quark-gluon matter, but are not so important for bulk properties such as pressure and energy density.[4]

Unfortunately, the conventional wisdom about the phase diagram has never been explicitly verified. It is quite difficult to disentangle the assumptions built into various model computations from the hard predictions of QCD.[5] Lattice Monte-Carlo computations are extremely difficult at high baryon density.[3]

In this talk, I will review the recent results concerning the phase diagram at finite T and μ in the limit of a large number of colors.[6] In this limit, it is possible to extract model independent results. The surprise result of these considerations is that in addition to the confined and de-confined phases of QCD, there is a third phase. It turns out that the pressure and energy

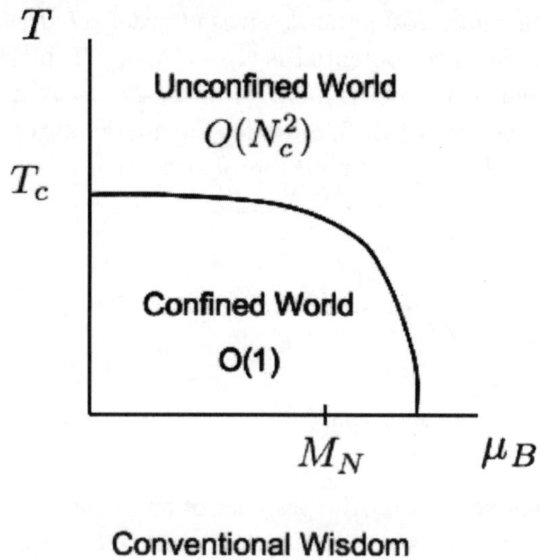

Fig. 1. The conventional wisdom about the QCD phase transition at finite temperature and density.

density of this phase behave like a gas of quarks at very high baryon density, but nevertheless is confined. Confinement is important for properties of the matter near the Fermi surface, where excitations are required to be bound into color singlets Rob Pisarski and I named this phase the quarkyonic phase, since it has properties of both high density baryonic matter and de-confined quark matter.

To understand how such a new phase of matter might come about, we need to understand that dynamical quarks do not modify the potential between heavy test quarks at large N_c. We shall first consider the case that $N_c \rightarrow \infty$ but that the number of quark flavors is held fixed. We will later turn to the case where N_F/N_c is held fixed. This limit with finite number of flavors is easiest to consider since there is a confined and a de-confined phase. For finite N_F/N_c, there is no distinction between a confined and de-confined phase, although there might perhaps be a remnant of these phases associated with a cross over. In Figure 2, the gluon loop and quark loop modifications of the potential are shown. At finite temperature, the first diagram Debye screens the potential at large distances. This can short out the linear potential when the temperaute is high enough. The second diagram corresponds to a quark loop and is suppressed by $1/N_c$ at large N_c. When expressed in terms of the t'Hooft coupling, the temperature T

and the quark chemical pottential μ_Q, it is of order $\alpha T^2 F(\mu_Q/T)/N_c$ Note that the baryon chemical potential is $\mu_B = N_c\mu_Q$. Typically, the baryon chemical potential is of order the baryon mass $M_B \sim N_c\Lambda_{QCD}$. The high density limit is $\mu_Q >> \Lambda_{QCD}$. In order for the quark loops to be important, the quark chemical potential must be of order $\mu_Q \sim \sqrt{N_c}\Lambda_{QCD}$ which approaches infinity in the large N_c limit

Fig. 2. Diagrams which modify the potential at high energy density .

This argument shows that the presence of baryons at finite density does not affect the values of the confinement-deconfinement transition temperature. How can there be any non-trivial physics due to finite baryon number density? It turns out that at large N_c there is another order parameter, which is the baryon number density. Remember that the baryon number density is

$$\rho_B \sim e^{(\mu_B - M_B)/T} \tag{1}$$

Since both μ_B and M_B are of order N_c, $\rho_B \sim e^{-\kappa N_c}$ so long as $\mu_B \leq M_B$. Here κ is a constant of order 1. Therefore there is a finite region in the $\mu_B - T$ plane where the baryon number density is zero. This is the confined-baryonless world. At high temperature, in the de-confined world, the quarks are the correct degrees of freedom in which to measure baryon number and the baryon number is finite. As one increases the baryon number density at temperatures below the confinement temperature, there is a phase which is confined, but the baryon number chemical potential is large enough that the baryon number density is finite.

The bottom line of these arguments is that there are two order parameters corresponding to confinement and to baryon number. This in principle allows four possible phases. The phase where there is de-cofninement and zero baryon number density is apparently not realized in nature, but the other three may be.

We can compute the dependence of the pressure and energy density on N_c for the quarkyonic phase. To do this, assume the quark chemical potential is large compared to Λ_{QCD}. The pressure and energy density are computable in this limit and are of order N_c. It is remarkable that the bulk properties of quarkyonic matter may be computed using perturbation theory at very high density. This is despite that such matter is confined. This is not unprecedented, since computations of scattering processes at short distances may be done in perturbation theory, even though the processes take place in the confined vacuum. The perturbation expansion works for bulk thermodynamic quantities because the major contribution arises from quarks deep inside the Fermi sea, where short distance interactions dominate. Pairing processes near the Fermi surface are sensitive to long distance affects, and presumably pairs which form near the Fermi surface must be confined.

The fact that the energy density and pressure are proportional to N_c may also be seen in a Skyrmyonic description of baryons for two flavors. Such a description follows from the large N_c limit.[7] The action for Skyrmions is

$$S = \int d^4x \ (f_\pi^2 \ tr \ V^\mu V_\mu^\dagger + \kappa \ tr \ [V^\mu, V^\nu]^2) \tag{2}$$

In this equation, V^μ is a derivative of a n $SU(2)$ group element, and both f_π and κ are constants of order N_c. We see that the action is therefore of order N_c, so that bulk quantities such as pressure and energy density computed in this theory should also be of order N_c

These arguments show that the confined-baryonless phase has bulk properties of order 1 in N_c, the de-coniined phase of order N_c^2 and the quarkyonic phase is of order N_c The phase diagram for QCD in the large N_c limit is shown in Fig. 3.

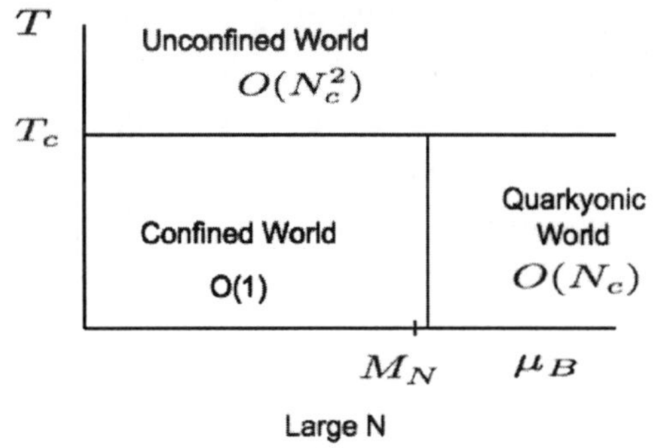

Fig. 3. The phase diagram for large N_c QCD.

The change in the bulk properties of the system when μ_B crosses the confinement-quarkyonic transition is of order N_c. We can also estimate the width of the transition region. Recall that in the large N_c limit baryons are very strongly interacting. When the number density of baryons becomes of order $1/\Lambda_{QCD}^3$ or when the Fermi momentum is $k_F \sim \Lambda_{QCD}$,, one will no doubt have made a transition to quarkyonc matter. This density is controlled by the Fermi momentum. For baryons with large masses of order N_c, we see that the baryon chemical potential is of order, $\mu_B \sim M_B + k_F^2/2M_B$. This means that in a width of order $1/N_c$ in μ_B or of order $1/N_c^2$ in μ_Q, the transition is achieved. In the large N_c limit, this width shrinks to zero.

2. Finite N_F/N_c

The finite ratio of N_F/N_c in the large N_c limit implies that there is no distinction between the confined and de-confined worlds at finite temperature and density. The baryon number density remains a valid order parameter. The transition is driven because of the huge degeneracy of the lowest mass baryon states. There are of order $e^{N_c G(N_F/N_c)}$ such states, where G is a function determined by Young-Tableau which counts such states. The baryon density is

$$\rho_B \sim e^{N_c G(N_F/N_c)} e^{-M_B/T + \mu_B/T} \tag{3}$$

This has a transition when $\mu_B = M_B - TN_c G(N_c/N_F)$ In this case the world is divided into that of mesons without baryons, and a world with finite baryon number density.

3. Phenomenology and Speculation

It is tempting to speculate on the nature of the phase transition in QCD for $N_c = 3$ and realistic numbers of flavors. It is difficult a priori to know whether phase transitions remain or whether they become cross overs. I have little to say about this. It is clear that the diagram drawn in Fig. 3 becomes smoothed due to finite N_c and N_F effects. This means that the transition line at small μ_B and finite T has some small decrease as μ_B increases. When $\mu_B \sim \sqrt{N_c}\Lambda_{QCD}$, the line of finite temperature transitions will have either disappeared or have dipped near $T = 0$. If it dips down, it probably is a weak transition because the effect of de-confinement is to liberate gluons and at low T and high μ, and there the ratio of gluons to quarks is very small. Most likely, the finite temperature line of transitions ends in a second order point, if it was ever a first order transition. Presumably, there is still a tricritical region as in the large N_c limit. If the transition becomes a cross over, then perhaps a critical point slides along the line of quarkyonic-confined transition.

The large N_c considerations must be further developed in order to say anything about the chiral transition. Chiral symmetry breaking should be a Fermi surface effect in the quarkyonic phase.[8] Whether or not it is restored in the quarkyonic phase, or only approximately restored is not resolved.

A hypothetical phase diagram for QCD is shown in Fig. 4.

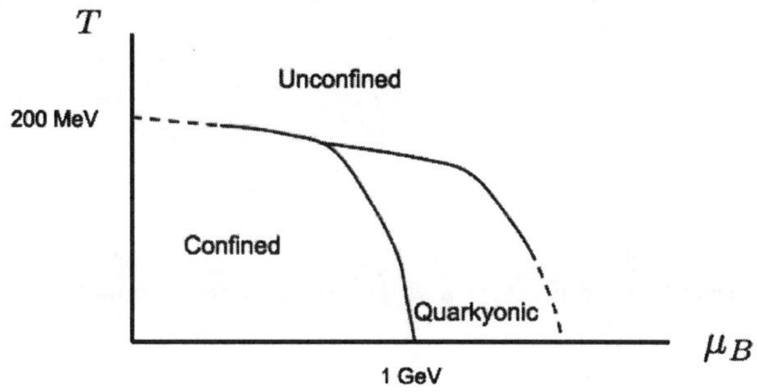

Fig. 4. A hypothetical phase diagram for QCD.

Perhaps the strongest hints for the existence of a quarkyonic phase come from the ratios of integrated yields of particles produced in heavy ion experiments.[9] This is supposed to give information about the decoupling temperature and chemical potential of matter produced in heavy ion collisions. Since the energy density jumps by of order $N_c N_F \Lambda_{QCD}^4$ across the quarkyonic transition, and since in large N_c, particle cross sections do not change their N_c dependence,[10] one might expect that the freeze out occurs at the quarkyonic phase transition.

In Figure 5, the supposed decoupling temperature and baryon chemical potential are computed for various energies of experiments. Note that the line goes to zero temperature at about the nucleon mass. This is easy to interpret in terms of the quarkyonic phase transition, but difficult to understand if the line corresponded to the confinement transition.

Also shown on the plot are results of computations from the bag model and from lattice gauge theory which show the weak dependence on baryon chemical potential of the confinement transition line.

There is some lattice data which argues in favor of the existence of the quarkyonic phase. In the computations of Fodor et. al., micro-canonical techniques were used on lattices of very small size.[11] They found the phase

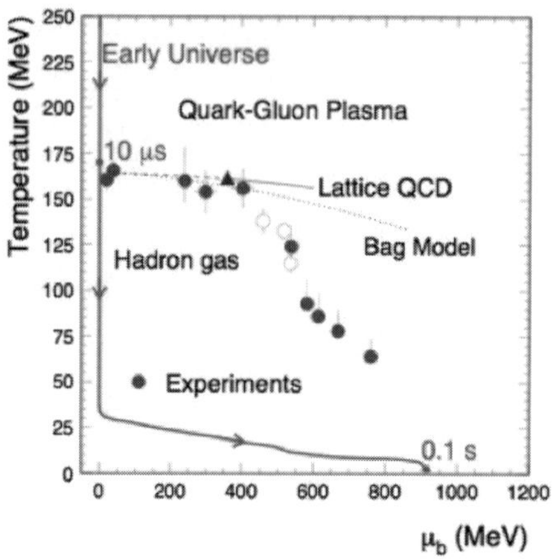

Fig. 5. The decoupling temperature and density of matter produced in heavy ion collisions.

diagram shown in Figure 6.

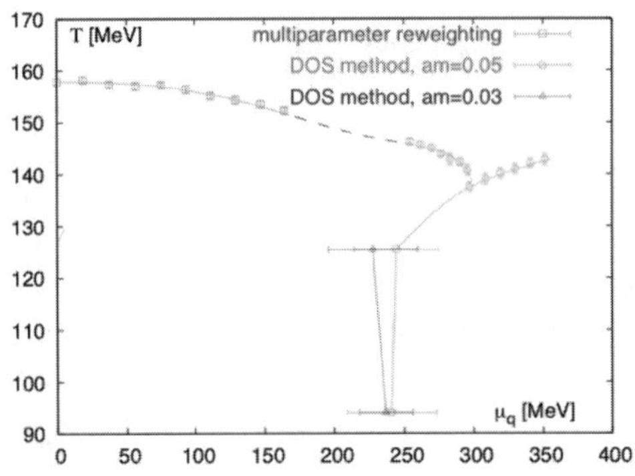

Fig. 6. The phase diagram found by Fodor and colleagues.

4. Conclusions

Quarkyonic matter forces us to revise our conception of the phase diagram of QCD. There are of course many unanswered questions: How does the chiral transition interplay with the quarkyonic transition? How is quarkyonic matter related to Skyrmionic crystals?[7] What is the nature of the Fermi surface of quarkyonic matter? How is the liquid gas phase transition related to the quarkyonic phase transitions? These are of course many more.

5. Acknowledgements

I thank Arkady Vainshtein and Misha Voloshin for inviting me to Continuous Advances in QCD, where this talk was presented. Their gracious hospitality is greatly appreciated. I thank Rob Pisarski and Yoshimasa Hidaka with whom many of these idea were developed. I thank my colleagues at Rob Pisarski and Yoshimasa Hidaka for their clever insights, and with whom these ideas were developed. I also thank Kenji Fukushima for his intuitive insights concerning such matter, and Kzysztof Redlich for his many insights concerning the properties of matter at finite baryon number density. My research is supported under DOE Contract No. DE-AC02-98CH10886.

References

1. G. 't Hooft, *Nucl. Phys.* **B72**, 461 (1984); **B75**, 461 (1974)
2. C. B. Thorn, *Phys. Lett. B* **99** 458 (1981)
3. For a review see F. Karsch, *Prog. Theor. Phys. Suppl.* **168**, 237 (2007).
4. M. G. Alford, K. Rajagopal and F. Wilczek, *Phys. Lett. B* **422** 247 (1998); R. Rapp, T. Shafer, E. Shuryak and M. Velkovsky, *Phys. Rev. Lett.* **81**, 53 (1998)
5. K. Fukushima, *Phys. Rev.* **D77**, 114028 (2008).
6. L. McLerran and R. Pisarski, *Nucl. Phys.* **A796**, 83 (2007); Y. Hidaka, L. McLerran and R. Pisarski, *Nucl. Phys.* **A808** 117 (2008).
7. M. Kutschera, C. J. Pethick and D. G. Ravenhall, *Phys. Rev. Lett. 53*, 1041 1984; I. R. Klebanov, *Nucl. Phys.* **B262**, 133 1985; A. S. Goldhaber and N. S. Manton, *Phys. Lett. B* **198**, 2311987; N. S. Manton, *Comm. Math. Phys.* **111**, 4691987; A. D. Jackson, A. Wirzba and N. S. Manton, *Nucl. Phys.* **B495**, 4991989; H. Forkel, A. D. Jackson, M. Rho, C. Weiss, A. Wirzba and H. Bang, *Nucl. Phys.* **B504**, 8181989; M. Kugler and S. Shtrikman, *Phys. Lett. B 208*, 4911988; *Phys. Rev.* **D40**,34211989; R. A. Battye and P. M. Sutcliffe, *Phys. Rev. Lett.* **79**,3631997 *Rev. Mod. Phys. 14*, 292002 *Nucl. Phys. B705*, 3842005 *Phys. Rev.* **C73**, 0552052006
8. L. Ya. Glozman and R. F. Wagenbrunn, *Phys. Rev.* **D77**, 054927 (2008), arXiv:0709.3080arXiv 0805.4799;L. Ya. Glozman, arXiv:0803.1636
9. A. Adronic, P. Braun-Munzinger, K. Redlich and J. Stachel, *Phys. Lett.* **B571**, 36 (2003); P. Braun-Munzinger, D, Magestro, K. Redlich and J. Stachel, *Phys. Lett.* **B518**, 41 (2001); J. Cleymans and K. Redlich, *Phys. Rev. Lett.* **52**, 84 (1998); J. Cleymans and K. Redlich, *Phys. Rev.* **C60**

054908 (1999); F. Beccatini, J. Cleymans, A. Keranen, E. Suhonen and K. redlich, *Phys. Rev.* **C64**, 024901 (2001)
10. L. Csernai, J. Kapusta and L. McLerran, *Phys. Rev. Lett.* **97**, 152303 (2006).
11. Z. Fodor, S. Katz and C. Schmidt, *JHEP* **0703**, 121 (2007).

Statistical QCD with Non-positive Measure

J.C. Osborn

Argonne Leadership Computing Facility
9700 S. Cass Avenue, Argonne, IL 60439, USA

Center for Computational Science, Boston University
Boston, MA 02215, USA

K. Splittorff

The Niels Bohr Institute
Blegdamsvej 17, DK-2100, Copenhagen Ø, Denmark
*speaker, * E-mail: split@nbi.dk*

J.J.M. Verbaarschot

Department of Physics and Astronomy, SUNY
Stony Brook, New York 11794, USA
E-mail: verbaarschot@cs.physics.sunysb.edu

In this talk we discuss the microscopic limit of QCD at nonzero chemical potential. In this domain, where the QCD partition function is under complete analytical control, we uncover an entirely new link between the spectral density of the Dirac operator and the chiral condensate: violent complex oscillations on the microscopic scale give rise to the discontinuity of the chiral condensate at zero quark mass. We first establish this relation exactly within a random matrix framework and then analyze the importance of the individual modes by Fourier analysis.

Keywords: nonperturbative QCD, sign problem, random matrix theory

1. Introduction

Many firm results about QCD are based on the fact that the Euclidean partition function has a real and positive measure. Prime examples are QCD inequalities and the evaluation of the lattice QCD partition function by Monte Carlo simulations. For this reason much of the intuition gained about nonperturbative QCD is based on a probabilistic interpretation. If we introduce a chemical potential in order to favor the presence of quarks over anti-quarks the measure in the Euclidean partition function is no longer positive definite. In this lecture we will analyze the link between the spectral density of the QCD Dirac operator and the order parameter for spontaneous chiral symmetry breaking. We will see that the intuition based on zero chemical potential results fails completely at nonzero chemical potential.

At zero chemical potential the eigenvalues of the anti-Hermitean Dirac operator are located on the imaginary axis. The chiral condensate can be thought of as the electric field at the position on the real axis given by the quark mass, m, created by positive charges located at the purely imaginary eigenvalues.[1] The chiral condensate is the jump in the electric field as the quark mass passes through zero. Clearly such a jump will occur if the density of the eigenvalues/charges near the origin of the imaginary axis scales with the space-time volume.[2]

At nonzero chemical potential the anti-Hermiticity of the Dirac operator is lost and the eigenvalues are located homogeneously in a strip parallel to the imaginary axis with width proportional to μ^2 (for sufficiently large μ). The constant of proportionality is such that, at sufficiently low temperatures, the quark mass hits the eigenvalue strip for $\mu = m_\pi/2$.[3–5] From our intuition at zero chemical potential we are lead to the conclusion that at any (however small) nonzero value of the chemical potential the chiral condensate has no discontinuity when the quark mass crosses the imaginary axis. This is in sharp contrast to the expected phase diagram, see Fig. 1. (Notice however, that in phase quenched QCD the chiral condensate rotates into a pion condensate so that chiral symmetry remains broken spontaneously in spite of a vanishing chiral condensate.) Increasing the temperature tends to decrease the width of the strip of eigenvalues so that the value of the chemical potential for which the quark mass is inside the domain of the eigenvalues increases as well (see Fig. 1 for illustration).

At nonzero chemical potential the eigenvalue density of the Dirac operator is defined as

$$\rho_{N_f}(x, y) = \left\langle \sum_k \delta(x - x_k)\delta(y - y_k) \right\rangle_{N_f} . \qquad (1)$$

Since the measure includes the fermion determinants there is no guarantee that the density is real or positive. To completely destroy the electrostatic picture it turns out that the eigenvalue density depends strongly on the quark mass.[7,8] Hence we can not interpret the quark mass as a test charge which, in a harmless manner, measures the chiral condensate/electric field. To study the eigenvalue density we turn to the microscopic scale of QCD[9]

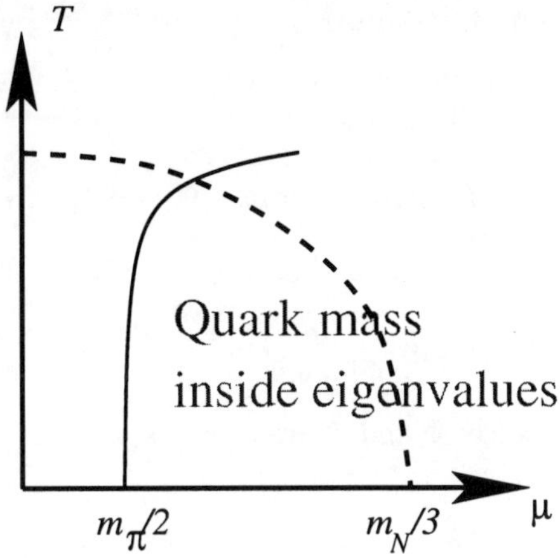

Fig. 1. Schematic phase diagram of QCD and region where the quark mass is inside the support of the eigenvalue density. The dashed curve indicates the chiral transition be it first order or crossover. To the right of the full line the quark mass is inside the support of the Dirac spectrum. The solid curve does *not* indicate a phase transition in QCD.

where the dimensionless combinations

$$\hat{m} \equiv m\Sigma V \qquad \text{and} \qquad \hat{\mu}^2 \equiv \mu^2 F_\pi^2 V \tag{2}$$

are kept fixed as the volume V is taken to infinity (Σ is the chiral condensate and F_π the pion decay constant). In this limit the Compton wavelength of the pions is much larger than the linear extent of the volume and the zero momentum mode of the pions dominates the low energy effective theory of QCD[10,11] resulting in a partition function that is given by a group integral. Here we will access the microscopic limit of QCD by analyzing the chiral random matrix model introduced previously.[7] We will show that the unquenched eigenvalue density is complex and oscillating when $\mu > m_\pi/2$ and that these oscillations are responsible for the discontinuity of the chiral condensate.[12] The complex eigenvalue density of the Dirac operator for QCD at nonzero chemical potential was also studied for gauge fields given by a liquid of instantons[6] and strong fluctuations in the spectral density could be identified.

2. The random matrix model

The random matrix partition function with N_f quark flavors with mass m and chemical potential μ is defined as[7]

$$\mathcal{Z}_N^{N_f}(m;\mu) \equiv \int d\Phi d\Psi \; w_G(\Phi) w_G(\Psi) \det{}^{N_f}(\, \mathcal{D}(\mu) + m \,), \tag{3}$$

where the non-Hermitian Dirac operator is given by

$$\mathcal{D}(\mu) = \begin{pmatrix} 0 & i\Phi + \mu\Psi \\ i\Phi^\dagger + \mu\Psi^\dagger & 0 \end{pmatrix} . \tag{4}$$

The $(N+\nu) \times N$ matrices Φ and Ψ are complex, ν is the topological index, and w_G is a Gaussian weight function

$$w_G(X) \;=\; \exp(-N \,\mathrm{Tr}\, \mathrm{X}^\dagger \mathrm{X}) \,. \tag{5}$$

In this random matrix model, the microscopic limit is defined as the limit $N \to \infty$ where

$$\hat{m} = 2Nm \quad \text{and} \quad \hat{\mu}^2 = 2N\mu^2 \tag{6}$$

are kept fixed. The random matrix partition function becomes identical to the microscopic QCD partition function provided that we identify, see for example,[8]

$$\begin{aligned} \hat{m} &= 2Nm \to m\Sigma V, \\ \hat{\mu}^2 &= 2N\mu^2 \to \mu^2 F_\pi^2 V. \end{aligned} \tag{7}$$

The main advantage of this matrix model as compared to the one originally introduced in[13] is that an eigenvalue representation is known[7]

$$\mathcal{Z}_N^{N_f}(m;\mu) \sim m^{\nu N_f} \int_{\mathbb{C}} \prod_{k=1}^{N} d^2 z_k \, \mathcal{P}^{N_f}(\{z_i\},\{z_i^*\},m;\mu), \tag{8}$$

where the joint probability distribution reads

$$\mathcal{P}^{N_f}(\{z_i\},\{z_i^*\},m;\mu) = \frac{1}{\mu^{2N}} \left|\Delta_N(\{z_l^2\})\right|^2 \prod_{k=1}^{N} w(z_k,z_k^*;\mu)(m^2 - z_k^2)^{N_f}.$$

The Vandermonde determinant is defined as

$$\Delta_N(\{z_l^2\}) \equiv \prod_{i>j=1}^{N} (z_i^2 - z_j^2), \tag{9}$$

and the weight function includes a modified Bessel function,

$$w(z_k,z_k^*;\mu) = |z_k|^{2\nu+2} K_\nu\left(\frac{N(1+\mu^2)}{2\mu^2}|z_k|^2\right) e^{-\frac{N(1-\mu^2)}{4\mu^2}(z_k^2 + z_k^{*2})}. \tag{10}$$

Given the eigenvalue representation we can employ the powerful method of orthogonal polynomials in the complex plane developed in.[14–19]

2.1. *Orthogonal polynomials, partition function and chiral condensate*

The complex orthogonal polynomials corresponding to the weight function (10) can be expressed in terms of complex Laguerre polynomials[7]

$$p_k(z;\mu) = \left(\frac{1-\mu^2}{N}\right)^k k! L_k^\nu\left(-\frac{Nz^2}{1-\mu^2}\right). \tag{11}$$

They satisfy the orthogonality relations

$$\int_{\mathbb{C}} d^2z\; w(z,z^*;\mu)\; p_k(z;\mu)\; p_l(z;\mu)^* \;=\; \delta_{kl}\, r_k^\nu\,, \tag{12}$$

with the norm r_k^ν given by

$$r_k^\nu \;=\; \frac{\pi\,\mu^2\;(1+\mu^2)^{2k+\nu}\;k!\;(k+\nu)!}{N^{2k+2+\nu}}\,. \tag{13}$$

The partition function for one fermion can be expressed in terms orthogonal polynomials as

$$Z_N^{N_f=1}(m;\mu) = m^\nu p_N(m;\mu). \tag{14}$$

Using this partition function (14) we immediately find the chiral condensate

$$\Sigma_N^{N_f=1}(m) = \frac{d}{dm}\log Z_N^{N_f=1}(m;\mu) = \frac{dp_N(m)/dm}{p_N(m)} + \frac{\nu}{m}. \tag{15}$$

Since we have not yet taken the microscopic limit the chiral condensate depends on μ.

3. The chiral condensate from the eigenvalue density using orthogonal polynomials

The quenched eigenvalue density given by

$$\rho_N^{N_f=0}(z,z^*;\mu) = w(z,z^*;\mu)\sum_{k=0}^{N-1}\frac{p_k(z^*)p_k(z)}{r_k}. \tag{16}$$

is real and positive. The unquenched spectral density[7] which can be expressed as

$$\rho_N^{N_f=1}(z,z^*,m;\mu) = w(z,z^*;\mu)\sum_{k=0}^{N-1}\frac{p_k(z^*)(p_k(z)-p_N(z)p_k(m)/p_N(m))}{r_k} \tag{17}$$

is not invariant under complex conjugation. The asymmetric second term causes complex valued oscillations, see figure 2.

Here we will compute the chiral condensate (15) starting from the complex and oscillating eigenvalue density given in (17),

$$\Sigma_N^{N_f=1}(m) = \int_{\mathbb{C}} d^2z \frac{2m}{z^2-m^2} \rho_N^{N_f=1}(z, z^*, m; \mu). \tag{18}$$

We will show that all integrals can be carried out using the orthogonality condition (12). The first step is to insert the above expression for the unquenched density

$$\Sigma_N^{N_f=1}(m) = \int_{\mathbb{C}} d^2z \frac{2mw}{z^2-m^2} \sum_{k=0}^{N-1} \frac{p_k(z^*)(p_k(z) - p_N(z)p_k(m)/p_N(m))}{r_k}. \tag{19}$$

We now use the fact that

$$\int_{\mathbb{C}} d^2z \frac{w(z,z^*;\mu)p_k(z^*)p_k(z)}{z^2-m^2} = \int_{\mathbb{C}} d^2z \frac{w(z,z^*;\mu)p_k(z^*)p_k(m)}{z^2-m^2}, \tag{20}$$

which is easily proved by first rewriting (the c_k's are independent of z)

$$p_k(z) = (z^2 - m^2)[p_{k-1}(z) + c_{k-2}p_{k-2}(z) + ...] + p_k(m) \tag{21}$$

and then using orthogonality, and find

$$\Sigma_N^{N_f=1}(m) = \int_{\mathbb{C}} d^2z \frac{2mw}{z^2-m^2} \sum_{k=0}^{N-1} \frac{p_k(z^*)p_k(m)}{r_k} \frac{1}{p_N(m)} (p_N(m) - p_N(z)) \tag{22}$$

For $p_N(m) - p_N(z)$ we now insert (this is just Eq. (21) rearranged a bit)

$$p_N(m) - p_N(z) = -(z^2 - m^2)[p_{N-1}(z) + c_{j-2}p_{N-2}(z) + ...] \tag{23}$$

and get

$$\Sigma_N^{N_f=1}(m) = -\int_{\mathbb{C}} d^2z \; 2mw \sum_{k=0}^{N-1} \frac{p_k(z^*)p_k(m)}{r_k p_N(m)} [p_{N-1}(z) + c_{N-2}p_{N-2}(z) + ...]. \tag{24}$$

Using orthogonality we find (note that the μ-dependent norms drop out)

$$\Sigma_N^{N_f=1}(m) = \frac{2m}{p_N(m)} [p_{N-1}(m) + c_{N-2}p_{N-2}(m) + ...]. \tag{25}$$

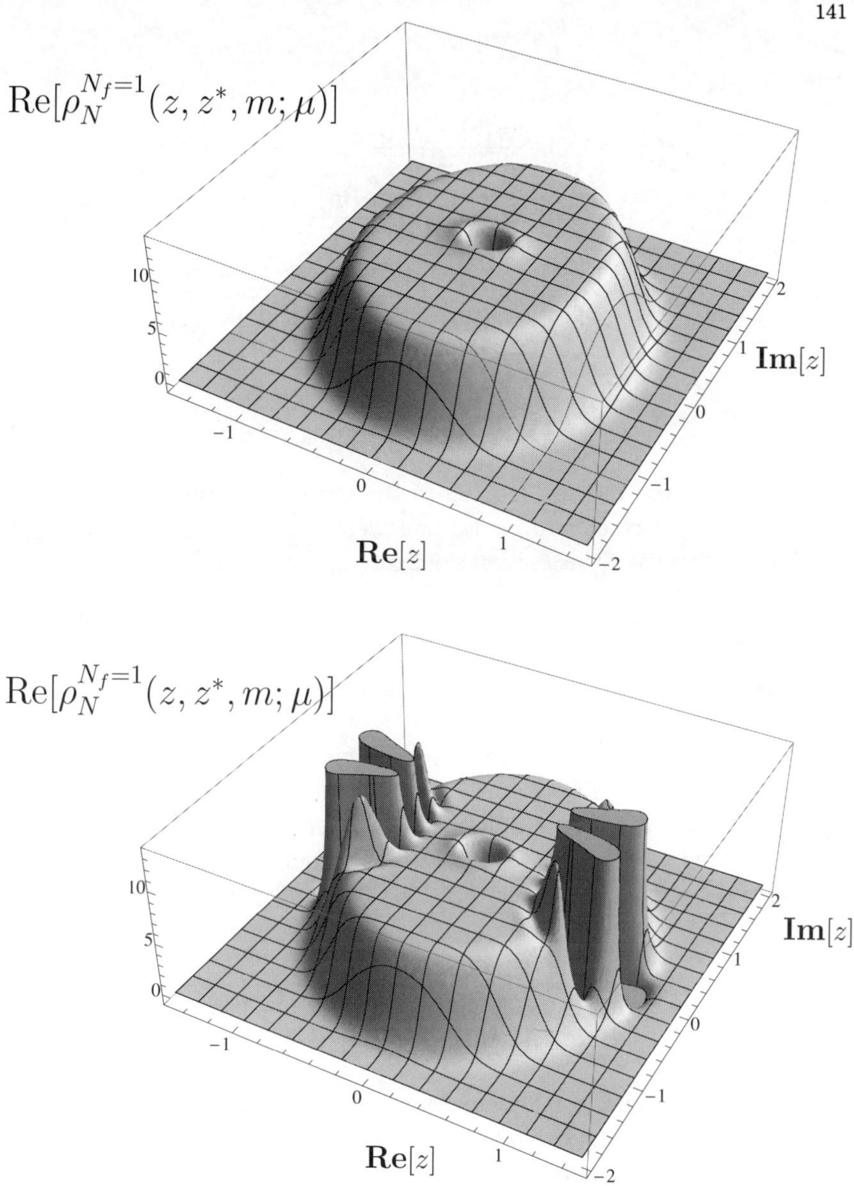

Fig. 2. The eigenvalue density of the random matrix model (3) for $N = 20$, $\mu = 0.8$ and $\nu = 0$. **Top:** The quark mass is well outside the support of the spectrum at $m = 2.0$, and the eigenvalue density is real and positive. **Bottom:** The quark mass is now inside the eigenvalue cloud at $m = 0.6$, and oscillations starting at $z = \pm m$ extend outward. Only the real part of the spectral density is shown – the imaginary part is nonzero inside the oscillating region.

Now,

$$\begin{aligned}\frac{dp_N(m)}{dm} &= \lim_{z\to m}\frac{p_N(z)-p_N(m)}{z-m},\\ &= \lim_{z\to m} 2m\frac{p_N(z)-p_N(m)}{z^2-m^2},\\ &= \lim_{z\to m} 2m[p_{N-1}(z)+c_{N-2}p_{N-2}(z)+...],\\ &= 2m[p_{N-1}(m)+c_{N-2}p_{N-2}(m)+...],\end{aligned} \tag{26}$$

so that

$$\Sigma_N^{N_f=1}(m) = \frac{p_N'(m)}{p_N(m)}. \tag{27}$$

This is the desired relation (15) up to the term ν/m which is due to the contribution of ν exact zero modes that were not included in the spectral density. The result of this section originally appeared in.[12]

Note that the proof relies entirely on orthogonality properties of polynomials. The detailed cancellations leading to the simple form of the chiral condensate are in this way linked to the orthogonality properties in the complex plane.

The above proof was given for finite N and is of course also valid in the universal microscopic limit. As shown in,[12] the proof can also be carried through in the microscopic limit without relying on the finite N chiral random matrix model. That is, one can start from the microscopic limit of the spectral density (17), which is[7,8]

$$\rho^{(\nu)}_{N_f=1}(\hat z,\hat z^*,\hat m;\hat\mu) = \frac{|\hat z|^2}{2\pi\hat\mu^2}K_\nu\left(\frac{|\hat z|^2}{4\hat\mu^2}\right)\mathrm{e}^{-\frac{\hat z^2+\hat z^{*2}}{8\hat\mu^2}} \tag{28}$$
$$\times\left(\int_0^1 dt\; t\;\mathrm{e}^{-2\hat\mu^2t^2}I_\nu(\hat z t)I_\nu(\hat z^* t) - \frac{I_\nu(\hat z)}{I_\nu(\hat m)}\int_0^1 dt\; t\;\mathrm{e}^{-2\hat\mu^2t^2}I_\nu(\hat m t)I_\nu(\hat z^* t)\right),$$

and compute the chiral condensate using complex contour integration techniques. In the microscopic limit the cancellations are even more dramatic than at finite N. The entire μ dependence of the eigenvalue density vanishes upon integration and leaves us with a μ independent chiral condensate

$$\Sigma(\hat m) = \frac{I_\nu'(\hat m)}{I_\nu(\hat m)}. \tag{29}$$

Rather than repeating the proof of[12] we now focus on the Fourier components of the eigenvalue density that are responsible for chiral symmetry breaking.

Note that the first term in (28) is the quenched eigenvalue density[20] and the effect of unquenching is in the second term. This motivates the notation

$$\rho^{(\nu)}_{N_f=1} = \rho^{(\nu)}_Q - \rho^{(\nu)}_U, \tag{30}$$

which we shall make use of below.

4. The Fourier transform of the eigenvalue density

To further expose the nature of the cancellations leading to a μ independent chiral condensate we now consider the Fourier transform of the eigenvalue density. The motivation to do so is that the discontinuity in the chiral condensate is due to the strongly oscillating part of the spectral density.[12] To evaluate the chiral condensate from the eigenvalue density

$$\Sigma_{N_f}(m) = \int dxdy \, \frac{\rho_{N_f}(x,y)}{x+iy-m} \tag{31}$$

illustrates the effects of the phase of the fermion determinant in QCD at nonzero chemical potential. In particular for $\mu > m_\pi/2$ where the sign problem is acute,[21] it would be desirable if we could find a real and positive reformulation of this integral.

The Fourier transform of the eigenvalue density along the imaginary axis is defined by

$$\begin{aligned}\tilde{\rho}_{N_f}(x,t) &\equiv \int_{-\infty}^{\infty} dy e^{-iyt} \rho_{N_f}(x,y), \\ &= \left\langle \sum_k \delta(x-x_k) e^{-iy_k t} \right\rangle_{N_f}.\end{aligned} \tag{32}$$

Since we integrate over all y the Fourier transform is necessarily real.

The chiral condensate becomes

$$\begin{aligned}\Sigma(m) &= \int dxdy \frac{\rho(x,y)}{x+iy-m}, \\ &= \int dxdy \frac{1}{x+iy-m} \frac{1}{2\pi} \int dt e^{iyt} \tilde{\rho}(x,t), \\ &= \frac{1}{2\pi} \int dx \int dt \tilde{\rho}(x,t) \int dy \frac{e^{iyt}}{x+iy-m}, \\ &= \int dx \int dt \tilde{\rho}(x,t) e^{t(m-x)} \theta(x-m).\end{aligned} \tag{33}$$

Now the integrand is real and positive if $\tilde{\rho}$ is. Below we will see that this is the case for the asymptotic limit of the unquenched microscopic density.

4.1. *The asymptotic limit of the microscopic spectral density*

In the limit where the microscopic variables $\hat{m}$ and $\hat{\mu}$ are much larger than unity, the expression for the eigenvalue density simplifies considerably. The quenched part of the microscopic eigenvalue distribution is simply constant and equal to $1/4\hat{\mu}^2$ for $\text{Re}[\hat{z}] < 2\hat{\mu}^2$. This leads to a chiral condensate that decreases linearly with the quark mass

$$\Sigma_Q(\hat{m}) = \frac{\hat{m}}{2\hat{\mu}^2} \qquad \text{for} \qquad |\hat{m}| < 2\hat{\mu}^2. \tag{34}$$

In the asymptotic limit the oscillating part of the microscopic eigenvalue density reads[12] (here and below $\hat{m} \geq 0$)

$$\rho_U(\hat{x}, \hat{y}) = \frac{1}{4\pi\hat{\mu}^2} e^{(\hat{m}-\hat{x}-i\hat{y})(\hat{m}+3\hat{x}-8\hat{\mu}^2-i\hat{y})/(8\hat{\mu}^2)}. \tag{35}$$

Obviously the amplitude grows exponentially with the volume whereas the period is inversely proportional to the volume.

By simple Gaussian integration it follows that the Fourier transform of ρ_U is

$$\tilde{\rho}_U(\hat{x}, \hat{t}) = \frac{1}{\sqrt{2\pi\hat{\mu}^2}} e^{-2\hat{\mu}^2(\hat{t}-1)^2-\hat{x}^2/2\hat{\mu}^2-\hat{m}\hat{t}-(\hat{t}-2)\hat{x}+2\hat{m}}. \tag{36}$$

It is a positive definite and nicely behaved function with maximum at $\hat{t} = 1 - (\hat{m} + \hat{x})/4\hat{\mu}^2$ (see Fig. 3). The contribution to the chiral condensate of the unquenched part of the spectral density thus becomes an integral over a positive definite function of the real part of the eigenvalue, x, and the Fourier coordinate, t,

$$\Sigma_U(\hat{m}) = \frac{1}{\sqrt{2\pi\hat{\mu}^2}} \int_{-2\hat{\mu}^2}^{2\hat{\mu}^2} d\hat{x} \int_{-\infty}^{\infty} d\hat{t} e^{-2\hat{\mu}^2(\hat{t}-1)^2-\hat{x}^2/2\hat{\mu}^2-2(\hat{t}-1)\hat{x}} \theta(\hat{x} - \hat{m}) \tag{37}$$

A simple computation yields

$$\begin{aligned}
\Sigma_U(\hat{m}) &= \frac{1}{\sqrt{2\pi\hat{\mu}^2}} \int_{-2\hat{\mu}^2}^{2\hat{\mu}^2} d\hat{x}\theta(\hat{x} - \hat{m}) e^{-\hat{x}^2/2\hat{\mu}^2} \int_{-\infty}^{\infty} d\hat{q} e^{-2\hat{\mu}^2\hat{q}^2-2\hat{q}\hat{x}}, \\
&= \frac{1}{\sqrt{2\pi\hat{\mu}^2}} \int_{-2\hat{\mu}^2}^{2\hat{\mu}^2} d\hat{x}\theta(\hat{x} - \hat{m}) e^{-\hat{x}^2/2\hat{\mu}^2} \sqrt{\frac{\pi}{2\hat{\mu}^2}} e^{\hat{x}^2/2\hat{\mu}^2}, \\
&= \frac{1}{2\hat{\mu}^2} \int_{-2\hat{\mu}^2}^{2\hat{\mu}^2} d\hat{x}\theta(\hat{x} - \hat{m}), \\
&= 1 - \frac{\hat{m}}{2\hat{\mu}^2}.
\end{aligned} \tag{38}$$

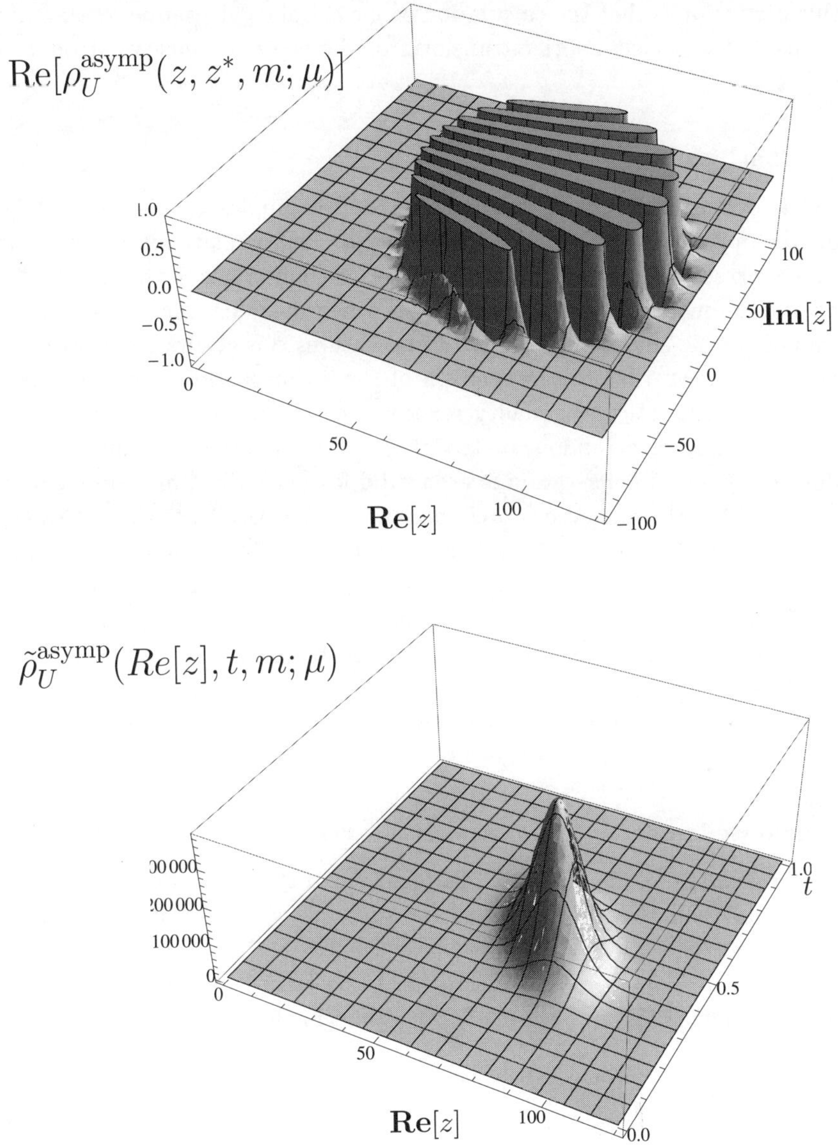

Fig. 3. **Top:** The asymptotic limit of the oscillating part of the unquenched eigenvalue density. **Bottom:** The Fourier transform of the asymptotic limit of the oscillating part of the unquenched eigenvalue density. In both cases $\hat{m} = 30$ and $\hat{\mu} = 7$.

The full chiral condensate is obtained by adding the quenched part (34), note that the μ dependent term drops out. The message to take away from this derivation is that the calculation of the chiral condensate may simplify by using the mixed Fourier transform of the spectral density introduced above.

5. Conclusions

Chiral symmetry breaking has been linked to complex oscillations on the microscopic scale. The cancellation of these violent oscillations, which result in the correct value of the chiral condensate, demonstrates the numerical difficulties in dealing with the sign problem when the quark mass is inside the support of the eigenvalues. At low temperatures this occurs for $\mu > m_\pi/2$. We have shown that the mechanism of these cancellations can be understood in terms of orthogonality relations for the orthogonal polynomials corresponding to a random matrix theory that describes the microscopic limit of QCD. The derivation is even valid for finite size random matrices. The fact that the oscillations predominantly occur parallel to the imaginary axis motivated us to study the Fourier transform of the eigenvalue density in the imaginary part of the eigenvalue. In the asymptotic limit the resulting Fourier transformed eigenvalue density is real and positive, and in this formulation a probabilistic interpretation of the contributions to the chiral condensate is possible.

It would be most interesting to generalize these results to the p-domain of chiral perturbation theory.

Acknowledgments

It is a pleasure to thank the FTPI faculty and staff for an exciting workshop and Poul Henrik Damgaard for stimulating discussions. This work was supported by U.S. DOE Grant No. DE-FG-88ER40388 (JV), the Villum Kann Rassmussen Foundation (JV), the Danish National Bank (JV) and the Danish Natural Science Research Council (KS).

References

1. I. Barbour, N. E. Behilil, E. Dagotto, F. Karsch, A. Moreo, M. Stone and H. W. Wyld, Nucl. Phys. B **275**, 296 (1986).
2. T. Banks and A. Casher, Nucl. Phys. **B169**, 103 (1980).
3. P. E. Gibbs, Glasgow University preprint PRINT-86-0389 (1986) unpublished.
4. D. Toublan and J. J. M. Verbaarschot, Int. J. Mod. Phys. B **15**, 1404 (2001) [hep-th/0001110].
5. M. P. Lombardo, J. B. Kogut and D. K. Sinclair, Phys. Rev. D **54** (1996) 2303 [arXiv:hep-lat/9511026].

6. T. Schafer, Phys. Rev. D **57**, 3950 (1998) [arXiv:hep-ph/9708256].
7. J. C. Osborn, Phys. Rev. Lett. **93**, 222001 (2004), [arXiv:hep-th/0403131].
8. G. Akemann, J. C. Osborn, K. Splittorff and J. J. M. Verbaarschot, Nucl. Phys. B **712**, 287 (2005), arXiv:hep-th/0411030.
9. E. V. Shuryak and J. J. M. Verbaarschot, Nucl. Phys. A **560**, 306 (1993) [arXiv:hep-th/9212088].
10. J. Gasser and H. Leutwyler, Phys. Lett. B **184**, 83 (1987).
11. H. Leutwyler and A. Smilga, Phys. Rev. D **46**, 5607 (1992).
12. J. C. Osborn, K. Splittorff and J. J. M. Verbaarschot, Phys. Rev. Lett. **94**, 202001 (2005), [arXiv:hep-th/0501210]; arXiv:0805.1303 [hep-th].
13. M.A. Stephanov, Phys. Rev. Lett. 76 (1996) 4472 [arXiv:hep-lat/9604003].
14. Y.V. Fyodorov and H.-J Sommers, J.Phys.A:Math.Gen. **36** (2003), 3303 [arXiv:nlin/0207051].
15. G. Akemann and G. Vernizzi, Nucl. Phys. B **660** (2003) 532 [hep-th/0212051].
16. G. Akemann, Acta Phys. Polon. B **34** (2003) 4653 [hep-th/0307116].
17. M.C. Bergère, [hep-th/0311227].
18. M.C. Bergère, [hep-th/0404126].
19. G. Akemann and A. Pottier, J. Phys. A **37**, L453 (2004) [math-ph/0404068].
20. K. Splittorff and J. J. M. Verbaarschot, Nucl. Phys. B **683**, 467 (2004) [arXiv:hep-th/0310271].
21. K. Splittorff, PoS **LAT2006**, 023 (2006) [arXiv:hep-lat/0610072]; K. Splittorff and J. J. M. Verbaarschot, Phys. Rev. Lett. **98**, 031601 (2007) [arXiv:hep-lat/0609076]; Phys. Rev. D **75**, 116003 (2007) [arXiv:hep-lat/0702011].

From Equilibrium to Transport Properties of Strongly Correlated Fermi Liquids

Thomas Schäfer

Department of Physics, North Carolina State University, Raleigh, NC 27695

We summarize recent results regarding the equilibrium and non-equilibrium behavior of cold dilute atomic gases in the limit in which the two body scattering length a goes to infinity. In this limit the system is described by a Galilean invariant (non-relativistic) conformal field theory. We discuss the low energy effective lagrangian appropriate to the limit $a \to \infty$, and compute low energy coefficients using an ϵ-expansion. We also show how to combine the effective lagrangian with kinetic theory in order to compute the shear viscosity, and compare the kinetic theory predictions to experimental results extracted from the damping of collective modes in trapped Fermi gases.

Keywords: cold atomic gases, conformal symmetry, shear viscosity

1. Introduction

Over the last ten years there has been remarkable progress in the study of "designer fluids", dilute, non-relativistic Bose and Fermi gases in which the scattering length between the Bosons or Fermions can be continuously adjusted. In the following we are particularly interested in Fermi gases, since these systems are stable for both positive and negative values of the scattering length, including the strongly correlated limit in which the scattering length is taken to infinity.

The scattering length is controlled through a Feshbach resonance. Alkali atoms such as ^{6}Li and ^{40}K have a single valence electron. When a dilute gas of atoms is cooled to very low temperatures, we can view the atoms as pointlike particles interacting via interatomic potentials which depend on the hyperfine quantum numbers. A Feshbach resonance arises if a molecular bound state in a closed hyperfine channel crosses near the threshold of an energetically lower "open" channel. Because the magnetic moments of the open and closed states are in general different, Feshbach resonances can be tuned using an applied magnetic field. At resonance the two-body scattering length in the open channel diverges, and the cross section σ is limited only by unitarity, $\sigma(k) = 4\pi/k^2$ for low momenta k. In the unitarity limit, details about the microscopic interaction are irrelevant, and the system displays universal properties.

Near a Feshbach resonance the scattering length behaves as

$$a = a_0 \left(1 + \frac{\Delta B}{B - B_0}\right) \tag{1}$$

where a_0 is the non-resonant value of the scattering length (typically on the order of the effective range of the interatomic potential), B is the magnetic field, B_0 the position of the resonance, and ΔB the width. A small negative scattering length corresponds to a weak attractive interaction between the atoms. This case is known as the BCS (Bardeen-Cooper-Schrieffer) limit. On the other side of the resonance the scattering length is positive. In the BEC (Bose-Einstein condensation) limit the interaction is strongly attractive and the fermions form deeply bound molecules. For this reason the unitarity limit $a \to \infty$ is also known at the BCS/BEC crossover.

The unitarity limit is of interest to QCD practitioners for a number of reasons:

- The unitarity limit provides an approximate description of dilute neutron matter. The neutron-neutron scattering length is $a_{nn} = 17$ fm, and the effective range is $r_{nn} = 2.7$ fm. This means that there is a range of densities, relevant to the outer layers of neutron stars, for which the interparticle spacing is large compared to the effective range, but small compared to the scattering length.
- The Fermi gas at unitarity is a high T_c superconductor. There is an attractive interaction in the spin singlet channel which leads to s-wave superconductivity below some critical temperature T_c. In the unitarity limit the only energy scale in the problem is the Fermi energy E_F, and we must have $k_B T_c = \alpha E_F$ with some numerical constant α. Quantum Monte Carlo calculations[1,2] (and experimental results) indicate that $\alpha \simeq 0.15$, much larger than in ordinary (or even high T_c) electronic superconductors, but comparable to what might be achieved in color superconducting quark matter.[3]
- The limit $a \to \infty$ corresponds to a non-relativistic conformal field theory.[4] In the unitarity limit there is no scale in the problem (other than the thermodynamic variables temperature and density). Indeed, one can show that the theory is not only scale invariant, but invariant under the full conformal group. This raises the question

whether there are any physical consequences of conformal symmetry that go beyond results that follow from scale invariance. It also raises the possibility that a holographic description, similar to the AdS/CFT correspondence, can be obtained.[5,6]

- Non-relativistic fermions at unitarity behave as a very good fluid and show interesting transport properties, including a very small shear viscosity. Kinetic theory suggests that the shear viscosity is inversely proportional to the scattering cross section, and reaches a minimum at unitarity. This expectation is confirmed by experiments that demonstrate large elliptic flow and a very small damping rate for collective oscillations.[7,8]

2. Equilibrium Properties

We begin by analyzing equilibrium properties of the dilute Fermi gas at unitarity. If the temperature is large, $k_BT > E_F$, then the scattering cross section is regularized by the thermal wave length, and the effective interaction is weak. Here the Fermi energy is defined by $E_F = (3\pi^2 n)^{2/3}/(2m)$, where n is the density, and m is the mass of the atoms. In the high temperature regime the equation of state is well described by the Virial expansion, and the system has single particle excitations with the quantum numbers of the fundamental fermions. In the regime $k_BT \sim E_F$ the interactions are strong. As noted above, superfluidity occurs at $k_BT_c \simeq 0.15E_F$. Below the critical temperature the excitations are Goldstone bosons. In following section we will discuss the effective theory of the Goldstone bosons, and relate the parameters in the effective lagrangian to static properties of the system.

2.1. *Low Energy Effective Theory and Density Functional*

The Goldstone boson field can be defined as the phase of the difermion condensate $\langle\psi\psi\rangle = e^{2i\varphi}|\langle\psi\psi\rangle|$. The effective Lagrangian at next-to-leading order (NLO) in derivatives of φ and the external potential is[9]

$$\mathcal{L} = c_0 m^{3/2} X^{5/2} + c_1 m^{1/2} \frac{(\vec{\nabla} X)^2}{\sqrt{X}} + \frac{c_2}{\sqrt{m}} \left[(\nabla^2 \varphi)^2 - 9m\nabla^2 V \right] \sqrt{X} \,, \quad (2)$$

where we have defined

$$X = \mu - V - \dot{\varphi} - \frac{(\vec{\nabla}\varphi)^2}{2m} \,. \quad (3)$$

Here, μ is the chemical potential and $V(\vec{x},t)$ is an external potential. The functional form of the effective lagrangian is fixed by the symmetries of the problem, Galilean invariance, $U(1)$ symmetry, and conformal symmetry. The NLO effective lagrangian is characterized by three dimensionless parameters, c_0, c_1, c_2. These parameters can be related to physical properties of the system. The first parameter, c_0, can be related to the equation of state. We have

$$c_0 = \frac{2^{5/2}}{15\pi^2\xi^{3/2}}\,, \tag{4}$$

where ξ determines the chemical potential in units of the Fermi energy, $\mu = \xi E_F$. The two NLO parameters c_1, c_2 are related to the momentum dependence of correlation functions. The phonon dispersion relation, for example, is given by

$$q_0 = v_s q \left[1 - \pi^2\sqrt{2\xi}\left(c_1 + \frac{3}{2}c_2\right)\frac{q^2}{k_F^2} + O(q^4\log(q^2))\right] \tag{5}$$

where $v_s = \sqrt{\xi/3}v_F$ and $v_F = k_F/m$. The static susceptibility

$$\chi(q) = -i\int dt\, d^3x\, e^{-i\vec{q}\cdot\vec{x}}\,\langle\psi^\dagger\psi(0)\psi^\dagger\psi(x)\rangle \tag{6}$$

involves a different linear combination of c_1 and c_2,

$$\chi(q) = -\frac{mk_F}{\pi^2\xi}\left[1 + 2\pi^2\sqrt{2\xi}\left(c_1 - \frac{9}{2}c_2\right)\frac{q^2}{k_F^2} + O(q^4\log(q^2))\right]. \tag{7}$$

Higher derivative terms in the effective lagrangian can also be used to compute the energy of inhomogeneous matter. At NLO in an expansion in derivatives of the density we find the following energy density functional[10]

$$\mathcal{E}(x) = n(x)V(x) + \frac{3\cdot 2^{2/3}}{5^{5/3}mc_0^{2/3}}n(x)^{5/3} - \frac{4}{45}\frac{2c_1+9c_2}{mc_0}\frac{(\nabla n(x))^2}{n(x)} - \frac{12}{5}\frac{c_2}{mc_0}\nabla^2 n(x)\,. \tag{8}$$

The first two terms correspond to the local density approximation (LDA) and the terms proportional to c_1 and c_2 are the leading correction to the LDA involving derivatives of the density.

2.2. *Epsilon Expansion*

At unitarity the determination of c_1 and c_2 is a non-perturbative problem, and we will perform the calculation using an expansion around $d = 4-\epsilon$ spatial dimensions.[11,12] Our starting point is the lagrangian

$$\mathcal{L} = \Psi^\dagger\left[i\partial_0 + \sigma_3\frac{\vec{\nabla}^2}{2m}\right]\Psi + \mu\Psi^\dagger\sigma_3\Psi + \left(\Psi^\dagger\sigma_+\Psi\phi + h.c.\right) - \frac{1}{C_0}\phi^\dagger\phi\,, \tag{9}$$

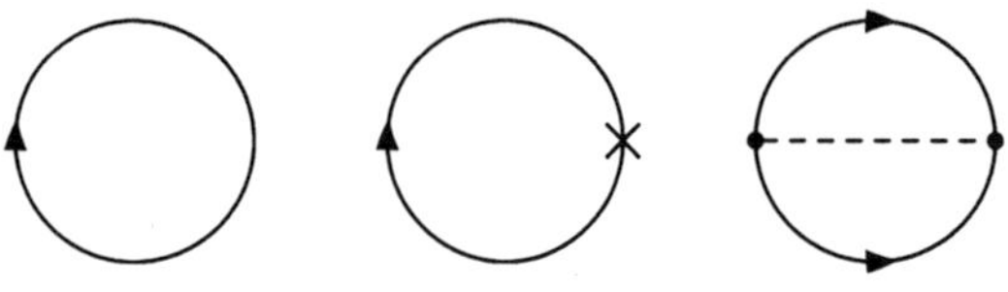

Fig. 1. Leading order contributions to the effective potential in the epsilon expansion. Solid lines denote fermions propagators, dashed lines denote boson propagators, and the cross is an insertion of the chemical potential.

where $\Psi = (\psi_\uparrow, \psi_\downarrow^\dagger)^T$ is a two-component Nambu-Gorkov field, σ_i are Pauli matrices acting in the Nambu-Gorkov space, $\sigma_\pm = (\sigma_1 \pm i\sigma_2)/2$, ϕ is a complex boson field, and C_0 is a coupling constant. In dimensional regularization the fermion-fermion scattering length becomes infinite for $1/C_0 \to 0$.

The epsilon expansion is based on the observation that the fermion-fermion scattering amplitude near $d = 4$ dimensions is saturated by the propagator of a boson with mass $2m$. The coupling of the boson to pairs of fermions is given by

$$g = \frac{\sqrt{8\pi^2\epsilon}}{m}\left(\frac{m\phi_0}{2\pi}\right)^{\epsilon/4}. \qquad (10)$$

In the superfluid phase ϕ acquires an expectation value $\phi_0 = \langle\phi\rangle$. We write the boson field as $\phi = \phi_0 + g\varphi$. The lagrangian is split into a free part

$$\mathcal{L}_0 = \Psi^\dagger \left[i\partial_0 + \sigma_3 \frac{\vec{\nabla}^2}{2m} + \phi_0(\sigma_+ + \sigma_-) \right] \Psi + \varphi^\dagger \left(i\partial_0 + \frac{\vec{\nabla}^2}{4m} \right) \varphi, \qquad (11)$$

and an interacting part $\mathcal{L}_I + \mathcal{L}_{ct}$, where

$$\mathcal{L}_I = g\left(\Psi^\dagger\sigma_+\Psi\varphi + h.c\right) + \mu\Psi^\dagger\sigma_3\Psi + 2\mu\varphi^\dagger\varphi, \qquad (12)$$

$$\mathcal{L}_{ct} = -\varphi^\dagger\left(i\partial_0 + \frac{\vec{\nabla}^2}{4m}\right)\varphi - 2\mu\varphi^\dagger\varphi. \qquad (13)$$

Note that the leading self energy corrections to the boson propagator generated by the interaction term $\mathcal{L}_I$ cancel against the counterterms in $\mathcal{L}_{ct}$. The chemical potential term for the fermions is included in $\mathcal{L}_I$ rather than in $\mathcal{L}_0$. This is motivated by the fact that near $d = 4$ the system reduces to a non-interacting Bose gas and $\mu \to 0$. We will count μ as a quantity of $O(\epsilon)$. The Feynman rules are quite simple. The fermion and boson propagators

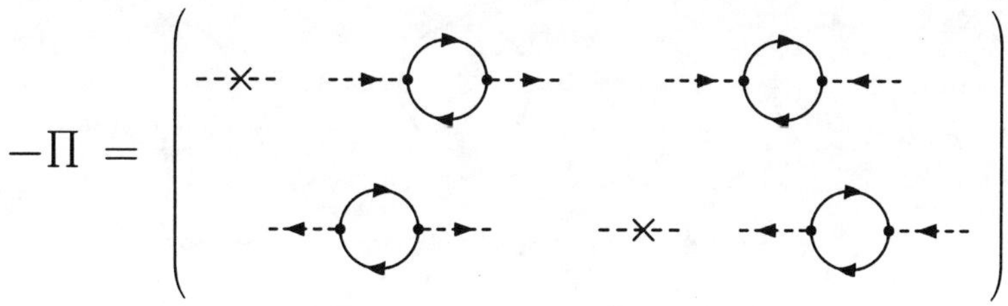

Fig. 2. Scalar self energy at LO in the epsilon expansion.

are

$$G(p_0, p) = \frac{i}{p_0^2 - E_p^2} \begin{bmatrix} p_0 + \epsilon_p & -\phi_0 \\ -\phi_0 & p_0 - \epsilon_p \end{bmatrix}, \tag{14}$$

$$D(p_0, p) = \frac{i}{p_0 - \epsilon_p/2}, \tag{15}$$

where $E_p^2 = \epsilon_p^2 + \phi_0^2$ and $\epsilon_p = p^2/(2m)$. The fermion-boson vertices are $ig\sigma^{\pm}$. Insertions of the chemical potential are $i\mu\sigma_3$. Both g^2 and μ are corrections of order ϵ.

In order to determine c_0, c_1, c_2 we have to compute three physical observables. We have studied $\xi = \mu/E_F$, and the curvature terms in the phonon dispersion relation and the static susceptibility. The universal parameter ξ was originally calculated by Nishida and Son. They computed the effective potential to NLO in the epsilon expansion, see Fig. 1. The derivative of the effective potential with respect to μ determines the density n, and the relation between n and μ fixes ξ. The result is

$$\xi = \frac{\epsilon^{3/2}}{2} \left[1 + \frac{1}{8}\epsilon \log(\epsilon) - \frac{1}{4}\left(12C - 5 + 5\log(2)\right)\epsilon + O(\epsilon^2) \right], \tag{16}$$

with $C = 0.144$. The phonon dispersion relation can be extracted from the scalar propagator. We introduce a two-component scalar field $\Phi = (\varphi, \varphi^*)$. The scalar propagator satisfies a Dyson-Schwinger equation[13]

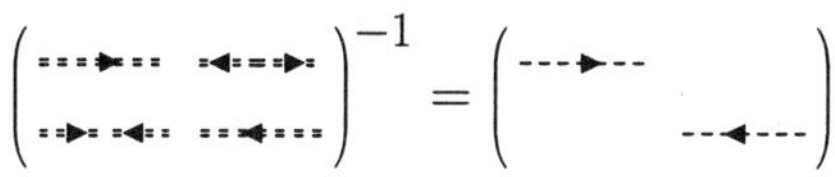

$- \Pi$ (17)

At LO in the epsilon expansion the self energy is determined by the diagrams shown in Fig. 2. NLO contributions were calculated in reference.[10] The phonon dispersion relation is

$$p_0 = \sqrt{\mu\epsilon_p}\left(1 + \frac{\epsilon}{8}\right)\left\{1 + \frac{\epsilon_p}{8\mu}\left(1 - \frac{\epsilon}{4}\right) + \ldots\right\} \tag{18}$$

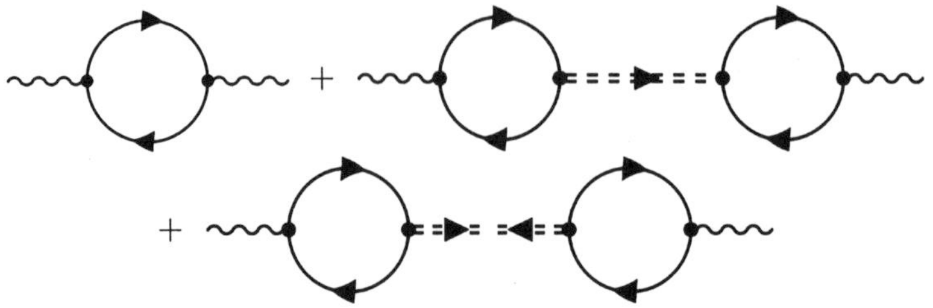

Fig. 3. Leading order contributions to the static susceptibility. The wiggly line denotes an external current. The double line is the scalar propagator defined in equ. (17).

We note that the dispersion relation curves up (unlike ^{4}He, but similar to weakly interacting Bose gases). This implies that there is $\varphi \to \varphi + \varphi$ decay. Finally, we can determine the static susceptibility. Computing the diagrams in Fig. 3 we get

$$\chi(q) = -\frac{2Z}{\epsilon\mu}\left\{1 - \frac{1}{8}\left(\frac{q^2}{m\mu}\right)\left(1-\frac{\epsilon}{4}\right) + O(q^4)\right\}\left(\frac{m\phi_0}{2\pi}\right)^{d/2}, \quad (19)$$
$$Z = 1 - \frac{1}{2}\left(\gamma - \log(2)\right)\epsilon\,.$$

The coefficient c_0 follows from the result for ξ using equ. (4). Matching equ. (18,19) against equ. (5,7) gives $c_2 = 0$ and $c_1/c_0 = 3/8 - \epsilon/4$. The corresponding energy density functional was studied in reference.[10] Compared to a free Fermi gas the local density term is reduced by a factor ~ 2 (the interaction is attractive), while the gradient correction proportional to $(\nabla n)^2/n$ is enhanced by a factor ~ 2.

3. Transport Properties

In the following we will discuss transport properties of the Fermi gas at unitarity. The interest in non-equilibrium properties arises from the observation that transport coefficients are much more sensitive to the strength of the interaction than thermodynamic quantities. A renewed interest in transport properties was also sparked the AdS/CFT correspondence and the experimental limits on the shear viscosity of the quark gluon plasma obtained at RHIC. In the following we shall focus on the shear viscosity of the Fermi gas at unitarity. Close to equilibrium the (coarse grained) energy

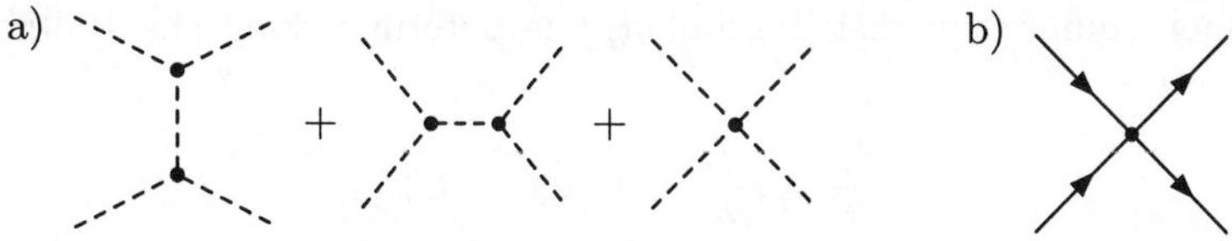

Fig. 4. Leading order processes that contribute to the shear viscosity at low temperature (Fig. a) and high temperature (Fig. b). Dashed lines are phonon propagators and solid lines are fermion propagators.

momentum tensor can be written as

$$T_{ij} = (P+\epsilon)v_i v_j - P\delta_{ij} + \delta T_{ij}\,, \qquad (20)$$
$$\delta T_{ij} = -\eta(\nabla_i v_j + \nabla_i v_j - \frac{2}{3}\delta_{ij}\vec{\nabla}\cdot\vec{v}) + \cdots\,,$$

where ϵ and P are the energy density and pressure, and v_i is the local flow velocity. The first term is the ideal gas contribution, and δT_{ij} is the leading order (in gradients of v_i) dissipative correction. The traceless part of δT_{ij} is proportional to the shear viscosity η.

3.1. *Kinetic Theory*

We first consider the case that the fluid is composed of weakly interacting quasi-particles. In the unitarity limited Fermi gas this is the case at $T \ll T_c$ (phonons) and $T \gg T_c$ (atoms). In these limits we can compute the shear viscosity using kinetic theory. In the following we will concentrate on the low temperature case discussed in reference.[14] In kinetic theory the stress-energy tensor is given by

$$T_{ij} = v_s^2 \int \frac{d^3p}{(2\pi)^3}\frac{p_i p_j}{E_p} f_p\,, \qquad (21)$$

where f_p is the distribution function of the phonons, v_s is the speed of sound, p_i is the momentum and E_p the quasi-particle energy. Close to equilibrium $f_p = f_p^{(0)} + \delta f_p$, where $f_p^{(0)}$ is the Bose-Einstein distribution and δf_p is a small departure from equilibrium. We write $\delta f_p = -\chi(p) f_p^{(0)}(1+f_p^{(0)})/T$. In the case of shear viscosity we can further decompose

$$\chi(p) = g(p)(p_i p_j - \frac{1}{3}\delta_{ij}p^2)(\nabla_i v_j + \nabla_j v_i - \frac{2}{3}\delta_{ij}\vec{\nabla}\cdot\vec{v})\,. \qquad (22)$$

Inserting equ. (22) into equ. (21) we get

$$\eta = \frac{4v^2}{15T}\int \frac{d^3p}{(2\pi)^3}\frac{p^4}{2E_p} f_p^{(0)}(1+f_p^{(0)})g(p)\,. \qquad (23)$$

The non-equilibrium distribution $g(p)$ is determined by the Boltzmann equation

$$\frac{df_p}{dt} = \frac{\partial f_p}{\partial t} + \vec{v}\cdot\vec{\nabla} f_p = C[f_p], \tag{24}$$

relating the rate of change of the distribution function f_p to the collision operator $C[f_p]$. The $2 \leftrightarrow 2$ collision integral is given by

$$C_{2\leftrightarrow 2}[f_p] = \frac{1}{2E_p}\int \frac{d^3k}{(2\pi)^3 2E_k}\frac{d^3k'}{(2\pi)^3 2E_{k'}}\frac{d^3p'}{(2\pi)^3 2E_{p'}} \times (2\pi)^4\delta^{(4)}(p+k-p'-k')|\mathcal{M}|^2 D_{2\leftrightarrow 2}, \tag{25}$$

where $D_{2\leftrightarrow 2}$ contains the distribution functions and $|\mathcal{M}|$ is the $2 \leftrightarrow 2$ scattering amplitude shown in Fig. 4. The three and four-phonon vertices are fixed by the effective lagrangian (2). Linearizing $D_{2\leftrightarrow 2}$ in δf_p one finds

$$D_{2\leftrightarrow 2} = \frac{1}{T} f_{k'}^{(0)} f_{p'}^{(0)} (1+f_k^{(0)})(1+f_p^{(0)}) \left(\chi(p)+\chi(k)-\chi(p')-\chi(k')\right) . \tag{26}$$

There are a variety of methods for solving the linearized Boltzmann equation. A standard technique is based on expanding $g(p)$ in a complete set of functions. A nice feature of this method is that the truncated expansion gives a variational estimate

$$\eta \geq \frac{4v^4}{25T^2}\frac{(b_0A_{00})^2}{\sum_{s,t} b_s b_t M_{st}} \tag{27}$$

where b_s is a set of expansion coefficients, A_{00} is a normalization integral, and M_{st} are matrix elements of the linearized collision operator. For the best trial function we find[14]

$$\eta/s = 7.7\times 10^{-6}\xi^5 \frac{T_F^8}{T^8}, \tag{28}$$

where ξ is the universal parameter introduced in Sect. 2.1 and we have normalized the result to the entropy density s of a weakly interacting phonon gas. A similar estimate can be obtained in the high temperature limit. In this case the relevant degrees of freedom are atoms, and the dominant scattering process is shown in Fig. 4b. The result is[15,16]

$$\eta/s = \frac{45\pi^{3/2}}{64\sqrt{2}}\left(\frac{T}{T_F}\right)^{3/2}\left[\log\left(\frac{3\sqrt{\pi}}{4}\frac{T^{3/2}}{T_F^{3/2}}\right)+\frac{5}{2}\right]^{-1} . \tag{29}$$

The high and low temperature limits of η/s are shown in Fig. 5, together with the proposed lower bound[17] $\eta/s = 1/(4\pi)$ and experimental data which we will discuss in the next section.

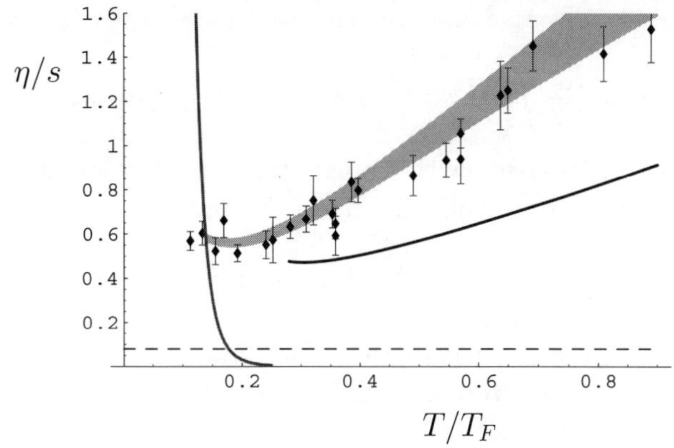

Fig. 5. Viscosity to entropy density ratio of a cold atomic gas in the unitarity limit. This plot is based on the damping data published in ref.[8] and the thermodynamic data in ref.[18,19] The dashed line shows the conjectured viscosity bound $\eta/s = 1/(4\pi)$, and the solid lines show the high and low temperature limits.

3.2. *Hydrodynamics*

Hydrodynamics describes the evolution of long-wavelength, slow-frequency modes. The hydrodynamic description remains valid even if there is no underlying kinetic theory. The hydrodynamic equations follow from conservation of mass (particle number), energy and momentum. In a non-relativistic system the equations of continuity and of momentum conservation are given by

$$\frac{\partial n}{\partial t} + \vec{\nabla} \cdot (n\vec{v}) = 0, \tag{30}$$

$$mn\frac{\partial \vec{v}}{\partial t} + mn\left(\vec{v} \cdot \vec{\nabla}\right)\vec{v} = -\vec{\nabla}P - n\vec{\nabla}V, \tag{31}$$

where n is the number density, m is the mass of the atoms, $\vec{v}$ is the fluid velocity, P is the pressure and V is the external potential. In an ideal fluid the equation of energy conservation can be rewritten as conservation of entropy,

$$\frac{\partial ns}{\partial t} + \vec{\nabla} \cdot (ns\vec{v}) = 0\,. \tag{32}$$

A non-zero shear viscosity leads to dissipation, converting kinetic energy to heat and increasing the entropy. The shear viscosity of the dilute Fermi gas in the unitarity limit can be measured by studying the damping of collective modes in trapped systems.[20] The frequency of these modes agrees well the prediction of ideal hydrodynamics. The dissipated energy is given by

$$\dot{E} = -\frac{1}{2}\int d^3x\,\eta(x)\left(\partial_i v_j + \partial_j v_i - \frac{2}{3}\delta_{ij}\partial_k v_k\right)^2 . \tag{33}$$

The damping rate is given by the ratio of the energy dissipated to the total energy of the collective mode. The kinetic energy is

$$E_{kin} = \frac{m}{2}\int d^3x\, n(x)\vec{v}^{\,2} . \tag{34}$$

If the damping rate is small both $\dot{E}$ and E_{kin} can be computed using the solution of ideal hydrodynamics. We recently performed an analysis[21] which is based on measurements of the damping rate of the lowest radial breathing mode performed by the Duke group.[8] We showed that can relate the dimensionless ratio Γ/ω, where Γ is the damping rate and ω is the trap frequency, to the shear viscosity to entropy density ratio

$$\frac{\eta}{s} = \frac{3}{4}\xi^{1/2}(3N)^{1/3}\left(\frac{\bar{\omega}\Gamma}{\omega_\perp^2}\right)\left(\frac{E}{E_{T=0}}\right)\left(\frac{N}{S}\right). \tag{35}$$

Here N is the total number of particles in the trap ($2\cdot 10^5$ in ref.[8]), ξ is the universal parameter defined in Sec. 2.1, $E/E_{T=0}$ is the ratio of the total energy to the energy at $T=0$ (which can be extracted using a Virial theorem from the measured cloud size), and S/N is the entropy per particle (which is measured using adiabatic sweeps to the BCS limit[19]). The results are compared to theoretical prediction in the high and low temperature limit in Fig. 5. The data show a minimum near $T/T_F \simeq 0.2$. At the minimum $\eta/s \sim 1/2$. This should probably be considered as an upper bound, since dissipative mechanism other than shear viscosity may be present. In the high T limit there is fairly good agreement with kinetic theory. The temperature dependence implied by the low T prediction is not seen in the data. This is maybe not very surprising, since the mean free path in the low T regime quickly exceeds the system size.

4. Outlook

There are many promising directions for further study. Clearly, it is desirable to obtain additional experimental constraints on the shear viscosity, and to improve the theoretical analysis of the existing data sets. It would also be interesting to confirm that the bulk viscosity vanishes in the normal phase, and to measure the thermal conductivity. We would also like to improve the theoretical tools for computing transport properties in the interesting regime near T_c. There are some recent ideas for applying holography and the AdS/CFT correspondence to Galilean invariant conformal field theories,[5,6] but there are also many purely field theoretic methods (ϵ expansions, large N methods) that have yet to be pursued.

Acknowledgments: Much of this work was carried out in collaboration with G. Rupak. The work is supported in part by the US Department of Energy grant DE-FG02-03ER41260.

References

1. E. Burovski, N. Prokof'ev, B. Svistunov, M. Troyer, Phys. Rev. Lett. **96**, 160402 (2006) [cond-mat/0602224].
2. A. Bulgac, J. E. Drut, P. Magierski, arXiv:0803.3238 [cond-mat.stat-mech].
3. M. G. Alford, A. Schmitt, K. Rajagopal and T. Schäfer, Rev. Mod. Phys., in press, arXiv:0709.4635 [hep-ph].
4. T. Mehen, I. W. Stewart and M. B. Wise, Phys. Lett. B **474**, 145 (2000) [arXiv:hep-th/9910025].
5. D. T. Son, arXiv:0804.3972 [hep-th].
6. K. Balasubramanian and J. McGreevy, arXiv:0804.4053 [hep-th].
7. K. M. O'Hara, S. L. Hemmer, M. E. Gehm, S. R. Granade, J. E. Thomas, Science **298**, 2179 (2002) [cond-mat/0212463].
8. J. Kinast, A. Turlapov, J. E. Thomas, Phys. Rev. Lett. **94**, 170404 (2005) [cond-mat/0502507].
9. D. T. Son and M. Wingate, Annals Phys. **321**, 197 (2006) [arXiv:cond-mat/0509786].
10. G. Rupak and T. Schäfer, arXiv:0804.2678 [nucl-th].
11. Z. Nussinov and S. Nussinov, Phys. Rev. A **74**, 053622 (2006), cond-mat/0410597.
12. Y. Nishida and D. T. Son, Phys. Rev. Lett. **97**, 050403 (2006) [arXiv:cond-mat/0604500].
13. Y. Nishida, Phys. Rev. A **75**, 063618 (2007) [arXiv:cond-mat/0608321].
14. G. Rupak and T. Schäfer, Phys. Rev. A **76**, 053607 (2007) [arXiv:0707.1520 [cond-mat.other]].
15. P. Massignan, G. M. Bruun, and H. Smith, Phys. Rev. A **71**, 033607 (2005).
16. G. M. Bruun and H. Smith, Phys. Rev. A **72**, 043605 (2005).
17. P. Kovtun, D. T. Son and A. O. Starinets, Phys. Rev. Lett. **94**, 111601 (2005) [hep-th/0405231].
18. J. Kinast, A. Turlapov, J. E. Thomas, Q. Chen, J. Stajic, and K. Levin, Science **307**, 1296 (2005) [cond-mat/0502087].
19. L. Luo, B. Clancy, J. Joseph, J. Kinast, J. E. Thomas, preprint, cond-mat/0611566.
20. J. Kinast, S. L. Hemmer, M. E. Gehm, A. Turlapov, and J. E. Thomas, Phys. Rev. Lett. **92**, 150402 (2004).
21. T. Schäfer, Phys. Rev. A **76**, 063618 (2007) [arXiv:cond-mat/0701251].

LESSONS FROM RANDOM MATRIX THEORY FOR QCD AT FINITE DENSITY

K. Splittorff*

The Niels Bohr Institute,
Blegdamsvej 17, Dk-2100, Copenhagen, Denmark
**E-mail: split@nbi.dk*

J.J.M. Verbaarschot** (*speaker*)

Stony Brook University,
Stony Brook, NY 11749-3800, USA
***E-mail: jacobus.verbaarschot@stonybrook.edu*

In this lecture we discuss various aspects of QCD at nonzero chemical potential, including its phase diagram and the Dirac spectrum, and summarize what chiral random matrix theory has contributed to this subject. To illustrate the importance of the phase of the fermion determinant, we particularly highlight the differences between QCD and phase quenched QCD.

1. Introduction

QCD at nonzero baryon chemical potential has turned out to be particularly challenging, and as of today, the phase diagram of QCD in the chemical potential temperature plane is far from being understood. There is not even agreement of its gross features such as the existence of a critical endpoint.[1–3] The reason for these difficulties is that the Dirac operator at nonzero chemical potential is nonhermitean resulting in a complex determinant so that the partition function cannot be evaluated by Monte-Carlo simulations. In this lecture we compare the QCD partition function to the QCD like partition function that only differs by the absence of the phase of the fermion determinant. This theory, known as phase quenched QCD, can be studied by means of lattice QCD simulations and has a phase diagram that is certainly completely different from QCD. This makes it clear that the phase of the fermion determinant is absolutely essential for the physics of QCD at nonzero chemical potential.

In the second half of this lecture we will discuss various aspects of the phase of the fermion determinant and analyze its connections with the spectrum of the Dirac operator. We will do this in the context of chiral random matrix theory[4,5] at nonzero chemical potential.[6–8] In the microscopic domain,[5] this theory is equivalent to QCD and, in spite of the fact that it can be solved analytically, it shares the sign problem and the physical effects of the phase of the fermion determinant with QCD.

2. Phase Diagram of QCD

The phase diagram of QCD in the temperature-chemical potential plane can only be obtained by means of nonperturbative methods. The method that has had the most impact is lattice QCD, but because it requires a real action, only the temperature axis has been studied extensively. The consensus is that, in the temperature range of 150-200 MeV, QCD undergoes a crossover transition from a chirally broken phase to a chirally symmetric phase. The picture at nonzero chemical potential is much less clear. Although significant progress has been made for small chemical potentials, questions about the structure of the phase diagram along the chemical potential axis could not be answered based on first principles (except for asymptotically large chemical potentials[9]). That is why most of our knowledge of QCD at nonzero baryon density is based on models and general properties of the phases of QCD.

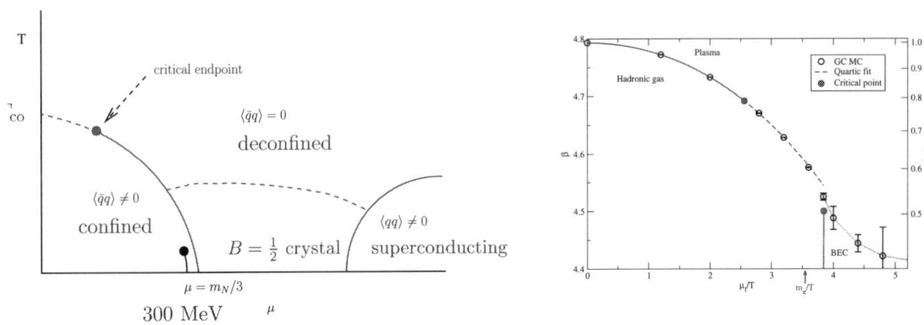

Fig. 1. The phase diagram of QCD in the temperature baryon chemical potential plane (left) and in the temperature isospin chemical potential plane (right). The data points in the right figure are from lattice QCD simulations.[10]

A tentative phase diagram of QCD in the chemical potential temperature plane is shown in Fig. 1 (left). Among others we conjecture a chirally symmetric crystalline phase made out of $B = \frac{1}{2}$ objects.[11] The best evidence for the existence of such phase is based on the Skyrme model where this phase exists as a strongly bound crystalline state with a binding energy of 135 MeV per nucleon (see Fig. 2 (left)). The chiral restoration phase transition to this phase is of second order for a face centered cubic crystal[12,13] but is of first order for a cubic crystal.[14] The restoration of chiral symmetry in the dense phase is illustrated by the vanishing of the average σ-field over a unit cell (see Fig. 2 (right)). When the temperature is increased, we expect that this crystal will melt at a temperature that is of the order of the binding energy. This phase has also been identified as the quarkyonic phase.[15,16]

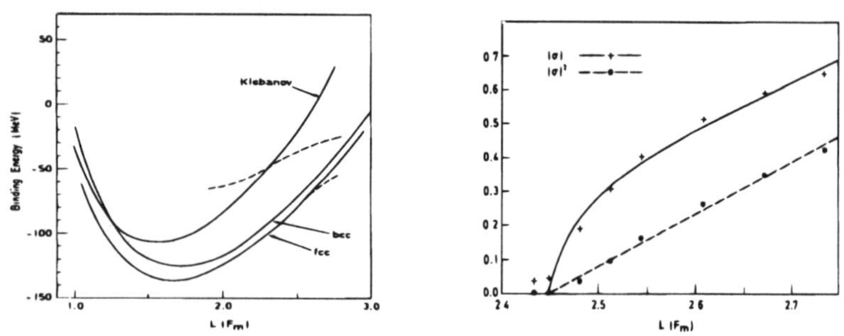

Fig. 2. Energy of a unit cell of a Skyrmion crystal (left) for a cubic crystal with cubic boundary conditions (obtained in[14]) and for a bcc and fcc crystal (obtained in[12]) versus the size of the unit cell. In the right figure we show the average value of the σ-field for an fcc crystal.[12]

Our aim is to understand QCD at $\mu \neq 0$ from first principles, i.e. starting from the partition function

$$Z_{\rm QCD} = \langle \prod_f \det(D + m_f + \mu\gamma_0) \rangle, \tag{1}$$

but the nonhermiticity of $D + \mu\gamma_0$ makes this a very challenging problem indeed. Let us first discuss the effect of the phase of the fermion determinant by studying the theory where this phase has been quenched. For $N_f = 2$ this partition function, known as phase quenched QCD, is given by

$$Z_{|\rm QCD|} = \langle |\det(D + m + \mu\gamma_0)|^2 \rangle = \langle \det(D + m + \mu\gamma_0) \det(D + m - \mu\gamma_0) \rangle \tag{2}$$

Therefore, the chemical potential of phase quenched QCD can be interpreted as an isospin chemical potential,[17] and at low enough temperatures, pions will Bose condense for $\mu > m_\pi/2$.[18–20] This partition function can be simulated by Monte-Carlo methods[10,21–24] (see Fig. 1) and its phases are in agreement with theoretical expectations based on a mean field treatment of a chiral Lagrangian.[18,25,26]

We conclude that for $\mu > m_\pi/2$ the phase factor quark determinant completely changes the phase diagram. The would be pion condensate is nullified by this phase factor after averaging. To better understand the effect of nonhermiticity, we will analyze the QCD partition function in the microscopic domain where it is equivalent to a chiral random matrix theory. This theory is analytically solvable in the nonperturbative domain of QCD, but the sign problem has all complications of QCD.

3. Random Matrix Model at $\mu \neq 0$

A chiral Random Matrix Theory (chRMT) is obtained by replacing the matrix elements of the Dirac operator by Gaussian random numbers[4,5]

$$D = \begin{pmatrix} 0 & iW^\dagger + \mu^\dagger \\ iW + \mu & 0 \end{pmatrix} \tag{3}$$

with W a complex $n \times (n+\nu)$ matrix, and μ is a multiple of the identity[6] or an arbitrary random matrix.[8] The random matrix partition function in the sector of topological charge ν is given by

$$Z_\nu(m_f;\mu) = \langle \prod_f \det(D + m_f) \rangle, \tag{4}$$

where the average is over the Gaussian distribution of W. This partition function has the global symmetries and transformation properties of QCD. At fixed θ-angle it is given by

$$Z(m_f,\theta;\mu) = \sum_\nu \mathcal{N}_\nu(\mu) e^{i\nu\theta} Z_\nu(m_f;\mu) \tag{5}$$

with $\mathcal{N}_\nu(\mu)$ s normalization constant that may depend on μ and ν.

In the microscopic domain of QCD, defined by

$$m_\pi^2 \ll \frac{1}{\sqrt{V}}, \qquad \mu^2 \ll \frac{1}{\sqrt{V}}, \qquad V\Lambda_{\rm QCD}^4 \gg 1 \tag{6}$$

the mass and the chemical potential dependence of the QCD partition function is given by the random matrix partition function

$$Z_\nu^{\rm QCD}(m;\mu) \sim Z_\nu^{\rm chRMT}(m;\mu).$$

The reason for this equivalence is that in this domain, because the Compton wave length of the Goldstone modes is much larger than the size of the box, both theories reduce to the same static chiral Lagrangian. The Goldstone mass may either refer to physical quarks masses or to complex parameters (ghost quark masses) to probe the Dirac spectrum.

Since the kinetic term of the chiral Lagrangian does not contribute in a mean field analysis, chRMT correctly describes mean field results for the low-energy limit of QCD beyond the microscopic domain provided that $m \ll \Lambda_{\rm QCD}$ and $\mu \ll \Lambda_{\rm QCD}$.

4. Lessons from Random Matrix Theory

4.1. *Lesson 1: Homogeneity of Dirac Eigenvalues*

At nonzero chemical potential, the Dirac operator is nonhermitean with eigenvalues that are scattered in the complex plane. On inspection of quenched lattice QCD Dirac spectra (see Fig. 3) two properties stand out: i) The Dirac spectrum has a sharp edge, and ii) the distribution of the eigenvalues is more or less homogeneous. Both properties can be understood in terms of random matrix theory.[6,7] In essence they follow from the fact that eigenvalues of nonhermitean random matrices behave as repulsive electric charges in the plane. Because the chiral condensate is given by

$$\Sigma(m) = \frac{1}{V}\sum_{\lambda_k}\frac{1}{m + i\lambda_k},$$

it can be interpreted as the electric field at m of charges at λ_k. Therefore, a direct consequence of the homogeneity of the Dirac spectrum is that the quenched chiral condensate increases linearly with m when m is inside the domain of the eigenvalues.

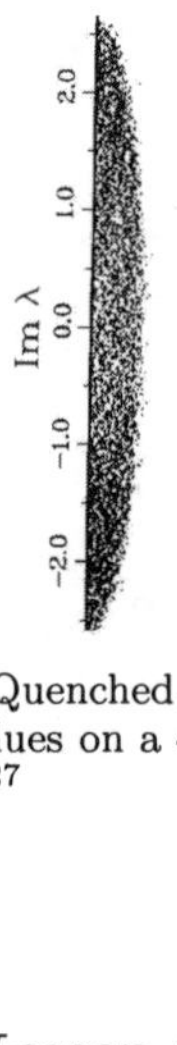

Fig. 3. Quenched Dirac eigenvalues on a $4^3 \times 8$ lattice.[27]

4.2. *Lesson 2: Quenched Limit*

The quenched limit is the limit where the fermion determinant is ignored in generating the statistical ensemble. For zero chemical potential this has been a reasonable approximation, but at nonzero chemical potential, the limit of no fermion determinant is not given by

$$\lim_{n\to 0}\langle(\det(D + \mu\gamma_0 + m))^n\rangle,$$

but rather by

$$\lim_{n\to 0}\langle|\det(D + \mu\gamma_0 + m)|^n\rangle,$$

i.e. the quenched limit of phase quenched QCD.[6] Although similar ideas were proposed before,[28,29] this result was first demonstrated convincingly for random matrix theory where the averages can be evaluated analytically.

Because phase quenched QCD is QCD at nonzero isospin chemical potential, a phase transition to a pion condensate occurs at $\mu = m_\pi/2$ for low temperatures. This phase is absent in full QCD where, at low temperatures a phase transition to a phase with nonzero baryon density occurs at $\mu = m_N/3$. We conclude that quenching completely fails for $\mu \neq 0$.

4.3. *Lesson 3: Width of the Dirac Spectrum*

The phase quenched partition function undergoes a phase transition at $\mu = m_\pi/2$. In terms of Dirac eigenvalues the only critical point occurs when the quark mass hits the domain of eigenvalues (remember that it has a sharp edge). Therefore these two points have to coincide.[28] This leads to the following condition for the half-width m_c of the Dirac spectrum:

$$m_\pi^2 = \frac{2m_c\Sigma}{F^2} = 4\mu^2. \tag{7}$$

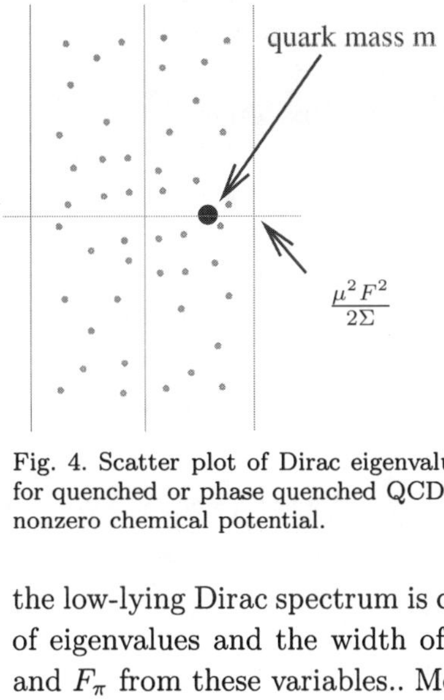

Fig. 4. Scatter plot of Dirac eigenvalues for quenched or phase quenched QCD at nonzero chemical potential.

This result has also been derived from the static limit of the chiral Lagrangian for phase quenched QCD[19] which contains two low-energy constants, Σ and F_π, as parameters. Since, at nonzero chemical potential the low-lying Dirac spectrum is characterized by two quantities: the density of eigenvalues and the width of the spectrum, it is possible to extract Σ and F_π from these variables.. More sophisticated methods to extract these constants from the low lying Dirac spectrum have been proposed and have been successfully applied to lattice QCD.[30–38] For sufficiently small chemical potential, when perturbation theory applies, the width of the Dirac spectrum increases linearly with μ rather than quadratically.

4.4. *Lesson 4: Infrared Dominance*

The question we wish to address in this section is whether the fluctuations of the fermion determinant given by

$$\det(D + m + \mu\gamma_0) = \prod_k (\lambda_k + m) \tag{8}$$

are dominated by the infrared part of the Dirac spectrum. At $\mu = 0$, because the low-energy limit of QCD is given by chiral perturbation theory for sufficiently small quark masses, this is apparently the case. This is less clear at $\mu \neq 0$ when the eigenvalues are complex. In particular, because the eigenvalue density for large λ increases as $\sim V\lambda^3$ one would expect that also the ultraviolet part of the Dirac spectrum contributes to the phase of the determinant. Although a nonzero chemical potential does not introduce new infinities once the theory is regularized at $\mu = 0$, one cannot exclude finite contributions from large Dirac eigenvalues.

When the chemical potential and the quark masses are in the microscopic domain of QCD, though, we can use random matrix theory to show that the average determinant is determined by the small Dirac eigenvalues. Further evidence for the infrared dominance of this quantity comes from lattice QCD at imaginary chemical potential, where a small number of low-lying eigenvalues reproduce the random matrix result.[39]

Because of agreement with lattice QCD we know that chiral perturbation theory can be applied to phase quenched QCD when both the chemical potential and m_π are of order of F_π or less. This shows that the magnitude of the quark determinant is infrared dominated in this domain. Since the full QCD partition function in this domain does not depend on the chemical potential at low temperatures, in agreement with chiral perturbation theory, we conclude that the μ-dependence of the the phase of the fermion determinant resides in the infrared part of the Dirac spectrum.

4.5. *Lesson 5: Failure of Banks-Casher at* $\mu \neq 0$

The Banks-Casher relation[40] states that

$$\Sigma = \lim_{m \to 0} \lim_{V \to \infty} \frac{\pi \rho(m)}{V}, \tag{9}$$

where $\rho(\lambda)$ is the density of Dirac eigenvalues and V is the volume of space-time. Although originally intended for a Hermitian Dirac operator, this relation correctly gives a vanishing chiral condensate for phase quenched QCD at $\mu \neq 0$. However, for full QCD at $\mu \neq 0$, the chiral condensate has a discontinuity when the quark mass crosses the imaginary axis, but it does so without the occurrence of an accumulation of eigenvalues. The alternative mechanism that is at work has been understood in detail in random matrix theory.[41,42] For a comprehensive review we refer to the talk by Splittorff also in this volume.[43] Below we will illustrate this mechanism for QCD in one dimension which can also be viewed as a random matrix model.

Fig. 5. Schematic plot of eigenvalues of the Dirac operator for lattice QCD in one dimension (left). The yellow dots denote the position of the eigenvalues for a single gauge field configuration, whereas the green ellipse shows the support of the spectrum in the thermodynamic limit. The chiral condensate for one flavor versus the quark mass is shown in the right figure.

The partition function of lattice QCD in one dimension is given by

$$Z = \int_{U \in U(N_c)} dU \det D, \tag{10}$$

where the integral is over the Haar-measure of $U(N_c)$. The Dirac operator is given by the $N \times N_c$ matrix (for N lattice points)

$$D = \begin{pmatrix} mI & e^{\mu} & \dots & e^{-\mu}U^{\dagger} \\ -e^{-\mu} & mI & \cdots & 0 \\ \vdots & & & \vdots \\ 0 & \cdots & mI & e^{\mu} \\ -e^{\mu}U/2 & \cdots & -e^{-\mu} & mI \end{pmatrix}. \tag{11}$$

The chiral condensate for one flavor is defined by

$$\Sigma(m) = \frac{\left\langle \frac{1}{N}\sum_k \frac{1}{\lambda_k+m}\prod_k(\lambda_k+m)\right\rangle}{\langle\prod_k(\lambda_k+m)\rangle}. \tag{12}$$

Notice that the determinant has a complex phase. For $U(1)$ the chiral condensate can be evaluated analytically[44] with the result that is shown in the right figure of Fig. 5. The amazing phenomenon, also known as the "Silver Blaze Problem",[45] is that the chiral condensate is continuous when m crosses the ellipse of eigenvalues, but shows a discontinuity at $m = 0$ where there are no eigenvalues. This can happen because the chiral condensate is determined by exponentially large (in the number of lattice points) contributions which cancel to give a finite result for the chiral condensate for $N \to \infty$.

4.6. *Lesson 6: Quark Mass and Average Phase Factor*

The severity of the sign problem can be measured through the expectation value of the average phase factor. A physical interpretation is obtained by defining the phase with respect to the phase quenched partition function:

$$\langle e^{2i\theta} \rangle_{\rm pq} = \frac{\langle {\det}^2(D + m + \mu\gamma_0) \rangle}{\langle |\det(D + m + \mu\gamma_0)|^2 \rangle} \equiv \frac{Z_{\rm QCD}}{Z_{|\rm QCD|}}.$$

At nonzero temperature the free energies of QCD and |QCD| are different so that the average phase factor vanishes in the thermodynamic limit.

At zero temperature, the free energy of QCD and phase quenched QCD are the same for $\mu < m_\pi/2$. Therefore, in the thermodynamic limit, the average phase factor is one for $\mu < m_\pi/2$. At finite volume, a nontrival μ-dependence is obtained from the one-loop corrections which give rise to an $O(1)$ contribution.

In the microscopic limit the average phase is given by the random matrix result that can be evaluated analytically (see Fig. 6 (right)). In the left part of the same figure we show a scatter plot of the Dirac eigenvalues of 40 200×200 random matrices. We conclude that, in the microscopic domain of QCD, the average phase factor vanishes if the quark mass is inside the support of the Dirac spectrum.

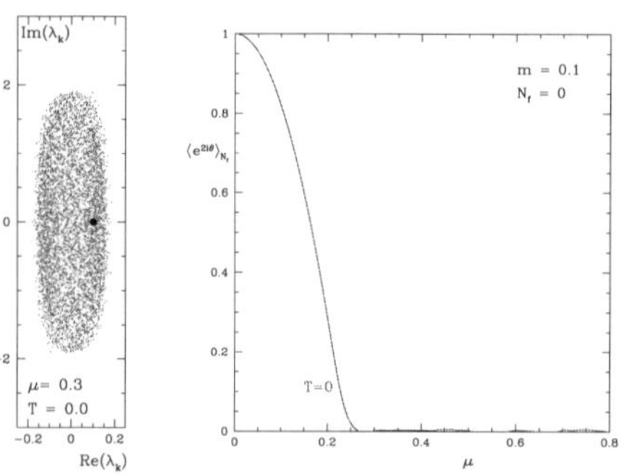

Fig. 6. Scatter plot of Dirac eigenvalues for chiral random matrix theory at nonzero chemical potential (left) and the average phase factor as a function of the chemical potential (right) for $m = 0.1$. In the left figure the mass is indicated by a black dot.

The average phase factor has been evaluated to one-loop order in chiral perturbation theory,[46] and the results are in good agreement with lattice QCD.[47] The mean field limit of this result agrees with random matrix theory in the limit where the microscopic variables are large. Not surprisingly, also in the domain of chiral perturbation theory, the sign problem is most severe when the quark mass is inside the domain of eigenvalues.

4.7. *Lesson 7: Distribution of Small Dirac Eigenvalues*

One of the greatest successes of chiral random matrix theory has been the exact description of the distribution of the low-lying eigenvalues of the Dirac operator[4,5,48] both at zero (see review[49]) and at nonzero chemical potential (see review[50]). In this section we show recent results for the Dirac spectrum at nonzero chemical potential and topology. Such lattice calculations have become possible because of the introduction of an overlap Dirac operator at nonzero chemical potential.[51] In Fig. 7, the radial distribution of the lowest Dirac eigenvalue is shown for different values of the topological charge in comparison with analytical results derived in.[52]

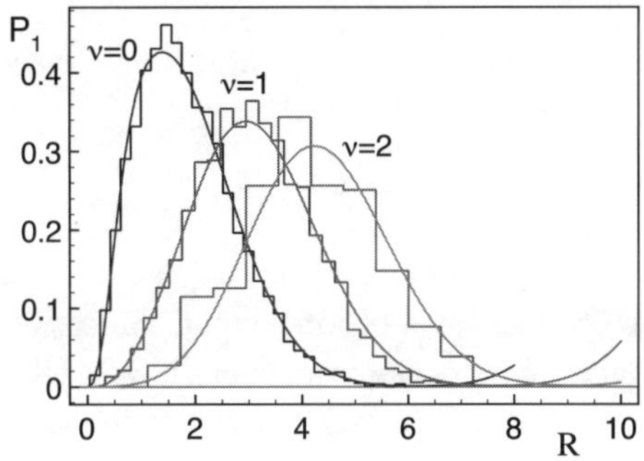

Fig. 7. Radial distribution of smallest Dirac eigenvalue for $\mu \neq 0$ in different topological charge sectors using a Dirac operator that satisfies the Ginsparg-Wilson relation (histograms)[52] compared to the prediction of chiral random matrix theory (smooth curve).

4.8. *Lesson 8: Equality Two Condensates*

The chiral condensate can be calculated in two ways

$$\Sigma^{(1)} = \lim_{m\to 0}\lim_{V\to\infty}\frac{1}{V}\left\langle\sum_k \frac{1}{\lambda_k + m}\right\rangle, \Sigma^{(2)} = \lim_{V\to\infty}\lim_{m\to 0}\frac{1}{V}\left\langle\sum_k \frac{1}{\lambda_k + m}\right\rangle, \tag{13}$$

where $\Sigma^{(1)}$ is nonzero because of spontaneous symmetry breaking, and its value does not depend on the total topological charge. The second chiral condensate can be expressed as

$$\Sigma^{(2)} = \lim_{V\to\infty}\frac{1}{V}\frac{\langle\prod_{\lambda_k\neq 0}\lambda_k\rangle_{\nu=1}}{\langle\prod_{\lambda_k}\lambda_k\rangle_{\nu=0}}. \tag{14}$$

The reason is that in the sector of topological charge ν, there are ν exact zero modes. The equality $\Sigma^{(1)} = \Sigma^{(2)}$ requires a subtle reshuffling of the eigenvalues: For $\nu = 1$ the eigenvalues are shifted by approximately half a level spacing w.r.t. $\nu = 0$ in order to satisfy the chiral Ward identity. Therefore, $\lambda_k^{\nu=1} \approx \sqrt{\lambda_k^{\nu=0}\lambda_{k+1}^{\nu=0}}$. To correctly normalize the ratio in (14) we evaluate it for a finite Dirac operator with an $N \times (N+\nu)$ nonzero off-diagonal block. Such Dirac matrix has ν exact zero eigenvalues (and perhaps additional paired zero modes which, being a set of measure zero, can be safely ignored). As it stands, the ratio (14) is dimensionally incorrect. A dimensionally correct ratio is obtained by replacing the largest squared eigenvalue pair for $\nu = 0$ by its square root, $[\lambda_N^{\nu=0}]^2 \to \lambda_N^{\nu=0}$. For the chiral condensate we then obtain the approximate expression

$$\Sigma^{(2)} = \frac{1}{V\lambda_{\min}}. \tag{15}$$

Using the Banks-Casher relation, the smallest nonzero eigenvalue is approximately given by

$$\lambda_{\min} \approx \frac{\pi}{2\Sigma V}, \tag{16}$$

resulting in $\Sigma^{(2)} \approx \frac{\pi}{2}\Sigma^{(1)}$ This calculation can be made rigorous (see[53]) with the result that both condensates become equal. Since we compare the ratio of two partition functions, it is essential that they are normalized correctly. This has been studied in a random matrix framework[53] confirming the above results.

Also within a random matrix framework, it turns out that for $\mu \neq 0$ the two condensate become only equal after the partition functions have been correctly normalized.[54]

4.9. *Lesson 9: Test of Algorithms*

Because random matrix models are exactly solvable and show all essential features due to nonhermiticity, they are an ideal tool to test algorithms for lattice QCD at nonzero chemical potential. We mention two examples: i) The density of states method was analyzed and tested in a chiral random matrix model at nonzero chemical potential[55] and has been applied successfully to lattice QCD.[56,57] ii) The radius of convergence has been determined[58] for algorithms that rely on Taylor expansion[47,59] or analytical continuation in μ.[60,61]

5. Conclusions

We have shown that the phase of the fermion determinant dramatically affects the physics of the QCD partition function. For example, the phase diagram of QCD and |QCD| are completely different. This is our main motivation for studying the behavior of the average phase factor and try to understand its relation with the spectrum of the Dirac operator. In particular, this has been done in the microscopic domain where QCD is given by a chiral random matrix theory that can be solved analytically. In this lecture we have discussed nine questions where random matrix theory has contributed significantly to our understanding of QCD at $\mu \neq 0$. Perhaps the most important contributions are the failure of the quenched approximation and the explanation of the mechanism that results in a discontinuity of the chiral condensate when the quark mass crosses the imaginary axis.

Acknowledgments: We wish to thank the organizer's of CAQCD 2008 for an inspiring week in Minneapolis as well as all of our colleagues who has contributed to the work reported here. This work was supported in part by U.S. DOE Grant No. DE-FAG-88ER40388. We thank the Institute for Nuclear Theory at the University of Washington for its hospitality and the Department of Energy for partial support during the completion of this work.

References

1. M. A. Stephanov, PoS **LAT2006**, 024 (2006) [arXiv:hep-lat/0701002].
2. Z. Fodor and S. D. Katz, JHEP **0404**, 050 (2004) [arXiv:hep-lat/0402006].
3. P. dE Forcrand and O. Philipsen, arXiv:0808.1096 [hep-lat].
4. J. J. M. Verbaarschot, Phys. Rev. Lett. **72**, 2531 (1994) [arXiv:hep-th/9401059].
5. E. V. Shuryak and J. J. M. Verbaarschot, Nucl. Phys. A **560**, 306 (1993) [arXiv:hep-th/9212088].
6. M. A. Stephanov, Phys. Rev. Lett. **76**, 4472 (1996) [arXiv:hep-lat/9604003].
7. A. M. Halasz, J. C. Osborn and J. J. M. Verbaarschot, Phys. Rev. D **56**, 7059 (1997) [arXiv:hep-lat/9704007].
8. J. C. Osborn, Phys. Rev. Lett. **93**, 222001 (2004), [arXiv:hep-th/0403131].
9. D. T. Son, Phys. Rev. D **59**, 094019 (1999) [arXiv:hep-ph/9812287].
10. P. dE Forcrand, M. A. Stephanov and U. Wenger, PoS **LAT2007**, 237 (2007) [arXiv:0711.0023 [hep-lat]].
11. A. S. Goldhaber and N. S. Manton, Phys. Lett. B **198**, 231 (1987).
12. L. Castillejo, P. S. J. Jones, A. D. Jackson, J. J. M. Verbaarschot and A. Jackson, Nucl. Phys. A **501**, 801 (1989).
13. A. D. Jackson and J. J. M. Verbaarschot, Nucl. Phys. A **484**, 419 (1988).
14. I. R. Klebanov, Nucl. Phys. B **262**, 133 (1985).
15. L. McLerran and R. D. Pisarski, Nucl. Phys. A **796**, 83 (2007) [arXiv:0706.2191 [hep-ph]].
16. Y. Hid aka, L. D. McLerran and R. D. Pisarski, Nucl. Phys. A **808**, 117 (2008) [arXiv:0803.0279 [hep-ph]].
17. M. G. Alford, A. Kapustin and F. Wilczek, Phys. Rev. D **59**, 054502 (1999) [arXiv:hep-lat/9807039].
18. J. B. Kogut, M. A. Stephanov, D. Toublan, J. J. M. Verbaarschot and A. Zhitnitsky, Nucl. Phys. B **582**, 477 (2000) [arXiv:hep-ph/0001171].
19. D. Toublan and J. J. M. Verbaarschot, Int. J. Mod. Phys. B **15**, 1404 (2001)

[hep-th/0001110].
20. D. T. Son and M. A. Stephanov, Phys. Rev. Lett. **86**, 592 (2001) [arXiv:hep-ph/0005225].
21. S. Hands, I. Montvay, S. Morrison, M. Oevers, L. Scorzato and J. Skullerud, Eur. Phys. J. C **17**, 285 (2000) [arXiv:hep-lat/0006018].
22. J. B. Kogut and D. K. Sinclair, Phys. Rev. D **77**, 114503 (2008) [arXiv:0712.2625 [hep-lat]].
23. J. B. Kogut and D. K. Sinclair, Phys. Rev. D **77**, 114503 (2008) [arXiv:0712.2625 [hep-lat]].
24. J. B. Kogut and D. K. Sinclair, Phys. Rev. D **66**, 014508 (2002) [arXiv:hep-lat/0201017].
25. K. Splittorff, D. Toublan and J. J. M. Verbaarschot, Nucl. Phys. B **639**, 524 (2002) [arXiv:hep-ph/0204076].
26. G. V. Dunne and S. M. Nishigaki, Nucl. Phys. B **670**, 307 (2003) [arXiv:hep-ph/0306220].
27. I. Barbour, N. E. Behilil, E. Dagotto, F. Karsch, A. Moreo, M. Stone and H. W. Wyld, Nucl. Phys. B **275**, 296 (1986).
28. P. E. Gibbs, Glasgow preprint 86-0389 (unpublished).
29. A. Gocksch, Phys. Rev. D **37**, 1014 (1988).
30. M. E. Berbenni-Bitsch, A. D. Jackson, S. Meyer, A. Schafer, J. J. M. Verbaarschot and T. Wettig, Nucl. Phys. Proc. Suppl. **63**, 820 (1998) [arXiv:hep-lat/9709102].
31. J. C. Osborn and T. Wettig, PoS **LAT2005**, 200 (2006) [arXiv:hep-lat/0510115].
32. G. Akemann, J. C. Osborn, K. Splittorff and J. J. M. Verbaarschot, Nucl. Phys. B **712**, 287 (2005) [arXiv:hep-th/0411030].
33. G. Akemann, P. H. Damgaard, J. C. Osborn and K. Splittorff, Nucl. Phys. B **766**, 34 (2007) [Erratum-ibid. B **800**, 406 (2008)] [arXiv:hep-th/0609059].
34. G. Akemann and P. H. Damgaard, JHEP **0803**, 073 (2008) [arXiv:0803.1171 [hep-th]].
35. G. Akemann and P. H. Damgaard, JHEP **0803**, 073 (2008) [arXiv:0803.1171 [hep-th]].
36. P. H. Damgaard, U. M. Heller, K. Splittorff, B. Svetitsky and D. Toublan, Phys. Rev. D **73**, 105016 (2006) [arXiv:hep-th/0604054].
37. P. H. Damgaard, U. M. Heller, K. Splittorff, B. Svetitsky and D. Toublan, Phys. Rev. D **73**, 074023 (2006) [arXiv:hep-lat/0602030].
38. P. H. Damgaard, U. M. Heller, K. Splittorff and B. Svetitsky, Phys. Rev. D **72**, 091501 (2005) [arXiv:hep-lat/0508029].
39. K. Splittorff and B. Svetitsky, Phys. Rev. D **75**, 114504 (2007) [arXiv:hep-lat/0703004].
40. T. Banks and A. Casher, Nucl. Phys. **B169**, 103 (1980).
41. J. C. Osborn, K. Splittorff and J. J. M. Verbaarschot, Phys. Rev. Lett. **94**, 202001 (2005), [arXiv:hep-th/0501210].
42. J. C. Osborn, K. Splittorff and J. J. M. Verbaarschot, arXiv:0805.1303 [hep-th].
43. J. C. Osborn, K. Splittorff and J. J. M. Verbaarschot, this volume,

arXiv:0808.1982 [hep-lat].
44. L. Ravagli and J. J. M. Verbaarschot, Phys. Rev. D **76**, 054506 (2007) [arXiv:0704.1111 [hep-th]].
45. T. D. Cohen, Phys. Rev. Lett. **91**, 222001 (2003) [arXiv:hep-ph/0307089].
46. K. Splittorff and J. J. M. Verbaarschot, Phys. Rev. D **77**, 014514 (2008) [arXiv:0709.2218 [hep-lat]].
47. C. R. Allton *et al.*, Phys. Rev. D **66**, 074507 (2002) [arXiv:hep-lat/0204010].
48. J.J.M. Verbaarschot and I. Zahed, Phys. Rev. Lett. **70**, 3852 (1993) [arXiv:hep-th/9303012].
49. J. J. M. Verbaarschot and T. Wettig, Ann. Rev. Nucl. Part. Sci. **50**, 343 (2000) [arXiv:hep-ph/0003017].
50. G. Akemann, Int. J. Mod. Phys. A **22**, 1077 (2007) [arXiv:hep-th/0701175].
51. J. Bloch and T. Wettig, Phys. Rev. Lett. **97**, 012003 (2006) [arXiv:hep-lat/0604020].
52. G. Akemann, J. Bloch, L. Shifrin and T. Wettig, Phys. Rev. Lett. **100**, 032002 (2008) [arXiv:0710.2865 [hep-lat]].
53. L. Shifrin and J. J. M. Verbaarschot, Phys. Rev. D **73**, 07400 [arXiv:hep-th/0507220].
54. C. Lehner, M. Otani, J.J.M. Verbaarschot and T. Wettig, in progress.
55. J. Ambjorn, K. N. Anagnostopoulos, J. Nishimura and J. J. M. Verbaarschot, JHEP **0210**, 062 (2002) [arXiv:hep-lat/0208025].
56. Z. Fodor, S. D. Katz and C. Schmidt, JHEP **0703**, 121 (2007) [arXiv:hep-lat/0701022].
57. S. Ejiri, arXiv:0804.3227 [hep-lat].
58. M. A. Stephanov, Phys. Rev. D **73**, 094508 (2006) [arXiv:hep-lat/0603014].
59. R. V. Gavai, S. Gupta and P. Majumdar, Phys. Rev. D **65**, 054506 (2002) [arXiv:hep-lat/0110032].
60. M. D'Elia and M. P. Lombardo, Phys. Rev. D **67**, 014505 (2003) [arXiv:hep-lat/0209146].
61. P. de Forcrand and O. Philipsen, Nucl. Phys. B **642**, 290 (2002) [arXiv:hep-lat/0205016].

SECTION 4.

METHODS AND MODELS OF HOLOGRAPHIC CORRESPONDENCE

SOFT-WALL DYNAMICS IN AdS/QCD

BRIAN BATELL*

School of Physics and Astronomy, University of Minnesota,
Minneapolis, MN 55455, USA
** E-mail: batell@physics.umn.edu*

Linear confinement in holographic QCD can be obtained with a soft-wall quadratic dilaton background. We present a solution to the five-dimensional Einstein's equations which dynamically realizes the soft-wall model. We discuss aspects of the solution and comment on the possible relation to a noncritical string theory.

1. Introduction

The anti-de Sitter/conformal field theory (AdS/CFT) correspondence[1–3] states that strongly coupled gauge theories can be equivalently described by higher-dimensional gravitational theories. One obvious application of this duality is to study Quantum Chromodynamics (QCD). AdS/QCD models represent a phenomenological twist to these studies, using known facts about QCD to gain insight into the string/gravity dual theory.[4,5]

In AdS/QCD models, one starts from a weakly coupled five-dimensional (5D) anti-de Sitter background, dual to conformal symmetry of the 4D field theory. Bulk fields in the 5D theory map to QCD operators through the AdS/CFT correspondence. Confinement can be modeled with a hard cutoff in the extra dimension, or with a dynamical cutoff.[6,7] These models do a fairly good job of reproducing the lowest lying meson spectra and decay constants. One of the main criticisms leveled at hard-wall models is in regards to the absence of linear Regge trajectories associated with higher radial or spin quantum numbers, expected from the experimental data[8] and semiclassical QCD arguments.[9]

The so-called *soft wall* AdS/QCD models correctly produce the Regge behavior.[10,11] In the model with a quadratic dilaton,[10] the mesons follow a mass spectrum $m^2_{n,S} \propto n + S$ for excitation number n and spin S (see also[12]).

Below we will present a dynamical soft-wall model with a pure AdS metric and quadratic in z "dilaton" Φ.[13] Certain aspects of the scalar potential will resemble a 5D non-critical string theory. Our solution provides a bottom-up holographic model of linear confinement and dynamical realization of the proposals in Refs.[10,11]

2. A dynamical soft-wall model

We start from the following 5D local action:

$$S = M^3 \int d^5x\sqrt{-g}\Big[e^{-2\Phi}\Big(R + 4g^{MN}\partial_M\Phi\partial_N\Phi - \frac{1}{2}g^{MN}\partial_M T\partial_N T - V(\Phi,T)\Big) + e^{-\Phi}\mathcal{L}_{meson}\Big], \quad (1)$$

where Φ is the dilaton, T is the tachyon, and M is the 5D Planck scale. We will search for a solution with a pure AdS metric $g_{MN} = z^{-2}\eta_{MN}$ and $\Phi(z) \sim z^2$.

The action (1) is defined in the string frame, but we will find it is easier to obtain solutions in the Einstein frame, which can be reached with the conformal transformation

$$g_{MN} = e^{4\Phi/3}\tilde{g}_{MN}. \quad (2)$$

The gravity-dilaton-tachyon action in this frame becomes

$$S = M^3 \int d^5x\sqrt{-\tilde{g}}\Big[\tilde{R} - \frac{1}{2}\tilde{g}^{MN}\partial_M\phi\partial_N\phi - \frac{1}{2}\tilde{g}^{MN}\partial_M T\partial_N T - \tilde{V}(\phi,T)\Big], \quad (3)$$

where the tilde distinguishes the two frames, $\tilde{V} = e^{4\Phi/3}V$, and we have canonically normalized the dilaton $\phi = \sqrt{8/3}\,\Phi$.

The so-called superpotential method[14,15] will be used to look for a solution. As an ansatz, we take

$$ds^2 = e^{-2A(y)}dx^2 + dy^2, \quad (4)$$

$$\phi = \phi(y), \quad (5)$$

$$T = T(y). \quad (6)$$

Note that the metric (4) is parametrized with the (x^μ, y) coordinates. The equations of motion become

$$3A'' - 6A'^2 = \frac{1}{4}\phi'^2 + \frac{1}{4}T'^2 + \frac{1}{2}\widetilde{V}, \tag{7}$$

$$6A'^2 = \frac{1}{4}\phi'^2 + \frac{1}{4}T'^2 - \frac{1}{2}\widetilde{V}, \tag{8}$$

$$\phi'' - 4A'\phi' = \frac{\partial \widetilde{V}}{\partial \phi}, \tag{9}$$

$$T'' - 4A'T' = \frac{\partial \widetilde{V}}{\partial T}, \tag{10}$$

where ($'$) denotes differentiation with respect to y. Next, we introduce a "superpotential" $W(\phi, T)$, leading to the first-order system of equations:

$$A' = W, \tag{11}$$

$$\phi' = 6\frac{\partial W}{\partial \phi}, \tag{12}$$

$$T' = 6\frac{\partial W}{\partial T}. \tag{13}$$

The scalar potential is then

$$\widetilde{V}(\phi, T) = 18\left[\left(\frac{\partial W}{\partial \phi}\right)^2 + \left(\frac{\partial W}{\partial T}\right)^2\right] - 12W^2. \tag{14}$$

We will specify the metric function $A(y)$ and work backwards to determine ϕ, T, W and $\widetilde{V}$ which support the metric solution. For a quadratic dilaton $\Phi \sim z^2$, the conformal transformation (2) indicates that the Einstein frame metric must have the form $\widetilde{g}_{MN} = z^{-2}e^{-2az^2}\eta_{MN}$, where a is an arbitrary constant. To use the superpotential method we transform to the y coordinates:

$$y = \int dz\, \frac{e^{-az^2}}{z} = \frac{1}{2}\mathrm{Ei}\left(-az^2\right), \tag{15}$$

where $\mathrm{Ei}(x)$ is the exponential integral function. We define an inverse function I through the relation

$$z^2 = -\frac{1}{a}\mathrm{Ei}^{(-1)}(2y) \equiv -\frac{1}{a}I(2y), \tag{16}$$

which obeys the differentiation rule:

$$\frac{dI}{dy} = \frac{d}{dy}(-az^2) = -2az^2e^{az^2} = 2Ie^{-I}. \tag{17}$$

Using (16) the function $A(y)$ becomes

$$A(y) = az^2 + \frac{1}{2}\log z^2 = -I + \frac{1}{2}\log\left(-\frac{I}{a}\right). \tag{18}$$

Differentiating this expression, using (11), (12) and (13), we find

$$W' = -6Ie^{-2I} + 4I^2e^{-2I} = \frac{\phi'^2}{6} + \frac{T'^2}{6}, \tag{19}$$

and can make the identification

$$\phi'^2 = 24I^2e^{-2I}, \tag{20}$$

$$T'^2 = -36Ie^{-2I}. \tag{21}$$

These equations can be integrated to obtain two possible solutions for ϕ and T:

$$\phi = \pm\sqrt{6}I = \mp\sqrt{6}az^2, \tag{22}$$

$$T = \pm 6\sqrt{-I} = \pm 6\sqrt{a}z, \tag{23}$$

where we have set the integration constants to zero. Note the dilaton ϕ has a quadratic profile as desired.

With the background solutions for the scalar fields, it is straightforward to find the superpotential:

$$W_\phi = -2\left(1 \pm \frac{\phi}{\sqrt{6}}\right)e^{\mp\phi/\sqrt{6}}, \tag{24}$$

$$W_T = 3e^{T^2/36}. \tag{25}$$

These expressions can be substituted into (14) to find the scalar potential

$$\begin{aligned}\widetilde{V}_\pm(\phi, T) = {} & \frac{T^2}{2}e^{T^2/18} + 2\phi^2e^{\mp 2\phi/\sqrt{6}} \\ & -12\left[3e^{T^2/36} - 2\left(1 \pm \frac{\phi}{\sqrt{6}}\right)e^{\mp\phi/\sqrt{6}}\right]^2.\end{aligned} \tag{26}$$

Going back to the string frame (1) and choosing $\phi(z) = \sqrt{8/3}\,z^2$ in (22) with the normalization $a = 2/3$ gives the solution

$$\Phi(z) = z^2, \tag{27}$$

$$T(z) = \pm 2\sqrt{6}z, \tag{28}$$

$$g_{MN} = \frac{\eta_{MN}}{z^2}, \tag{29}$$

$$V_-(\Phi, T) = e^{-4\Phi/3}\widetilde{V}_-(\phi, T). \tag{30}$$

The metric is pure AdS_5 and the dilaton is quadratic in z realizing the soft-wall model.[10]

3. Discussion

The action (1) is somewhat similar to the low energy effective action arising from a non-critical string theory, with T identified as the closed string tachyon and the meson action arising from a D-brane action. Closed string tachyon condensation was suggested as a possible way of screening the metric from the backreaction of the dilaton and maintaining AdS space.[7,10] A typical dilaton-tachyon potential arising from string theory is of the form $e^{c\phi}V(T)$, where c is a constant. Notice for the solution with $\Phi = z^2$ there are terms containing $e^{2\phi/\sqrt{6}} = e^{4\Phi/3}$ in the potential, matching what is found in a 5D non-critical string theory.[16] However, the terms linear and quadratic in ϕ in (26) are peculiar and perhaps not expected from string theory.

To see if our scalar fields map to QCD operators, we expand the potential to quadratic order in the fields:

$$\widetilde{V}_{\pm}(\phi, T) \simeq -12 - 2\phi^2 - \frac{3}{2}T^2 + \cdots . \tag{31}$$

As $z \to 0$ the metric is asymptotically AdS, and from the background solutions (22),(23) all higher-order terms in the potential can be neglected. The operator dual to ϕ has dimension $\Delta_\phi = 2 + \sqrt{4 + m_\phi^2} = 2$, and does not correspond to any local, gauge-invariant operator in 4D QCD. Some discussion in the literature has been given to the possible relevance of a dimension two condensate in the form of a gluon mass term,[17] though it is not obvious that we are to associate ϕ with this operator as the AdS/CFT correspondence dictates that bulk fields are dual to gauge-invariant local operators.

To obtain asymptotically linear trajectories, we only require $\phi \sim z^2$ in the IR (large z). At small z we should expect the z-dependence to change in a more complete description. For example, the running gauge coupling should be incorporated, which requires $\phi \sim -\log\log(-z)$ for small z.[7,18]

The operator dual to the tachyon T has dimension $\Delta_T = 2 + \sqrt{4 + m_T^2} = 3$, tempting us to associate this operator with the $\overline{q}q$ operator. Though perhaps not well motivated from a string perspective, the linear in z behavior (23) is precisely that which is required for the constant mass-squared splittings between chiral partners along Regge trajectories[19,20] and implies the chiral symmetry is not restored in the highly excited states.[21]

It still remains to check the stability of the model by examining fluctuations of the background associated with the scalar sector. The general perturbation analysis for the case of one bulk scalar field has been performed,[22] and we might expect based on these analyses that neither tachyonic nor massless scalar modes exist in our two-field model.

The five-dimensional solution can be generalized to an arbitrary spacetime dimension and with an Einstein frame metric $\tilde{g}_{MN} = z^{-2}\, e^{-2az^{\nu}}\, \eta_{MN}$, where ν is a real constant. In particular it is interesting to analyze the case of 2D QCD (i.e. $\nu = d = 2$), or the 't Hooft model. In the limit of large number of colors, the 't Hooft model is soluble and the mesons follow asymptotically linear trajectories.[23] In our 3D holographic model the bulk field ϕ corresponds to the dimension two gluon field strength squared G^2 while T is dual to the dimension one $\overline{q}q$ operator. The connection between holography and 2D QCD has recently been explored.[24]

We have discussed a dynamical soft-wall AdS/QCD model that leads to Regge trajectories for the high spin mesons, realizing the proposals of Refs.[10,11] The model contains scalars dual to operators with simple integer dimensions. It is not clear how our particular solution might be connected to string theory, but it is encouraging that such a confining soft-wall background exists.

Acknowledgments

We thank Andreas Karch, Misha Shifman, Arkady Vainshtein, and Valya Zakharov for helpful discussions. This work was supported in part by a Department of Energy grant DE-FG02-94ER40823 at the University of Minnesota, and an award from Research Corporation.

References

1. J. M. Maldacena, Adv. Theor. Math. Phys. **2** 231 (1998) [Int. J. Theor. Phys. **38** 1113 (1999)] [arXiv:hep-th/9711200].
2. S. S. Gubser, I. R. Klebanov and A. M. Polyakov, Phys. Lett. B **428** 105 (1998) [arXiv:hep-th/9802109].
3. E. Witten, Adv. Theor. Math. Phys. **2** 253 (1998) [arXiv:hep-th/9802150].
4. J. Erlich, E. Katz, D. T. Son, M. A. Stephanov, Phys. Rev. Lett. **95** 261602 (2005) [arXiv:hep-ph/0501128].
5. L. Da Rold, A. Pomarol, Nucl. Phys. B **721** 79 (2005), [arXiv:hep-ph/0501218].
6. K. Ghoroku, N. Maru, M. Tachibana and M. Yahiro, Phys. Lett. B **633**, 602 (2006) [arXiv:hep-ph/0510334].
7. C. Csaki and M. Reece, JHEP **0705** 062 (2007) [arXiv:hep-ph/0608266].
8. S. Eidelman *et al.* [Particle Data Group] , Phys. Lett. B **592**, 1 (2004).
9. M. Shifman, arXiv:hep-ph/0507246.
10. A. Karch, E. Katz, D. T. Son, M. A. Stephanov, Phys. Rev. D **74**, 015005 (2006) [arXiv:hep-ph/0602229].

11. O. Andreev, Phys. Rev. D **73**, 107901 (2006) [arXiv:hep-ph/0603170].
12. R. R. Metsaev, arXiv:hep-th/0002008.
13. B. Batell and T. Gherghetta, Phys. Rev. D **78**, 026002 (2008) [arXiv:0801.4383 [hep-ph]].
14. K. Skenderis and P. K. Townsend, Phys. Lett. B **468**, 46 (1999) [arXiv:hep-th/9909070].
15. O. DeWolfe, D. Z. Freedman, S. S. Gubser, A. Karch, Phys. Rev. D **62**, 046008 (2000) [arXiv:hep-th/9909134].
16. J. Polchinski, "String theory. Vol. 1: An introduction to the bosonic string," *Cambridge, UK: Univ. Pr. (1998) 402 p.*
17. F. V. Gubarev, L. Stodolsky, V. I. Zakharov, Phys. Rev. Lett. **86** 2220 (2001) [arXiv:hep-ph/0010057]; F. V. Gubarev, and V. I. Zakharov, Phys. Lett. B **501** 28 (2001) [arXiv:hep-ph/0010096].
18. U. Gursoy and E. Kiritsis, JHEP **0802**, 032 (2008) [arXiv:0707.1324 [hep-th]]. U. Gursoy, E. Kiritsis and F. Nitti, JHEP **0802**, 019 (2008) [arXiv:0707.1349 [hep-th]].
19. R. Casero, E. Kiritsis and A. Paredes, Nucl. Phys. B **787**, 98 (2007) [arXiv:hep-th/0702155].
20. O. Bergman, S. Seki and J. Sonnenschein, JHEP **0712**, 037 (2007) [arXiv:0708.2839 [hep-th]].
21. R. L. Jaffe, D. Pirjol and A. Scardicchio, Phys. Rept. **435**, 157 (2006) [arXiv:hep-ph/0602010]; O. Cata, M. Golterman and S. Peris, Phys. Rev. D **74**, 016001 (2006) [arXiv:hep-ph/0602194]; M. Shifman and A. Vainshtein, Phys. Rev. D **77**, 034002 (2008) [arXiv:0710.0863 [hep-ph]].
22. L. Kofman, J. Martin, and M. Peloso, Phys. Rev. D **70**, 085015 (2004) [arXiv:hep-ph/041189]; E. Kiritsis, and F. Nitti, Nucl. Phys. B **772**, 67 (2007) [arXiv:hep-th/0611344].
23. G. 't Hooft, Nucl. Phys. B **75**, 461 (1974).
24. E. Katz and T. Okui, arXiv:0710.3402 [hep-th].

Holographic QCD

Nick Evans & Ed Threlfall

School of Physics and Astronomy, University of Southampton, Southampton, SO17 1BJ, UK

** E-mail: evans@phys.soton.ac.uk, ejt@phys.soton.ac.uk*

A simple holographic model, derived from the AdS/CFT Correspondence, is described that displays many features of QCD's dynamics. The geometry has a non-trivial dilaton flow that breaks supersymmetry but preserves the SO(6) R-symmetry of $\mathcal{N} = 4$ Yang Mills. We speculate that this flow describes the super-Yang Mills theory with a scalar mass term (which is highly irrelevant and would enter as a sharp UV cut off). The geometry describes confinement and chiral symmetry breaking at low temperatures and a first order phase transition to a deconfined plasma with melted mesons at a finite critical temperature.

Keywords: QCD; AdS/CFT Correspondence; holography

1. Introduction

The AdS/CFT Correspondence[1–3] has provided a new tool for the study of strongly coupled gauge theories through weakly coupled supergravity/string duals. The initial correspondence was for a highly supersymmetric and conformal gauge theory which did not confine nor dynamically generate a scale like Λ_{QCD}. Progress has been made in deforming the AdS space in a way to break both supersymmetry and conformal symmetry. Quark fields can also be introduced - the original gauge theory only has matter in the adjoint representation. Here we will introduce a simple example[4] of such a gravity dual and show how it encodes confinement and chiral symmetry breaking. We will also show that the model has a finite temperature phase transition to a quark-gluon plasma in which bound states melt. These models are not directly QCD but gauge theories with related behaviour from which we hope to learn about generic features of strong dynamics.

2. The AdS/CFT Correspondence

The correspondence[1–3] emerged from type IIB string theory - the massless open string states with ends on N D3 brane describe the 3+1 dimensional U(N) $\mathcal{N}=4$ super-Yang Mills theory. In the infinite tension limit other string states decouple.

The closed string sector of the same string theory is described by the supergravity background to the stack of D3 branes. In the limit of a large number, N, of D3s the curvature becomes weak and is well described by classical supergravity. In the decoupling limit this geometry becomes $\mathrm{AdS}_5 \times S^5$

$$ds^2 = \frac{u^2}{L^2} dx_{3+1}^2 + \frac{L^2}{u^2} du^2 + L^2 d\Omega_5^2 \tag{1}$$

with L the AdS radius. The correspondence states that the gauge theory and the supergravity/string modes in this geometry are two equivalent descriptions of the same physics. The dilaton, a scalar in the supergravity theory which is dual to the gauge coupling, is constant, reflecting the conformal nature of the $\mathcal{N}=4$ theory.

A useful caricature of the correspondence is as follows: if one takes a slice though AdS at a particular value of the radial direction, $\tilde{u}$, then the resulting 3+1 dimensional space (parallel to the D3) is where the gauge theory lives. The five sphere provides a global SO(6) symmetry that acts on the matter fields of the model as an R-symmetry. Now, we can see from the metric that if we rescale spatial lengths then we must rescale u as an inverse length for the metric to remain invariant. The radial direction is therefore an energy scale in the gauge theory. By choosing different values of $\tilde{u}$ one is looking at the gauge theory at different energy scales.

The supergravity fields live throughout the AdS space and can have non-trivial u-dependence. These fields must couple to the gauge theory on each $\tilde{u}$ slice in a symmetry-preserving fashion. In other words the classical value of these fields enters the field theory as some coupling or other source (they can also describe the vevs of operators). The u-dependence tells us the renormalization group flow of these quantities - the magic is that the supergravity equations of motion are classical yet yield the non-perturbative quantum theory's renormalization dynamics.

3. Dilaton Flow Geometries

We now turn to constructing an example of a solution of the supergravity equations of motion which breaks supersymmetry and conformal invariance. We will allow the dilaton, a scalar in the supergravity theory, to deviate from it's constant value in pure AdS4−.[12] We will work in $\mathcal{N} = 8$ supergravity in five dimensions which is a truncation of IIB string theory on $AdS_5 \times S^5$ and it is known that any solution can be lifted to a complete ten dimensional geometry. We will neglect the 40 scalars which participate in the superpotential - these fields are R-charged and switching them on would break the SO(6) symmetry or "crush" the 5-sphere. The only motivation for preserving the SO(6) is to do something simple!

The effective five-dimensional action is (we use the normalization conventions of Gubser[6])

$$S = \frac{1}{4\pi G_5} \int d^5x\sqrt{-g}\left(\frac{1}{4}R - \frac{1}{8}g^{ab}\nabla_a\phi\nabla_b\phi + 3\right) \tag{2}$$

We will make an ansatz of the form

$$ds_5^2 = e^{2A}\left(-e^{2B}dt^2 + dx_3^3\right) + dr^2 \tag{3}$$

The presence of A allows the dilaton to have a non-trivial r dependence to which the geometry will backreact and that of B allows for non-zero temperature which we will discuss later.

The field equations are

$$\frac{1}{4}R_{ab} = \frac{1}{8}\partial_a\phi\partial_b\phi - g_{ab}, \qquad \nabla^2\phi = 0 \tag{4}$$

Using the linear combinations $\bar{A} \equiv A + \frac{1}{4}B$ and $\bar{B} \equiv \frac{\sqrt{3}}{4}B$ as in? these equations take the form

$$\phi'' + 4\bar{A}'\phi' = 0 \tag{5}$$

$$\bar{B}'' + 4\bar{A}'\bar{B}' = 0 \tag{6}$$

$$6(\bar{A}')^2 = \frac{1}{4}(\phi')^2 + 2(\bar{B}')^2 + 6 \tag{7}$$

These equations are analytically solvable with solutions

$$e^{4\bar{A}} = \frac{c_4^2e^{8r} - c_3^2}{2c_4e^{4r}} \tag{8}$$

$$\bar{B} = \frac{c_2}{4c_3}\ln\left(\frac{c_4e^{4r} - c_3}{c_4e^{4r} + c_3}\right) + B_0 \tag{9}$$

$$\phi = \frac{c_1}{4c_3}\ln\left(\frac{c_4e^{4r} - c_3}{c_4e^{4r} + c_3}\right) + \phi_0 \tag{10}$$

where the c_i are integration constants. For any solution that returns to AdS asymptotically $B_0 = 0$ and ϕ_0 is the dilaton value in the AdS limit.

3.1. *Solution with no event horizon*

Let us first take the solution above and set $B \equiv 0$ to find a zero temperature dilaton flow with manifest 4D Lorentz invariance. The solution can be recast (by setting $c_3 = c_4\zeta$) in the form

$$e^{2A} = \sqrt{\frac{c_4}{2}}\,\sqrt{e^{4r} - \zeta^2 e^{-4r}}, \qquad \phi = \sqrt{\frac{3}{2}}\ln\left(\frac{e^{4r}-\zeta}{e^{4r}+\zeta}\right) + \phi_0 \quad (11)$$

To match to other results in the literature[12] one can rescale the x_4 coordinates to effectively set $c_4 = 1/2$, set $2u^2 = e^{2r}$ and $\zeta = -4u_0^4$. Reinstating the five sphere and moving to string frame one arrives at the 10D metric

$$ds^2 = c^{\phi/2}\left(\frac{u^2}{L^2}\mathcal{A}^2(u)dx_4^2 + \frac{L^2}{u^2}du^2 + L^2 d\Omega_5^2\right) \quad (12)$$

with

$$\mathcal{A}(u) = \left(1 - \left(\frac{u_0}{u}\right)^8\right)^{\frac{1}{4}}, \qquad e^{\phi} = \left(\frac{(u/u_0)^4+1}{(u/u_0)^4-1}\right)^{\sqrt{3/2}} \quad (13)$$

The four form is just that in pure AdS. This metric clearly becomes $AdS_5 \times S^5$ at large u and has a deformation parameter u_0^4 which has dimension four and no R-charge - this parameter is naturally identified with TrF^2 in the gauge theory. Since TrF^2 is the F-term of a chiral superfield supersymmetry is therefore broken in this gauge theory.

A crucial aspect of the geometry is that it is singular at $u = u_0$ with the dilaton blowing up. A singularity should be a source of unease and we do not have a full explanation of it but we wish to argue there are number of ideas that suggest such geometries are worthy of study none the less. The presence of D3 branes in the geometry provides sources that, in some non-supersymmetric configuration, might complete the geometry. The N=2* geometry[13]is also singular at a point where the effective gauge coupling diverges - this geometry has been matched to the expected field theory solution at a particular point on moduli space.[14] That model provides evidence that a divergent gauge coupling can show up as a divergence in the geometry.

Our real motivation for the continued study here though is the phenomenological successes of the geometry. It has been shown to be confining in[6,10] and to break chiral symmetries when quarks are introduced[12] as we will discuss below. In this sense we can think of it as a back reacted hard wall (a sharp cut off on AdS has long been used as a primitive method for including confinement) with the correct properties to describe QCD-like physics.

3.2. *Black hole geometry*

The high temperature phase of the dilaton flow geometry just explored should correspond to a geometry with a black hole horizon. Black holes have the correct thermodynamic properties to describe a finite temperature bath.

To have a solution with a horizon we will choose constants such that the function B goes as a constant plus $\ln r$ near $r = 0$ ($\bar{B} \sim \sqrt{3}/4 \ln r$). From (9) this gives the constraints $c_2 = \sqrt{3} c_3$ and $c_4 = c_3$. At this point we note that, with the definition of c_3 ($6c_3^2 \equiv (\frac{1}{4}c_1^2 + 2c_2^2)$), c_1 vanishes so the dilaton profile in the non-extremal solution is just a constant.

We have learnt that there is no black hole solution with a radially dependent dilaton. This means there is not a gravity dual of a high temperature theory with TrF^2 switched on. The vev for TrF^2 is not the fundamental perturbation but an operator induced by the dynamics. We will discuss what the true perturbation might be in the next section.

We are left with a unique black hole solution which in the unbarred quantities is thus

$$e^{2A} = \frac{c_4}{2}\left(e^{2r} + e^{-2r}\right), \qquad B = \ln\left(\frac{e^{4r}-1}{e^{4r}+1}\right), \qquad \phi = \phi_0 \quad (14)$$

In fact by rescaling x_4 to set $c_4 = 2u_h^2$ and defining $e^{2r} = (u^2 + \sqrt{u^4 - u_h^4})/u_h^2$ this solution can be reduced to the usual Poincare coordinate form of five-dimensional AdS-Schwarzschild with Hawking temperature $T_H = \frac{u_h}{\pi}$.

3.3. *Interpretation*

We now turn to the question of the origin of supersymmetry breaking[4] in the dilaton flow geometry if TrF^2 is an induced operator as it appears it must be from the above analysis.

Since supersymmetry is broken, yet $SO(6)_R$ preserved, in the T=0 geometry, we expect all SO(6) invariant operators to be present. Amongst these SO(6) invariant operators is an equal mass for each of the six scalar fields - one would expect the scalar masses to rise to the scale of the supersymmetry breaking. Such an SO(6) invariant mass is a tricky beast though! It is associated with an SO(6) invariant vev for the scalars of the $\mathcal{N} = 4$

theory - that theory is understood as the geometry where the D3 branes are spread evenly over the surface of a 5-sphere. The supergravity solution around the 5-sphere is just AdS_5 by a simple Gauss' law argument - so the vev only shows up as a hard cut off at the D3-sphere surface. These operators are therefore invisible in the supergravity solution.

A frequently argued interpretation for the fact that this source and vev are not described by a supergravity mode is that they are super-irrelevant/relevant operators. The vev for the scalar operator would then be super-relevant; that is it would have no impact on the UV of the theory until suddenly at some point in the IR it would dominate it - this can certainly be matched to it's appearance as a sharp IR cut off on the geometry. In this language one would expect the mass term to show up as a sharp cut off on the space at some large radius or UV scale. Below that scale it would naively have no impact on the dynamics. This is not quite true though because it would define the symmetries of the theory below that UV cut off and, in particular, leave a non-supersymmetric but SO(6) invariant flow at lower energies. The dilaton flow geometry is the natural candidate for this flow. In this interpretation one should cut off the dilaton flow at some point in the UV, although this point could presumably be set at an arbitrarily high scale. The point where the cut off appears would be undetectable in the low energy flow.

4. QCD-like Dynamics

Let us now review the dynamics of the gauge theory described by the dilaton flow geometry.

4.1. *Glueballs at T=0*

The simplest sign of confinement is to look for a discrete glueball spectrum. Glueballs correspond to small fluctuations of TrF^2 about the vacuum value of the form $f(u)e^{ik\cdot x_4}$. The linearized equation for the dilaton has regular solutions for $f(u)$ only for discrete values of $k^2 = -M^2$ which pick out the glueball states.[6] Table I shows the masses of the lowest five scalar glueball states, in units of u_0/L^2.

4.2. *D7 Branes and Quarks at T=0*

The $N = 4$ gauge theory only has adjoint matter fields - the original construction realized the gauge theory through open string modes with both

n	1	2	3	4	5
M_n	4.1	7.2	10.2	13.2	16.2

Table I: Lowest five glueball masses in the zero temperature dilaton flow geometry in units of the deformation scale u_0/L^2.

ends tied to a D3 brane (they transform as $N_c, \bar{N}_c$). To generate fundamental representation quarks one must detach one of the string's ends from the D3 - it is useful to tie it to a D7 brane[15–19] as shown in Figure 1.

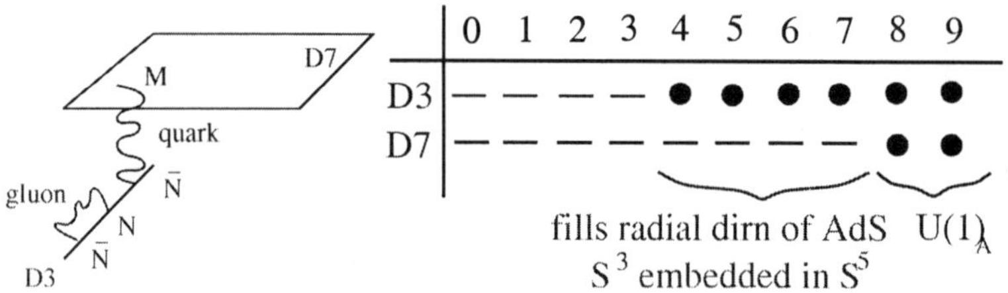

Fig. 1. The D3 D7 configuration that introduces quarks into the AdS/CFT Correspondence.

The D3 and the D7 share the 0-3 directions, the D7 are in addition extended in the 4-7 directions (we will call the radial coordinate in this space ρ), and finally the D3 and D7 can be separated in the 8-9 directions (w_5 and w_6 below). This configuration preserves $N = 2$ supersymmetry and in its simplest form corresponds to the $N = 4$ gauge theory with a fundamental representation quark hypermultiplet.

The minimum length D7-D3 string indicates (length $\times$ tension) the mass of the quark. If the D7 brane lies along the ρ axis then the quarks are massless and there is an SO(2) symmetry in the $w_5 - w_6$ plane. If the D7 lies off axis there is a non-zero quark mass and the SO(2) symmetry is explicitly broken. This indicates that the SO(2) symmetry is a geometric realization of the U(1) axial symmetry of the gauge theory (in the supersymmetric case that symmetry is part of a $U(1)_R$ symmetry). Note that at large N_c we neglect anomalies.

Using these techniques we can next include quarks into the dilaton deformed geometry above.[12] We will work in the approximation where the D7 brane is a probe (so there is no backreaction on the geometry[17]) - this is the quenched limit where the number of flavours $N_f \ll N_c$. One simply embeds the D7 brane so as to minimize its world volume via it's Dirac Born

Infeld action

$$S_{D7} = -T_7 \int d^8\xi \sqrt{P[G_{ab}]}, \qquad P[G_{ab}] = G_{MN} \frac{dx^M}{d\xi^a} \frac{dx^N}{d\xi^b} \tag{15}$$

where T_7 is the tension, ξ the coordinates on the D7, x^M are the spacetime coordinates and G_{MN} the metric.

In the dilaton flow geometry one finds that the D7 brane is repelled by the singular core of the geometry and the regular embeddings of interest are those shown in Figure 2.

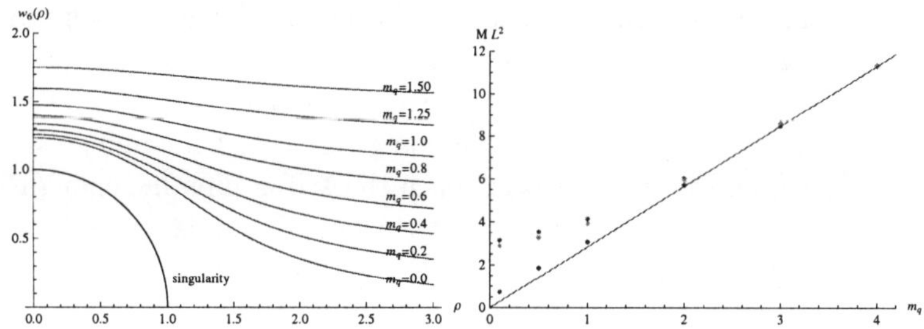

Fig. 2. Embedding solutions for a D7 probe in the dilaton flow geometry and a plot of the meson masses vs quark mass in that model.

At large ρ the solutions become flat as the gauge theory returns to AdS. The solution is of the form $w_6 = m + c/\rho^2 + ..$ where m corresponds to the quark mass and c to the $\bar{q}q$ condensate - we can read off the condensate as a function of the quark mass in this theory. A more intuitive understanding of the embedding results from interpreting the separation of the D7 brane from the ρ axis as the effective quark mass. As one moves in ρ one is changing RG scale - at large ρ one sees a small bare quark mass but in the IR (small ρ) a dynamical mass is generated.

In particular we can see that the solution exhibits chiral symmetry breaking. If we try to lie a D7 along the ρ axis, so $m = 0$, it is repelled from the origin and there is a non-zero value of the quark condensate. In fact the D7 may be deflected to any point on a circle in the $w_5 - w_6$ plane. We explicitly see the breaking of the SO(2) symmetry in that plane and the circle is the vacuum manifold. There should be a Goldstone boson associated with fluctuations of the D7 along the vacuum manifold. One can seek solutions to the equations of motion from the DBI action for those angular fluctuations of the form $\theta(\rho, x) = f(\rho)e^{-ikx}, \quad k^2 = -M^2$. Only for particular values of M is $f(\rho)$ regular and hence the mesonic $\bar{q}q$ bound state masses are picked out. In figure 3 the meson masses as a function of quark mass are shown. There is a massless Goldstone at $m = 0$ and it's mass grows as $\sqrt{m}$ as chiral perturbation theory predicts. The mass of the meson associated with radial fluctuations and the ρ meson state are also shown - they always have a mass gap.

Finally we can make another test of confinement using the D7 brane probe. If we place a D7 brane far from the singularity at u_0 it describes very heavy quarks (the embedding is one of those towards the top of figure 2 which is just flat). A quark anti-quark state is described by a string with both ends on the D7 brane but dipping down into centre of the space in the middle. The embedding has been performed in[6,10] and the core of the geometry is again repulsive. In the asymptotic AdS part of the space the string dips to smaller u the further the ends are separated (we can call the quark anti-quark separation L) - the action gives a $1/L$ Coulomb behaviour for the interaction energy. However, once the string dips far enough into the space to encounter the singularity it is repelled and lies parallel to the singularity a little away from u_0 - the quark anti-quark potential becomes linear with their separation, the classic sign of confinement.

4.3. *Thermodynamic computation*

We can compute the scale of the thermal phase transition[4] in the dilaton flow theory by comparing the action of the T=0 geometry, but now with a compact Euclidean time direction, and the action of the Schwarzschild black hole. We will see there is a first order phase transition between the two geometries.

To make the comparison fair we must set the parameter c_4 equal in the two geometries so they have the same large-r AdS limit. We will perform the calculation in the Schwarzschild-type coordinates, rescaling the Euclidean time coordinate to keep the temperature of the two geoemtries equal at some UV cut off Λ. Both geometries asymptote to AdS_5 so we can set the same cut-off in both cases, before taking the limit $\Lambda \to \infty$.

The Euclidean action density per unit spatial volume for the black hole solution is

$$S_{BH} = -\frac{1}{4\pi G_5}\int_0^{\frac{\pi L^2}{u_h}} d\tau \int_{u_h}^{\Lambda} \sqrt{-g}\left(\frac{1}{4}R + \frac{3}{L^2}\right) dr \tag{16}$$

The trace of the Einstein equation gives $R = -\frac{20}{L^2}$ so

$$S_{BH} = \frac{1}{2G_5 u_h L^3}\int_{u_h}^{\Lambda} u^3 du = \frac{1}{8G_5 u_h L^3}\left(\Lambda^4 - u_h^4\right) \tag{17}$$

The Euclidean action density per unit spatial volume for the dilaton flow solution is, having used the trace of the equation of motion to remove the scalar gradient term and allowing for the rescaling of Euclidean time, simply

$$S_{DF} = \frac{1}{2G_5L^2}\int_0^{\frac{\pi L^2}{u_h}\sqrt{1-\frac{u_h^4}{\Lambda^4}}} d\tau \int_{u_0}^{\Lambda} \sqrt{-g}\, dr \tag{18}$$

This is

$$\begin{aligned} S_{DF} &= \frac{1}{2G_5u_hL^3}\sqrt{1-\frac{u_h^4}{\Lambda^4}}\int_{u_0}^{\Lambda}\left(u^3-\frac{u_0^8}{u^5}\right)dr \\ &= \frac{1}{8G_5u_hL^3}\sqrt{1-\frac{u_h^4}{\Lambda^4}}\left(\Lambda^4+\frac{u_0^8}{\Lambda^4}-2u_0^4\right) \end{aligned} \tag{19}$$

Hence, in the $\Lambda \to \infty$ limit, the difference in the actions is simply

$$S_{BH} - S_{DF} = \frac{1}{16G_5u_hL^3}\left(4u_0^4 - u_h^4\right) \tag{20}$$

For a deformation scale $u_0 > \frac{u_h}{\sqrt{2}}$ the dilaton flow solution is thermodynamically favoured whereas for a deformation scale $u_0 < \frac{u_h}{\sqrt{2}}$ the AdS-Schwarzschild solution is favoured. The transition temperature is clearly $T_c = \frac{\sqrt{2}u_0}{\pi L^2}$.

We can perform an estimate of the deconfinement temperature in our model. The mass of the lowest-lying vector state for zero quark mass can be compared to the mass of the ρ meson, experimentally 776MeV. The vector mass is $2.80u_0/L^2$ and the deconfinement temperature is $T_c = \frac{\sqrt{2}u_0}{\pi L^2}$. This gives an estimate for the deconfinement temperature of $T_c \sim 124$ MeV.

4.4. *The Deconfined Phase*

Above the critical temperature, the theory is in a deconfined phase. There is no longer a spectrum of glueball normal modes, rather the gravity dual admits a spectrum of unstable quasinormal modes which was calculated in.[22] The field theory interpretation of the quasinormal spectrum is to give the mass and decay width for a glueball excitation embedded in a thermal bath of SYM plasma. The finite decay timescale can be viewed as the timescale for the 'melting' of the glueball state. The breaking of Lorentz symmetry means there is a nontrivial dispersion relation $\omega(k)$ for a scalar glueball excitation.[22] Table II shows the lowest five quasinormal frequencies, which are measured in units of $\frac{u_h}{L^2}$ which is $\propto T$ - the natural scale of the lowest quasinormal frequency is the temperature.

n	ω_n
1	$\pm$ 2.1988 - 1.7595 i
2	$\pm$ 4.2119 - 3.7749 i
3	$\pm$ 6.2155 - 5.7773 i
4	$\pm$ 8.2172 - 7.7781 i
5	$\pm$ 10.2181- 9.7785 i

Table 2: the scalar mesonic quasinormal frequencies in the high T phase ($m_q = 0$) - in units of $\frac{u_h}{L^2}$.

Flavour degrees of freedom can be included by embedding D7 branes as discussed above. Above the critical temperature the physically relevant D7 embeddings in the black hole geometry[19] give $\langle \bar{q}q \rangle = 0$ for $m_q = 0$ - there is no chiral symmetry breaking and hence no pion-like meson. The D7 can either end on the black hole horizon (small m_q) or for large enough m_q it has sufficient tension to support itself away from the black hole - there is a first-order phase transition in the behaviour of (quenched) quark matter as one raises the quark mass in the plasma background. In the former case there are quasinormal spectra representing the melting of scalar and vector mesonic excitations in the hot background.[20] In the latter case there are discrete spectra of scalar and vector meson masses with scale set by m_q which tend to the values in[18] as $m_q \gg T$. Our transition behaves in the same way as long as the quarks are sufficiently light ($m_q < 0.92\frac{\sqrt{\lambda}T}{2}$).[20] In the undeformed theory this bound implies that the mesons melt once the temperature of the background becomes of order the meson mass since the meson masses are $\sim \frac{m_q}{\sqrt{\lambda}}$. In our case the pion-like meson is an exception to this - one can have a massless pion that does not 'melt' until the background reaches some finite temperature.

There are recent results concerning a lower bound for the ratio of viscosity to entropy density of a strongly-coupled field theory,[21] $\frac{\eta}{s} \geq \frac{1}{4\pi}$, where the equality is for the deconfined phase of $\mathcal{N} = 4$ SYM theory. Our findings show that strict equality also applies to this non-supersymmetric theory in its deconfined phase.

5. Conclusion

Holographic techniques are now beginning to shed light on the dynamics of non-conformal and non-supersymmetric gauge theories as we have seen. New descriptions of confinement, chiral symmetry breaking, meson masses and finite temperature transitions have been found. Whether these techniques can ever be turned into a precise calculator for QCD remains unclear - understanding the transition from the perturbative QCD regime (where the gravity dual is strongly coupled) to the strong coupling regime remains a serious challenge. Nevertheless important new insights into gauge theory and string theory have already been made.

Acknowledgements: ET would like to thank STFC for his studentship funding.

References

1. J. M. Maldacena, Adv. Theor. Math. Phys. **2**, 231 (1998) Int. J. Theor. Phys. **38**, 1113 (1999).
2. E. Witten, Adv. Theor. Math. Phys. **2** (1998) 253.
3. S. S. Gubser, I. R. Klebanov and A. M. Polyakov, Phys. Lett. B **428** (1998) 105.
4. N. Evans and E. Threlfall, arXiv:0805.0956 [hep-th].
5. A. Kehagias and K. Sfetsos, Phys. Lett. B **454** (1999) 270.
6. S. S. Gubser, arXiv:hep-th/9902155.
7. S. Nojiri and S. D. Odintsov, Phys. Lett. B **449** (1999) 39.
8. S. Nojiri and S. D. Odintsov, Phys. Lett. B **458** (1999) 226.
9. K. Ghoroku, M. Tachibana and N. Uekusa, Phys. Rev. D **68** (2003) 125002.
10. I. Brevik, K. Ghoroku, A. Nakamura, Int. J. Mod. Phys. **D15**:57-68, 2006.
11. N. R. Constable and R. C. Myers, JHEP **9911** (1999) 020.
12. K. Ghoroku and M. Yahiro, Phys. Lett. B **604** (2004) 235.
13. K. Pilch and N. P. Warner, Nucl. Phys. B **594** (2001) 209.
14. A. Buchel, A. W. Peet and J. Polchinski, Phys. Rev. D **63** (2001) 044009.
15. M. Grana and J. Polchinski, Phys. Rev. **D65** (2002) 126005, [arXiv: hep-th/0106014].
16. M. Bertolini, P. Di Vecchia, M. Frau, A. Lerda and R. Marotta, Nucl. Phys. B **621**, 157 (2002).
17. A. Karch and E. Katz, JHEP **0206**, 043 (2002).
18. M. Kruczenski, D. Mateos, R. C. Myers and D. J. Winters, JHEP **0307** 049, 2003.
19. J. Erdmenger, N. Evans, I. Kirsch and E. Threlfall, Eur. Phys. J. A **35** (2008) 81.
20. C. Hoyos-Badajoz, K. Landsteiner and S. Montero, JHEP **0704** (2007) 031.
21. G. Policastro, D. T. Son and A. O. Starinets, Phys. Rev. Lett. **87** (2001) 081601.
22. A. O. Starinets. Phys. Rev. **D66** (2002) 124013.

QCD GLUEBALL SUM RULES AND VACUUM TOPOLOGY

HILMAR FORKEL*

Institute for Theoretical Physics, University of Heidelberg
D-69102 Heidelberg, Germany
and
IFT-UNESP
01405-900 - São Paulo, SP, Brazil
** E-mail: forkel@ift.unesp.br*

Several key problems of QCD sum rules in the spin-0 glueball channels are resolved by implementing nonperturbative short-distance physics from direct instantons and topological charge screening. A lattice-based instanton size distribution and the IR renormalization of the nonperturbative Wilson coefficients are also introduced. Results of a comprehensive quantitative sum rule analysis are reviewed and their implications discussed.

1. Introduction

The gluonium states of QCD have remained intriguing for almost four decades.[1] Their "exotic" nature reflects itself not least in several long-standing problems which the QCD sum rule approach faces in the spin-0 glueball channels.[2] In the scalar (0^{++}) glueball correlator, in particular, the departure from asymptotic freedom sets in at unusually small distances[3] and the perturbative Wilson coefficients of the standard operator product expansion (OPE) proved inadequate to establish consistency both among the 0^{++} glueball sum rules and with an underlying low-energy theorem.[4] Although nonperturbative contributions due to direct (i.e. small) instantons[5] were early candidates for the missing short-distance physics,[6] insufficient knowledge of the instanton size distribution[5,7] prevented their quantitative implementation at the time. Only recently, the derivation of the exact instanton contributions (to leading order in $\hbar$), their duality continua and the corresponding Borel sum-rule analysis[4] showed that direct instantons indeed solve the mentioned key problems in the scalar glueball channel. Below we will outline a more thorough and systematic treatment[9] which eliminates artefacts of earlier approximations, significantly modifies the sum-rule results in both spin-0 channels and improves their reliability. Moreover, we implement for the first time topological charge screening contributions to the OPE in the 0^{-+} channel[9] and show how those resolve earlier sum-rule inconsistencies.[10] Finally, we review several new predictions for spin-0 glueball properties.

2. Correlators and sum rules

The spin-0 glueball sum rules are based on the scalar (0^{++}) and pseudoscalar (0^{-+}) correlations functions

$$\Pi_G(-q^2) = i \int d^4x \, e^{iqx} \left\langle 0 | T \, O_G(x) \, O_G(0) \, | 0 \right\rangle \tag{1}$$

where O_G with $G \in \{S, P\}$ are the gluonic interpolating fields (of lowest mass dimension) $O_S = \alpha_s G^a_{\mu\nu} G^{a\mu\nu}$ and $O_P = \alpha_s G^a_{\mu\nu} \tilde{G}^{a\mu\nu}$ with $\tilde{G}_{\mu\nu} \equiv (1/2) \, \varepsilon_{\mu\nu\rho\sigma} G_{\rho\sigma}$ (the definition in Ref. 9 contains a typo). The zero-momentum limit of the correlator (1) is governed by the low-energy theorems (LETs)

$$\Pi_S \left(q^2 = 0\right) = \frac{32\pi}{b_0} \left\langle \alpha G^2 \right\rangle \tag{2}$$

in the scalar[11] and (for three light flavors and $m_{u,d} \ll m_s$)

$$\Pi_P \left(q^2 = 0\right) = (8\pi)^2 \, \frac{m_u m_d}{m_u + m_d} \left\langle \bar{q} q \right\rangle \tag{3}$$

in the pseudoscalar channel.[12] (Note that Eq. (3) vanishes in the chiral limit.) Consistency with the low-energy theorems places stringent constraints on the sum rules which cannot be satisfied without nonperturbative short-distance physics.[4,9]

Contact with the hadronic information in the glueball correlators is established by means of the dispersive representation

$$\Pi_G \left(Q^2\right) = \frac{1}{\pi} \int_0^\infty ds \frac{\operatorname{Im} \Pi_G(-s)}{s + Q^2} \tag{4}$$

where the necessary number of subtractions is implied but not written explicitly. The standard sum-rule description of the spectral functions

$$\operatorname{Im} \Pi_G^{(ph)}(s) = \operatorname{Im} \Pi_G^{(pole)}(s) + \operatorname{Im} \Pi_G^{(cont)}(s) \tag{5}$$

contains one or two resonance poles in zero-width approximation, $\operatorname{Im} \Pi^{(pole)}(s) = \pi \sum_{i=1}^{2} f_{Gi}^2 m_{Gi}^4 \delta\left(s - m_{Gi}^2\right)$, and the local-duality continuum $\operatorname{Im} \Pi_G^{(cont)}(s) = \theta\left(s - s_0\right) \operatorname{Im} \Pi_G^{(IOPE)}(s)$ from the IOPE discontinuities in the "duality range" which starts at an effective threshold s_0.

In order to write down QCD sum rules, the Borel-transformed IOPE and dispersive representations of the correlators - weighted by k powers of $-Q^2$ - are matched in the fiducial τ-region (τ is the Borel variable) and rearranged in terms of the continuum-subtracted Borel moments $\mathcal{R}_{G,k}$ as

$$\mathcal{R}_{G,k}(\tau; s_0) \equiv \frac{1}{\pi}\int_0^{s_0} ds s^k \operatorname{Im}\Pi_G^{(IOPE)}(s) e^{-s\tau} \tag{6}$$
$$= \sum_{i=1}^{2} f_{Gi}^2 m_{Gi}^{4+2k} e^{-m_{Gi}^2\tau} - \delta_{k,-1}\Pi_G^{(ph)}(0).$$

The pole contributions of interest (and for $k = -1$ the crucial subtraction terms) are then isolated on the RHS, and the hadronic parameters m_{Gi}, f_{Gi} and s_0 can be determined numerically.

3. IOPE

Our theoretical framework, the instanton-improved operator product expansion (IOPE), factorizes the correlators at large, spacelike momenta $Q^2 \equiv -q^2 \gg \Lambda_{QCD}$ into contributions from "hard" field modes (with momenta $|k| > \mu$) in the Wilson coefficients $\tilde{C}_D(Q^2)$ and "soft" field modes (with $|k| \le \mu$) in the "condensates" $\left\langle \hat{O}_D \right\rangle_\mu$ of operators $\hat{O}_D$ with increasing dimension D. Previous glueball sum rules based on the OPE with purely perturbative Wilson coefficients were plagued by notorious inconsistencies between the predictions of different moment sum rules and by massive LET violations. Moreover, the soft nonperturbative condensate contributions were exceptionally small.

We have therefore analyzed *hard* nonperturbative contributions to the Wilson coefficients. They are strongly channel dependent and due to direct instantons and topological charge screening. The instanton contributions to the spin-0 coefficients,[9]

$$\Pi_G^{(I+\bar{I})}(x^2) = \frac{2^8 3}{7}\int d\rho n(\rho)\frac{1}{\rho^4}\,{}_2F_1\left(4, 6, \frac{9}{2}, -\frac{x^2}{4\rho^2}\right), \tag{7}$$

are large (while those in the 2^{++} tensor channel vanish) and add to the unit-operator coefficients $\tilde{C}_0^{(G)}$. The imginary part of their Fourier transform at timelike momenta generates the Borel moments[4]

$$\mathcal{R}_k^{(I+\bar{I})}(\tau) = -2^7\pi^2\delta_{k,-1}\int d\rho n(\rho) - 2^4\pi^3\int d\rho \tag{8}$$
$$\times n(\rho)\rho^4\int_0^{s_0} ds s^{k+2} J_2(\sqrt{s}\rho) Y_2(\sqrt{s}\rho) e^{-s\tau}$$

which are similar or larger in size than the perturbative ones. The evaluation of these moments requires as the sole input the (anti-) instanton distribution $n(\rho)$ which is implemented by means of a lattice-based Gaussian-tail parametrization with the correct small-ρ behavior,

$$n_g(\rho) = \frac{2^{18}}{3^6\pi^3}\frac{\bar{n}}{\bar{\rho}}\left(\frac{\rho}{\bar{\rho}}\right)^4 \exp\left(-\frac{2^6}{3^2\pi}\frac{\rho^2}{\bar{\rho}^2}\right) \tag{9}$$

(for $N_c = N_f = 3$), which was introduced in Ref. 9 and shown to prevent several artefacts of the oversimplified "spike" approximation $n(\rho) = \bar{n}\delta(\rho - \bar{\rho})$ on which all previous direct-instanton calculations had relied.

As an additional benefit, the realistic size distribution allows for a gauge-invariant IR renormalization which excludes large instantons with size $\rho > \mu^{-1}$ from the Wilson coefficients,

$$n(\rho) \to \tilde{n}(\mu;\rho) \equiv \theta_\beta\left(\rho - \mu^{-1}\right) n(\rho) \tag{10}$$

(θ_β is a soft step function). The instanton-induced μ dependence turns out to be relatively weak for $\mu < \bar{\rho}^{-1}$, as necessary to compensate its perturbative counterpart. Neglect of this renormalization, although common practice in perturbative Wilson coefficients, would significantly contaminate the results, e.g. by missing the reduction of the direct-instanton density

$$\bar{n} = \int_0^\infty d\rho n(\rho) \to \int_0^\infty d\rho \tilde{n}(\mu;\rho) \equiv \bar{n}(\mu). \tag{11}$$

Another important renormalization effect is the reduction of the instanton contributions to the pseudoscalar relative to the scalar sum rules.

The instanton's self-duality causes a strongly repulsive contribution to the 0^{-+} channel, with seemingly detrimental impact on the sum rules:[10] the glueball signal disappears and both unitarity and the LET (3) are badly violated. The origin of these problems can be traced to the neglect of topological charge screening.[9] Due to their high channel selectivity and a small screening length $\lambda_D \sim m_{\eta'}^{-1} \sim 0.2$ fm, the model-independent screening correlations[14]

$$\Pi_P^{(scr)}(x) \simeq -2^8\pi^2\left(\xi\gamma_{\eta_0}\right)^2 \langle\eta_0(x)\,\eta_0(0)\rangle \tag{12}$$

(ξ is the overall topological charge density ($= \bar{n}$ for instantons) and η_0 the flavor-singlet part of the η') affect almost exclusively the 0^{-+} Wilson coefficients. They arise from the axial-anomaly induced attractive (repulsive) interaction between topological charge lumps of opposite (equal) sign due to η_0 exchange[13] and, after correcting for $\eta_0 - \eta_8$ mixing, add the terms

$$\mathcal{R}_{P,k}^{(scr)}(\tau) = -\delta_{k,-1}\left(\frac{F_{\eta'}^2}{m_{\eta'}^2} + \frac{F_\eta^2}{m_\eta^2}\right) + F_{\eta'}^2 m_{\eta'}^{2k} e^{-m_{\eta'}^2\tau} + F_\eta^2 m_\eta^{2k} e^{-m_\eta^2\tau} \tag{13}$$

to the pseudoscalar IOPE moments.

They strongly reduce the direct-instanton induced repulsion and resolve the disastrous problems mentioned above:[9] positivity of the spectral function is restored, the four 0^{-+} Borel sum rules (6) are stable and contain consistent pseudoscalar glueball information. (All previous analyses had discarded the $k = -1$ sum rule and thereby missed valuable first-principle information and LET consistency checks.)

4. Results and discussion

We have implemented direct instanton and topological charge screening contributions into the OPE coefficients of the spin-0 glueball correlators, evaluated their duality continuua and demonstrated how these contributions resolve the problems of previous QCD sum rule analyses.[9] A lattice-based instanton size distribution and the gauge-invariant IR renormalization of the nonperturbative Wilson coefficients were also introduced. Quark admixtures, and thereby quarkonium mixing effects, enter through quark loops, the instanton size distribution and the condensates.

In the scalar channel, the sizeable direct instanton contributions are indispensable for mutually and LET consistent sum rules. Their improved treatment reduces our earlier (spike-distribution based) result for the 0^{++} glueball mass to $m_S = 1.25 \pm 0.2$ GeV. (The mass stays well beyond 1 GeV, however, in contrast to obsolete predictions based on purely perturbative coefficients.) This value is somewhat smaller than the quenched lattice results[15] (which will probably be reduced by light-quark effects) and consistent with the broad glueball state found in a comprehensive K-matrix analysis.[16] The systematics among our different Borel moments likewise indicates a rather large width of the scalar glueball, $\Gamma_S \gtrsim 0.3$ GeV.

Our prediction for the scalar glueball decay constant, $f_S = 1.05 \pm 0.1$ GeV, is several times larger than the value obtained when ignoring the nonperturbative Wilson coefficients. This implies an exceptionally small 0^{++} glueball size, in agreement with several lattice results.[17] Another stringent, OPE- and sum-rule-independent consistency check of the instanton contributions and their f_S enhancement provide numerical simulations in an instanton ensemble[18] (ILM). For $|x| \lesssim 0.5$ fm the scalar ILM and IOPE correlators are indeed very similar, and the ILM prediction $f_S^{(ILM)} = 0.8$ GeV is similarly large. This indicates a robust instanton effect and rules out that the large f_S "may signal some eventual internal inconsistencies in the treatment of the instanton contributions".[19] A subsequent model-independent confirmation of the f_S enhancement was supplied by the first direct (quenched) lattice result[20] $f_S = 0.86 \pm 0.18$ GeV which is consistent both with our prediction and the ILM value (the latter is practically independent of quark quenching[18]).

Our prediction for f_S implies enhanced partial widths for radiative J/ψ and Υ decays into scalar glueballs and is therefore relevant for experimental glueball searches, e.g. in the CLEO and BES data on $\Upsilon \to \gamma f_0$ and other decay branches. Since the exceptionally small size and large decay constant are particular to the *scalar* glueball and since only part of the scalar decay constant contributes to the radiative production rates,[21] however, the above results are not ruled out by experimental data on decays into 0^{-+} glueballs (cf. Ref. 19).

In the pseudoscalar (0^{-+}) glueball correlator we have identified and implemented a new type of nonperturbative contributions to the Wilson coefficients, due to topological charge screening. Roughly speaking, the screening effects "unquench" the direct instanton contributions, thereby restoring unitarity, the axial Ward identity and the resonance signals. Consistency among all moment sum rules and with the underlying LET is also achieved, and the resulting mass prediction $m_P = 2.2 \pm 0.2$ GeV lies inside the range of quenched and unquenched lattice data. The coupling $f_P = 0.6 \pm 0.25$ GeV is somewhat enhanced by the topological short-distance physics, affecting radiative production rates, the $\gamma\gamma \to G_P \pi^0$ cross section at high momentum transfers and other glueball signatures.

The crucial impact of the nonperturbative Wilson coefficients on both spin-0 glueball correlators is particularly evident in the interplay between their subtraction constants. Indeed, the notorious consistency problems which plagued previous 0^{++} glueball sum rules were primarily caused by the large LET-induced subtraction constant $\Pi_S(0) \simeq 0.6$ GeV^4 (cf. Eq. (2)): it cannot be matched by perturbative Wilson coefficients (since $\Pi_S^{(pert)}(0) = 0$) and requires the direct instanton contribution $\Pi_{S/P}^{(I,\bar{I})}(0) = \pm 2^7 \pi^2 \bar{n}_{dir} \simeq 0.63$ GeV^4. At first sight this seems to imply a conflict with the much smaller LET subtraction constant $\Pi_P(0) \simeq -0.02$ GeV^4 (cf. Eq. (3)) in the 0^{+-} channel, however, since the instanton contributions to both spin-0 correlators are equal (up to a sign). Here the topological screening contributions $\Pi_P^{(scr)}(0) \simeq 0.59$ GeV^4 from Eq. (13) prove indispensable: they restore consistency by canceling most (and in the chiral limit, where $\Pi_P(0) \to 0$, all) of the instanton contributions: $\Pi_P^{(scr)}(0) + \Pi_P^{(I,\bar{I})}(0) \simeq \Pi_P(0)$.

To summarize: contrary to naive expectation, the nonperturbative contributions to the OPE of the spin-0 glueball correlators reside primarily in the Wilson coefficients (i.e not in the condensates) and are closely related to the topological vacuum structure. The nonperturbative short-distance physics resolves long-standing consistency problems of the associated QCD sum rules and generates a rather diverse set of new glueball predictions. A large part of the 0^{++} glueball mass and binding originates from direct instantons, for example, while their net effects in the 0^{+-} channel are smaller and more subtle, due to cancellations between instanton and topological charge screening contributions.

This work was supported by FAPESP and CNPq of Brazil.

References

1. M. Gell-Mann, Acta Phys. Aust. Suppl. **9**, 733 (1972); H. Fritzsch and M. Gell-Mann, 16th Int. Conf. High-Energy Phys., Chicago, Vol. 2, 135 (1972).
2. S. Narison, Nucl. Phys. **B509**, 312 (1998) and references therein.
3. V.A. Novikov, M.A. Shifman, A.I. Vainsthein, and V.I. Zakharov, Nucl. Phys. **B191**, 301 (1981).
4. H. Forkel, Phys. Rev. D **64**, 034015 (2001).
5. T. Schäfer and E.V. Shuryak, Rev. Mod. Phys. **70**, 323 (1998). For an introduction see H. Forkel, *A Primer on Instantons in QCD*, hep-ph/0009136.
6. V.A. Novikov, M.A. Shifman, A.I. Vainsthein, and V.I. Zakharov, Nucl. Phys. **B165**, 67 (1980).
7. A. Ringwald and F. Schremmp, Phys. Lett. **B 459**, 249 (1999).
8. D. Harnett and T.G. Steele, Nucl. Phys. **A 695**, 205 (2001).
9. H. Forkel, Phys. Rev. D **71**, 054008 (2005); Braz. J. Phys. **34**, 875 (2004); AIP Conf. Proc. **739**, 434 (2004).
10. A. Zhang and T.G. Steele, Nucl. Phys. **A 728**, 165 (2003).
11. V.A. Novikov, M.A. Shifman, A.I. Vainshtein and V.I. Zakharov, Nucl. Phys. **B191**, 301 (1981).
12. H. Leutwyler and A. Smilga, Phys. Rev. D **46**, 5607 (1992).
13. P. Di Vecchia and G. Veneziano, Nucl. Phys. **B 171**, 253 (1980).
14. N.J. Dowrick and N.A. McDougall, Phys. Lett. **B 285**, 269 (1992); H. Kikuchi and J. Wudka, Phys. Lett. **B 284**, 111 (1992).
15. W. Lee and D. Weingarten, Phys. Rev. D **61**, 014015 (2000) and references therein; C. Morningstar and M. Peardon, Phys. Rev. D **60**, 034509 (1999).
16. V.V. Anisovich, AIP Conf. Proc. **717**, 441 (2004), arXiv:hep-ph/0310165.
17. N. Ishii, H. Suganuma, and H. Matsufuru, Phys. Rev. D **66**, 94506 (2002); P. de Forcrand and K.-F. Liu, Phys. Rev. Lett. **69**, 245 (1992); R. Gupta et al., Phys. Rev. D **43**, 2301 (1991).
18. T. Schäfer and E.V. Shuryak, Phys. Rev. Lett. **75**, 1707 (1995).
19. S. Narison, arXiv:hep-ph/0512256.
20. Y. Chen et al., Phys. Rev. D **73**, 014516 (2006).
21. X.-G. He, H.-Y. Jin and J. P. Ma, Phys. Rev. D **66**, 74015 (2002).

THE PION FORM FACTOR IN AdS/QCD

HERRY J. KWEE* and RICHARD F. LEBED†
Department of Physics, Arizona State University,
Tempe, AZ 85287-1504, USA
** E-mail: Herry.Kwee@asu.edu*
† E-mail: Richard.Lebed@asu.edu
http://phy.asu.edu

Holographic QCD provides a unique framework in which to compute QCD observables. In this talk we summarize recent numerical work on computing the pion electromagnetic form factor using an AdS/QCD action that includes both spontaneous and explicit chiral symmetry breaking. We consider both hard- and soft-wall model results and develop an intermediate background that supports the best features of both. We also begin to see possible evidence in the fit for the presence of $1/N_c$ corrections.

1. Introduction and Background

Studies of the duality between strongly-coupled Yang-Mills gauge theories and weakly-coupled gravity on curved backgrounds, originating with the anti-de Sitter/conformal field theory (AdS/CFT) correspondence,[1] have become prominent in studies of strongly-coupled field theories. In QCD, the most accessible of these theories, the approach is dubbed "AdS/QCD." In contrast to the exact conformality of the original AdS/CFT example $\mathcal{N}=4$ SUSY, the approximate conformality of QCD is broken by explicit mass scales such as $\Lambda_{\rm QCD}$ and quark masses (as evidenced by confinement and chiral symmetry breaking), which must in some manner be incorporated into the theory if one hopes to achieve a satisfactory picture for the rich spectrum and dynamics of QCD.

Since a great deal of AdS/QCD phenomenology has already been studied, even in just the meson sector,[2–28] we provide only a summary of the most salient features of the approach, leading eventually to a discussion of our work on the pion electromagnetic form factor $F_\pi(Q^2)$.[27,28] In the holographic approach one begins with the 5-dimensional AdS "sliced" metric,

$$ds^2 = g_{MN}\,dx^M dx^N = \frac{1}{z^2}(\eta_{\mu\nu}dx^\mu dx^\nu - dz^2)\,, \tag{1}$$

where $\eta_{\mu\nu} = \text{diag}(+,-,-,-)$ is distinguished from the full nontrivial 5D metric g_{MN} obtained from Eq. (1). One conjectures that weakly-coupled gravity on this background corresponds to strongly-coupled QCD. Crudely speaking, the z "bulk" coordinate corresponds to an inverse momentum scale: $z \equiv \epsilon \to 0$ corresponds to the UV limit, while $z > 0$ probes the IR behavior of the theory. Every QCD operator $\mathcal{O}(x)$ is sourced by a 5D operator $\Psi(x,z)$ that is uniquely determined by its boundary value $\Psi(x,\epsilon)$, hence the term "holographic." The behavior of $\Psi(x,z)$ for $z > 0$ then encapsulates the IR dynamics. Lastly, the global symmetry of isospin is promoted to a gauged symmetry in the bulk.

2. Formalism

The application of these assertions provides a holographic dictionary between QCD and the 5D theory. In particular, to the QCD quark bilinear operators $\bar{q}_L \gamma^\mu t^a q_L$, $\bar{q}_R \gamma^\mu t^a q_R$, and $\bar{q}_R^\alpha q_L$, one associates the gauge fields $A^a_{L\mu}$ and $A^a_{R\mu}$ (with coupling g_5), and the bifundamental field $(2/z)X^{\alpha\beta}$, respectively. From these one defines polar- and axial-vector fields $V^M, A^M \equiv \frac{1}{2}(A^M_L \pm A^M_R)$, from which one defines field strengths $F^{MN}_V \equiv \partial^M V^N - \partial^N V^M - i([V^M, V^N] + [A^M, A^N])$ and $F^{MN}_A \equiv \partial^M A^N - \partial^N A^M - i([V^M, A^N] + [A^M, V^N])$ and the covariant derivative $D^M X \equiv \partial^M X - i[V^M, X] - i\{A^M, X\}$. One further decomposes $X = X_0 \exp(2i\pi^a t^a)$, where the modulus field $X_0 = \frac{1}{2}v(z)$ carries information on the form of chiral symmetry breaking (as discussed below), and the exponent is the usual nonlinear representation of pseudoscalar pion fields (in this notation, we take $\pi \equiv \tilde{\pi}/f_\pi$, where $\tilde{\pi}$ is the canonically-normalized pion field and $f_\pi = 92.1$ MeV). The lowest-order 5D action then reads

$$S = \int d^5x\, e^{-\Phi(z)} \sqrt{g}\, \text{Tr}\left\{ |DX|^2 + 3|X|^2 - \frac{1}{2g_5^2}(F_V^2 + F_A^2) \right\}\,, \tag{2}$$

where $e^{-\Phi(z)}$ represents a background dilaton coupling. Working in the axial-like gauges $V_z = A_z = 0$, resolving $A_\mu = A_{\mu\perp} + \partial_\mu \varphi$ into transverse and longitudinal parts, and working in momentum space, one obtains the Euler-Lagrange equations:

$$\partial_z\left(\frac{e^{-\Phi(z)}}{z}\,\partial_z V^a_\mu\right) + \frac{q^2 e^{-\Phi(z)}}{z} V^a_\mu = 0\,, \tag{3}$$

$$\left[\partial_z\left(\frac{e^{-\Phi(z)}}{z}\,\partial_z A^a_\mu\right)+\frac{q^2e^{-\Phi(z)}}{z}A^a_\mu-\frac{g_5^2\,v(z)^2e^{-\Phi(z)}}{z^3}A^a_\mu\right]_\perp=0\,, \quad (4)$$

$$\partial_z\left(\frac{e^{-\Phi(z)}}{z}\,\partial_z\varphi^a\right)+\frac{g_5^2\,v(z)^2e^{-\Phi(z)}}{z^3}(\pi^a-\varphi^a)=0\,, \quad (5)$$

$$-q^2\partial_z\varphi^a+\frac{g_5^2\,v(z)^2}{z^2}\,\partial_z\pi^a=0\,, \quad (6)$$

$$\partial_z\left(\frac{e^{-\Phi(z)}}{z^3}\partial_z X_0\right)+\frac{3e^{-\Phi(z)}}{z^5}X_0=0\,. \quad (7)$$

The meson masses/wave functions are then obtained as the eigenvalues/eigenfunctions of the equations of motion treated as Sturm-Liouville systems, leading to Kaluza-Klein towers of meson states reminiscent of old-fashioned Regge trajectories. Meanwhile, the source currents that create or destroy mesons appear as the free-field solutions to the equations of motion. Meson form factors are then overlap integrals (in z) of the source solutions with the external-state eigenmode solutions. Our case of interest is the pion electromagnetic form factor $F_\pi(Q^2)$; the pion wave function is the lowest-mass mode (massless in the limit $m_q\to 0$) of the field $\pi(q^2,z)$, which is seen from Eqs. (5)–(6) to be coupled to the solution for $\varphi(q^2,z)$. With $V(q,z)$ being the source current from Eq. (3) normalized by $V(0,z)=1$, one obtains

$$F_\pi(q^2)=\int dz\,e^{-\Phi(z)}\,\frac{V(q,z)}{f_\pi^2}\left\{\frac{1}{g_5^2 z}[\partial_z\varphi(z)]^2+\frac{v(z)^2}{z^3}\left[\pi(z)-\varphi(z)\right]^2\right\}\,, \quad (8)$$

which is most useful for spacelike Q^2. In the timelike region, the large-N_c nature of the holographic approach gives the form factor as a sum over zero-width vector meson poles (the ρ and its excitations):

$$F_\pi(q^2)=-\sum_{n=1}^{\infty}\frac{f_n g_{n\pi\pi}}{q^2-M_n^2}\,, \quad (9)$$

where $g_{n\pi\pi}$ is given by

$$g_{n\pi\pi}=\frac{g_5}{f_\pi^2}\int dz\,\psi_n(z)\,e^{-\Phi(z)}\left\{\frac{1}{g_5^2 z}[\partial_z\varphi(z)]^2+\frac{v(z)^2}{z^3}\left[\pi(z)-\varphi(z)\right]^2\right\}\,, \quad (10)$$

where ψ_n are the eigenmodes for the vector states. While Ref. 27 considered both $F_\pi(Q^2)$ curves and the pattern of $g_{n\pi\pi}$ values, in this talk we focus exclusively on the spacelike pion form factor $F_\pi(Q^2)$ obtained from Eq. (8).

3. Background and Chiral Symmetry-Breaking Fields

The discussion to this point is just a straightforward application of the basic holographic scheme. To continue from this point, however, one must make two choices: for the field $v(z)$ encapsulating the chiral symmetry breaking, and for the background field $e^{-\Phi(z)}$.

Addressing first the background field, we note that two popular choices permeate the literature: the hard-wall model[29] with step function $e^{-\Phi(z)} = H(z_0 - z)$ and the soft-wall model[2] with Gaussian $e^{-\Phi(z)} = e^{-\kappa^2 z^2}$. We introduce[28] the interpolating "semi-hard" option, inspired by the Saxon-Woods model of nuclear charge density:

$$e^{-\Phi(z)} = \frac{e^{\lambda^2 z_0^2} - 1}{e^{\lambda^2 z_0^2} + e^{\lambda^2 z^2} - 2}, \tag{11}$$

which, like the hard-wall profile, has a drop-off at $z = z_0$, but like the soft-wall profile decreases as $e^{-\lambda^2 z^2}$ for large z. The hard-wall model was introduced for its simplicity: The fields simply permeate a fixed distance z_0 into the bulk, and the resulting meson trajectories as a function of excitation quantum number n scale as $m_n^2 \sim n^2$. On the other hand, semiclassical flux-tube QCD reasoning leads[30] to the conclusion $m_n^2 \sim n^1$; the soft-wall model was developed precisely to accommodate this behavior. Unfortunately, hard-wall models tend to give more accurate predictions for QCD observables than soft-wall models; as an example,[16] the experimental value for the ratio $m_\rho^2/f_\rho = 5.02 \pm 0.04$ compares favorably with the hard-wall result 5.55, but rather poorly with the soft-wall prediction 8.89 (other examples appear in Table 1). Nevertheless, as seen below, the linear trajectory of the soft-wall model is preserved simply by the exponential tail of the background, which motivates the hybrid choice in Eq. (11).

Turning now to the choice of chiral symmetry breaking represented by the field $v(z)$, we begin by noting that the two solutions to Eq. (7) in the hard wall case are z^1 and z^3. Since the standard gauge/gravity techniques identify the operator source as the non-normalizable (more singular) solution and the state and associated vev with the normalizable (less singular) as $z \to 0$, one identifies[3] the coefficient of z^1 with the quark mass and z^3 with the quark condensate: $v(z) = m_q z + \sigma z^3$. In the soft-wall model, the exact solutions turn out to be Kummer (confluent hypergeometric) functions, which have the unfortunate feature that only one solution satisfies the appropriate boundary conditions[2] by vanishing asymptotically as $z \to \infty$. Taken literally, this unique solution would give an unphysical fixed ratio for m_q to σ; as argued in Ref. 2, however, neglected higher-order terms in the quark potential permit independent coefficients for the z^1 and z^3 terms in the low-z expansion of the solution for $v(z)$. Lacking exact forms for the higher-order terms, one may implement this fact in two ways: One may note[27] that the soft-wall background $e^{-\kappa^2 z^2}$ suppresses the distinction between the exact solution for $v(z)$ and $m_q z + \sigma z^3$ at large z, or one may choose[28] a modified form for $v(z)$ that behaves like $m_q z + \sigma z^3$ for small z and like the asymptotic form of the appropriate Kummer function for large z. In either case, one finds that the numerical solutions for static observables and $F_\pi(Q^2)$ are never better than those of the hard-wall model.

4. Numerical Solutions

This brings us to the question of how to solve Eqs. (3)–(7) in practice. In particular, Eqs. (5) and (6) are coupled, and the whole set depends upon three adjustable parameters, z_0 or κ, m_q, and σ. Analytic solutions exist only in certain limits,[27] particularly, as studied in Ref. 20, $m_q \to 0$. If $m_q \neq 0$, one must resort to a numerical approach, using standard techniques[31] such as the "shooting method" and properly convergent numerical integrations to solve the equations. Such calculations, carried out in hard, soft, and semi-hard backgrounds, form the core of our work.[27,28]

Figure 1 compares our $F_\pi(Q^2)$ hard- and soft-wall model results to data. The value of z_0 (hard wall) or κ (soft wall) is completely fixed by the value of m_ρ; once this primary parameter is fixed, the pion decay constant f_π is determined by adjusting σ. Finally, the Gell-Mann–Oakes–Renner formula $m_\pi^2 f_\pi^2 = 2 m_q \sigma$ uses m_π to fix a value of m_q. Empirically, the shape of $F_\pi(Q^2)$ is driven primarily by σ. Using the experimental values $m_\rho = 775.3$ MeV, $m_\pi = 139.6$ MeV, and $f_\pi = 92.1$ MeV, one obtains the hard-wall parameters $1/z_0 = 322$ MeV, $\sigma^{1/3} = 326$ MeV, and $m_q = 2.30$ MeV, which in turn generate the solid line in Fig. 1. The same experimental values for m_ρ and m_π (but taking[a] $f_\pi = 87.0$ MeV) in the soft-wall model give $\kappa = 389$ MeV, $\sigma^{1/3} = 368$ MeV, and $m_q = 1.45$ MeV, and produce the dashed line. Both models predict a value of $F_\pi(Q^2)$ clearly more shallow than data, an effect even more pronounced when one views the same plot using the dependent variable $Q^2 F_\pi(Q^2)$ [Fig. 2]. Interestingly, the discrepancy with $F_\pi(Q^2)$ data could easily be cured if f_π were smaller: $f_\pi = 64.2$ MeV ($\sigma^{1/3} = 254$ MeV) (hard-wall) gives the dash-dot curve in Fig. 1, and $f_\pi = 52.2$ MeV ($\sigma^{1/3} = 262$ MeV) (soft-wall) gives the dash-dot-dot curve. We return presently to the question of the meaning of these anomalously small f_π values.

[a]The slightly smaller value used for f_π allows for a much improved fit to $F_\pi(Q^2)$.

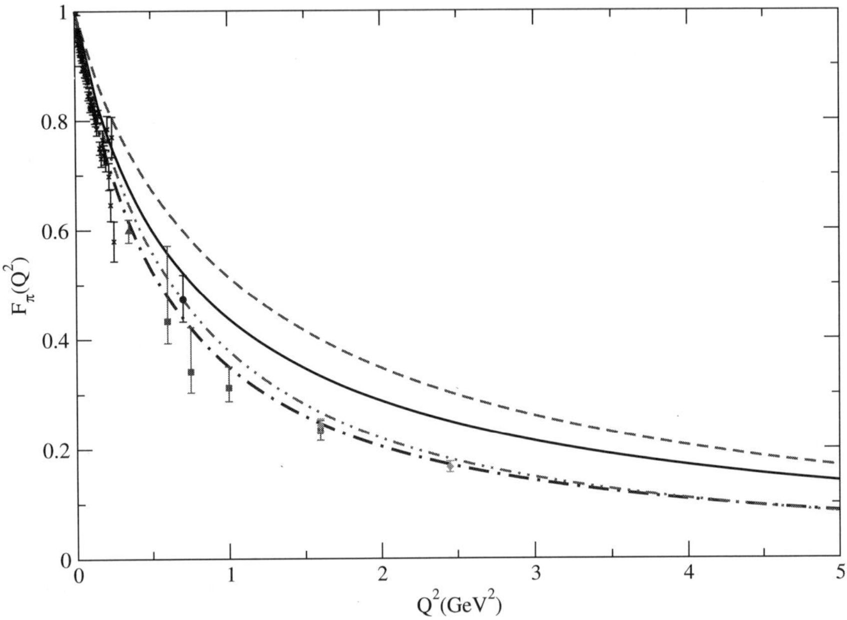

Fig. 1. The pion form factor $F_\pi(Q^2)$ prediction in hard- and soft-wall models compared to data.[32–37] The solid and dash-dot lines are hard-wall model predictions whose input parameters differ only by use of a smaller value of f_π than experiment in the latter, and analogously for the dashed and dash-dot-dot lines in the soft-wall model. The input values appear in the text.

Figure 2 is also interesting because it seems to suggest near-asymptotic values for $Q^2F_\pi(Q^2)$. For example, if one adopts the standard abbreviation $s_0 \equiv 8\pi^2 f_\pi^2 = 0.67$ GeV^2, then the original hard-wall model appears to approach at least $Q^2F_\pi(Q^2) \approx 1.2 s_0$ as $Q^2 \to \infty$. In fact, the analytic $m_q = 0$ hard-wall results of Ref. 20 for $Q^2F_\pi(Q^2)$, which appear to conform closely with our numerical $m_q \neq 0$ results, predict that $Q^2F_\pi(Q^2) \to s_0$ as $Q^2 \to \infty$, but also that $Q^2F_\pi(Q^2)$ overshoots its asymptote and does not return to it until values of $Q^2 \gg 5$ GeV^2, at which partonic effects (absent in this holographic approach) are expected to become relevant. Note that the perturbative QCD result[38] for $Q^2F_\pi(Q^2)$ scales not as a constant, but rather falls off as $\alpha_s(Q^2)f_\pi^2$.

We have argued that the semi-hard background in Eq. (11), for suitable values of λ (or dimensionless variable λz_0), can be made to simulate either hard-wall or soft-wall backgrounds. This effect is illustrated in Fig. 3, which again presents the data and original hard- (solid) and soft-wall (dashed) re-

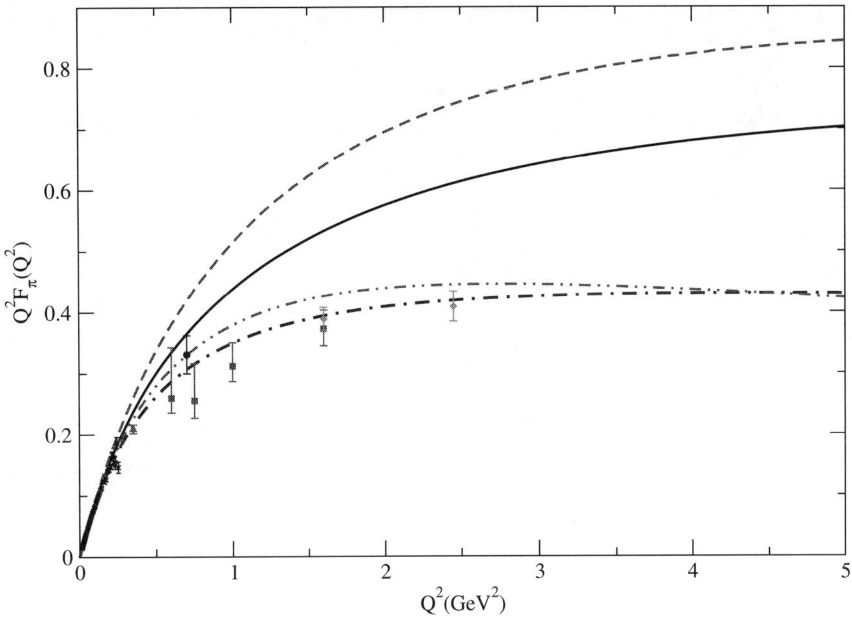

Fig. 2. The same as Fig. 1, for the combination $Q^2F_\pi(Q^2)$.

sults from Fig. 1, superimposed with the result of the semi-hard model for $\lambda z_0 = 2.1$ (crosses) and $\lambda z_0 = 1.0$ (pluses). While the agreement between the semi-hard wall and original hard- and soft-wall models for $F_\pi(Q^2)$ is impressive, one must check that the meson static observables (masses, decay constants, *etc.*) also agree; this is confirmed by a glance at Table 1. Despite agreeing so well with so many hard-wall quantities, the semi-hard wall model with $\lambda z_0 = 2.1$ nevertheless generates a very different meson trajectory, as illustrated in Table 2: One finds that the exponential tail is sufficient, even for the modest value $\lambda z_0 = 2.1$, to turn the hard-wall $m_n^2 \sim n^2$ trajectory into one that is $\sim n^1$, as a careful examination of the numbers confirms. Thus, the semi-hard wall model carries all the best features of both hard- and soft-wall models.

One mystery remains, namely, why all the models considered here using only experimental inputs predict curves for $F_\pi(Q^2)$ too shallow compared to data (predicting, *e.g.*, too small a pion charge radius $\langle r_\pi^2 \rangle$). As mentioned above, partonic degrees of freedom have not entered into the holographic calculation in any essential way; indeed, the only place that the fundamental QCD gauge theory appears is through matching[3] the vector-current

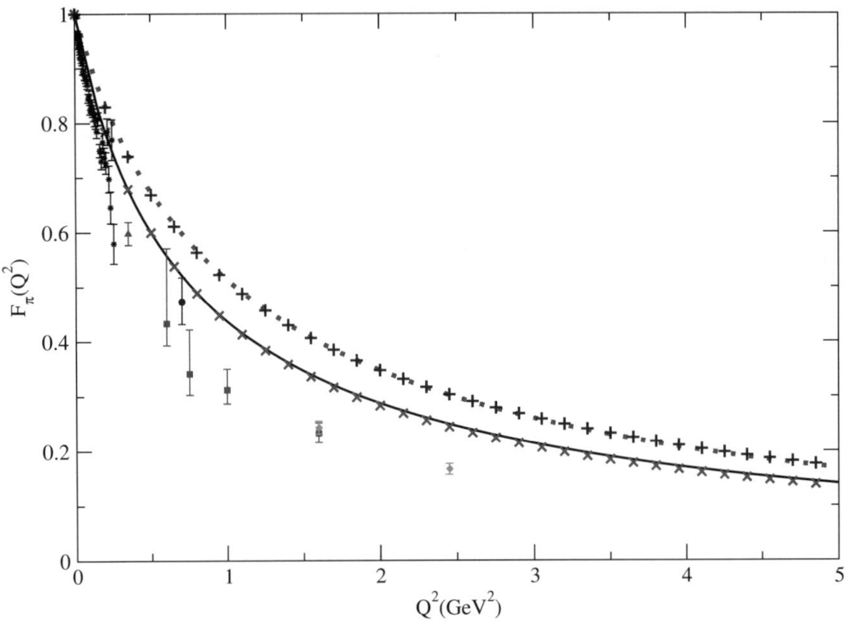

Fig. 3. Data and hard- and soft-wall model results for $F_\pi(Q^2)$ (same symbols as in Fig. 1), superimposed with semi-hard model with $\lambda z_0 = 2.1$ (crosses) and $\lambda z_0 = 1.0$ (pluses).

Table 1. Observables in soft- and hard-wall models compared to those from the semi-hard model of Eq. (11) with $\lambda z_0 = 1$ and $\lambda z_0 = 2.1$, respectively; values in MeV (except for $g_{\rho\pi\pi}$, which is dimensionless).

Observable	Experiment	Soft-wall	$\lambda z_0 = 1$	Hard-wall	$\lambda z_0 = 2.1$
m_π	139.6±0.0004[39]	139.6	139.6	139.6	139.6
m_ρ	775.5±0.4[39]	777.4	779.2	775.3	777.5
m_{a_1}	1230±40[39]	1601	1596	1358	1343
f_π	92.4±0.35[39]	87.0	92.0	92.1	88.0
$f_\rho^{1/2}$	346.2±1.4[40]	261	283	329	325
$f_{a_1}^{1/2}$	433±13[41,42]	558	576	463	474
$g_{\rho\pi\pi}$	6.03±0.07[39]	3.33	3.49	4.48	4.63

two-point function calculated both in the 5D theory and in (perturbative) QCD, from which one determines the 5D gauge coupling $g_5 = 2\pi$. It is natural to suppose that the holographic model is at best incomplete, due to the absence of partonic degrees of freedom; however, the accurate $F_\pi(Q^2)$

Table 2. Comparison of vector meson masses (in MeV) from the hard-wall model and the semi-hard wall model using $e^{-\Phi(z)}$ of Eq. (11) with $\lambda z_0 = 2.1$.

	$n=1$	$n=2$	$n=3$	$n=4$	$n=5$
Hard wall	775.6	1780.2	2790.8	3802.8	4815.2
Semi-hard wall	777.5	1608.1	2226.8	2637.5	2986.6

data[b] extends only out to ~ 3 GeV^2, far below the regime where one would expect the partonic expression for $F_\pi(Q^2)$ to dominate. Alternately, one might argue that the treatment of chiral symmetry used here is inadequate; while this is certainly possible, it appears to be the most realistic treatment available. These were the possible explanations proffered in Ref. 28.

However, we present here one further possible explanation: The holographic method implicitly assumes both large N_c and large 't Hooft coupling $g_s^2 N_c$. Can the shallow slope of $F_\pi(Q^2)$ be a $1/N_c$ correction?[c] To test this idea, we ask which quantities in the analysis are most sensitive to $1/N_c$ corrections. While meson masses m_π and m_ρ are $O(N_c^0)$, their decay constants are $O(N_c^{1/2})$ and thus are fractionally more sensitive to variations in N_c. Recall that the original hard-wall model gave a perfect account of the $F_\pi(Q^2)$ data if f_π had been 64.2 MeV. This number is, interestingly, very close to a factor $1-1/N_c$ smaller than the experimental value 92.1 MeV, so that their difference is easily attributable as a $1/N_c$ correction.

While such an explanation may seem a bit glib, a similar effect has been seen long ago in the literature: The pioneering soliton model work of Adkins, Nappi, and Witten.[43] Using our normalization for f_π, their model values inserted into the Goldberger–Treiman relation $f_\pi = M_N g_A / g_{\pi\pi N}$ predict[d] $f_\pi = 61$ MeV, a result traditionally attributed to being due to a $1/N_c$ correction.

These considerations suggest a supposedly "perfect" holographic model for low-energy hadronic phenomena that should include two features: A semi-hard wall background with an exponential tail extending to large z, and $1/N_c$ corrections of a natural size, particularly for f_π. The semi-hard wall model has been seen to give a fit to low-energy observables just as good as that of the hard-wall model, but nevertheless generates the desired linear trajectories for excited mesons. Meanwhile, allowing f_π to be smaller

[b]Some $F_\pi(Q^2)$ data points extend out to about 10 GeV^2, but the uncertainties are sufficiently large as to accommodate almost any model.
[c]That the discrepancy can be a $1/g_s^2 N_c$ correction was suggested to us by O. Andreev.
[d]Reference 43 actually used the Goldberger–Treiman relation to predict $g_{\pi\pi N}$.

by a relative $1/N_c$ correction is sufficient to correct the shallowness of the $F_\pi(Q^2)$ curves compared to experiment. Indeed, it is remarkable that these modifications are sufficient to cure the discrepancies with data but still make no mention of the partonic degrees of freedom in QCD. A truly perfect holographic model would of necessity incorporate dynamical quarks as well.

5. Conclusions

Since their inception, holographic methods have provided a compelling theoretical framework in which to study hadronic quantities, including $F_\pi(Q^2)$. In this talk we have seen that the choice of background field behavior has a strong effect on low-energy observables, but it is possible to retain many of the best features of each model while adjusting this background; the semi-hard wall background proves ideal for accomplishing this goal. In passing, we note that the treatment of chiral symmetry breaking advocated by Ref. 3 appears completely suitable for this purpose. However, we begin to see in the precise value of the slope of $F_\pi(Q^2)$ possible evidence for the necessity of including $1/N_c$ corrections in order to achieve completely satisfactory agreement with data. The great remaining challenge appears to be how to knit together the promising first results of this all-hadronic approach with the fundamental QCD partonic degrees of freedom.

Acknowledgments

RFL thanks the organizers for their kind invitation and an interesting scientific program. This presentation benefited from discussions with H. Grigoryan and A. Radyushkin. This work was supported by the NSF under Grant No. PHY-0456520.

References

1. J.M. Maldacena, Adv. Theor. Math. Phys. **2**, 231 (1998) [Int. J. Theor. Phys. **38**, 1113 (1999)] [arXiv:hep-th/9711200]; S.S. Gubser, I.R. Klebanov, and A.M. Polyakov, Phys. Lett. B **428**, 105 (1998) [arXiv:hep-th/9802109]; E. Witten, Adv. Theor. Math. Phys. **2**, 253 (1998) [arXiv:hep-th/9802150].
2. A. Karch, E. Katz, D.T. Son, and M.A. Stephanov, Phys. Rev. D **74**, 015005 (2006) [arXiv:hep-ph/0602229].
3. J. Erlich, E. Katz, D.T. Son, and M.A. Stephanov, Phys. Rev. Lett. **95**, 261602 (2005) [arXiv:hep-ph/0501128].
4. H. Boschi-Filho and N.R.F. Braga, JHEP **0305**, 009 (2003) [arXiv:hep-th/0212207].
5. G.F. de Teramond and S.J. Brodsky, Phys. Rev. Lett. **94**, 201601 (2005) [arXiv:hep-th/0501022].

6. N. Evans and A. Tedder, Phys. Lett. B **642**, 546 (2006) [arXiv:hep-ph/0609112].
7. P. Colangelo, F. De Fazio, F. Jugeau and S. Nicotri, Phys. Lett. B **652**, 73 (2007) [arXiv:hep-ph/0703316].
8. H. Forkel, M. Beyer and T. Frederico, JHEP **0707**, 077 (2007) [arXiv:0705.1857 (hep-ph)].
9. L. Da Rold and A. Pomarol, Nucl. Phys. B **721**, 79 (2005) [arXiv:hep-ph/0501218]; JHEP **0601**, 157 (2006) [arXiv:hep-ph/0510268].
10. J. Hirn and V. Sanz, JHEP **0512**, 030 (2005) [arXiv:hep-ph/0507049]; J. Hirn, N. Rius and V. Sanz, Phys. Rev. D **73**, 085005 (2006) [arXiv:hep-ph/0512240].
11. K. Ghoroku, N. Maru, M. Tachibana and M. Yahiro, Phys. Lett. B **633**, 602 (2006) [arXiv:hep-ph/0510334].
12. T. Huang and F. Zuo, arXiv:0708.0936 [hep-ph].
13. H. Boschi-Filho, N.R.F. Braga and C.N. Ferreira, Phys. Rev. D **73**, 106006 (2006) [Erratum-ibid. D **74**, 089903 (2006)] [arXiv:hep-th/0512295]; O. Andreev and V.I. Zakharov, Phys. Rev. D **74**, 025023 (2006) [arXiv:hep-ph/0604204]; C.D. White, Phys. Lett. B **652**, 79 (2007) [arXiv:hep-ph/0701157].
14. T. Hambye, B. Hassanain, J. March-Russell and M. Schvellinger, Phys. Rev. D **74**, 026003 (2006) [arXiv:hep-ph/0512089]; Phys. Rev. D **76**, 125017 (2007) [arXiv:hep-ph/0612010].
15. H.R. Grigoryan and A.V. Radyushkin, Phys. Lett. B **650**, 421 (2007) [arXiv:hep-ph/0703069].
16. H.R. Grigoryan and A.V. Radyushkin, Phys. Rev. D **76**, 095007 (2007) [arXiv:0706.1543 (hep-ph)].
17. S. Hong, S. Yoon and M.J. Strassler, JHEP **0604**, 003 (2006) [arXiv:hep-th/0409118].
18. A.V. Radyushkin, Phys. Lett. B **642**, 459 (2006) [arXiv:hep-ph/0605116].
19. S.J. Brodsky and G.F. de Téramond, Phys. Rev. D **77**, 056007 (2008) [arXiv:0707.3859 [hep-ph]].
20. H.R. Grigoryan and A.V. Radyushkin, Phys. Rev. D **76**, 115007 (2007) (arXiv:0709.0500 [hep-ph]).
21. H. R. Grigoryan and A. V. Radyushkin, Phys. Rev. D **77**, 115024 (2008) [arXiv:0803.1143 [hep-ph]].
22. P. Colangelo, F. De Fazio, F. Giannuzzi, F. Jugeau and S. Nicotri, arXiv:0807.1054 [hep-ph].
23. W. de Paula, T. Frederico, H. Forkel and M. Beyer, arXiv:0806.3830 [hep-ph].
24. S.S. Agaev and M.A.G. Nobary, Phys. Rev. D **77**, 074014 (2008) [arXiv:0805.0993 [hep-ph]].
25. D. f. Zeng, arXiv:0805.2733 [hep-th].
26. B. Batell and T. Gherghetta, Phys. Rev. D **78**, 026002 (2008) [arXiv:0801.4383 [hep-ph]].
27. H.J. Kwee and R.F. Lebed, JHEP **0801**, 027 (2008) (arXiv:0708.4054 [hep-ph]).
28. H. J. Kwee and R. F. Lebed, Phys. Rev. D **77**, 115007 (2008) [arXiv:0712.1811

[hep-ph]].
29. J. Polchinski and M.J. Strassler, Phys. Rev. Lett. **88**, 031601 (2002) [arXiv:hep-th/0109174].
30. M. Shifman, arXiv:hep-ph/0507246.
31. William H. Press, Brian P. Flannery, Saul A. Teukolsky, William T. Vetterling, *Numerical Recipes in FORTRAN 77: The Art of Scientific Computing*, (Cambridge University Press, 1992).
32. S.R. Amendolia *et al.*, Phys. Lett. B **138**, 454 (1984); Nucl. Phys. **B277**, 168 (1986).
33. P. Brauel *et al.*, Phys. Lett. B **69**, 253 (1977); P. Brauel *et al.*, Z. Phys. C **3**, 101 (1979).
34. V. Tadevosyan *et al.* [Jefferson Lab F(pi) Collaboration], Phys. Rev. C **75**, 055205 (2007) [arXiv:nucl-ex/0607007].
35. H. Ackermann *et al.*, Nucl. Phys. **B137**, 294 (1978).
36. T. Horn *et al.* [Fpi2 Collaboration], Phys. Rev. Lett. **97**, 192001 (2006) [arXiv:nucl-ex/0607005].
37. C.J. Bebek *et al.*, Phys. Rev. D **17**, 1693 (1978).
38. G.P. Lepage and S.J. Brodsky, Phys. Lett. B **87**, 359 (1979); A.V. Efremov and A.V. Radyushkin, Theor. Math. Phys. **42**, 97 (1980) [Teor. Mat. Fiz. **42**, 147 (1980)].
39. S. Eidelman *et al.* (PDG), Phys. Lett. B **592**, 1 (2004).
40. J.F. Donoghue, E. Golowich, and B.R. Holstein, *Dynamics of the Standard Model* (Cambridge University Press, Cambridge 1992)
41. N. Isgur, C. Morningstar, and C. Reader, Phys. Rev. D **39**, 1357 (1989).
42. D.T. Son and M.A. Stephanov, Phys. Rev. D **69**, 065020 (2004) [arXiv:hep-ph/0304182].
43. G.S. Adkins, C.R. Nappi and E. Witten, Nucl. Phys. B **228**, 552 (1983).

The Fast Life of Holographic Mesons

R. C. Myers[1,2] and A. Sinha[1]

[1] *Perimeter Institute for Theoretical Physics, Waterloo, Ontario N2L 2Y5, Canada.*
[2] *Department of Physics and Astronomy, University of Waterloo, Waterloo, Ontario N2L 3G1, Canada.*

Holographic techniques are used to study the dispersion relation of meson quasiparticles moving through a thermal plasma in $\mathcal{N} = 2$ super-Yang-Mills theory with N_c colours and coupled to N_f flavours of fundamental matter in the regime $N_f/N_c \ll 1$. The meson states are destabilized by introducing a small quark density n_q. Holographic spectral functions are used to investigate the dispersion relations and widths of these quasiparticles. In a low momentum regime $\mathfrak{q} < \mathfrak{q}_{\text{max}}$, the quasiparticles approach a limiting velocity that can be significantly less than the speed of light. In this regime, the widths of the quasiparticles also rise dramatically as the momentum approaches $\mathfrak{q}_{\text{max}}$. The spectral functions do not display isolated resonances for $\mathfrak{q} > \mathfrak{q}_{\text{max}}$. However, by studying the dual quasinormal modes one can extend the dispersion relation into this regime. We give a qualitative argument suggesting that the group velocity rises to the speed of light for $\mathfrak{q} \gg \mathfrak{q}_{\text{max}}$. This contribution provides a summary of the results in [1].

Keywords: AdS/CFT correspondence, thermal field theory, strongly coupled gauge theories

1. Introduction

The AdS/CFT correspondence has been used extensively to gain insights about strongly coupled gauge theories at finite temperature. The gauge theories in question have a large number of colours N_c and a large 't Hooft coupling $\lambda = g_{YM}^2 N_c$. The motivation for these studies is the hope that QCD just above deconfinement falls into the same universality class and so may have certain dynamical features in common with the holographic theories. Evidence in favour of this argument can be found in examining the ratio of the shear-viscosity to the entropy-density: One finds $\eta/s = 1/4\pi$ for a very broad class of gauge theories with a gravity dual.[2,3] At the same time, evidence is emerging from the experiments at RHIC that the strongly coupled quark gluon plasma also has $\eta/s \sim 1/4\pi$.[4]

The best studied holographic theory is $U(N_c)$ super-Yang-Mills theory in $d = 3+1$ with $\mathcal{N} = 4$ supersymmetry. The gravity dual is type IIb string theory in $AdS_5 \times S^5$. This holographic background is an effective approach to study the SYM theory when both the 't Hooft coupling and the number of colours N_c are large. Introducing N_f D7-branes into the $AdS_5 \times S^5$ background incorporates fundamental matter in the gauge theory.[5] Working in the limit $N_f/N_c \ll 1$ allows us to ignore the backreaction of the D7-branes, *i.e.*, we work in the probe brane approximation. The resulting gauge theory is $\mathcal{N} = 2$ $U(N_c)$ super-Yang-Mills theory with $N_f + 1$ hypermultiplets. Here N_f hypermultiplets are in the fundamental representation of the gauge group and we refer to these fields as the "quarks". The remaining hypermultiplet is in the adjoint representation and combines with the vector multiplet to give the field content of the original $\mathcal{N} = 4$ SYM theory. The $\mathcal{N} = 2$ theory constructed in this way is not confining. Rather it has free quarks with mass m_q. Further, it also has deeply bound quark-antiquark states or "mesons" with masses of the order $m_q/\sqrt{\lambda}$.

In the gravity dual, the free quarks correspond to strings stretching from the D7-branes down to the Poincaré horizon of the AdS_5 geometry. The mesons, *i.e.*, bound states of fundamental and antifundamental fields, are dual to open string states supported on the D7-branes. The lowest-lying mesons then correspond to excitations of the world-volume gauge theory. For example, the modes of the D7-brane gauge field can be thought of as vector mesons while fluctuations moving the brane in the two transverse coordinates provide a scalar and a pseudoscalar meson. The quadratic action for these mesons can be derived using the DBI action describing the D7-brane. The meson spectrum was found analytically at $T = 0$ in [6].

To study the theory at finite temperature using holographic techniques, the AdS_5 geometry is replaced with an asymptotically AdS black hole. The Hawking temperature of the event horizon then corresponds to the gauge theory temperature, T. At low temperatures, the spectrum of the theory is only slightly modified with free quarks having masses $m_q - \Delta m(T)$ and mesons having masses $m_q/\sqrt{\lambda} - \Delta m'(T)$. One finds there is a phase transition at $T_{\text{fun}} \sim m_q/\sqrt{\lambda}$ in this strongly coupled theory.[7,8] At higher temperatures above this transition, there are no quark or meson quasi-particles. Rather the quarks can be thought to have dissolved in the strongly coupled plasma. These two phases of the theory are reflected in the holographic dual with two distinct embeddings of the D7-branes:

(1) **Minkowski embeddings**: At low temperatures below T_{fun}, the D7-brane tension is able to support the branes against the gravitational

pull of the black hole. Hence in these embeddings, the D7-branes remain entirely outside the horizon.

(2) **Black hole embeddings**: These embeddings have the D7-branes extending down through the black hole horizon and happen at temperatures above T_{fun} where the attraction of gravity overcomes the brane tension.

An interesting physical effect, discovered for the low temperature Minkowski phase,[8,9] is that the mesons appear to have an asymptotic group velocity that can be significantly less than that of light. This effect is understood in the gravitational background as a redshift effect for excitations traveling near the event horizon. In a more realistic model, one would like to examine the same physics for mesons with finite widths. However, in the model described above, the mesons in the low temperature phase are absolutely stable, *i.e.*, these states have zero width, as a result of the large-N_c limit.[8] Mesons can also be studied in the high temperature phase using spectral functions,[10] however, they show no evidence of metastable quasiparticles.[11] Hence the holographic model needs to be extended in order to have mesons with controllably small widths. In fact, this is easily achieved by introducing a small quark density into the construction.[12,13] In the following, we will study the dispersion relations of the resulting quasiparticles using spectral function techniques. This discussion below summarizes the detailed analysis presented in [1].

2. Dispersion relations and widths

Quasiparticles appear as simple poles in thermal correlators:

$$G^R \simeq \frac{A}{\omega - \Omega(q,\alpha) + i\Gamma(q,\alpha)} \,. \tag{1}$$

The spectral function is defined as

$$\mathfrak{R}(\mathfrak{w},\mathfrak{q}) = -2\mathrm{Im}G^R(\mathfrak{w},\mathfrak{q}) \,. \tag{2}$$

Hence given an isolated pole, the spectral function exhibits a peak with a characteristic Breit-Wigner form

$$\mathfrak{R}(\mathfrak{w}) \simeq \frac{2A\Gamma}{(\mathfrak{w}-\Omega)^2+\Gamma^2} \,, \tag{3}$$

A clear peak with a quasiparticle interpretation appears if the Landau criterion $\Gamma \ll \Omega$ is satisfied. We also note that the spectral function $\mathfrak{R}(\omega)$

has a characteristic form in the 'high-frequency' limit. This behaviour is determined by the leading short-distance singularity

$$\lim_{(t^2-\mathbf{x}^2)\to 0} \langle \mathcal{O}(t,\mathbf{x})\,\mathcal{O}(0)\rangle = \frac{\mathcal{A}}{|t^2-\mathbf{x}^2|^{\Delta}} + \cdots\,, \tag{4}$$

where Δ denotes the dimension of the operator $\mathcal{O}$ and $\mathcal{A}$ is a dimensionless constant. A Fourier transform then leads to the following contribution to the spectral function

$$\aleph(\omega) \sim \mathcal{A}\,(\omega^2-q^2)^{\Delta-2}\,. \tag{5}$$

Note that this high-frequency tail is Lorentz invariant and shows no indications of a limiting velocity $v_{\rm lim} < 1$. This should be expected as it describes the very high-energy/short-distance behaviour, which is independent of temperature.[11] Thermal spectral functions were used to study holographic mesons in the high temperature phase of the holographic model described above[11] (at zero quark density). As commented before, the mesons in this phase are too unstable for an appreciable peaks to appear in the spectral functions, as illustrated in fig. 2.

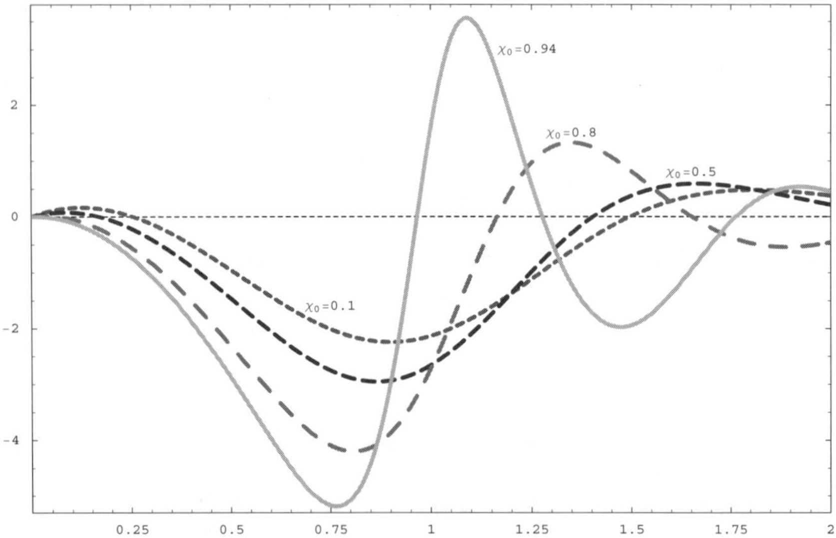

Fig. 1. Finite temperature part of the spectral function for vector mesons, *i.e.*, spectral function minus the high-frequency tail. Various values of χ_0 correspond to different values of $T \geq T_{\rm fun}$ with $\chi_0 = 0.94 \leftrightarrow T = T_{\rm fun}$ and $\chi_0 = 0 \leftrightarrow T = \infty$. (See [11] for details.)

We wish to study (the real and imaginary parts of) the dispersion relations of meson quasiparticles with small (tunable) finite widths. This is achieved by introducing a small quark density[a] to the thermal plasma.[12,13] In the gravity dual, this is accomplished by turning on an electric field on the world-volume of the D7-branes which deforms their geometry to have a narrow neck extending down to the horizon at all temperatures.[12] Hence all excitations on the branes are slowly absorbed by the black hole and this results in the mesons having a finite lifetime. Now the most direct approach to studying the dispersion relations would be to identify the poles in the thermal correlators by studying the quasinormal modes[14] in the induced black hole geometry on the D7-branes. However, this is technically a very demanding problem in this set-up.[15] An alternative approach to access the same physics is studying the corresponding spectral functions $\mathfrak{R}$ as a function of $\mathfrak{w}, \mathfrak{q}$. Essentially examining the locations and widths of the peaks in the spectral function, we can reconstruct the quasinormal frequencies for the modes dual to the mesons.

The simplest case to study are the transverse components of the vector meson. The holographic spectral function $\mathfrak{R}$ is given by

$$\mathfrak{R}(\mathfrak{w}, \mathfrak{q}) = -\frac{N_f N_c T^2}{4}\, \mathfrak{w}^2 \,\mathrm{Im}\mathfrak{S}(\rho \to \infty)\,, \tag{6}$$

where $\mathfrak{S}$ obeys the non-linear differential equation

$$\partial_\rho \mathfrak{S} + \frac{\mathfrak{w}^2}{F}\mathfrak{S}^2 + G\left(\frac{\tilde{f}^2}{\Delta f^2} - \frac{\mathfrak{q}^2}{\mathfrak{w}^2}\right) = 0\,. \tag{7}$$

Here ρ is the holographic radial coordinate, with the black hole horizon at $\rho = 1$ and the asymptotic boundary at $\rho \to \infty$. The boundary condition is $\mathfrak{S}(1) = -4i\sqrt{(1-\chi_0^2)^3 + \tilde{d}^2}/\mathfrak{w}$ where χ_0 specifies the size of the induced horizon on the D7-brane and so naturally controls the lifetimes of the meson quasiparticles. Here $\tilde{d}$ determines the quark density n_q while the quark mass m_q implicitly depends on both χ_0 and $\tilde{d}$. Hence, as we will see below, by varying both of these parameters together, the meson widths can be tuned while keeping m_q fixed. The four-momentum of the meson is given by $k = (\mathfrak{w}, \mathfrak{q}, 0, 0)$. The functions $f, \tilde{f}$ are inherited from the black hole geometry while the functions F, G, Δ depend on the profile of the D7-brane embedded in the black hole background. For more details, the reader is referred to [1]. This non-linear differential equation (7) can be solved numerically. Typical spectral functions with $\mathfrak{q} = 0$ are depicted in fig. 2.

[a]Note that this density consists of both fermions and bosons because of the underlying supersymmetry of the theory.

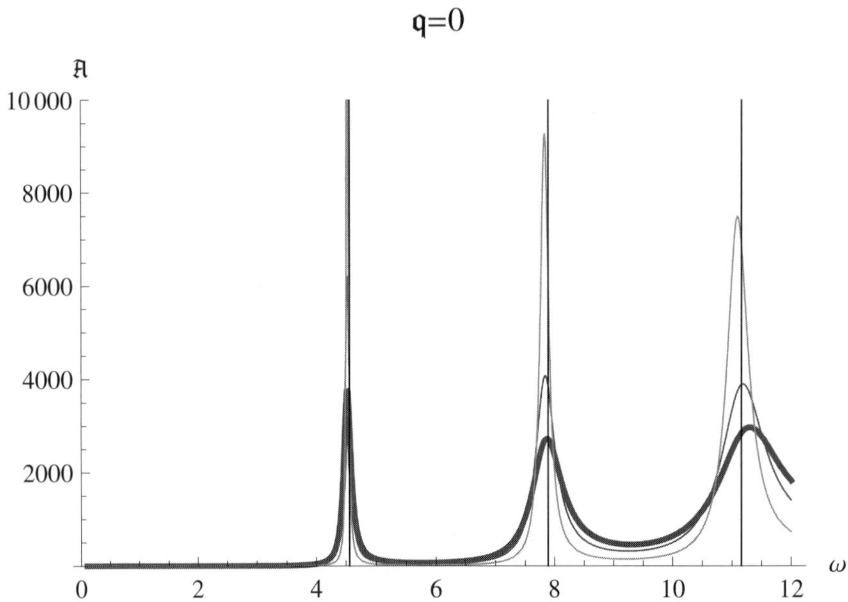

Fig. 2. Spectral function for mesons with non-zero quark density and $\mathfrak{q} = 0$. Different lines correspond to different quark densities but a fixed quark mass. The quasiparticles acquire an increasing width as the quark density increases.

Now to study the dispersion relations, we introduce a finite momentum $\mathfrak{q}$. First following the location of the peaks with increasing $\mathfrak{q}$, we found that the curves $\Omega_i(\mathfrak{q})$ appear to asymptote to the straight-line form:

$$\Omega_i(\mathfrak{q}) = v_{\rm lim}\,\mathfrak{q} + a_i + O(1/\mathfrak{q}) \tag{8}$$

Examples are shown in fig. 3. This illustrates that $v_{\rm lim}$ can be much less than the speed of light, as we found for the stable messons. This effect arises because even though the dual excitations locally travel at the speed of light in the bulk, the speed is $v_{\rm lim} < 1$ in the gauge theory due to the redshift near the black hole horizon. In fact, we found that $v_{\rm lim}$ for the quasiparticles accurately matches that for the stable mesons with the same m_q at $n_q = 0$:

$$v_{\rm lim}^2 \simeq 1 - \frac{\lambda^2}{4}\left(\frac{T}{m_q}\right)^4 . \tag{9}$$

Next we wish to consider the behaviour of the widths Γ as a function of the momentum $\mathfrak{q}$. As noted above, for an isolated pole in the thermal correlator, we expect the spectral function to exhibit a peak of the form

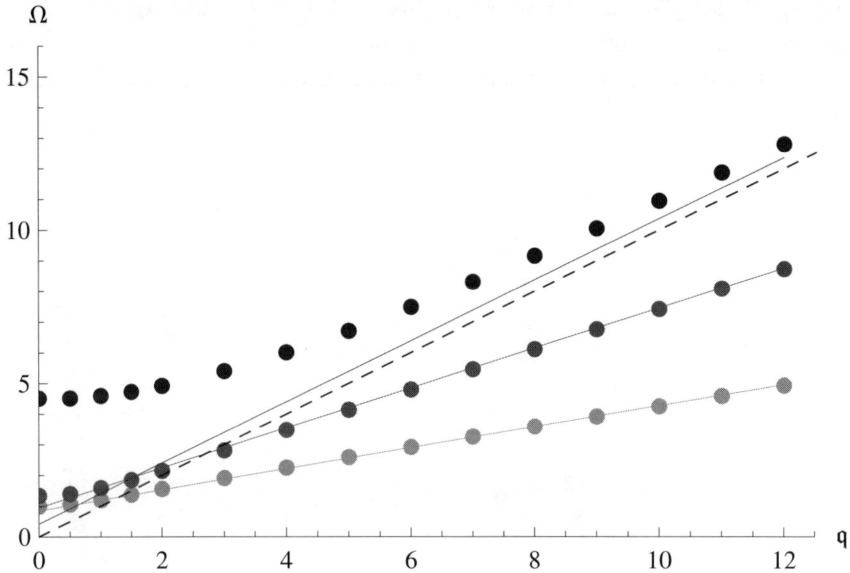

Fig. 3. The plot shows the positions of the first peak in the spectral function for three different quark densities. The dashed line is $\Omega = \mathfrak{q}$.

given in eq. (3). However, the full spectral function will also include contributions from the high-frequency tail (5). In the present case, $\Delta = 3$ and so we observe that by taking the third derivative, $\partial^3_{\mathfrak{w}}\mathfrak{R}$, the latter contribution will be eliminated. Further the resulting expression will have zeros at $\mathfrak{w}_0 \equiv \Omega$ and $\mathfrak{w}_{0,\pm} \equiv \Omega \pm \Gamma$ and so in principle, the zeros of $\partial^3_{\mathfrak{w}}\mathfrak{R}$ can be used to determine Γ. However, rather than working directly with these zeros, we note that there are also a maximum and a minimum to either side of the central zero at Ω. A straightforward calculation shows that these extrema occur at

$$\mathfrak{w}_{m,\pm} \equiv \Omega \pm \delta\,\Gamma \qquad \text{with } \delta = \sqrt{1 - \frac{2}{\sqrt{5}}} \simeq 0.3249\,. \tag{10}$$

We can use these expressions to estimate both the width of a given peak but also an error because our form (3) is not perfect. In particular, we have

$$\bar{\Gamma} = \frac{\mathfrak{w}_{m,+} - \mathfrak{w}_{m,-}}{2\delta} \qquad \Delta\Gamma = \frac{1}{2\delta}\left(\mathfrak{w}_{m,+} + \mathfrak{w}_{m,-} - 2\mathfrak{w}_0\right)\,. \tag{11}$$

Examples of the resulting widths using these expressions are shown in fig. 4. As is clear, the widths of the quasiparticles increase with increasing $\mathfrak{q}$ and there appears a $\mathfrak{q}_{\max}$ beyond which we cannot reliably use the notion of quasiparticles. The latter notion requires both $\Gamma \ll \Omega$ and $\Gamma \ll \Delta\Omega$ where $\Delta\Omega$ is the spacing between neighbouring peaks. However, subsequent peaks appear to coalesce at roughly the point where they vanish from the spectral function.

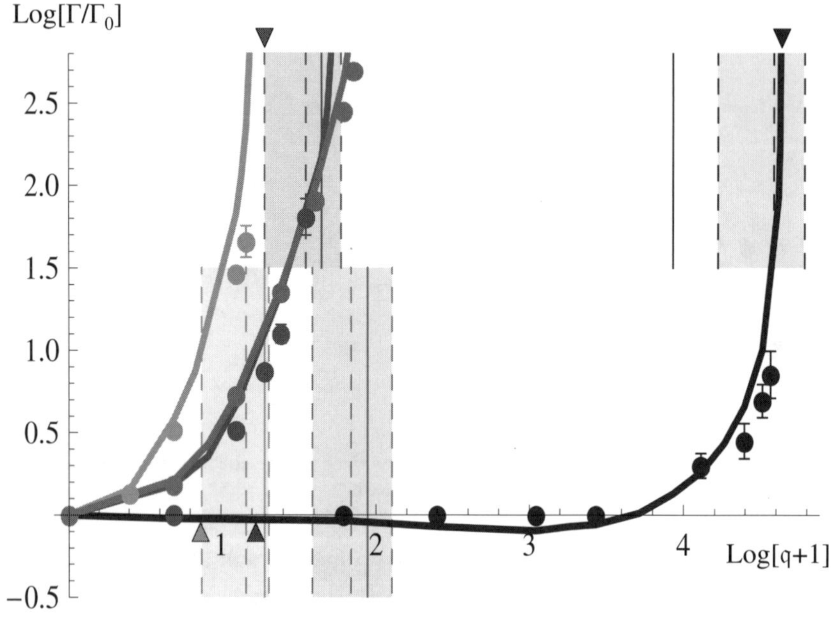

Fig. 4. This plot shows the widths of the quasiparticles. The continuous curves were obtained from (3) ignoring the high-frequency tail. The error bars were obtained using (11). The shaded regions illustrate the range of $\mathfrak{q}$ over which the peak is washed away.

As described in [1,11], some qualitative information can be inferred about the quasinormal modes by casting the relevant radial equations of motion in the form of a Schrödinger equation:

$$-\partial_{R_*}^2 \psi + V\,\psi = \mathfrak{w}^2 \psi\,. \tag{12}$$

Here the new radial $R_*(\rho)$ has $R_* \to \infty$ at the horizon (*i.e.*, $\rho \to 1$) and $R_* \to 0$ asymptotically (*i.e.*, $\rho \to \infty$). Some examples of the effective potential V for varying values of $\mathfrak{q}$ are illustrated in fig. 5a. We note that this potential has the form $V = V_0 + \mathfrak{q}^2\, V_1$ where V_0 and V_1 are independent of both $\mathfrak{q}$ and $\mathfrak{w}$. Hence the Schrödinger problem has a particularly simple dependence on the frequency and wave-number of the mesons. That is, $\mathfrak{w}^2$ plays the role of the effective energy while $\mathfrak{q}^2$ appears as the coefficient of the term V_1 in the effective potential. Further we observe that in the limit that $\rho \to \infty$, $V_1 \to 1$ and hence in this UV regime, these two contributions to (12) can be brought together in the Lorentz invariant combination $\mathfrak{w}^2 - \mathfrak{q}^2$.

To summarize, given the effective Schrödinger equation (12), we develop an intuitive picture with the following key observations: For small $\mathfrak{q}$, the effective potential has a large barrier which is responsible for the metastability of the quasiparticles. That is, the low-lying quasiparticles correspond to metastable states which are bound in the potential well and which slowly leak through the barrier to be absorbed at the horizon. However, this barrier shrinks with increasing $\mathfrak{q}$ which then explains why the decay rates of the metastable states should increase as $\mathfrak{q}$ increases. Further this barrier actually disappears above some momentum which again points to a limiting momentum $\mathfrak{q}_{\rm max}$ beyond which the quasiparticles disappear. Since our intuition is that metastable states exist because of the potential barrier appearing V, we propose to estimate $\mathfrak{q}_{\rm max}$ as the value of the momentum at which the barrier disappears and is reduced to a point of inflexion.

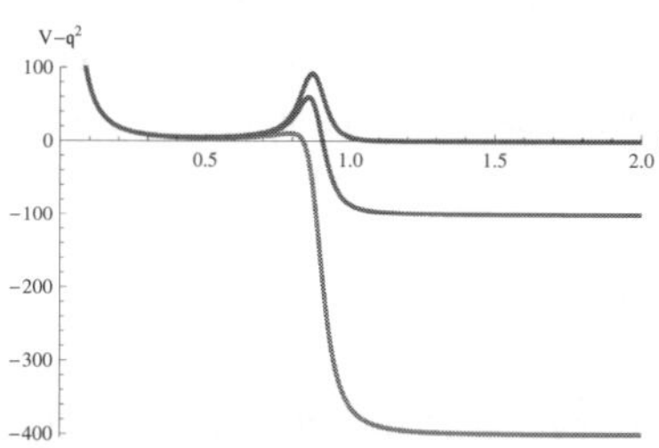

(a)

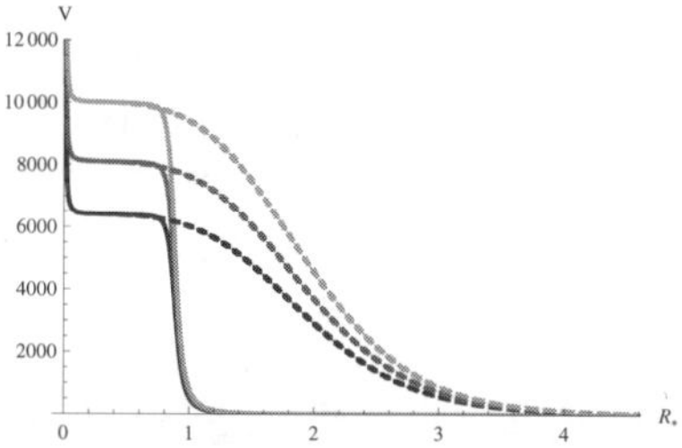

(b)

Fig. 5. (a) Effective Schrödinger potential for non-zero n_q. With increasing $\mathfrak{q}$ the barrier decreases and eventually disappears. (b) The solid lines correspond to non-zero n_q while the dashed lines correspond to the analogous supergravity potential for R-currents for $\mathfrak{q} > \mathfrak{q}_{\rm max}$.

An interesting question is to ask what happens to the quasinormal modes beyond $\mathfrak{q}_{\rm max}$. In order to answer this question qualitatively, we again look at the effective Schrödinger equation for the modes. One can compare the nature of the potentials with that for R currents in supergravity[14] which are similar to the case we are investigating. This is illustrated in fig. 5b, from which we see that for large momenta and small Schrödinger coordinate R_*, the supergravity potential and the meson potential for $n_q \neq 0$ are the essentially same. This suggests that the quasinormal modes in the two cases will also have the same behaviour. Hence we infer from the results of

[14] that in the regime $\mathfrak{q} \gg \mathfrak{q}_{max}$, the group velocity for the meson modes should rise to the speed of light. Further Γ should also decrease with growing $\mathfrak{q}$ in this regime. These results can be understood by arguing that the primary support for the corresponding wavefunctions is on the ledge near the $R_* = 0$. The latter corresponds to the asymptotic region of the gravity background where any redshift effects are minimal.

3. Discussion

Using holographic techniques, we investigated the spectral functions of meson quasiparticles. From the spectral functions, we deduced the behaviour of the corresponding quasinormal modes. We found that for momenta below some $\mathfrak{q}_{max}$, the limiting velocity is substantially smaller than the speed of light. Since this is the same limiting velocity found in the case of stable mesons, our results suggest that the limiting velocity will be a universal feature in any holographic model. The widths also rise dramatically so that the notion of quasiparticles breaks down eventually. We gave a qualitative argument that for $\mathfrak{q} \gg \mathfrak{q}_{max}$ the group velocity will rise to the speed of light. The physical picture that arises from our analysis is the following: In the low-momentum regime, the quasiparticles are strong coupled to the deconfined plasma of the adjoint fields. The gravitational redshift leading to $v_{lim} < 1$ is the geometric realization of this strong coupling effect. The energy density of this halo must be rapidly increasing to maintain the meson's velocity at $v_{lim} < 1$. However, in the high momentum regime, the quasiparticles and the adjoint plasma are no longer strongly coupled so that the meson excitations can achieve the speed of light. This restores the intuitive picture that at high momentum, the meson excitations should be largely unaffected by the surrounding plasma. It is an interesting open problem to verify this picture by a direct calculation of the quasinormal frequencies. Recently, the fact that the widths increases with momenta has also been observed as a strong coupling effect in [16]. In this case, the net quark density was fixed at zero but the presence of free quarks (and anti-quarks) in the thermal bath catalyzed the decay of the mesons. Of course, if a limiting velocity and limiting momentum were realized in the strongly coupled quark-gluon plasma at the RHIC or the LHC, this would provide interesting experimental signatures, *e.g.*, in the suppression of J/ψ production.[9]

Acknowledgments

Research at Perimeter Institute is supported by the Government of Canada through Industry Canada and by the Province of Ontario through the Ministry of Research & Innovation. RCM also acknowledges support from an NSERC Discovery grant and funding from the Canadian Institute for Advanced Research. RCM thanks the organizers of CAQCD-08 for the opportunity to speak on this work.

References

1. R. C. Myers and A. Sinha, JHEP **0806**, 052 (2008) arXiv:0804.2168 [hep-th].
2. P. Kovtun, D.T. Son and A.O. Starinets, Phys. Rev. Lett. **94**, 111601 (2005) [arXiv:hep-th/0405231];
   P. Kovtun, D.T. Son and A.O. Starinets, JHEP **0310**, 064 (2003) [arXiv:hep-th/0309213].
3. A. Buchel and J. T. Liu, Phys. Rev. Lett. **93**, 090602 (2004) [arXiv:hep-th/0311175].
4. D. Teaney, Phys. Rev. C **68** (2003) 034913 [arXiv:nucl-th/0301099];
   A. Adare *et al.* [PHENIX Collaboration], Phys. Rev. Lett. **98** (2007) 172301 [arXiv:nucl-ex/0611018];
   P. Romatschke and U. Romatschke, Phys. Rev. Lett. **99** (2007) 172301 [arXiv:0706.1522 [nucl-th]];
   H. Song and U. W. Heinz, Phys. Lett. B **658** (2008) 279 [arXiv:0709.0742 [nucl-th]];
   K. Dusling and D. Teaney, Phys. Rev. C **77** (2008) 034905 [arXiv:0710.5932 [nucl-th]];
   H. Song and U. W. Heinz, arXiv:0712.3715 [nucl-th];
   M. Luzum and P. Romatschke, arXiv:0804.4015 [nucl-th];
   H. Song and U. W. Heinz, arXiv:0805.1756 [nucl-th].
5. O. Aharony, A. Fayyazuddin and J. M. Maldacena, JHEP **9807**, 013 (1998) [arXiv:hep-th/9806159];
   A. Karch and E. Katz, JHEP **0206**, 043 (2002) [arXiv:hep-th/0205236].
6. M. Kruczenski, D. Mateos, R. C. Myers and D. J. Winters, JHEP **0307**, 049 (2003) [arXiv:hep-th/0304032].
7. J. Babington, J. Erdmenger, N.J. Evans, Z. Guralnik and I. Kirsch, Phys. Rev. D **69** 066007 (2004) [arXiv:hep-th/0306018];
   I. Kirsch, Fortsch. Phys. **52** 727 (2004) [arXiv:hep-th/0406274];
   T. Albash, V.G. Filev, C.V. Johnson and A. Kundu, Phys. Rev. D **77**, 066004 (2008) [arXiv:hep-th/0605088];
   D. Mateos, R.C. Myers and R.M. Thomson, Phys. Rev. Lett. **97** 091601 (2006) [arXiv:hep-th/0605046].
8. D. Mateos, R.C. Myers and R.M. Thomson, JHEP **0705**, 067 (2007) [arXiv:hep-th/0701132].
9. Q.J. Ejaz, T. Faulkner, H. Liu, K. Rajagopal and U.A. Wiedemann, arXiv:0712.0590 [hep-th].

10. D.T. Son and A.O. Starinets, JHEP **0209**, 042 (2002) [arXiv:hep-th/0205051].
11. R.C. Myers, A.O. Starinets and R.M. Thomson, JHEP **0711**, 091 (2007) [arXiv:0706.0162 [hep-th]].
12. S. Kobayashi, D. Mateos, S. Matsuura, R. C. Myers and R. M. Thomson, JHEP **0702**, 016 (2007) [arXiv:hep-th/0611099];
    D. Mateos, S. Matsuura, R. C. Myers and R. M. Thomson, JHEP **0711**, 085 (2007) [arXiv:0709.1225 [hep-th]].
13. J. Erdmenger, M. Kaminski and F. Rust, arXiv:0710.0334 [hep-th].
14. A. Nunez and A.O. Starinets, Phys. Rev. D **67** 124013 (2003) [arXiv:hep-th/0302026];
    A.O. Starinets, Phys. Rev. D **66** 124013 (2002) [arXiv:hep-th/0207133].
15. C. Hoyos-Badajoz, K. Landsteiner and S. Montero, JHEP **0704**, 031 (2007) [arXiv:hep-th/0612169].
16. T. Faulkner and H. Liu, arXiv:0807.0063 [hep-th].

Properties of Baryons from D-branes and Instantons *

Shigeki Sugimoto

Institute for the Physics and Mathematics of the Universe, University of Tokyo, Chiba 277-8568, Japan
** E-mail: shigeki.sugimoto@ipmu.jp*
www.ipmu.jp

We give a brief summary of our recent works[1,2] on the baryons in holographic QCD.[3] In the holographic description, meson effective action is given by the five dimensional Yang-Mills - Chern-Simons theory in a certain curved background. The baryon is described as a soliton carrying an instanton number on the four dimensional space. By quantizing the soliton, we obtain various baryon states and calculate static quantities such as charge radii, magnetic moments, axial coupling, etc.

1. Introduction

In the early 60's, Skyrme proposed that baryons are solitons (Skyrmion) in a pion effective theory.[4] This remarkable idea was further investigated by Adkins, Nappi and Witten (ANW) in 1983.[5] They succeeded in analyzing the static properties of the baryons by quantizing the collective modes of the Skyrmion. Their results roughly agree with the experimental data, suggesting that the Skyrme model does catch the essential ingredients of the baryon physics, though this description of baryon looks completely different from the quark model picture.

Recently, based on gauge/string duality, we obtained a holographic description of the effective theory of mesons in $U(N_c)$ QCD with N_f massless quarks.[3] The action is given by a five dimensional $U(N_f)$ Yang-Mills (YM)

*Talk given at Continuous Advances in QCD 2008, William I. Fine Theoretical Physics Institute, Minnesota, May 15-18, 2006

- Chern-Simons (CS) theory in a curved space-time:

$$S_{5\mathrm{dim}} \simeq S_{\mathrm{YM}} + S_{\mathrm{CS}} \ ,$$
$$S_{\mathrm{YM}} = \kappa \int d^4x dz \, \mathrm{Tr} \left(\frac{1}{2} h(z) F_{\mu\nu}^2 + k(z) F_{\mu z}^2 \right) \ ,$$
$$S_{\mathrm{CS}} = \frac{N_c}{24\pi^2} \int_{5\mathrm{dim}} \omega_5(A) \ , \tag{1}$$

where $\mu, \nu = 0, \cdots, 3$ are the Lorentz indices for the four dimensional world and z is the coordinate of the fifth dimension. The warp factors in the YM action are given by $k(z) = 1 + z^2$ and $h(z) = (1 + z^2)^{-1/3}$. $\kappa = a\lambda N_c$ ($a = 1/(216\pi^3)$) is a constant proportional to N_c and the 't Hooft coupling λ, and $\omega_5(A)$ is the Chern-Simons five form. This system is shown to be equivalent to four dimensional meson effective theory that contains infinitely many mesons, such as $\pi, \rho, a_1, \rho', a_1', \cdots$. Using this five dimensional YM-CS theory, masses and couplings of these mesons are calculated and it was found that they are roughly in agreement with the experimental data.

Given a meson effective theory, it is natural to expect that the Skyrme's idea can be applied to this system and a baryon can be realized as a kind of soliton solution. Actually, it has been shown that the baryons are realized as instantons localized on the four dimensional space. Now the question is whether we can apply the idea of ANW to this system. Since our model contains not only pion but also various vector and axial-vector mesons with reasonably simple action (1), we expect that the results should be improved compared with the Skyrme model (ANW). The purpose of this article is to show that this is indeed the case. We explain how to calculate the static quantities, such as charge radii, magnetic moments, axial coupling etc., generalizing the strategy of ANW, and show that most of them get closer to the experimental values. This article is mainly based on our recent papers Refs. 1 and 2, in which more details and further results can be found.[a]

2. Brief summary of the model

Here we briefly summarize the model.[3] If you accept the action (1) as our starting point, you can skip this section and go directly to section 3.

2.1. *Construction of QCD*

Recently, we proposed to consider D4-D8-$\overline{\mathrm{D8}}$ system in type IIA string theory to obtain a holographic dual of four dimensional QCD.[3] Let us

[a]See also closely related works Refs. 6–8

consider N_c D4-branes extended along $x^{0\sim4}$ directions and N_f D8-$\overline{\text{D8}}$ pairs extended along $x^{0\sim3}$ and $x^{5\sim9}$ directions. To obtain a four dimensional non-supersymmetric gauge theory, the x^4 direction is compactified to S^1 with the anti-periodic boundary condition for all the fermions in the system. The radius of the S^1 is denoted by M_{KK}^{-1}, although we mainly work in the $M_{\text{KK}} = 1$ unit. Then the gluon field is created by the open strings attached on the D4-branes and from the open strings connecting between the D4-brane and the D8-brane ($\overline{\text{D8}}$-brane), we find left (right) handed component of the quark field. In this way, it can be shown that the massless field content in this system is the four dimensional $U(N_c)$ QCD with N_f massless quarks. Note that the gauge symmetry for the D8-$\overline{\text{D8}}$ pairs is $U(N_f)_{\text{D8}} \times U(N_f)_{\overline{\text{D8}}}$ and this is interpreted as the chiral symmetry $U(N_f)_L \times U(N_f)_R$ of massless QCD.

On the other hand, if the number of color N_c and the 't Hooft coupling λ are large, the system can be well described by replacing D4-brane with the corresponding supergravity solution. The D4-brane solution is explicitly known[9] and it has been used to analyze four dimensional pure Yang-Mills theory. (See Ref. 10 for a review.) Assuming $N_c \gg N_f$, the D8-$\overline{\text{D8}}$ pairs are treated as probe D8-branes embedded in the D4-brane background. Now the topology of the background is $\mathbf{R}^{1,3} \times \mathbf{R}^2 \times S^4$, where $\mathbf{R}^{1,3}$ is the four dimensional space-time parameterized by $x^{0\sim3}$, the angular direction of the $\mathbf{R}^2$ is the S^1 parameterized by x^4, the radial direction of the $\mathbf{R}^2$ corresponds to the distance from the D4-brane, and the S^4 corresponds to the angular directions of the five dimensional plane parameterized by $x^{5\sim9}$. In this background, the D8-brane and $\overline{\text{D8}}$-brane should be smoothly connected and the system becomes a string theory with a single connected component of D8-brane embedded in the D4-brane background. The D8-brane is extended along $\mathbf{R}^{1,3} \times S^4$ and one dimensional subspace of $\mathbf{R}^2$ parameterized by $z \in (-\infty, +\infty)$.

Now, applying the idea of the gauge/string duality to the above construction, we conjecture that the four dimensional QCD with N_f massless quarks is dual to type IIA string theory in the D4-brane background with N_f probe D8-branes (at low energy). The former gauge theory description is better when the 't Hooft coupling is small, while the latter string theory description is better when the 't Hooft coupling is large.

Note however that in the gauge theory description, the gluons live on the D4-brane world-volume wrapped on an S^1 of radius M_{KK}^{-1} and hence the system becomes five dimensional if the energy scale is much higher than M_{KK}. In order to get rid of the Kaluza-Klein modes associated with the S^1, we should in principle consider a limit $M_{\rm KK} \to \infty$, keeping physical quantities, such as ρ meson mass, finite. This is analogous to the continuum limit in the lattice QCD. However, if we start with the string theory description and rely on the supergravity approximation, this limit is unfortunately not accessible, since the asymptotic freedom of QCD implies that the 't Hooft coupling becomes small when the "cut-off" scale $M_{\rm KK}$ becomes large, and we have to go beyond the supergravity approximation. This is one of the long standing problems in this kind of approach. In this article, we use the supergravity approximation assuming that the 't Hooft coupling λ is large and do not attempt to take the "continuum limit" $M_{\rm KK} \to \infty$. Therefore, we should keep in mind that our "QCD" deviates from real QCD at the energy scale higher than $M_{\rm KK}$.

2.2. *5 dim YM-CS theory as a theory of mesons*

In the string theory description of QCD constructed in the previous subsection, the open strings attached on the probe D8-branes are interpreted as mesons, since they carry two flavor indices. The low energy effective theory of the open strings on the D8-brane is given by a nine dimensional $U(N_f)$ gauge theory. Here recall that there is an $SO(5)$ isometry that acts as rotation of the S^4 of the background. Since the gluon and quark fields are invariant under this $SO(5)$, we restrict our attention to the $SO(5)$ invariant sector. The $SO(5)$ non-invariant particles are interpreted as bound states that involve unwanted Kaluza-Klein modes which are expected to decouple if we take the "continuum limit" $M_{\rm KK} \to \infty$. Then, the effective theory reduces to a five dimensional $U(N_f)$ gauge theory. Using the supergravity solution of the D4-brane background, we can explicitly calculate the effective action which turns out to be the five dimensional $U(N_f)$ YM-CS theory (1).

In order to extract four dimensional physics from the five dimensional gauge theory, it is convenient to expand the five dimensional gauge field as

$$A_\mu(x^\mu, z) = \sum_{n\geq 1} B_\mu^{(n)}(x^\mu)\psi_n(z) \ , \quad A_z(x^\mu, z) = \sum_{n\geq 0} \varphi^{(n)}(x^\mu)\phi_n(z) \ , \quad (2)$$

where $\{\psi_n(z)\}_{n\geq 1}$ and $\{\phi_n(z)\}_{n\geq 0}$ are complete sets of the functions of z. We choose the complete sets so that the kinetic terms and mass terms of the four dimensional fields $B_\mu^{(n)}(x)$ and $\varphi^{(n)}(x)$ are diagonalized. The functions $\psi_n(z)$ $(n = 1, 2, 3, \cdots)$ are eigenfunctions of the eigenequation

$$-h(z)^{-1}\partial_z(k(z)\partial_z\psi_n) = \lambda_n\psi_n \ , \quad (3)$$

where λ_n are the eigenvalues. The functions $\phi_n(z)$ $(n = 0, 1, 2, \cdots)$ are given by $\phi_n(z) = \lambda_n^{-1/2}\partial_z\psi_n(z)$ for $n = 1, 2, 3, \cdots$ and $\phi_0(z) = (\kappa\pi)^{-1/2}k(z)^{-1}$.

Inserting this mode expansion (2) into the five dimensional action (1), we obtain a four dimensional action of mesons. It turns out that $B_\mu^{(n)}$ become massive vector and axial-vector meson fields for odd and even n, respectively, whose masses m_n are related to the eigenvalues λ_n by $m_n^2 = \lambda_n M_{\mathrm{KK}}^2$. The scalar fields $\varphi^{(n)}$ $(n = 1, 2, 3, \cdots)$ are eaten by $B_\mu^{(n)}$ to make them massive, while $\varphi^{(0)}(x)$ remain massless. We interpret $\varphi^{(0)}$, $B_\mu^{(1)}$, $B_\mu^{(2)}$, etc., as pion, ρ meson, a_1 meson, etc., respectively. In this way, various meson fields π, ρ, a_1, etc. are beautifully unified in the five dimensional gauge field.

3. Baryons as instantons

Let us consider a static gauge configuration in the five dimensional YM-CS theory (1) with non-trivial instanton number in the four dimensional space parameterized by $x^M = (\vec{x}, z) \in \mathbf{R}^4$ $(M = 1, 2, 3, z)$. Since the energy density is localized along the spatial directions, it behaves as a point-like particle and we interpret it as a baryon. In fact, it is not difficult to show that the baryon number N_B is related to the instanton number as

$$N_B = \frac{1}{8\pi^2}\int \mathrm{Tr}\, F \wedge F\ . \tag{4}$$

A few comments are in order. Note that this instanton configuration is equivalent to a D4-brane embedded in the D8-brane world-volume.[11] This D4-brane is wrapped on the S^4 and is nothing but the baryon vertex considered in the context of AdS/CFT.[12,13] It is also interesting to note that the realization of baryon as an instanton configuration was introduced by Atiyah and Manton in 1989,[14] who proposed to use an instanton configuration to construct a Skyrmion. Our description of baryon naturally realizes their old idea and gives a definite physical interpretation of the fifth coordinate z as one of the coordinates on the D8-brane world-volume.

In order to make quantitative predictions, we need to find a classical solution with non-trivial instanton number. From now we consider the case with $N_f = 2$. If we ignore the CS-term, one can easily show that the instanton solution shrinks to zero size and located at $z = 0$, because of the warp factors $h(z)$ and $k(z)$. On the other hand, the CS-term has an effect to make the size of the instanton larger. This is because the CS-term contains a term like

$$S_{\mathrm{CS}} \sim \frac{N_c}{16\pi^2}\int d^4x dz\, A_0^{U(1)}\epsilon^{ijk}\mathrm{Tr}F_{ij}F_{kz} + \cdots\ , \tag{5}$$

where $A^{U(1)}$ is the $U(1)$ part of the $U(2)$ gauge field. This term acts as a source of the $U(1)$ charge for non-trivial instanton configuration. The repulsive force due to this $U(1)$ charge enlarge the size of the instanton and it is stabilized at a finite value.

The size of the instanton is estimated as $\mathcal{O}(\lambda^{-1/2})$ and hence it becomes small if λ is large. If the instanton is small enough, the five dimensional space-time can be approximated by the flat space-time, since we can use $h(z) \simeq k(z) \simeq 1$ for $|z| \ll 1$. Then, the leading order classical solution for large λ is given by the instanton solution in the flat space-time. The $SU(2)$ part is given by

$$A_M^{\rm cl} = -i\frac{\xi^2}{\xi^2+\rho^2} g\partial_M g^{-1} \tag{6}$$

with

$$g = \frac{(z-Z) - i(\vec{x}-\vec{X})\cdot\vec{\tau}}{\xi}\ , \quad \xi = \sqrt{(\vec{x}-\vec{X})^2 + (z-Z)^2}\ . \tag{7}$$

and the $U(1)$ part is

$$A_0^{U(1)} = \frac{1}{8\pi^2 a\lambda}\cdot\frac{1}{\xi^2}\left(1 - \frac{\rho^4}{(\rho^2+\xi^2)^2}\right)\ . \tag{8}$$

Here ρ is the size of the instanton and $(\vec{X}, Z)$ represents the position of the instanton in the four dimensional space.

4. Quantization

Next we consider a slowly moving (rotating) baryon configuration. We use the moduli space approximation method to quantize the system. The basic idea is to consider the classical solution (6) with the instanton moduli parameters promoted to time dependent variables and insert them into the action (1). Then we obtain an action for the quantum mechanics on the instanton moduli space. The $SU(2)$ one instanton moduli space is parameterized by $\{(\vec{X}, Z, \rho, \mathbf{a})\} \equiv \{(X^\alpha)\}$ where $\mathbf{a} \in SU(2)$ represent the $SU(2)$ orientation of the instanton. Then the Lagrangian of the quantum mechanics obtained in the above procedure is

$$L_{\rm QM} = \frac{1}{2}G_{\alpha\beta}\dot{X}^\alpha\dot{X}^\beta - U(X^\alpha)\ , \tag{9}$$

where $G_{\alpha\beta}$ is the metric of the instanton moduli space and

$$U(X^\alpha) = 8\pi^2\kappa\left(1 + \left(\frac{\rho^2}{6} + \frac{3^6\pi^2}{5\lambda^2\rho^2} + \frac{Z^2}{3}\right) + \cdots\right)\ . \tag{10}$$

The minimum of the potential $U(X^\alpha)$ is given by $\rho^2 = \rho_{\rm cl}^2 \equiv \frac{N_c}{8\pi^2\kappa}\sqrt{\frac{6}{5}}$ and $Z = 0$.

Here $(\vec{X}, \mathbf{a})$ are genuine moduli that also appear in the analysis of ANW for the Skyrme model. On the other hand, (ρ, Z) are not moduli parameters in the usual sense, since they have a non-trivial potential (10). We keep these new degrees of freedom, since one can show that they are light compared with the other massive modes for large λ.

Solving the Schrödinger equation for this quantum mechanics, we obtain the baryon states. For example, the wave function of the spin up proton state $|p\uparrow\rangle$ is given by

$$\psi(\vec{X}, \mathbf{a}, \rho, Z) \propto e^{i\vec{p}\cdot\vec{X}} R(\rho)\psi_Z(Z)T(\mathbf{a}) \tag{11}$$

where

$$R(\rho) = \rho^{-1+2\sqrt{1+N_c^2/5}} e^{-\frac{8\pi^2\kappa}{\sqrt{6}}\rho^2} , \quad \psi_Z(Z) = e^{-\frac{8\pi^2\kappa}{\sqrt{6}}Z^2} , \quad T(\mathbf{a}) = a_1 + ia_2 . \tag{12}$$

In this way, we can explicitly write down the wave function for various baryon states including not only nucleon and Δ, but also $N(1440)$, $N(1535)$, etc.

5. Currents

Here we calculate the currents (J_L^μ, J_R^μ) associated with the chiral symmetry $U(N_f)_L \times U(N_f)_R$. First, note that we can introduce the external gauge fields $(A_{L\mu}(x), A_{R\mu}(x))$ for the chiral symmetry $U(N_f)_L \times U(N_f)_R$, considering the gauge field with non-trivial boundary values as

$$A_{L\mu}(x) = \lim_{z\to+\infty} A_\mu(x,z) , \quad A_{R\mu}(x) = \lim_{z\to-\infty} A_\mu(x,z) . \tag{13}$$

The currents are obtained by inserting the gauge configuration with this boundary condition into the action (1) and picking up the terms linear with respect to the external gauge fields as

$$S_{\rm 5dim}\big|_{\mathcal{O}(A_L,A_R)} = -2\int d^4x \,{\rm Tr}\,(A_{L\mu}J_L^\mu + A_{R\mu}J_R^\mu) . \tag{14}$$

As a result, we obtain

$$J_{L\mu} = -\kappa(k(z)F_{\mu z})\big|_{z=+\infty} , \quad J_{R\mu} = +\kappa(k(z)F_{\mu z})\big|_{z=-\infty} . \tag{15}$$

The vector and axial-vector currents are given by

$$\begin{aligned} J_{V\mu} &= J_{L\mu} + J_{R\mu} = -\kappa\,[k(z)F_{\mu z}]_{z=-\infty}^{z=+\infty} , \\ J_{A\mu} &= J_{L\mu} - J_{R\mu} = -\kappa\,[\psi_0(z)k(z)F_{\mu z}]_{z=-\infty}^{z=+\infty} , \end{aligned} \tag{16}$$

with $\psi_0(\pm\infty) = \pm 1$.

In order to calculate the current (15), we need to know how $F_{\mu z}(x, z)$ behaves at $z \to \pm\infty$. For this purpose, we can no longer use the solution (6) which is valid only for $|z| \ll 1$. The key observation is that the gauge field (in a suitable gauge choice) becomes small when $\rho \ll |z|$ and the non-linear terms in the equations of motion can be neglected. Then, what we have to do is to solve the linearized equations of motion that agrees with (6) in the intermediate region $\rho \ll |z| \ll 1$.

As an illustrative example, let us consider the time component of the $U(1)$ gauge field whose solution for $\xi \ll 1$ is given by (8). For $\rho \ll \xi \ll 1$, it can be approximated as

$$A_0^{U(1)} \simeq \frac{1}{8\pi^2 a\lambda} \cdot \frac{1}{\xi^2} \equiv -\frac{1}{2a\lambda} G^{\rm flat}(\vec{x}, z; \vec{X}, Z) \tag{17}$$

where $G^{\rm flat}$ is the Green's function in the flat space satisfying

$$\partial_M \partial^M G^{\rm flat} = \delta^3(\vec{x} - \vec{X})\delta(z - Z) \ . \tag{18}$$

The solution for $\rho \ll \xi$ is then given by replacing flat space Green's function $G^{\rm flat}$ with the Green's function for curved space-time G:

$$A_0^{U(1)} \simeq -\frac{1}{2a\lambda} G(\vec{x}, z; \vec{X}, Z) \tag{19}$$

where G is the Green's function in the curved space satisfying

$$\left(h(z)\partial_i^2 + \partial_z k(z) \partial_z\right) G = \delta^3(\vec{x} - \vec{X})\delta(z - Z) \ . \tag{20}$$

Using (3), it can be shown that the Green's function G can be written as

$$G = \kappa \sum_{n=1}^{\infty} \psi_n(z)\psi_n(Z) \left(-\frac{1}{4\pi} \frac{e^{-m_n r}}{r} \right) , \tag{21}$$

where $r = |\vec{x} - \vec{X}|$.

6. Exploration

Now we are ready to calculate various physical quantities. We are going to show some numerical results that can be compared with experimental values. But, please keep in mind that you should not trust them too much, since our approximation is still very crude. For example, we have assumed λ is large, but it may not be large enough in the realistic situation. Although we ignored the higher derivative terms in the action (1), they may also contribute. $N_c = 3$ is unfortunately not large enough. We know that our model deviates from real QCD at high energy $\sim M_{\rm KK}$. We use $M_{\rm KK} \simeq$

949 MeV with which various quantities in the meson sector are consistent with the experimental values,[3] but we know this is too large to fit the baryon mass differences.[1]

6.1. *Baryon number (isoscalar) current*

Baryon number current is proportional to the $U(1)$ part of the vector current (16) and given by

$$J_B^\mu = -\frac{2}{N_c}\kappa \left[k(z)F_{U(1)}^{\mu z}\right]_{z=-\infty}^{z=+\infty} . \tag{22}$$

Using the trick explained in the previous section, we obtain

$$J_B^0 \simeq [k(z)\partial_z G]_{z=-\infty}^{z=+\infty} , \quad J_B^i \simeq -\frac{J^j}{16\pi^2\kappa}\epsilon^{ijk}\partial_k J_B^0 + \cdots , \tag{23}$$

where $J^j = -i4\pi^2\kappa\rho^2 \mathrm{Tr}(\tau^j \mathbf{a}^{-1}\dot{\mathbf{a}})$ is the spin operator. The terms in '$\cdots$' is the part irrelevant to the following calculation. It can be shown that the baryon number current is non-zero finite, since $k(z)$ and $\partial_z G$ behaves as z^2 and $1/z^2$ at $z \to \pm\infty$, respectively.

Using this expression, the isoscalar mean square radius for the nucleon is evaluated as

$$\langle r^2 \rangle_{I=0} = \int d^3x\, r^2 \left\langle J_B^0 \right\rangle \simeq (0.742 \text{ fm})^2 . \tag{24}$$

Here the expectation value $\langle J_B^0 \rangle$ is taken with respect to the nucleon wave function (12). This result is compared with the the experimental value $\langle r^2 \rangle_{I=0}^{1/2}|_{\text{exp}} \simeq 0.806$ fm. The result of ANW is $\langle r^2 \rangle_{I=0}^{1/2}|_{\text{ANW}} \simeq 0.59$ fm.

The isoscalar magnetic moment is given by

$$\mu_{I=0}^i = \frac{1}{2}\epsilon^{ijk}\int d^3x\, x^j J_B^k \simeq \frac{J^i}{16\pi^2\kappa} . \tag{25}$$

For a spin up proton state $|p\uparrow\rangle$, we obtain

$$\langle p\uparrow|\mu_{I=0}^i|p\uparrow\rangle = \frac{\delta^{i3}}{32\pi^2\kappa} \equiv \frac{g_{I=0}}{4M_N}\delta^{i3} \tag{26}$$

where $M_N \simeq 940$ MeV is the nucleon mass, and the isoscalar g-factor is calculated as

$$g_{I=0} = \frac{M_N}{8\pi^2\kappa M_{\text{KK}}} \simeq 1.68 . \tag{27}$$

The experimental value is $g_{I=0}|_{\text{exp}} \simeq 1.76$ and the result of ANW is $g_{I=0}|_{\text{ANW}} \simeq 1.11$.

6.2. *Isovector current*

The isovector current is obtained by applying the formula (16) to the $SU(2)$ part of the gauge field. The result is

$$J_V^{a\,0} \simeq I^a J_B^0 + \cdots , \quad J_V^{a\,i} \simeq (2\pi^2\kappa)\rho^2 \mathrm{Tr}(\tau^a \mathbf{a}\tau^j \mathbf{a}^{-1})\epsilon^{ijk}\partial_k J_B^0 + \cdots ,\tag{28}$$

where $I^a = -i4\pi^2\kappa\rho^2\mathrm{Tr}(\tau^a \mathbf{a}\dot{\mathbf{a}}^{-1})$ is the isospin operator.

Then, the isovector magnetic moment is obtained as

$$\mu_{I=1}^i = \epsilon^{ijk} \int d^3x\, x^j J_V^{3\,k} \simeq -4\pi^2\kappa\rho^2 \mathrm{Tr}(\tau^3 \mathbf{a}\tau^i \mathbf{a}^{-1}) \, .\tag{29}$$

Using the formula

$$\left\langle \mathrm{Tr}(\tau^a \mathbf{a}\tau^i \mathbf{a}^{-1}) \right\rangle = -\frac{2}{3}\left\langle \tau^a \sigma^i \right\rangle ,\tag{30}$$

the expectation value with respect to the spin up proton state is obtained as

$$\langle p\uparrow | \mu_{I=1}^i | p\uparrow \rangle = \frac{8\pi^2\kappa}{3}\left\langle \rho^2 \right\rangle \delta^{i3} \equiv \frac{g_{I=1}}{4M_N}\delta^{i3} \, .\tag{31}$$

If we approximate $\langle \rho^2 \rangle$ by its classical value ρ_{cl}^2we obtain $g_{I=1} \simeq 4.34$. If we evaluate $\langle \rho^2 \rangle$ by using the nucleon wave function (12), we obtain

$$\left\langle \rho^2 \right\rangle = \rho_{\mathrm{cl}}^2 \left(\sqrt{1 + \frac{5}{N_c^2}} + \frac{\sqrt{5}}{2N_c} \right) \simeq 1.62\, \rho_{\mathrm{cl}}^2 ,\tag{32}$$

which implies $g_{I=1} \simeq 7.03$. This value can be compared with $g_{I=1}|_{\mathrm{exp}} \simeq 9.4$ and $g_{I=1}|_{\mathrm{ANW}} \simeq 6.38$

If we use the above results $g_{I=0} \simeq 1.68$ and $g_{I=1} \simeq 7.03$, the magnetic moments for proton and neutron (in the unit of Bohr magneton $\mu_N = 1/(2M_N)$) are evaluated as

$$\mu_p = \frac{1}{4}(g_{I=0} + g_{I=1}) \simeq 2.18 , \quad \mu_n = \frac{1}{4}(g_{I=0} - g_{I=1}) \simeq -1.34 ,\tag{33}$$

The experimental values and the results of ANW are

$$\mu_p|_{\mathrm{exp}} \simeq 2.79 , \ \mu_n|_{\mathrm{exp}} \simeq -1.91 , \ \mu_p|_{\mathrm{ANW}} \simeq 1.87 , \ \mu_n|_{\mathrm{ANW}} \simeq -1.31 \tag{34}$$

Note however that these values may not be meaningful, since $g_{I=0} = \mathcal{O}(N_c^0)$ and $g_{I=1} = \mathcal{O}(N_c^2)$.

6.3. *Axial coupling*

The axial coupling g_A for nucleons can be read from the axial-vector current as

$$\int d^3x \langle J_A^{a\,i} \rangle = \frac{g_A}{3} \langle \sigma^i \tau^a \rangle \ , \tag{35}$$

where σ^i and τ^a in the right hand side are Pauli matrices acting on the spin and isospin space, respectively. Our result for the axial current (16) is

$$J_A^{a\,i} \simeq -2\pi^2 \kappa \rho^2 \mathrm{Tr}(\tau^a \mathbf{a} \tau^j \mathbf{a}^{-1}) \left[\psi_0(z) k(z) (\partial_i \partial_j - \delta_{ij} \partial_k^2) H \right]_{z=-\infty}^{z=+\infty} + \cdots \tag{36}$$

where

$$H = \kappa \sum_{n=0}^{\infty} \phi_n(z) \phi_n(Z) \left(-\frac{1}{4\pi} \frac{e^{-m_n r}}{r} \right) . \tag{37}$$

This is total derivative and hence most of the terms in the integral (35) vanish. But $n = 0$ component in H survives and we obtain

$$g_A = \frac{16\pi\kappa}{3} \left\langle \frac{\rho^2}{k(Z)} \right\rangle . \tag{38}$$

If we approximate $\langle \rho^2 / k(Z) \rangle$ by its classical value ρ_{cl}^2 we obtain $g_A \simeq 0.697$ and if we evaluate $\langle \rho^2 / k(Z) \rangle$ by using the nucleon wavefunction (12), we have $g_A \simeq 0.734$. This value is not in good agreement with the experimental value $g_A|_{\mathrm{exp}} \simeq 1.27$, as it is also the case for the Skyrme model $g_A|_{\mathrm{ANW}} \simeq 0.61$. It seems to suggest that the $1/N_c$ correction is relatively large for this quantity.

7. Conclusion and Discussion

We proposed a new method to analyze static properties of baryons generalizing the idea of ANW. The advantage of our model is that it automatically includes the contributions from various massive vector and axial-vector mesons. Compared with the results in the Skyrme model (ANW), the agreement with the experimental values are improved in most of the cases. In this article, we only present our results for nucleons, but other baryons, such as Δ, $N(1440)$, $N(1535)$, etc., can also be treated in a similar way. (See Ref. 2 for details.)

The main message here is that the Skyrme's original proposal "Baryons are solitons (Skyrmion) in a pion effective theory" can be extended to a statement in string theory as "Baryons are solitons (D-brane) in string theory". and if one looks at the low energy effective theory, it can be rephrased as "Baryons are solitons (instanton) in five dimensional Yang-Mills - Chern-Simons theory".

Acknowledgements

It is a pleasure to thank the organizers of the conference "Continuous Advances in QCD 2008" for giving us the opportunity to present and discuss our recent works. The author is especially grateful to K. Hashimoto, H. Hata, T. Sakai, S. Yamato for pleasant collaboration. This work is supported in part by JSPS Grant-in-Aid for Creative Scientific Research No. 19GS0219 and also by World Premier International Research Center Initiative (WPI Initiative), MEXT, Japan.

References

1. H. Hata, T. Sakai, S. Sugimoto and S. Yamato, "Baryons from instantons in holographic QCD," Prog. Theor. Phys. **117**, 1157 **(2007)** [arXiv:hep-th/0701280].
2. K. Hashimoto, T. Sakai and S. Sugimoto, "Holographic Baryons : Static Properties and Form Factors from Gauge/String Duality," arXiv:0806.3122 [hep-th].
3. T. Sakai and S. Sugimoto, "Low energy hadron physics in holographic QCD," Prog. Theor. Phys. **113**, 843 (2005) [arXiv:hep-th/0412141]; "More on a holographic dual of QCD," Prog. Theor. Phys. **114**, 1083 (2005) [arXiv:hep-th/0507073].
4. T. H. R. Skyrme, "A Nonlinear field theory," Proc. Roy. Soc. Lond. A **260**, 127 (1961); "Particle states of a quantized meson field," Proc. Roy. Soc. Lond. A **262**, 237 (1961); "A Unified Field Theory Of Mesons And Baryons," Nucl. Phys. **31**, 556 (1962).
5. G. S. Adkins, C. R. Nappi and E. Witten, "Static Properties Of Nucleons In The Skyrme Model," Nucl. Phys. B **228**, 552 (1983).
6. D. K. Hong, M. Rho, H. U. Yee and P. Yi, "Chiral dynamics of baryons from string theory," Phys. Rev. D **76**, 061901 (2007) [arXiv:hep-th/0701276]; "Dynamics of Baryons from String Theory and Vector Dominance," JHEP **0709**, 063 (2007) [arXiv:0705.2632 [hep-th]]; "Nucleon Form Factors and Hidden Symmetry in Holographic QCD," Phys. Rev. D **77**, 014030 (2008) [arXiv:0710.4615 [hep-ph]].
7. H. Hata, M. Murata and S. Yamato, "Chiral currents and static properties of nucleons in holographic QCD," arXiv:0803.0180 [hep-th].
8. K. Y. Kim and I. Zahed, "Electromagnetic Baryon Form Factors from Holographic QCD," arXiv:0807.0033 [hep-th].
9. E. Witten, "Anti-de Sitter space, thermal phase transition, and confinement in gauge theories," Adv. Theor. Math. Phys. **2**, 505 (1998) [arXiv:hep-th/9803131].
10. O. Aharony, S. S. Gubser, J. M. Maldacena, H. Ooguri and Y. Oz, "Large N field theories, string theory and gravity," Phys. Rept. **323**, 183 (2000) [arXiv:hep-th/9905111].
11. M. R. Douglas, "Branes within branes," arXiv:hep-th/9512077.

12. E. Witten, "Baryons and branes in anti de Sitter space," JHEP **9807**, 006 (1998) [arXiv:hep-th/9805112].
13. D. J. Gross and H. Ooguri, "Aspects of large N gauge theory dynamics as seen by string theory," Phys. Rev. D **58**, 106002 (1998) [arXiv:hep-th/9805129].
14. M. F. Atiyah and N. S. Manton, "SKYRMIONS FROM INSTANTONS," Phys. Lett. B **222**, 438 (1989).

The Master Space of $\mathcal{N} = 1$ Quiver Gauge Theories: Counting BPS Operators

A. ZAFFARONI

Physics Department, University of Milano-Bicocca,
and INFN, sezione di Milano-Bicocca
Milano 20126, Italy
**E-mail: alberto.zaffaroni@mib.infn.it*

We present general results about the chiral ring of quiver $\mathcal{N} = 1$ gauge theories, arising on the world-volume of D3-branes at conical Calabi-Yau singularities. We discuss the structure of the moduli space and the computation of generating functions for the chiral operators in the theory. The paper summarizes results from published work in Refs 1–5.

Keywords: Supersymmetric gauge theories; D-branes; AdS/CFT correspondence.

1. Introduction

Counting problems in supersymmetric field theories are rather interesting in many respects. Supersymmetric partition functions can be written for BPS states preserving a certain amount of the original supersymmetry. There exists many different types of partition functions, in this paper we will focus on 1/2 BPS states in $\mathcal{N} = 1$ gauge theories. The corresponding partition function reveals a rich structure of the chiral ring, its generators and their relations, and shares information about the dimension of the moduli space of vacua and the effective number of degrees of freedom in the system.

Partition functions for chiral operators in four dimensional supersymmetric gauge theories have been extensively studied in the past years, ranging from SQCD[6–9] to quiver gauge theories living on branes at singularities.[1–4,10–15] The computation of such generating functions is a very hard problem but it can be simplified in particular circumstances. In the case of SQCD the simplification arises from the fact that there is no superpotential for the chiral fields. In the case of superconformal quiver gauge theories, one can exploit the fact that they have an $AdS_5 \times H$ dual.[16–18] Information from the field theory and from the holographic dual can be combined to give a better understanding of the superconformal theory.

For the case of branes at toric Calabi-Yau singularities [a] the combination of the Plethystic Program[12] with algebraic tools in complex geometry allows to write quite explicit formulae for the partition functions. We will demonstrate the importance of the concept of master space, which is roughly the moduli space for one single brane.

2. Generating function for chiral operators

Before attacking the case of branes at singularities, let us briefly review some basic notions and introduce some useful concepts. In the case of $\mathcal{N} = 1$ gauge theories, the chiral operators form a ring (see for example Ref. 22). Chiral operators are those annihilated by the supercharges of one chirality $\bar{Q}_\alpha$ and it is useful to define them modulo operators of the form $\{\bar{Q}_\alpha, ...\}$ [b]. The product of two chiral operators is defined by their OPE which is not singular. In a theory with chiral superfields Φ_i and a superpotential $W(\Phi_i)$, the chiral operators involving scalar fields are obtained by gauge invariant products of the lowest component ϕ_i of the superfield Φ_i subject to the constraint $\partial_{\phi_i} W(\phi) = 0$. This follows from the equation of motion

$$\bar{D}_\alpha \bar{D}^\alpha \bar{\Phi}_i = \partial_{\Phi_i} W(\Phi) \tag{1}$$

which implies that all derivatives of the superpotential are descendants and vanish in the chiral ring. In addition, the superfield W_α can be used to construct other chiral operators containing fermions. For simplicity, in this paper we will only consider operators made with scalars. We refer to Ref. 23 for a general discussion of the role of W_α and to Ref. 15 for the case of superconformal quivers. Summarizing, we will consider the scalar part of the chiral ring consisting of all chiral gauge invariant operators modulo F terms.

Given a $\mathcal{N} = 1$ supersymmetric gauge theory with a collection of $U(1)$ global symmetries $\prod_{i=1}^{G} U(1)_i$, we can introduce a set of auxiliary parameters (fugacities) $\{t_i\}_{i=1}^{G}$ and define the generating function for the chiral

---

[a] Toric Calabi-Yaus are those having three isometries. They are simpler from the geometrical point of view because of toric geometry and its tools and from the physical point of view because of the language of **dimer models/brane tilings**[19,20] a generalization of brane boxes.[21] In particular there is a complete dictionary relating a toric Calabi-Yau to the corresponding quiver gauge theory.

[b] Chiral operators are the lowest component of chiral supermultiplets, which are protected by the BPS condition and short. Descendant operators of the form $\{\bar{Q}_\alpha, ...\}$ are not protected and belong to long multiplets.

ring

$$g(\{t_i\}) = \sum n_{k_1,...,k_G} t_1^{k_1} ... t_G^{k_G} \tag{2}$$

where $n_{k_1,...,k_G}$ is the number of operators with charge $\{k_1, ..., k_G\}$ under the global $U(1)^G$ symmetry. In mathematical language, we compute the **Hilbert series** of a graded ring. In the superconformal case we always have at least one symmetry, the $U(1)_R$ R-symmetry. By restricting to just one parameter t for the R-symmetry we can write an unrefined Hilbert series

$$g(t) = \sum_k n_k t^k \tag{3}$$

counting the chiral operators with given R-charge. Since, in a superconformal field theory, dimension and R-charge are related by $\Delta = 3R/2$, we are equivalently counting chiral operators with given dimension.

We can take a complementary point of view and consider the chiral operators as function on the moduli space of the supersymmetric gauge theory. The vacuum moduli space, $\mathcal{M}$, is given by the vanishing of the scalar potential as a function of the scalar components of the superfields of the field theory. This, in turn, is the set of zeros of D-terms and F-terms modulo gauge transformations. It is standard fact that we can dispose simultaneously of the D-term condition and the gauge freedom by modding by the complexified gauge group (see for example Ref.24). $\mathcal{M}$ can be phrased in a succinct mathematical language: it is the symplectic quotient of the space of F-flatness, by the gauge symmetries provided by the D-flatness. We will denote the space of F-flatness by $\mathcal{F}^\flat$ and symmetries prescribed by D-flatness as $G_{D^\flat}$, then we have

$$\mathcal{M} \simeq \mathcal{F}^\flat // G_{D^\flat} \ . \tag{4}$$

Chiral operators are invariant under the complexified gauge group and are subject to the F-terms. They are therefore holomorphic functions on the moduli space. Viceversa, with any holomorphic function on the moduli space we can associate a chiral operator. We can then reinterpret Eq. (2) as counting holomorphic functions on the moduli space. It is a general mathematical fact that $g(t)$ is a rational function which can be written as

$$g(t) = \frac{P(t)}{(1-t)^d} \tag{5}$$

for some polynomial $P(t)$. It is a well known and interesting fact that $g(t)$ has a pole at $t \to 1$ of order equal to the dimension of the moduli space.

The generating function (2) is hard to compute even at the classical level. In addition to the F-terms, we have other algebraic constraints satisfied by the elementary fields. For example, in superconformal quiver gauge theories, the elementary fields transform in the adjoint representation of an $SU(N)$ group or in the bifundamental representation of a pair of $SU(N)$ groups. The fields can be regarded as $N \times N$ matrices X_i. A gauge invariant operator made with, say, an adjoint $\mathrm{Tr}X^k$, for $k > N$ is a linear combination of the traces with $k < N$. Similarly, determinants of matrices X can be sometime written as linear combinations of products of traces. F-terms and complicated algebraic constraints make difficult the computation of the generating function. In principle, thanks to the results of commutative algebra, generating functions like (2) can be computed in an algorithmical way. We first compute the Hilbert series of an extended ring consisting of all the elementary fields (the entries of the matrices X_i) modded by the ideal of F-terms and graded with the parameter t_i and an extra set of gauge charges z_k labeling the Cartan part of the gauge group. This computation is algorithmical and can be performed with computer algebra programs like Macaulay2.[25] We then project on the set of gauge invariants by averaging on the parameters z_i with the Haar-Weyl measure. We are not entering in the details of this construction and we refer to Ref. 3 for a discussion and examples. This procedure works well in the case where there are no F-terms (for example SQCD, see Refs. 6,8) and for theories with small number of fields and moderate values of N. It is a very difficult task to find closed formulae for generic N in theories with F-terms. We will propose a general formula for superconformal quivers in the following.

3. The moduli space of superconformal quivers

The structure of the moduli space of a supersymmetric gauge theory is usually complicated, and should be best cast in the language of algebraic varieties. Typically, $\mathcal{M}$ consists of a union of various branches, such as the mesonic branch or the baryonic branch, the Coulomb branch or the Higgs branch; the names are chosen according to some characteristic property of the specific branch.

In the context of certain string theory backgrounds, $\mathcal{M}$ has an elegant geometrical realization. When D3-branes are transverse to an affine (non-compact) three-fold Calabi-Yau space $\mathcal{X}$, a supersymmetric gauge theory exists on the world-volume of the branes. When we consider N physical branes at the singularity we obtain a superconformal gauge theory of quiver type with $U(N)$ gauge groups and bi-fundamental or adjoint fields.[16,18] Our main interest is the IR physics of this system, where all Abelian symmetry decouples and the gauge symmetry is fully non-Abelian, typically given by products of $SU(N)$ groups. The Abelian factors are not gauged but rather appear as global baryonic symmetries of the gauge theory.

Under these circumstances, the moduli space $\mathcal{M}$ is a combined mesonic branch and baryonic branch. These branches are not necessarily separate (irreducible) components of $\mathcal{M}$ but are instead in most cases intrinsically merged into one or more components in $\mathcal{M}$. Even when mesonic and baryonic directions are mixed, it still makes sense to talk about the more familiar mesonic moduli space ${}^{\text{mes}}\mathcal{M}$, as the sub-variety of $\mathcal{M}$ parameterized by mesonic operators only. Since mesonic operators have zero baryonic charge, and thus invariant under the $U(1)$ Abelian factors, the mesonic moduli space can be obtained as a further quotient of $\mathcal{M}$ by the Abelian symmetries:

$$ {}^{\text{mes}}\mathcal{M} \simeq \mathcal{M}//U(1)_{D^\flat} \ . \tag{6} $$

It is of particular interest to consider the case of a single D3-brane transverse to the Calabi-Yau three-fold $\mathcal{X}$, which will enlighten the geometrical interpretation of the moduli space. It is common lore that, since the motion of the D-brane is parameterized by this transverse space, the moduli space $\mathcal{M}$ should coincide with the non-compact Calabi-Yau threefold $\mathcal{X}$ transverse to the D3-brane. This is not the full story. $\mathcal{X}$ is actually only the mesonic part of the moduli space. In general, even for one brane, $\mathcal{M}$ will be an algebraic variety of dimension greater than three. Since there is no gauge group in the IR, $\mathcal{M}$ is given by the space of F-flatness $\mathcal{F}^\flat$. Geometrically, $\mathcal{F}^\flat$ is a $\mathbb{C}^{dim\mathcal{F}^\flat-3}$ fibration over the mesonic moduli space $\mathcal{X}$ given by relaxing the $U(1)$ D-term constraints in Eq. (6). Physically, $\mathcal{F}^\flat$ is obtained by adding *baryonic* directions to the mesonic moduli space [c].

For $N > 1$, the situation is more subtle. Again, the common lore says that the moduli space, probed by a collection of N physical branes which are mutually BPS states and thus can arbitrarily located in the internal manifold, is given by the symmetrized product of N copies of $\mathcal{X}$. Actually this is true for the mesonic moduli space (this can be checked in many

[c]Of course, we can not talk about baryons for $N = 1$ but we can alternatively interpret these directions as Fayet-Iliopoulos (FI) parameters in the stringy realization of the $N = 1$ gauge theory. Indeed on the world-volume of a single D-brane there is a collection of $U(1)$ gauge groups, each giving rise to a FI parameter, which relax the D-term constraints.

examples and it is believed to be true in general [d]). The full moduli space $\mathcal{M}$ is a bigger algebraic variety of more difficult characterization. One of the purposes of this paper is to elucidate this situation and to show how the properties of $\mathcal{M}$ for arbitrary number of branes are encoded in the moduli space for a single brane. In view of the importance of the moduli space for one brane even for larger N, we will adopt the important convention that, in the rest of the paper, the the symbol $\mathcal{F}^\flat$ will refer to the $N = 1$ case and the $N = 1$ moduli space will be dubbed **master space**.

4. The case of one brane: the master space

The moduli space for one brane is the space of solutions $\mathcal{F}^\flat$ of the F-terms where the fields are considered as c-numbers. The generating function for one brane is just the Hilbert series of the quotient ring generated by the elementary fields modulo the ideal of F-terms. This generating function can be easily computed for all reasonable (=not too large) quivers by using computer algebra programs. In the case of a toric Calabi-Yau $\mathcal{X}$ we can give a quite general description of the master space $\mathcal{F}^\flat$. Many specific examples are discussed in details in Ref. 4. Some of the general properties of the master space are:[4]

- $\mathcal{F}^\flat$ is a toric variety of complex dimension $g + 2$, where g is the number of gauge groups. This is so because $\mathcal{X} \simeq \mathcal{F}^\flat//U(1)^g$, and an overall $U(1)$ decouples; thus $3 = \dim(\mathcal{F}^\flat) - (g - 1)$. It is toric since it is acted upon by exactly $g + 2$ $\mathbb{C}^*$-actions corresponding to the classical global symmetries of the gauge theory: one R and two flavor, coming from the isometries of the toric threefold $\mathcal{X}$, as well as $g - 1$ baryonic, IR relic symmetries of the non-trivial $U(1)$ factors, some of which are anomalous.
- The moduli space of gauge theories is well-known to have many branches; this is reflected by the fact that $\mathcal{F}^\flat$ *is typically a reducible algebraic*

[d]For example, for the simple and familiar case of $\mathcal{X}$ being the conifold,[16] a quiver with two gauge groups $SU(N) \times SU(N)$ and four chiral bi-fundamental fields, $A_{1,2}$ transforming in the $(N, \bar{N})$ representation and $B_{1,2}$ transforming in the $(\bar{N}, N)$ representation, interacting with the quartic superpotential $\epsilon_{ij}\epsilon_{pq}A_iB_pA_jB_q$, we have four basic mesonic operators $z_{ij} = \mathrm{Tr}(A_iB_j)$. By relaxing the trace in this expression, we can consider the $z_{ij} = A_iB_j$ as $N \times N$ matrices transforming in the adjoint of the first gauge group. It is easy to check that the F-terms implies that the z_{ij} commute and satisfy the defining equation of the conifold $z_{11}z_{22} - z_{12}z_{21} = 0$. Since the mutually commuting z_{ij} can be simultaneously diagonalized, up a remaining Weyl permutation symmetry, we see that the mesonic moduli space is indeed $\mathrm{Sym}^N\mathcal{X}$. A similar argument works for generic toric quivers.

variety. It contains a top-dimensional irreducible component of the same dimension, dubbed the **coherent component** and denoted as ${}^{\mathrm{Irr}}\mathcal{F}^{\flat}$, as well as many smaller dimensional irreducible (generically) linear pieces, realized as coordinate hyperplanes.

- The coherent component ${}^{\mathrm{Irr}}\mathcal{F}^{\flat}$ is itself an affine Calabi-Yau[e] of dimension $g+2$. This is related to an intriguing property of the Hilbert Series for ${}^{\mathrm{Irr}}\mathcal{F}^{\flat}$. The numerator $P(t)$ of $g(t; {}^{\mathrm{Irr}}\mathcal{F}^{\flat})$, which we recall to be an integer polynomial of degree, say, n, has a palindromic symmetry for its coefficients $a_{j=0,\ldots,n}$: $P(t)$ is invariant under the exchange $a_j \leftrightarrow a_{n-j}$. This is a consequence of a theorem by Stanley[26] and the fact ${}^{\mathrm{Irr}}\mathcal{F}^{\flat}$ is Calabi-Yau. Interestingly, the same palindromic property of the Hilbert series is shared by the classical moduli space of SQCD for all values of N_c and N_f;[9] indeed the moduli space of SQCD is Calabi-Yau. It seems that the Calabi-Yau property, although not universal (the quiver moduli space for $N > 1$ is not Calabi-Yau), is quite common in $\mathcal{N} = 1$ gauge theories.
- A detailed description of ${}^{\mathrm{Irr}}\mathcal{F}^{\flat}$ as algebraic variety is provided in Ref. 4. $\mathcal{F}^{\flat}$ and ${}^{\mathrm{Irr}}\mathcal{F}^{\flat}$ can be written as a set of algebraic equations in $\mathbb{C}^E$, where E is the number of elementary fields. Alternatively, one can provide a symplectic quotient description of the toric variety which is useful for computing the Hilbert series. The latter description is directly related to the Dimer model of the quiver and the intriguing fact that perfect matchings generate the coherent component. We refer to Ref. 4 for details on the algebraic properties of $\mathcal{F}^{\flat}$ and ${}^{\mathrm{Irr}}\mathcal{F}^{\flat}$, the computation of their Hilbert series and the relation to Dimers.
- It is known that a given toric Calabi-Yau corresponds to many different quivers which are related by Seiberg duality. The set of such quivers where all gauge groups have the same number of colors is known as the set of toric phases of the given theory. For $N > 1$, all these different non-abelian gauge theories are equivalent under Seiberg duality and flow to the same IR fixed point. For $N = 1$ there is no precise sense in which Seiberg duality can be defined, since there is no non-abelian gauge group. Nevertheless we expect to see some relic of this duality. And in fact, although not isomorphic, the coherent component of the master space for different toric phases are closely related: their generating functions coincide, not only as a function of the R-symmetry parameter t but also when refined with all the non anomalous charge parameters. This follows from

[e]We should better say that it is Gorenstein.

the fact that, as we discuss in the following, the coherent component of the master space determines the BPS spectrum for $N > 1$. Dual theories have the same non anomalous symmetries and the same spectrum of BPS operators.

Many other properties of the master space including the structure of the linear components, the relation to RG flows, and the existence of hidden symmetries are discussed in Ref. 4. We will see now that perhaps the most important property of the master space is the fact that it determines the generating function for chiral operators for any value of N.

5. The case of N branes

The case for an arbitrary number N of D3-branes is much more subtle. As we already said, the world-volume theory for N D3-branes is a quiver theory with product $U(N)$ gauge groups and in the IR, the $U(1)$ factors decouple since only the special unitary groups exhibit asymptotic freedom and are strongly coupled in the IR. Thus the moduli space of interest is the space of solutions to the F-flatness, modded out by a non-Abelian gauge group

$$\mathcal{M}_N = \mathcal{F}_N^\flat/(SU(N_1) \times \ldots \times SU(N_g)). \tag{7}$$

where the index N recalls that we are dealing with N branes. The moduli space $\mathcal{M}_N$ is of difficult characterization since the quotient is fully non-Abelian and it can not be described by toric methods, as in the $N = 1$ case.

The mesonic moduli space is simpler and, as already discussed, given by the symmetric product of N copies of $\mathcal{X}$. It is obtained from the full moduli space by performing a further quotient by the Abelian symmetries:

$$^{\text{mes}}\mathcal{M}_N \equiv \text{Sym}^N \mathcal{X} \simeq \mathcal{M}_N // U(1)^{g-1} . \tag{8}$$

We see that the mesonic moduli space, for $\mathcal{X}$ a Calabi-Yau threefold, is of dimension $3N$. The dimension of the moduli space $\mathcal{M}_N$ is thus $3N + g - 1$ for general N.

Now it is easy to write the generating function for a variety $\text{Sym}^N \mathcal{Y}$ once the generating function for $\mathcal{Y}$ is given. This is done with some Plethysm.[12] From the partition function $g_1(t)$ for $\mathcal{Y}$ we can obtain the partition function $g_N(t)$ for $\text{Sym}^N \mathcal{Y}$ as follows. Let us define the **plethystic exponential** of a multi-variable function $g(t_1, \ldots, t_n)$ that vanishes at the origin, $g(0,\ldots,0)=0$, to be

$$\mathrm{PE}[g(t_1,\ldots,t_n)] := \exp\left(\sum_{r=1}^{\infty}\frac{g(t_1^r,\ldots,t_n^r)}{r}\right). \tag{9}$$

The generating function g_N for $\mathrm{Sym}^N\mathcal{Y}$ is found by the series expansion of the ν - inserted plethystic exponential as

$$\mathrm{PE}[\nu g_1(t_1,\ldots,t_n)] = \exp\left(\sum_{r=1}^{\infty}\frac{\nu^r g_1(t_1^r,\ldots,t_n^r)}{r}\right) = \sum_{N=0}^{\infty} g_N(t_1,\ldots,t_n)\nu^N . \tag{10}$$

The plethystic program can be efficiently applied to the study of the coherent component of the moduli space as follows.[3] The generating function for the coherent component of the master space can be decomposed in sectors of definite baryonic charge. We write schematically

$$g_1(t_i;{}^{\mathrm{Irr}}\mathcal{F}^\flat) = \sum_B g_{1,B}(t_i) \tag{11}$$

where B generically denotes a set of complete quantum numbers for the $g-1$ baryonic symmetries and we introduced the subscript 1 to recall that we are dealing for the moment with $N=1$. The generating function $g_{1,0}$ counts the $N=1$ operators with zero baryonic charge and it is therefore the generating function for the mesonic moduli space for one brane. It can be determined from the knowledge of $g_1(t;{}^{\mathrm{Irr}}\mathcal{F}^\flat)$ by averaging over the baryonic quantum numbers. It can be alternatively considered as the generating function of holomorphic functions for the variety $\mathcal{X}$.

Now it is very simple to write the generating function $g_{N,0}(t_i)$ for the mesonic moduli space using the Pletystic Exponential.[12] Since ${}^{\mathrm{mes}}\mathcal{M}_N \equiv \mathrm{Sym}^N\mathcal{X}$, we conclude from the previous discussion that $g_{N,0}$ is obtained from $g_{1,0}$ by applying the Pletystic Exponential:

$$\mathrm{PE}[\nu g_{1,0}(t_i)] = \sum_{N=0}^{\infty} g_{N,0}(t_i)\nu^N . \tag{12}$$

It is quite remarkable that a similar formula holds for the entire coherent component of the moduli space. We just need to apply the Pletystic Exponential to each sector of definite baryonic charge. The final formula is:

$$\sum_B PE\left[\nu g_{1,B}(t_i)\right] = \sum_{N=0}^{\infty} \nu^N g_N(t_i) . \tag{13}$$

This quite remarkable formula is our proposal for the generating function of the chiral ring of superconformal quivers and one of the main results of our analysis. What is the rationale behind it? Notice that, while the formula for the mesonic branch is essentially proven, the formula for the entire moduli space stands as a conjecture. It is motivated by a computation in the gravitational dual where the BPS states can be represented as states of supersymmetric D3-branes.[1,27] A review of the dual computation will take us too far and we refer to Ref. 1 for details. Let us notice few things, however.

- BPS states in the gravitational AdS dual can be obtained by quantizing classical configurations of supersymmetric D3-branes. The latter fall in classes according to the homology of the cycle they are wrapping. It can be shown that the baryonic quantum numbers, that we generically denoted B, are in correspondence with the homology numbers. The decomposition (11) has therefore a natural interpretation in the gravitational dual. According to the analysis in Refs. 28,29, the classical supersymmetric D3-branes are in correspondence with zero loci of holomorphic functions or sections of line bundles over $\mathcal{X}$. Giant gravitons, branes wrapped on trivial cycles, correspond to $B = 0$ and can be described by zeros of holomorphic functions on $\mathcal{X}$. This perfectly agrees with the interpretation of $g_{1,0}$ as generating function for holomorphic functions on $\mathcal{X}$. Similarly $g_{1,B}$ can be interpreted as the generating function for sections of a given line bundle. As a consequence $g_{1,B}$ can be computed in a purely geometric way using the equivariant Atiyah-Singer theorem; this has been done for holomorphic functions in Ref. 30 and extended to $B \neq 0$ in Ref. 1. Formula (11) is then quite remarkable: the left hand side, which is computed in the quiver theory using F-terms, has an expansion as a sum of functions that can be computed using the geometry of the Calabi-Yau only.
- Quantization of classical supersymmetric D3-branes is done in each sector independently. Geometric quantization amounts to take the Pletystic Exponential in each sector.[1,27,29] This explains the conjecture in formula (13).
- We have been sloppy in writing the decomposition (11) for generic quivers. The sectors labelled by B have to be defined carefully in quivers with anomalous symmetries. The sum over B is in general replaced by a sum over a lattice of charges. We refer to Ref. 3 for examples and discussion. The general case still needs to be better understood.

In the case of the conifold the previous discussion gives a remarkably simple and neat answer [f]. We refer to Ref. 2 for the explicit form of the generating function. Based on the results in Refs. 1,2 we can view these results as a prediction of the gravity dual, valid at strong coupling and large N. For small values of N, it can be compared with the result in the classical theory computed using computer algebra programs and it turns out that there is perfect agreement. The partition function for a selected class of other quivers has been discussed in Ref. 3.

Acknowledgements

We thank Agostino Butti, Davide Forcella, Amihay Hanany, Yang-Hui He and David Vegh for collaboration on the research discussed in this paper and Nathan Broomhead and Alastair King for many enlightening discussions. A. Z. is supported in part by INFN and MIUR under contract 2005-024045-004 and 2005-023102 and by the European Community's Human Potential Program MRTN-CT-2004-005104.

References

1. A. Butti, D. Forcella and A. Zaffaroni, *JHEP* **0706**, 069 (2007).
2. D. Forcella, A. Hanany and A. Zaffaroni, *JHEP* **0712**, 022 (2007).
3. A. Butti, D. Forcella, A. Hanany, D. Vegh and A. Zaffaroni, *JHEP* **0711**, 092 (2007).
4. D. Forcella, A. Hanany, Y. H. He and A. Zaffaroni, *The Master Space of N=1 Gauge Theories,* arXiv:0801.1585.
5. D. Forcella, A. Hanany, Y. H. He and A. Zaffaroni, *Mastering the Master Space,* arXiv:0801.3477.
6. P. Pouliot, *JHEP* **9901**, 021 (1999).
7. C. Romelsberger, *Nucl. Phys. B* **747**, 329 (2006).
8. A. Hanany and N. Mekareeya, *Counting Gauge Invariant Operators in SQCD with Classical Gauge Groups,* arXiv:0805.3728.
9. J. Gray, A. Hanany, Y. H. He, V. Jejjala and N. Mekareeya, *JHEP* **0805**, 099 (2008).
10. J. Kinney, J. M. Maldacena, S. Minwalla and S. Raju, *Commun. Math. Phys.* **275**, 209 (2007).
11. D. Martelli and J. Sparks, *Nucl. Phys. B* **759**, 292 (2006).
12. S. Benvenuti, B. Feng, A. Hanany and Y. H. He, *JHEP* **0711**, 050 (2007).

---

[f]For the conifold all symmetries are non anomalous and there is a single baryonic charge B which takes integer values. The master space is simply $\mathbb{C}^4$ and the generating function is then simply $g_1(t) = 1/(1-t)^4$. From the index theorem it follows that the generating functions for branes branes wrapping the three cycle of homology B are $g_{1,B}(t) = \sum_{n=0}^{\infty}(n+1)(n+1+B)t^{2n+B}$. The result is consistent with equation (11).

13. B. Feng, A. Hanany and Y. H. He, *JHEP* **0703**, 090 (2007).
14. A. Hanany and C. Romelsberger, *Counting BPS operators in the chiral ring of N = 2 supersymmetric gauge theories or N = 2 braine surgery,* arXiv:hep-th/0611346.
15. D. Forcella, *BPS Partition Functions for Quiver Gauge Theories: Counting Fermionic Operators,* arXiv:0705.2989.
16. I. R. Klebanov and E. Witten, *Nucl. Phys. B* **536**, 199 (1998).
17. B. S. Acharya, J. M. Figueroa-O'Farrill, C. M. Hull and B. J. Spence, *Adv. Theor. Math. Phys.* **2**, 1249 (1999).
18. D. R. Morrison and M. R. Plesser, *Adv. Theor. Math. Phys.* **3**, 1 (1999).
19. S. Franco, A. Hanany, K. D. Kennaway, D. Vegh and B. Wecht, *JHEP* **0601**, 096 (2006).
20. A. Hanany and K. D. Kennaway, "Dimer models and toric diagrams," arXiv:hep-th/0503149.
21. A. Hanany and A. Zaffaroni, *JHEP* **9805**, 001 (1998).
22. D. Amati, K. Konishi, Y. Meurice, G. C. Rossi and G. Veneziano, *Phys. Rept.* **162**, 169 (1988).
23. F. Cachazo, M. R. Douglas, N. Seiberg and E. Witten, *JHEP* **0212**, 071 (2002).
24. M. A. Luty and W. Taylor, *Phys. Rev. D* **53**, 3399 (1996).
25. D. Grayson and M. Stillman, "Macaulay 2, a software system for research in algebraic geometry," Available at http://www.math.uiuc.edu/Macaulay2/
26. R. Stanley, *Adv. Math. 28* (1978), 57-83.
27. I. Biswas, D. Gaiotto, S. Lahiri and S. Minwalla, *JHEP* **0712**, 006 (2007).
28. A. Mikhailov, *JHEP* **0011**, 027 (2000).
29. C. E. Beasley, *JHEP* **0211**, 015 (2002).
30. D. Martelli, J. Sparks and S. T. Yau, *Commun. Math. Phys.* **280**, 611 (2008).

SECTION 5.

TOPOLOGICAL FIELD CONFIGURATIONS

SKYRMIONS IN THEORIES WITH MASSLESS ADJOINT QUARKS

R. AUZZI

Department of Physics, Swansea University,
Singleton Park, Swansea SA2 8PP, U.K.
E-mail: r.auzzi@swansea.ac.uk

Dynamics of SU(N_c) Yang–Mills theories with N_f adjoint Weyl fermions is quite different from that of SU(N_c) gauge theories with fundamental quarks. The symmetry breaking pattern is SU(N_f) $\to$ SO(N_f). The corresponding sigma model supports Skyrmions whose microscopic identification is not immediately clear. We address this issue as well as the issue of the Skyrmion stability. We discuss the coupling between the massless Goldstone bosons and massive composite fermions (with mass $O(N_c^0)$) from the standpoint of the low-energy chiral sigma model. We derive the Wess–Zumino–Novikov–Witten term and then determine Skyrmion statistics. We also determine their fermion number (mod 2) and observe an abnormal relation between the statistics and the fermion number. This explains the Skyrmion stability.

Keywords: Solitons; Skyrme Model

1. Introduction

Let us consider SU(N_c) gauge theories with N_f quarks in the adjoint representation. If the quarks are massless, the theory has the following pattern of chiral symmetry breaking (χSB):[1–3]

$$\mathrm{SU}(N_f) \times \mathbb{Z}_{2N_cN_f} \to \mathrm{SO}(N_f) \times \mathbb{Z}_2 \,, \tag{1}$$

where the discrete factors are the remnants of the anomalous singlet axial U(1). The low-energy spectrum is described by a nonlinear sigma model with the target space given by the coset

$$\mathcal{M}_{N_f} = \mathrm{SU}(N_f)/\mathrm{SO}(N_f). \tag{2}$$

This nonlinear sigma model is known to posses nontrivial topological solitons. In particular, it admits Skyrmions[4–7] which can be classified by topologically nontrivial maps from the compactified three-dimensional space to the coset space. The relevant topological winding number is given by the third homotopy group, which is $\mathbb{Z}$ for $N_f = 2$, $\mathbb{Z}_4$ for $N_f = 3$ and $\mathbb{Z}_2$ for $N_f \geq 4$.

Unlike QCD, in the theories with adjoint quarks the Skyrmion mass scales as N_c^2 (here N_c is the number of colors). A natural question arises whether these Skyrmions which are topologically stable in the low-energy (macroscopic) description, are stable in the full (microscopic) theory. Since the low-energy Lagrangian is insensitive to the ultraviolet behavior of the microscopic theory, it is *apriori* possible that these Skyrmions might decay into the lowest-energy states of Yang–Mills with adjoint quarks through tunneling into these states.

More precisely, we will consider N_f massless *Weyl* (or, which is the same, *Majorana*) fermions in the adjoint representation of the SU(N_c) gauge theory, to be denoted as $\lambda^a_{\alpha\, f}$ where a, α and f are the color, Lorentz and flavor indices, respectively. These Weyl fermions will be referred to as "quarks" or "adjoint quarks." To ensure flavor symmetry in the fundamental Lagrangian we take $N_f > 2$. At $N_f > 5$ asymptotic freedom is lost. The $N_f = 1$ theory is in fact $\mathcal{N} = 1$ super-Yang–Mills, it is gapped and has no Goldstone bosons in the physical spectrum.

Equation (1) can be elucidated as follows. In the vacuum the Lorentz-scalar bilinear $\lambda^a_{\alpha\, f}\, \lambda^{a\,\alpha}_g$ condenses,

$$\langle \lambda^a_{\alpha\, f}\, \lambda^{a\,\alpha}_g \rangle \sim \Lambda^3 \delta_{fg}\,. \tag{3}$$

The above order parameter stays intact under those transformations from SU(N_f) which are generated by matrices antisymmetric under transposition. In other words, the condensate (3) is invariant under transformations from the SO(N_f) subgroup. Thus, the low-energy pion Lagrangian is a nonlinear sigma model with the target space $\mathcal{M}_{N_f}$. Besides SO(N_f), the low-energy Lagrangian possesses a discrete symmetry: the Z_2 remnant of the axial flavor-singlet $U(1)$. We will call this the fermion number F since it counts the number of fermions modulo 2. Thus, $(-1)^F$ is well-defined.

The low-energy effective Lagrangian describes dynamics of the lightest particles. The massless particles are the Nambu–Goldstone bosons π, which are in the 2-index symmetric traceless representation of SO(N_f). In the fermionic sector the particle with the lowest mass is ψ, interpolated by the gauge invariant operator

$$\psi_{\beta\, f} = C\, \mathrm{Tr}\left(\lambda^\alpha_f\, F_{\alpha\beta}\right) \equiv C\, \mathrm{Tr}\left(\lambda^\alpha_f \sigma^{\mu\nu}_{\alpha\beta} F_{\mu\nu}\right), \tag{4}$$

were $F_{\alpha\beta}$ is the (anti)self-dual gluon field strength tensor (in the spinorial notation), and C is a normalizing factor, $C \sim (N\Lambda^2)^{-1}$. The above composite fermions are in the vectorial representation of SO(N_f).

For $N_f = 2$, the question of the skyrmion stability was addressed in Ref. 8. What is a crucial difference which makes the case of $N_f = 2$ easier than $N_f = 3$, 4 and 5? In this case a valid symmetry of the model surviving after χSB is SO(2) equivalent to U(1).[8] One can classify all physical states with respect to the U(1) charge (it was referred to as Q in Ref. 8).

The results of Ref. 8 can be summarized as follows. The Hilbert space accessible from the perturbative analysis is

$$\mathcal{H}^{(\text{hadronic})} = \mathcal{H}^{(+1,+1)} \oplus \mathcal{H}^{(-1,-1)} \tag{5}$$

containing the composite states with the even and odd U(1) charges, respectively. We have denoted the charges as $\left\{(-1)^Q, (-1)^F\right\}$. In particular, $\mathcal{H}^{(+1,+1)}$ contains the massless Nambu–Goldstone bosons $\pi^{\pm\pm}$ and, hence, there is no mass gap here. On the contrary, $\mathcal{H}^{(-1,-1)}$ has a mass gap m, the mass of the lightest composite fermion of the type (4). After analyzing the Skyrmion of the low-energy effective Lagrangian, it cab be argued that an *extra* sector to which the Skyrmions belong is

$$\mathcal{H}^{(\text{exotic})} = \mathcal{H}^{(+1,-1)} \oplus \mathcal{H}^{(-1,+1)} \,. \tag{6}$$

So the Hopf Skyrmion stability is not a low-energy artifact. This is due to the fact that all conventional mesons and baryons with $m = O(N_c^0)$ in the theory at hand have

$$(-1)^Q \cdot (-1)^F = 1\,,$$

while for the Hopf Skyrmion

$$(-1)^Q \cdot (-1)^F = -1\,.$$

The reason behind this is that the fermion number has an anomalous contribution that couples directly to the topological current of the Skyrmion. This means that the Skyrmion acquires a fermion number through the so called Goldstone–Wilczek mechanism.[9]

Now, what must be done to generalize this result to higher N_f? The generalization is not quite trivial. The residual symmetry which was U(1) in the $N_f = 2$ case is now replaced by SO(N_f) with $N_f = 3, 4, 5$. Correspondingly, all particles from the physical spectrum must be classified according to representations of SO(N_f). One can argue that for $N_f = 3, 4, 5$ the Goldstone–Wilczek mechanism provides the Skyrmion with an anomalous fermion number. Then we face a problem. For N_f odd (i.e. $N_f = 3$ and 5), the quantum number assignments used for $N_f = 2$ do *not* guarantee the stability of the Skyrmion. This is due to the existence of the antisymmetric tensor $\varepsilon^{i_1,i_2,\ldots,i_{N_f}}$ in SO(N_f). Using this tensor we can assemble N_f composite fermions ψ in a combination invariant under the flavor group SO(N_f), creating a baryonic final state. For N_f odd this state would have the same quantum numbers as the Skyrmion and, thus, we could conclude that the Skyrmion, being an object with mass $\propto N_c^2$ would decay into N_f composite fermions ψ with mass $O(N_c^0)$ in the flavor singlet configuration.

An important role in the spin and statistics determination for solitons belongs to the Wess–Zumino–Novikov–Witten (WZNW) term.[10] To make the question more explicit it is instructive to briefly review the situation in conventional QCD with fundamental quarks.[5,6] To begin with, let us consider two flavors, $N_f = 2$.

In this case the low-energy chiral Lagrangian does not admit the WZNW term. It does support Skyrmions, however. After quantization the Skyrmion quantum numbers $(I,\ J)$ form the following tower of possible values: $(0,0)$, $(1/2,1/2)$, $(1,1)$, $(3/2,3/2)$, etc. Here I, J stand for isospin and spin, respectively. In the absence of the WZNW term Skyrmions can be treated as both, bosons and fermions. This is due to the fact that we may or may not add an extra sign in the field configurations belonging to nontrivial maps of $\pi_4(SU(2))$.[11]

At $N_f \geq 3$ the choice of the Skyrmion statistics (i.e. boson vs. fermion) becomes unambiguous. The reason is well-known: at $N_f \geq 3$ it is possible (in fact, necessary) to introduce the WZNW term in the effective Lagrangian.[5,6] This term, which is absolutely essential in the anomaly matching between the ultraviolet (microscopic) and infrared (macroscopic) degrees of freedom, is responsible for the spin/statistics assignment for Skyrmions.

A similar situation takes place in adjoint QCD. With two flavors the WZNW term does not exist since $\pi_4(\mathrm{SU}(2)/\mathrm{U}(1)) = Z_2$. Quantization[12] gives us two possible tower of states: bosons with the U(1) charge and spin $(0,0)$, $(2,1)$, and so on, and fermions with the U(1) charge and spin $(1,1/2)$, $(3,3/2)$, and so on. In the effective low-energy theory it is impossible to decide in which of the two towers the Skyrmion lies. Only considering higher N_f can we answer this question. The answer will play a crucial role in the explanation of the Skyrmion stability.

As well known,[6] the SO(N_c) gauge theory with N_f Weyl fermions in the vectorial representation has the same as in Eq. (1) pattern of the global symmetry breaking, and is also described by a nonlinear sigma model with the target space $\mathcal{M}_{N_f}$. Witten proposed[6] that the Skyrmions of this theory must be identified with objects obtained by contracting the SO(N_c)

antisymmetric tensor $\varepsilon_{\alpha_1 \ldots \alpha_{N_c}}$ with the color indices of the vectorial quarks and/or the gluon field strength tensor. These objects are stable due to the quotient symmetry $\mathbb{Z}_2 = \mathrm{O}(N_c)/\mathrm{SO}(N_c)$, which acts as a global symmetry group. Our results for the Skyrmion statistics in adjoint QCD can be applied to the $\mathrm{SO}(N_c)$ gauge theory too. They give further evidence in favor of Witten's identification.

We find that the Skyrmions are stable since they are the only particles with an odd relation between statistics and fermion number. Namely, Skyrmions can be bosons with with fermion number one or fermions with the vanishing fermion number. Therefore, Skyrmions cannot decay to any final state consisting of "normal" or "perturbative" particles: pions and other similar mesons or baryons of the type (4).

In Sect. 2 we describe in detail the low-energy effective action, parametrization of the manifold $\mathcal{M}_{N_f}$, and introduce a coupling to baryons ψ of the type (4). Section 3 is devoted to determination of the WZNW term, to the calculation of its coefficient through the anomaly matching and to the effect of the WZNW term on the spin/statistics and fermion number of the Skyrmion. In Sect. 4 we discuss an anomalous term responsible for the shift of the Skyrmion fermion number, which, in turn, guarantees its stability.

This work has been done in collaboration with S. Bolognesi and M. Shifman and is published in Refs. 13,14. In this short presentation many details will be omitted.

2. Low-Energy Effective Action

We use the Cartan embedding to parametrize the coset (2). This parametrization is very useful, in particular, because it makes explicit the symmetries of the manifold. Then, coupling pions to baryons (4) becomes a straightforward task through the Cartan embedding.

The general element of the quotient $\mathcal{M}_{N_f} = \mathrm{SU}(N_f)/\mathrm{SO}(N_f)$ can be written in a compact form as $U \cdot \mathrm{SO}(N_f)$, where U is an $\mathrm{SU}(N_f)$ matrix (different U in $\mathrm{SU}(N_f)$ correspond to the same $\mathcal{M}_{N_f}$ element, modulo a product with an arbitrary $\mathrm{SO}(N_f)$ element). The map

$$U \cdot \mathrm{SO}(N_f) \to W = U \cdot U^t \,, \tag{7}$$

where the superscript t denotes transposition, is well-defined on the quotient because for the $\mathrm{SO}(N_f)$ matrices the inverse is equal to the transposed matrix. Equation (7) presents a one-to-one map between $\mathcal{M}_{N_f}$ and the submanifold of the matrices of $\mathrm{SU}(N_f)$ which are both unitary and symmetric.

In the mathematical literature it is called the Cartan embedding (see Ref. 15 for a review).

The quotient (2) can be parameterized using the matrix exponential of the SU(N_f) generators which do *not* belong to the unbroken subgroup SO(N_f). These generators are given by $N_f \times N_f$ *symmetric* Hermitian matrices with vanishing trace. These matrices necessarily have all real entries; they can be always diagonalized by virtue of an SO(N_f) change of basis. Thus, a convenient way of parameterization of the equivalence classes of $\mathcal{M}_{N_f}$ is as follows: one must choose an SU(N_f) representative

$$U = \exp(i\, V \cdot A \cdot V^\dagger), \tag{8}$$

where A is a Hermitian traceless diagonal matrix and V is an SO(N_f) element, which can be parameterized by N_f-dimensional Euler angles. Then we can construct the matrix W,

$$W = U \cdot U^t = \exp\left(2i\, V \cdot A \cdot V^\dagger\right). \tag{9}$$

The Lagrangian of the Skyrme model with the target space $\mathcal{M}_{N_f}$ can be computed by evaluating the Lagrangian of the SU(N_f) Skyrme model on the symmetric unitary matrix W,

$$\begin{aligned} \mathcal{L} &= \frac{F_\pi^2}{4}\mathcal{L}_2 + \frac{1}{e^2}\mathcal{L}_4 \\ &\equiv \frac{F_\pi^2}{4}\mathrm{Tr}\left(\partial_\mu W \partial^\mu W^\dagger\right) + \frac{1}{e^2}\mathrm{Tr}\left[(\partial_\mu W)W^\dagger, (\partial_\nu W)W^\dagger\right]^2 . \end{aligned} \tag{10}$$

We have also to include in the effective Lagrangian the fermions ψ, which have mass $O(N^0)$ in the large N limit. The fermion coupling is

$$-\frac{g}{2}\left\{W^{fg}\psi_{\alpha f}\psi_g^\alpha + \text{H.c.}\right\}. \tag{11}$$

To the lowest order, the effective Lagrangian which includes both pions and the fermions $\psi_{\alpha a}$ is

$$\mathcal{L} = \frac{F_\pi^2}{4}\mathrm{Tr}\left(\partial_\mu W \partial^\mu W^\dagger\right) + \bar{\psi}_{f\dot{\alpha}} i\partial^{\dot{\alpha}\alpha}\psi_{f\alpha} - \frac{g}{2}\left\{W^{fg}\psi_{\alpha f}\psi_g^\alpha + \text{H.c.}\right\}. \tag{12}$$

If we expand around the vacuum where W is given by the identity matrix, the fermionic part of the Lagrangian is given by

$$\mathcal{L}_{\text{ferm}} = \bar{\psi}_{f\dot{\alpha}} i\partial^{\dot{\alpha}\alpha}\psi_{f\alpha} - g\left\{\psi_f^\alpha\psi_{\alpha f} + \text{H.c.}\right\}. \tag{13}$$

The coupling constant g is identified with the mass of the fermion.

3. WZNW Term, Anomalies and Statistics

We can write the WZNW term for the $\mathcal{M}_{N_f}$ sigma model ($N_f \geq 3$) by virtue of evaluating the SU(N_f) Wess–Zumino–Novikov–Witten term on the symmetric unitary matrices W introduced in Eqs. (8) and (9). Namely,

$$\Gamma = -\frac{i}{240\pi^2}\int_{B_5} d\Sigma^{\mu\nu\rho\sigma\lambda}\mathrm{Tr}\left[(W^\dagger\partial_\mu W)\cdot(W^\dagger\partial_\nu W)\right.$$
$$\left.\cdot\ (W^\dagger\partial_\rho W)\cdot(W^\dagger\partial_\sigma W)\cdot(W^\dagger\partial_\lambda W)\right]. \tag{14}$$

In order to compute the WZNW term for the $\mathcal{M}_{N_f}$ sigma model, we need to take the result for SU(N_f) and restrict it to the submanifold of the unitary symmetric matrices.

There is a subtle difference regarding the possible coefficients allowed for Γ in the action. In the Lagrangian of the SU(N_f) sigma model, relevant for QCD Skyrmions, the WZNW term must have just integer coefficient k,

$$\mathcal{L} = \mathcal{L}_2 + k\,\Gamma + \text{Higher order terms}. \tag{15}$$

This is due to the fact that the integral of this term on an arbitrary S^5 submanifold of SU(N_f) must be an integer multiple of 2π. In the $\mathcal{M}_{N_f}$ sigma model relevant for adjoint QCD we need to use the same normalization prescription. The main difference is that if we integrate Γ on the minimal S^5 which we can build inside the SU(N_f) subspace of the symmetric Hermitian matrices, the result will be 4π rather than 2π, as we get for the generator of $\pi_5\,(\mathrm{SU}(N_f))$ (for an explicit calculation see Ref. 14). Therefore, if we restrict ourself to this subspace it is consistent to consider also half-integer values of k.

Let us consider a U(1) subgroup of SO(N_f) generated by the charge matrix Q. We have the following expression for the gauged WZNW term (see Refs. 5,18):

$$\tilde{\Gamma} = \Gamma + \int dx^4 x A_\mu J^\mu + \frac{i}{24\pi^2}\int dx^4 \epsilon^{\mu\nu\rho\sigma}(\partial_\mu A_\nu)A_\sigma$$
$$\times \mathrm{Tr}\left[Q^2(\partial_\sigma W)W^\dagger + Q^2 W^\dagger(\partial_\sigma W) + \frac{1}{2}QW^\dagger Q(\partial_\sigma W) - \frac{1}{2}QWQ(\partial_\sigma W^\dagger)\right], \tag{16}$$

where

$$J^\mu = \frac{1}{48\pi^2}\epsilon^{\mu\nu\rho\sigma}\,\mathrm{Tr}\Big[Q(\partial_\nu W)W^\dagger\cdot(\partial_\rho W)W^\dagger\cdot(\partial_\sigma W)W^\dagger$$

$$+\,QW^\dagger(\partial_\nu W)\cdot W^\dagger(\partial_\rho W)\cdot W^\dagger(\partial_\sigma W)\Big]. \tag{17}$$

The general expression for the anomalous current in four-dimensional gauge theories can be read off from the usual triangle graph,

$$\langle\partial_\mu J^\mu_{\kappa_1}\rangle = \frac{1}{32\pi^2}D_{\kappa_1\kappa_2\kappa_3}\epsilon^{\kappa\nu\lambda\rho}F^{\kappa_2}_{\kappa\nu}F^{\kappa_3}_{\lambda\rho}\,, \tag{18}$$

where

$$D_{\kappa_1\kappa_2\kappa_3} = \frac{1}{2}\mathrm{Tr}\Big(\{T_{\kappa_1},\,T_{\kappa_2}\}T_{\kappa_3}\Big)\,. \tag{19}$$

Let us gauge the U(1) subgroup generated by

$$Q = \begin{pmatrix} 0 & i & 0 \\ -i & 0 & 0 \\ 0 & 0 & 0 \end{pmatrix}\,. \tag{20}$$

We have $T_{\kappa_2} = T_{\kappa_3} = Q$. The expression for the anomaly coefficient of a generic element of the flavor group T_{κ_3} is

$$D_{\kappa_1 QQ} = D_{QQ\kappa_1} = (N_c^2-1)\,\mathrm{Tr}\left[\begin{pmatrix} 1 & 0 & 0 \\ 0 & 1 & 0 \\ 0 & 0 & 0 \end{pmatrix} T_{\kappa_1}\right]\,. \tag{21}$$

The fact that D is completely symmetric facilitates the calculation. Let us take

$$T_{\kappa_1} = \begin{pmatrix} 1 & 0 & 0 \\ 0 & 1 & 0 \\ 0 & 0 & -2 \end{pmatrix}\,, \tag{22}$$

which corresponds to the Golstone boson π_3. We then find

$$\langle\partial_\mu J^\mu_{\kappa_1}\rangle = \frac{N_c^2-1}{16\pi^2}\epsilon^{\kappa\nu\lambda\rho}F_{\kappa\nu}F_{\lambda\rho}\,. \tag{23}$$

At this point we can match this value with the one found from the low-energy theory in Eq. (16). We obtain in this way that the coefficient in front of the WZNW term is

$$k = \frac{N_c^2-1}{2}. \tag{24}$$

The crucial 1/2 factor comes from the fact that we consider a theory with the Weyl fermions rather than Dirac fermions as is the case in QCD. Note that k is half-integer for N_c even and integer for N_c odd.

In order to determine the Skyrmion statistics it is necessary to calculate the value of the WZNW functional Γ for a process of a 2π rotation of the soliton. This calculation in the case of QCD has been done in Ref.6. The explicit calculation for adjoint QCD can be found in Ref. 14. The result is:

$$\Gamma = 2\pi\, k. \tag{25}$$

With our conventions for the coefficient k, it can be integer or half-integer, depending on the number of colors N_c. As was discussed in Eq. 24, k is half-integer for N_c even and integer for N_c odd. It immediately follows that the Skyrmion is quantized as a fermion for N_c even and as a boson for N_c odd.

4. Skyrmion Stability

We introduce now a different, but equivalent, formulation that will be particularly useful for the purposes addressed in this section. In the $N_f = 2$ case there are two ways to parametrize the target space S^2. One can use a vector $\vec{n}$ subject to the constraint $|\vec{n}| = 1$. This is the so-called O(3) formulation. Another approach, which goes under the name of the gauged formulation of the CP^1 sigma model, is to use a complex doublet z_i subject to the constraint $z_i^* z_i = 1$. This leaves us with an S^3 sphere. We have to further reduce it by gauging the phase rotation $z_i \to e^{i\theta} z_i$. This Hopf fibration leaves us exactly with the sphere S^2. The map between the two formulations is

$$\vec{n} = z_i^* \vec{\tau} z_i \, .$$

The derivatives acting on the doublet z_i are the covariant derivatives

$$D_\mu = \partial_\mu - i A_\mu$$

where (see e.g. Ref. 16 for a review)

$$A_\mu = -\frac{i}{2} \left[z_i^* (\partial_\mu z_i) - (\partial_\mu z_i^*) z_i \right] . \tag{26}$$

To avoid confusion in the case of higher N_f we will refer to the latter formulations as the z formulation. The z_i formulation has a valuable advantage. It is possible to express the Hopf charge (the charge of $\pi_3(S^2) = Z$) as a local function of the gauge field A,

$$s = \frac{1}{4\pi^2} \int d^3x \epsilon^{\mu\nu\rho} A_\mu \partial_\nu A_\rho \, . \tag{27}$$

An equivalent local expression in terms of the $\vec{n}$ field is impossible.[17]

Generalization to higher N_f *is not* achieved by extending the doublet to a complex N_f-plet. For $N_f = 2$ this strategy works because SU(2) is equivalent to the sphere S^3. In order to generalize to higher N_f we need to start with an SU(N_f) sigma model and then gauge an SO(N_f) subgroup. Let us consider the exact sequence

$$\ldots \to \pi_3\left(\mathrm{SO}(k)\right) \to \pi_3\left(\mathrm{SU}(k)\right) \to \pi_3\left(\mathrm{SU}(k)/\mathrm{SO}(k)\right) \to \pi_2\left(\mathrm{SO}(k)\right) \to \ldots$$

For every k we have $\pi_2\left(\mathrm{SO}(k)\right) = 0$. Therefore, every non-zero element of $\pi_3\left(\mathrm{SU}(k)/\mathrm{SO}(k)\right)$ can be lifted to a non-zero element of $\pi_3\left(\mathrm{SU}(k)\right)$. Then we can calculate the S^3 winding number of the lifted 3-cycle, using the SU(N_f) result,

$$s = -\frac{i}{24\pi^2}\int_{S^3} \mathrm{Tr}\,(U^\dagger dU)^3\,. \tag{28}$$

It is possible to present the topological winding number as an SU(N_f) Chern–Simons current. Let us introduce

$$\mathcal{A}_\mu = iU^\dagger \partial_\mu U\,. \tag{29}$$

Then

$$s = \frac{1}{8\pi^2}\int d^3x K^0, \quad K^\mu = \epsilon^{\mu\nu\rho\sigma}\,\mathrm{Tr}\left(\mathcal{A}_\nu\partial_\rho\mathcal{A}_\sigma - \frac{2}{3}\,i\,\mathcal{A}_\nu\mathcal{A}_\rho\mathcal{A}_\sigma\right). \tag{30}$$

The topological winding number s is defined modulo 4 for $N_f = 3$ and modulo 2 for $N_f > 3$, due to a topological arbitrariness in the choice of U.

The main outcome of Ref. 8 in the $N_f = 2$ case, is that the Skyrmion, through the Goldstone–Wilczek mechanism,[9] acquires a fermion number 1. The reason is as follows. Equation (27) indicates how to express the Hopf charge as a function of the gauge field A_μ. On the other hand, we know that the fermion current has an axial anomaly,

$$\partial^\mu J^{F0}_\mu = \frac{1}{8\pi^2}F_{\mu\nu}\tilde{F}^{\mu\nu} = \frac{1}{4\pi^2}\partial_\mu(\epsilon^{\mu\nu\rho\sigma}A_\nu\partial_\rho A_\sigma), \tag{31}$$

where

$$\widetilde{F}_{\mu\nu} = \frac{1}{2}\epsilon_{\mu\nu\rho\sigma}F^{\rho\sigma}\,.$$

We have omitted the mass term which explicitly brakes $\mathrm{U}(1)_F \to Z_2$.

One must remember that the fermion number is well defined only modulo two. The last equality in Eq. (31) is the well-known expression of the topological density as a total derivative of the Chern–Simons term. Integrating the anomaly equation we arrive at

$$J_\mu^F = J_\mu^{F0} - \frac{1}{4\pi^2} K_\mu, \qquad K_\mu = \epsilon^{\mu\nu\rho\sigma} A_\nu \partial_\rho A_\sigma\,. \tag{32}$$

This expression is matched with the Hopf number in Eq. (27). We thus find that the fermion ψ transfers one unit of fermion number to the Hopf Skyrmion.

In order to generalize to higher N_f one must consider the triangle anomaly

$$\mathrm{U}(1) - \mathrm{SO}(N_f) - \mathrm{SO}(N_f)\,.$$

The U(1) corresponds to the fermion number. For SO(N_f) we introduce an auxiliary gauge field. The anomaly is

$$\partial_\mu J_\mu^{F0} = \frac{1}{16\pi^2} \mathrm{Tr}(F^{\mu\nu} \widetilde{F}_{\mu\nu}) = \frac{1}{8\pi^2} \partial_\mu K^\mu, \tag{33}$$

where $F_{\mu\nu} = F_{\mu\nu}^k T^k$, with T^k standing for the generators of SO(N_f) (with $\mathrm{Tr}(T_j T_k) = \delta_{ij}$), and K_μ is given in Eq. (30).

The net effect of the baryon ψ with mass $O(N_c^0)$ is to shift the Skyrmion fermion number by one unit, without changing its statistics. For N_c odd, the Skyrmion is a boson with an odd fermion number. For N_c even, it is a fermion with an even fermion number. The relation between the Skyrmion statistics and fermion number is abnormal. In both cases it is a $\mathbb{Z}_2$-stable object, because in the "perturbative" spectrum the normal relation between the fermion number and statistics takes place.

Acknowledgments

I wish to thank the organizers of the Workshop for their kind hospitality and for providing us with an occasion for interesting discussions.

References

1. S. Dimopoulos, Nucl. Phys. B **168**, 69 (1980); M. E. Peskin, Nucl. Phys. B **175**, 197 (1980).
2. Y. I. Kogan, M. A. Shifman and M. I. Vysotsky, Yad. Fiz. **42**, 504 (1985) [Sov. J. Nucl. Phys. **42**, 318 (1985)].
3. J. J. Verbaarschot, Phys. Rev. Lett. **72**, 2531 (1994) [hep-th/9401059]; A. Smilga and J. J. Verbaarschot, Phys. Rev. D **51**, 829 (1995) [hep-th/9404031]; M. A. Halasz and J. J. Verbaarschot, Phys. Rev. D **52**, 2563 (1995) [hep-th/9502096].

4. T. H. R. Skyrme, Proc. Roy. Soc. Lond. A **262**, 237 (1961); Nucl. Phys. **31**, 556 (1962). See also *Selected Papers of Skyrme*, Ed. G. Brown (World Scientific, Singapore, 1994); for a review see *Chiral Solitons*, Ed. Keh-Fei Liu (World Scientific, Singapore, 1987).
5. E. Witten, Nucl. Phys. B **223**, 422 (1983).
6. E. Witten, Nucl. Phys. B **223**, 433 (1983) [reprinted in S. Treiman et al., *Current Algebra and Anomalies*, (Princeton University Press, 1985), p. 515.]
7. G. S. Adkins, C. R. Nappi and E. Witten, Nucl. Phys. B **228**, 552 (1983).
8. S. Bolognesi and M. Shifman, Phys. Rev. D **75** (2007) 065020 [arXiv:hep-th/0701065].
9. J. Goldstone and F. Wilczek, Phys. Rev. Lett. **47**, 986 (1981).
10. J. Wess and B. Zumino, Phys. Lett. B **37**, 95 (1971); S. P. Novikov, *Multi-valued Functions and Functionals. An Analogue of the Morse Theory,* Dokl. Akad. Nauk SSSR **260**, 31 (1981) [Soviet Math. Doklady, **24**, 222 (1981)]; E. Witten, Commun. Math. Phys. **92**, 455 (1984).
11. D. Finkelstein and J. Rubinstein, J. Math. Phys. **9** (1968) 1762.
12. S. Krusch and J. M. Speight, Commun. Math. Phys. **264**, 391 (2006) [hep-th/0503067].
13. R. Auzzi and M. Shifman, J. Phys. A **40** (2007) 6221 [arXiv:hep-th/0612211].
14. R. Auzzi, S. Bolognesi and M. Shifman, Phys. Rev. D **77**, 125029 (2008) [arXiv:0804.0229 [hep-th]].
15. T. Puettmann and A. Rigas, Commentarii Mathematici Helvetici **78**, 648 (2003) [arXiv:math/0301192].
16. For a review see e.g. V. A. Novikov, M. A. Shifman, A. I. Vainshtein and V. I. Zakharov, Phys. Rept. **116**, 103 (1984), Sect. 5.2.
17. F. Wilczek and A. Zee, Phys. Rev. Lett. **51**, 2250 (1983).
18. O. Kaymakcalan, S. Rajeev and J. Schechter, Phys. Rev. D **30** (1984) 594.

DOMAIN WALLS, LOCALIZATION AND CONFINEMENT:
- What Binds Strings Inside Walls -

S. BOLOGNESI

William I. Fine Theoretical Physics Institute, University of Minnesota,
116 Church St. S.E., Minneapolis, MN 55455, USA

We study a mechanism that induces confinement inside domain walls. Condensates on the two vacua separated by the wall, if they are equivalent, can tunnel from one side to the other. They can create a current and trap a confining string inside the domain wall. Strings in the two vacua are attracted toward the domain wall and form a bound state with it.

We consider a simple toy model that realizes this mechanism. We also provide an example from the Seiberg-Witten theory where everything can be studied in the strong coupling regime using the dual effective action. We finally compare our findings with the previously studied wall with two different quark flavors and explain why here the discussed mechanism of confinement does not act.

Keywords: Topological solitons: Strings and Walls

1. Introduction

Elementary domain walls can localize U(1) fields.[1] In 2+1 dimensions the U(1) gauge field is dual to a phase field σ living on S_1. The U(1) gauge theory in 2+1 dimensions can exist in distinct phases: (*i*) Coulomb; (*ii*) Higgs; (*iii*) Topological massive; (*iv*) Confining. In this latter regime electric charges are connected by a flux line which plays the role of a confining string. Interaction between the electric charges grows linearly with the distance r. The domain line endpoints are the σ field vortices.

In a recent paper[2] a mechanism has been suggested that leads to confinement on domain walls. We describe the physical idea. Let us take a theory with a Confining vacuum and a Coulomb vacuum. Then create a composite domain wall Confining-Coulomb-Confining. In this way the Coulomb vacuum is constrained in a strip of thickness d, sandwiched between two confining vacua. In order to avoid a wall-antiwall instability, the two confining vacua must be different.

In the $3+1$ dimension, confinement can be understood as condensation of magnetic monopoles. In this dual variables the domain wall of interest can be schematized as in Figure 1. The monopole condensate is equal to v on the left and the same on the right. In the middle the monopole is a massive particle of mass M and thus the condensate thus drop to zero exponentially. The condensate in the center is roughly $ve^{-Md/2}$. *There is*

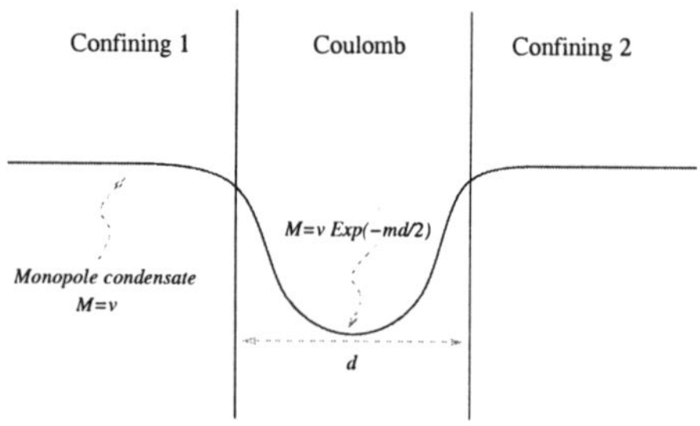

Fig. 1. The monopole condensation along the domain wall profile.

always a residue condensate in the middle.

We suggest two stable model examples: one at weak coupling (Section 2) and another using the Seiberg-Witten solution at strong coupling (Section 3). These models are demonstrated to be working examples of confinement on the domain wall. We finally discuss (Section 4) the theories with two flavors and explain why here the gauge field on the wall remain massless despite the presence of a residual condensate in the middle of the wall.

2. The Simplest Example at Weak Coupling

Consider a U(1) gauge theory with a charged scalar field Q. Our task is to study a domain wall interpolating between Higgs-Coulomb-Higgs vacua. To create a stable configuration we need at least an extra real neutral field a. The Lagrangian is:

$$L = -\frac{1}{4e^2}(F^{\mu\nu})^2 + \frac{1}{2e^2}(\partial_\mu a)^2 + |\nabla_\mu Q|^2 - \frac{(a-m)^2(a+m)^2|Q|^2}{m^2} - \frac{e^2}{2}(v^2 - |Q|^2)^2 \,. \quad (1)$$

We have two distinct vacua $a = \pm m$, $|Q| = v$. If we keep $m \gg ev$, there is a large intermediate region inside the domain wall where the VEV of Q is almost zero (see Fig. 2). Even if the theory has no Coulomb vacuum, if we look at the domain wall profile we the inside region of the wall is almost in the Coulomb phase.

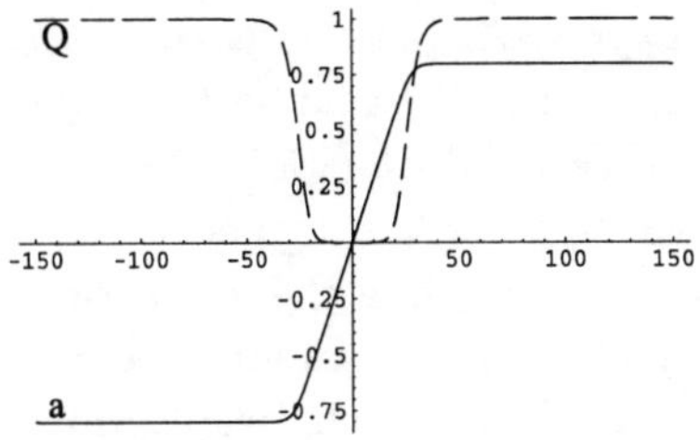

Fig. 2. Domain wall in the toy model (1). The profile of the $a(z)$ field is presented by the solid line while that of $Q(z)$ by dashed.

Let us estimate the thickness d. In the inside region the VEV of Q almost vanishes while a is linear in z. If $m >> ev$, the dominating contributions to the energy come from the potential and from the kinetic term for the field a:

$$T_{\rm wall} \sim \frac{e^2v^4}{2}d + \frac{(2m)^2}{de^2}. \tag{2}$$

The minimum is achieved at

$$d = \frac{2\sqrt{2}m}{e^2v^2}, \qquad T_{\rm wall} \sim 2\sqrt{2}\, m\, v^2. \tag{3}$$

Localization of gauge fields on lower-dimensional topological objects is, generally speaking, a nontrivial task. Massless scalars can be localized as Goldstone bosons of continuous symmetries spontaneously broken on the given topological defects. Massless fermions can be localized via index theorems. No such powerful tool exists for gauge fields.

Since the domain wall is an object with thickness d a low-energy effective action makes sense only up to energy scales $\sim 1/d$. At higher energies excitations of the wall internal structure become important. Fluctuations of the wall as a whole in the transverse direction are massless, they always belong to the low-energy effective action. Other - massive - excitations can be considered a part of the $2+1$ dimensional effective action as long as their mass is much smaller than $1/d$. The localization limit is $e\,v_0 \ll 1/d \ll e\,v$. Up to energies $\sim 1/d$ the gauge field can be considered as a field localized on the $2+1$ dimensional world volume.

The simplest solutions that describe localization of the gauge field are a constant magnetic field and a constant electric field. From now on we will always work in the gauge $A_z = 0$.

A constant magnetic field inside the wall is parallel to the wall surface. Let us assume the magnetic field to be aligned along the x axis, $\vec{B} = B\hat{x}$ where $\hat{x}$ is the unit vector along x. At negative z we take the field $Q = ve^{iky}$ and the gauge field $A_y = k$. At positive z we take $Q = v$ and $A_y = 0$. In this way in two vacua, to the left and to the right of the wall, the field configuration is pure gauge. Inside the wall, A_y linearly interpolates between $-k$ and 0 on the interval of size d. The magnetic field is $B_x = -\partial_z A_y = k/d$. The magnetic flux per unit length in the y direction is $\int dz B_x = k$.

The electric field inside the wall can only be perpendicular to the wall surface, aligned along $\hat{z}$. Thus, on the wall it must be interpreted as a pseudoscalar. To obtain such an electric field inside the wall consider $Q = ve^{i\omega t}$ at negative z and $A_t = \omega$. At positive z we have $Q = v$ and $A_t = 0$. Inside the wall A_t linearly interpolates between ω and 0. The electric field inside the wall is $E_z = -\partial_z A_t = \omega/d$.

Of course, from the $2+1$ dimensional point of view the picture must be dualized, since in 2+1 dimensions it is the electric field F_{0i} which is a vector while the magnetic field F_{12} is a scalar. Now we can further dualize $F_{\mu\nu}^{(2+1)}$ à la Polyakov,[3] expressing $F_{\mu\nu}^{(2+1)}$ in terms of a phase fields σ. Assembling everything together we have, with our gauge choice

$$\begin{aligned} B_x^{(3+1)} = -\partial_z A_y \quad &= \quad E_x^{(2+1)} = \frac{1}{d}\,\partial_y \sigma\,, \\ B_y^{(3+1)} = \partial_z A_x \quad &= \quad E_y^{(2+1)} = -\frac{1}{d}\,\partial_x \sigma\,, \\ E_z^{(3+1)} = -\partial_z A_t \quad &= \quad B^{(2+1)} = \frac{1}{d}\,\partial_t \sigma\,. \end{aligned} \tag{4}$$

Note that the angle field σ exactly corresponds to the phase of Q at negative z relative to that at positive z.

In the model at hand, the phase field σ is not an exact modulus. A potential $V(\sigma)$ is generated forcing $\sigma = 0$ in the true vacuum which is unique. In the localization limit of large d we can nevertheless speak of a $2+1$ dimensional effective theory for σ, since the σ field mass is much smaller than the excitation energy of the domain wall $\sim 1/d$. In the leading approximation the σ field Lagrangian will be of the sine-Gordon type,

$$\mathcal{L}_{2+1} = \frac{1}{2e^2\,d}\;\partial_\mu\sigma\partial_\mu\sigma - \Delta\cdot(\sin\sigma)^2 \tag{5}$$

where Δ is the difference between two tensions: the tension of the $\sigma = 0$ wall and and that of the $\sigma = \pi$ wall. The mass of σ is $m_\sigma \approx \sqrt{2\Delta\, e^2\, d}$. We can naturally guess that m_σ is of the same order of magnitude of ev_0.

Consider for example the solution with a constant magnetic field in the $\hat{y}$ direction (see Fig. 3). As a first approximation, we can choose $Q = ve^{ikx}$

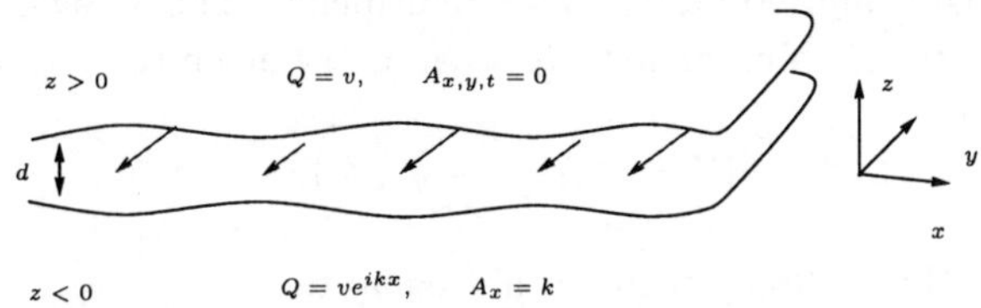

Fig. 3. An (almost) constant magnetic field inside the domain wall. When $\sigma \approx 0$, the wall is thinner and the magnetic field larger. At $\sigma \approx \pi$ the magnetic field is smaller and the wall thicker.

and $A_x = k$ at $z < 0$, and $Q = v$ and $A_x = 0$ at $z > 0$, similar to the discussion above. This gives a constant magnetic field. But it is clear that, due to a nonvanishing Q condensate inside the wall, this is *not* the exact solution. The latter requires modulations of k. Due to topological reasons, the field Q must exactly vanish at some x and $z \approx 0$ each time the relative phase σ rotates by 2π. The lines (in the y direction) on which Q vanishes are the lines where the magnetic field reaches its maximum.

At $\sigma = \pi$ the wall solution corresponds to the maximum of the potential $V(\sigma)$. It is stationary but unstable. In this solution the Q field vanishes exactly in the middle of the wall, at $z = 0$. We can easily get the solution if we impose two conditions: (i) the field Q is real; (ii) Q changes sign in passing from one side of the wall to the other, see Fig. 4. The difference between tensions of the unstable and stable walls will determine Δ in Eq. (5).

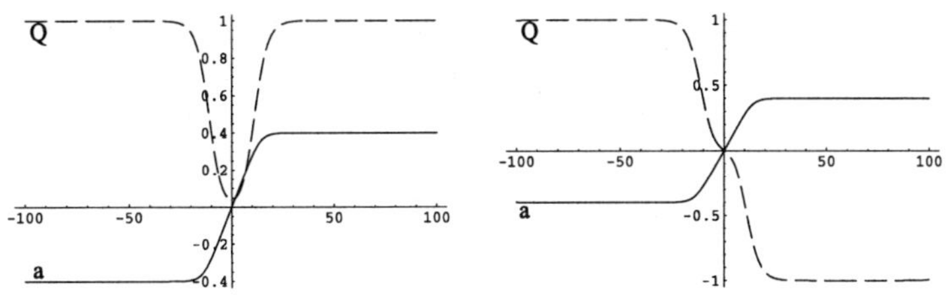

Fig. 4. Left: the stable wall ($\sigma = 0$). Right: the unstable wall ($\sigma = \pi$).

3. A Modified Seiberg–Witten framework

We now turn to a strong coupling example of the confinement phenomenon inside domain walls. The theory of interest is $\mathcal{N}=2$ gauge theory, with the gauge group U(2) and no matter hypermultiplets.[4] The following superpotential breaks then the extended supersymmetry down to $\mathcal{N}=1$:

$$W = \alpha \,\mathrm{Tr}\left(\frac{\Phi^3}{3} - \xi\Phi\right). \tag{6}$$

Classically, we have three vacua, with ϕ equal to

$$\begin{pmatrix} \sqrt{\xi} & 0 \\ 0 & \sqrt{\xi} \end{pmatrix}, \quad \begin{pmatrix} \sqrt{\xi} & 0 \\ 0 & -\sqrt{\xi} \end{pmatrix}, \quad \begin{pmatrix} -\sqrt{\xi} & 0 \\ 0 & -\sqrt{\xi} \end{pmatrix}. \tag{7}$$

The first and the last vacua preserve the non-Abelian SU(2) gauge symmetry. Strong coupling effects à la Seiberg-Witten will then split each of them into two vacua (the monopole and dyon vacua). The vacuum in the middle preserve only the U(1) × U(1) gauge symmetry, and is not split. We, thus, expect in total five vacua, for generic values of ξ.

If we set α at zero, the U(1) and SU(2) sectors get completely decoupled. Since there are no matter hypermultiplets, only a a nonvanishing superpotential can make the two sectors communicate with each other. Dynamics of the U(1) sector is trivial, while the SU(2) sector is described by the Seiberg–Witten solution.[4] If we parametrize the moduli space by

$$\Phi = \begin{pmatrix} a_0 + a_3 & 0 \\ 0 & a_0 - a_3 \end{pmatrix}, \tag{8}$$

the conventional SW solution is written in terms of the invariant $u = 2a_3^2$. Another gauge invariant way to parametrize the moduli space by $u_1 = \mathrm{Tr}\,\Phi$, $u_2 = \mathrm{Tr}\,\Phi^2/2$.

After the superpotential (6) is switched on, the moduli space is lifted. Five discrete vacua described above survive. The value of u_2 is ξ for all five vacua. It is not modified by quantum corrections. The Coulomb vacuum in the middle is not modified by quantum correction either; it lies at $u_1 = 0$, $u_2 = \xi$. The monopole-1 and dyon-1 vacua from $\phi = \mathrm{diag}\,(\xi, \xi)$ are

$$u_1 = 2\sqrt{\xi - \Lambda^2},\ u_2 = \xi \quad \text{and} \quad u_1 = 2\sqrt{\xi + \Lambda^2},\ u_2 = \xi\,, \tag{9}$$

respectively. The dyon-2 and monopole-2 vacua from $\phi = \mathrm{diag}\,(-\xi, -\xi)$ are at

$$u_1 = -2\sqrt{\xi - \Lambda^2},\ u_2 = \xi \quad \text{and} \quad u_1 = -2\sqrt{\xi + \Lambda^2},\ u_2 = \xi\,. \tag{10}$$

In the limit $\sqrt{\xi} \gg \Lambda$ the Coulomb vacuum is such that the electric coupling is small. As ξ decreases and becomes of order Λ, it enters a strong coupling regime and at the critical value $\xi = \Lambda^2$ it lies exactly in the monopole singularity and coalesces with two monopole vacua. Around this critical value, the Coulomb vacuum is such that the magnetic coupling is small, so we can use the same set of low-energy effective variables to describe both the Coulomb and the confining vacua (Fig. 5). We will work in this limit, but keeping three vacua separate.

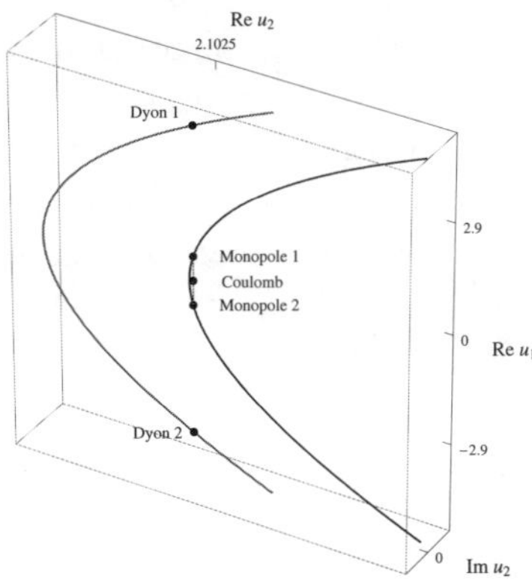

Fig. 5. As ξ becomes smaller, we reach a critical region where the Coulomb vacua lies near the the monopole singularity ($\Lambda = 1$, $\xi = 1.45$ in this plot).

The effect of the superpotential W in the infrared theory, near the monopole vacuum, is described by the following effective superpotential:

$$W_{\text{eff}} = \frac{A_{3D} M \widetilde{M}}{\sqrt{2}} + \frac{\alpha(2A_0^3 + 6u(A_{3D})A_0)}{3} - 2\alpha\xi A_0\,, \tag{11}$$

where we can use the expansion $u(A_{3D}) \approx \Lambda^2 - 2i\Lambda A_{3D} - A_{3D}^2/4 + \ldots$. The vacuum expectation values in the confining vacua are

$$A_{3D} = 0, \qquad A_0 = \pm\sqrt{\xi - \Lambda^2}, \qquad M\widetilde{M} = \pm 4\sqrt{2}\, i\, \Lambda^2 \alpha(\xi - \Lambda^2)\,. \tag{12}$$

The dual description is valid in these vacua provided that the monopole condensate is $\ll \Lambda^2$. The Coulomb vacuum is described by the following solution:

$$M = \widetilde{M} = 0\,, \qquad A_0 = 0\,, \qquad u(A_{3D}) = \xi\,. \tag{13}$$

The nice feature of the limit $|\xi - \Lambda^2| \ll \Lambda^2$ is that we can use the same weakly coupled effective description in all three vacua of interest. This unified description is also valid for the domain wall interpolating between them.

The value of the superpotential in the two confining vacua, monopole-1 and monopole-2, is

$$W_{1,2} = \mp\frac{4}{3}\alpha(\xi - \Lambda^2)^{3/2} \,. \tag{14}$$

In the Coulomb vacuum the superpotential vanishes. We look for the wall solution in the ansatz $|\widetilde{M}| = |M|$ which automatically guarantees vanishing of the D term. For the elementary wall we will assume the phases of $\widetilde{M}$ and M to be constant constant. It is convenient to choose $\widetilde{M} = iM$. For the elementary wall we can also take M and A_0 real and A_{3D} pure imaginary.

The tension of the BPS wall interpolating between monopole-1 and monopole-2 vacua, if it existed, would be twice the tension of the BPS wall interpolating between the Coulomb and confining vacua. A composite wall built of two elementary ones at a finite distance from each other, does not exist. There is a repulsive force that falls off exponentially with the walls separation.[5]

In order to avoid the problem and stabilize the composite domain wall an extra term is introduced in the superpotential,

$$W' = \alpha\left(\mathrm{Tr}\left(\frac{\Phi^3}{3} - \xi\Phi\right) + \frac{i\mu}{2}(\mathrm{Tr}\,\Phi)^2\right), \tag{15}$$

where ξ and μ are chosen real. The values of the superpotential in both confining vacua change being shifted in the complex plane,

$$W'_{1,2} = \frac{2}{3}\alpha\left[-2i\mu^3 + 3i(\xi - \Lambda^2)\mu \mp 2(\xi - \Lambda^2 - \mu^2)^{3/2}\right]. \tag{16}$$

The tension of the BPS domain wall is given by the absolute value of the difference of the superpotentials at two vacua between which it interpolates. Now the three values of the superpotential are no more allineated. For this reason, if the composite BPS walls exist at $\mu \neq 0$, the composite wall will be stable.

In order to write the BPS equations for the elementary wall, we need complex profile functions for each field, M, A_{3D} and A_0. The ansatz $\widetilde{M} = iM$ can still be used. The corresponding profiles are shown in Fig. 6. The boundary conditions are such that $M(\infty) = -iM(-\infty)$. The tensions of the BPS walls are given by the central charges, implying that the composite wall is stable.

We would like also to understand the localization of the (massive) gauge field on the wall as a quasimodulus σ localized on the wall world-volume. Previous consideration suggests us to look for an opposite direction rotation of the U(1) phase of the monopole field at $z < 0$ and $z > 0$, respectively.

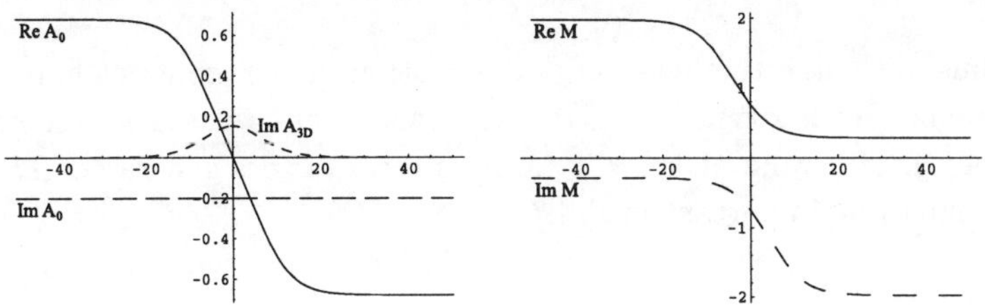

Fig. 6. The profile functions of the stable composite wall profiles at nonvanishing μ Left: Re A_0 (solid), Im A_0 (long dashes), Im A_3 (short dashes). Right: Re M (solid) and Im M (dashed).

There is an important difference between the toy model and that of Sect. 3. In the former case the Coulomb phase was not a true vacuum of the theory, while in the latter it is. This implies that imposing the condition $\sigma = \pi$ we get, in fact, two elementary walls, with no binding energy, separated by infinite distance. Needless to say, the condition that typical energies in the low-energy theory must be $\ll 1/d$, cannot be met then. In this formulation it makes no sense to speak of localization and reduction to $2+1$ dimensions.

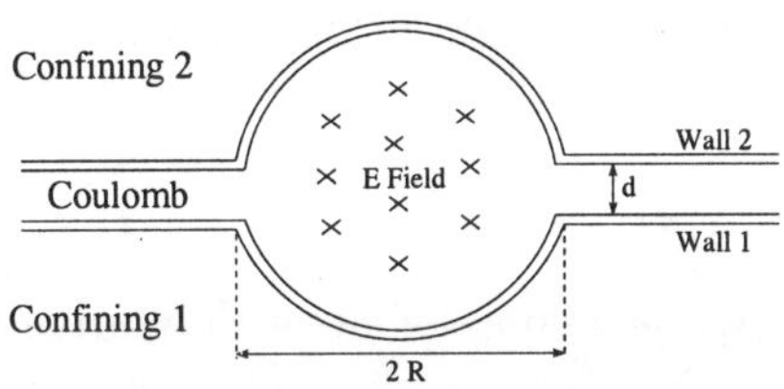

Fig. 7. An electric flux tube trapped in the middle of the composite wall.

Consider the field configuration shown in Fig. 7, which displays a flux tube trapped in the middle of the composite wall. Let us call R the radius of the region where the electric field is localized. The two-component wall is stabilized at a distance d_0 between the elementary walls. To separate them will cost a finite amount of energy per unit surface of the wall, to be referred to as $\delta T_{\rm w}$. In the model discussed in this section, at the first nonvanishing order in μ, we have $\delta T_{\rm w} = 6\,\alpha\,\sqrt{\xi - \Lambda^2}\,\mu^2$. Denote the electric flux of the vortex by f. An estimate for the tension of the flux tube is $T_{\rm ft} \approx \frac{1}{2e_{3D}^2}\,\frac{f^2}{\pi R^2} + 2R\,\delta T_{\rm w}$.. Minimizing with respect to R we arrive at $R_* = \left(f^2/(2\pi\, e_{3D}^2\,\delta T_w)\right)^{1/3}$. This shows that R_* is a finite quantity, and the flux is indeed squeezed into flux tubes inside the composite domain wall.

Thus, although the low-energy description in the model at hand is not of the sine-Gordon type (cf. (5)), the quasimodulus based low-energy description is still valid at $|\sigma| \ll \pi$: a mass term $m\sigma^2$ is generated. This fact turns out to be important in Ref.[6]

4. Two Flavor Model

At this point one is tempted to say that the fact that the gauge field localized on the wall weakly confines is due to residual exponentially small VEV's of the Higgs field inside the wall. In this case, the absence of the Coulomb regime on the wall would be a universal phenomenon. In fact, we know that this is not the case.[1]

Consider $\mathcal{N} = 2$ SQED with two flavors and the matter mass terms $m = m_1 = -m_2$. We will introduce the Fayet–Iliopoulos term ξ through the superpotential. Then the bosonic part of the action can be written as

$$S = \int d^4x \left\{ \frac{1}{4e^2} F_{\mu\nu}^2 + \frac{1}{e^2} |\partial_\mu a|^2 + \bar{\nabla}_\mu \bar{q}_A \nabla_\mu q^A + \bar{\nabla}_\mu \tilde{q}_A \nabla_\mu \bar{\tilde{q}}^A \right.$$
$$+ \frac{1}{2} \left| q^B (a + \sqrt{2} m_B) \right|^2 + \frac{1}{2} \left| \tilde{q}_B (a + \sqrt{2} m_B) \right|^2$$
$$\left. \frac{e^2}{2} \left| \tilde{q}_A q^A - \frac{\xi}{2} \right|^2 + \frac{e^2}{8} \left(|q^B|^2 - |\tilde{q}_B|^2 \right)^2 \right\}, \tag{17}$$

This theory has two vacua, and a domain wall interpolating between them.

This two-flavor model has a massless phase field σ localized on the domain wall. This is not related to supersymmetry. Any model with a global U(1) symmetry, two distinct vacua in which the global symmetry is unbroken, and a domain wall that spontaneously breaks this symmetry, automatically has a massless Goldstone boson localized on the domain wall.

Consider a string parallel to the domain wall (a grid of such strings is depicted in Fig. 8). If the distance between the wall and the string is very large we have an ordinary ANO string with thickness $\sim 1/(e\sqrt{\xi})$. As we move the string toward the wall, the quark condensate decreases and, in the center of the wall, the thickness of the string (in the directions parallel to the wall) becomes $\sim R = 1/(e\sqrt{\xi} e^{-dm/2})$, where d is the thickness of the wall. The thickness of the wall depends on the regime in which we find ourselves. In the limit $m/(e\sqrt{\xi}) \gg 1$ the thickness is entirely determined by the matter field and is $d \sim m/(e^2\xi)$. In the opposite limit $m/(e\sqrt{\xi}) \ll 1$ (the sigma-model limit) the thickness is is $d \sim 1/m$.

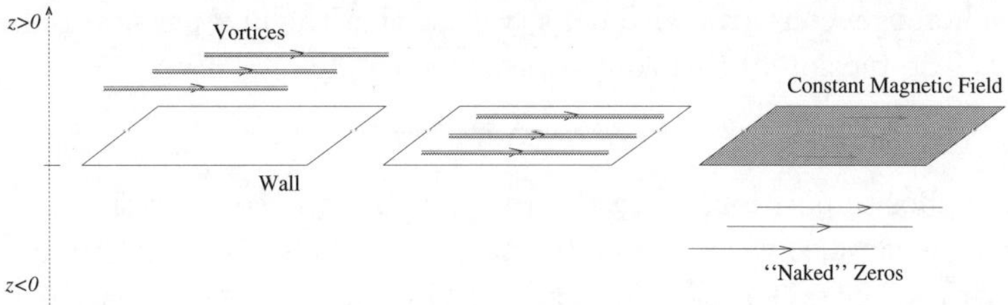

Fig. 8. A grid of flux tubes parallel to the domain wall. At $z > 0$ they are made out of the q_2 field condensate. The line of zeros of the q_2 field is surrounded by the magnetic field. At $z = 0$ the vortices lies in the middle of the wall. Proceeding further the zeros of q_2 go to z negative while the magnetic field remains on the domain wall surface.

We want to compare the thickness of the domain wall with that of the string in the middle of the wall. Multiplying both sides by $e\sqrt{\xi}$ we can express everything in terms of a dimensionless parameter $x = \frac{m}{e\sqrt{\xi}}$. Thus, we should compare

$$\text{Max}\left[\frac{1}{x}, x\right] \qquad \text{with} \qquad e^{x^2/2}. \tag{18}$$

In the $x \gg 1$ limit the thickness of the string is much larger than the thickness of the domain wall. This is qualitatively consistent with the existence of a Coulomb phase in the middle of the wall. The opposite limit $x \ll 1$ is problematic. The thickness of the string is much smaller that the thickness of the wall. We are thus tempted to conclude that in this regime the wall lives in a confining phase, which is clearly in contradiction with the rigorous proof above of the existence of a massless modulus and, hence, the Coulomb phase.

The fact is that wall world-volume dynamics reflects not only what happens at $z \sim 0$, but also what happens at $z \to \pm\infty$. Consider again a string parallel to the wall. When we reach $z = 0$ the string is inside the wall and has thickness $1/(e\sqrt{\xi}e^{-dm/2})$. Continue further and send the string to $z \to -\infty$ on the opposite side. In this limit the magnetic field remains trapped inside the domain wall and becomes constant (like in a Coulomb phase). All the energy remains trapped at $z = 0$. But the zeros of q_1 form a line that goes to $z \to -\infty$. In this sense we can say that the world-volume dynamics is dual to the $z \pm \infty$ in the bulk. In particular the q_2 flux tube, passing through the wall to the other side of the wall, *does not* become the q_1 flux tube.

Let us discuss energetics of this process. First of all, consider the string grid when it is far away from the wall in the $z > 0$ region. Assume we deal with a homogeneous grid with density f, so that the flux per unit of length is $4\pi f$. The tension of this configuration is simply

$$T_{z\gg 0} = \xi \Delta m + 2\pi \xi f\,. \tag{19}$$

In the opposite position, when the zeros of the strings are in the far $z < 0$ region, we can also easily compute the tension. It is just that of a domain wall with a constant magnetic flux inside it. An easy way to get the result is the thin edge approximation. The wall tension is now given by a sum of three terms

$$T\,(d) = \frac{1}{d}\left(\frac{2(\Delta m)^2}{g^2} + \frac{8\pi^2 f^2}{g}\right) + \frac{g^2\xi^2}{8} d\,. \tag{20}$$

Minimizing with respect to d we obtain

$$T_{z\ll 0} = \xi\sqrt{(\Delta m)^2 + 4\pi^2 f^2}\,. \tag{21}$$

At large positive z the tension of this configuration is given by the sum of the tensions of the wall and the grid of the flux tubes. The energy minimum is reached when the distance is $-\infty$. The Coulomb phase on the wall is energetically preferred to the confining phase

Acknowledgments

Is a pleasure to thank R. Auzzi, M. Shifman and A. Yung, with whom I enjoyed doing research on this problem. This proceeding is a short version of the original paper[7] where the reader can find all the details about the presented results and much more. This work is supported by DOE grant DE-FG02-94ER40823.

References

1. M. Shifman and A. Yung, Phys. Rev. D **67**, 125007 (2003) [arXiv:hep-th/0212293].
2. G. Dvali, H. B. Nielsen and N. Tetradis, Phys. Rev. D **77**, 085005 (2008) [arXiv:0710.5051 [hep-th]]
3. A. M. Polyakov, Nucl. Phys. B **120**, 429 (1977).
4. N. Seiberg and E. Witten, Nucl. Phys. B **426**, 19 (1994) [Erratum-ibid. B **430**, 485 (1994)] [arXiv:hep-th/9407087].
5. R. Portugues and P. K. Townsend, Phys. Lett. B **530**, 227 (2002) [arXiv:hep-th/0112077].
6. S. Bolognesi, arXiv:0710.5198 [hep-th].
7. R. Auzzi, S. Bolognesi, M. Shifman and A. Yung, arXiv:0807.1908 [hep-th].

Static Interactions of non-Abelian Vortices

Minoru Eto
Department of Physics, University of Pisa,
INFN, Sezione di Pisa
Largo Pontecorvo, 3, Ed. C, 56127 Pisa, Italy
E-mail: minoru@df.unipi.it

Interactions between non-BPS non-Abelian vortices are studied in non-Abelian $U(1)\times SU(N)$ extensions of the Abelian-Higgs model in four dimensional space-time. In addition to the usual type I/II Abelian superconductors, we find two new regime which we call type I*/II*.

Keywords: Soliton, vortex, superconductor

1. Introduction

Recently, a new type of BPS vortex was found in $U(N)$ gauge theories [1,2]. This is called a non-Abelian vortex and carries the non-Abelian charge $\mathbf{C}P^{N-1} = \frac{SU(N)_{\mathrm{C+F}}}{SU(N-1)_{\mathrm{C+F}}\times U(1)_{\mathrm{C+F}}}$. The reader can find good reviews in [3,4] and references of related works therein. In this talk, we are interested in studying interactions between *non-Abelian* vortices which are non-BPS. The non-BPS vortices are more natural than BPS ones in the sense that the BPS always requires a fine tuning or supersymmetry. It is well known that ANO vortices [5,6] in the type I system feel an attractive force while those in the type II model feel a repulsive force [7–10]. Specifically, we are interested in the interactions between vortices with different internal orientations, which is the distinct feature [11] from the ANO case.

This talk is based on [12] in collaboration with R.Auzzi and W.Vinci.

2. The model

2.1. *A fine-tuned model*

We start with a non-Abelian, $U(N)$, extension of the Abelian-Higgs model

$$\mathcal{L} = \mathrm{Tr}\left[-\frac{1}{2e^2}F_{\mu\nu}F^{\mu\nu} + \mathcal{D}_\mu H(\mathcal{D}^\mu H)^\dagger - \frac{\lambda^2 e^2}{4}\left(v^2\mathbf{1}_N - HH^\dagger\right)^2\right]. \quad (1)$$

Here, for simplicity we take the same gauge coupling e for both the $U(1)$ and $SU(N)$ groups, while $\lambda^2 e^2/4$ is a scalar coupling and v (> 0) determines the Higgs VEV. H is N Higgs fields in the fundamental representation of $U(N)$. The Higgs vacuum of the model is given by $HH^\dagger = v^2\mathbf{1}_N$. It breaks completely the gauge symmetry, although a global color-flavor locking symmetry $SU(N)_{\rm C+F}$ is preserved

$$H \to U_{\rm G} H U_{\rm F}, \quad U_{\rm G} = U_{\rm F}^\dagger, \quad U_{\rm G} \in SU(N)_{\rm G},\ U_{\rm F} \in SU(N)_{\rm F}. \qquad (2)$$

The trace part ${\rm Tr}H$ is a singlet under the color-flavor group and the traceless parts are in the adjoint representation. The $U(1)$ and the $SU(N)$ gauge vector bosons have both the same mass $M_{U(1)} = M_{SU(N)} = e\,v$. The N^2 real scalar fields in H are eaten by the gauge bosons and the other N^2 (one singlet and the rest adjoint) have same masses $M_{\rm s} = M_{\rm ad} = \lambda\, e\, v$. The critical coupling $\lambda = 1$ (BPS) allows an $\mathcal{N} = 2$ supersymmetric extension.

2.2. *Models with general couplings*

A generalization of (1) is to consider different gauge couplings, e for the $U(1)$ part and g for the $SU(N)$ part, and a general quartic scalar potential

$$\mathcal{L} = {\rm Tr}\left[-\frac{1}{2g^2}\hat{F}_{\mu\nu}\hat{F}^{\mu\nu} - \frac{1}{2e^2}f_{\mu\nu}f^{\mu\nu} + \mathcal{D}_\mu H(\mathcal{D}^\mu H)^\dagger\right] - V, \qquad (3)$$

where we have defined $\hat{F}_{\mu\nu} = \sum_{A=1}^{N^2-1} F_{\mu\nu}^A T_A$ and $f_{\mu\nu} = F_{\mu\nu}^0 T^0$ with ${\rm Tr}(T^A T^B) = \delta^{AB}/2$ and $T^0 = \mathbf{1}/\sqrt{2N}$ The scalar potential is:

$$V = \frac{\lambda_g^2 g^2}{4}{\rm Tr}\hat{X}^2 + \frac{\lambda_e^2 e^2}{4}{\rm Tr}\left(X^0 T^0 - v^2\mathbf{1}_N\right)^2, \qquad (4)$$

where $HH^\dagger = X^0 T^0 + \hat{X}$ and $\hat{X} = 2\sum_{A=1}^{N^2-1}\left(H^{i\dagger}T^A H_i\right)T^A$. The symmetries is the same in the previous fine-tuned model (1). In this model, the $U(1)$ and the $SU(N)$ vector bosons have different masses $M_{U(1)} = e\,v$, $M_{SU(N)} = g\,v$. Moreover, the singlet part of H has a mass $M_{\rm s}$ different from that of the adjoint part $M_{\rm ad}$ as $M_{\rm s} = \lambda_e\, e\, v$, $M_{\rm ad} = \lambda_g\, g\, v$. For the critical values $\lambda_e = \lambda_g = 1$, the Lagrangian again allows an $\mathcal{N} = 2$ susy extension.

2.3. *Vortex equations in the fine-tuned model*

Let us make the following rescaling of fields and coordinates:

$$H \to vH, \quad W_\mu \to evW_\mu, \quad x_\mu \to \frac{x_\mu}{ev}. \qquad (5)$$

The masses of vector and scalar bosons are rescaled to

$$M_{U(1)} = M_{SU(N)} = 1, \qquad M_{\rm s} = M_{\rm ad} = \lambda. \tag{6}$$

In order to construct non-BPS non-Abelian vortex solutions, we have to solve the equation of motion derived from the Lagrangian (1),

$$\mathcal{D}_\mu F^{\mu\nu} - \frac{i}{2}\left[H(\mathcal{D}^\nu H)^\dagger - (\mathcal{D}^\nu H)H^\dagger\right] = 0, \tag{7}$$

$$\mathcal{D}_\mu \mathcal{D}^\mu H + \frac{\lambda^2}{4}\left(1 - HH^\dagger\right)H = 0. \tag{8}$$

From now on, we restrict ourselves to static configurations only depending on the coordinates x^1, x^2. Here we introduce a complex notation $z = x^1 + ix^2, \partial = \frac{\partial_1 - i\partial_2}{2}$, $W = \frac{W_1 - iW_2}{2}$, $\mathcal{D} = \frac{\mathcal{D}_1 - i\mathcal{D}_2}{2} = \partial + iW$. Instead of the equations of motion itself, it will prove useful to study gauge invariant quantities. For that purpose, let us define

$$\bar{W}(z,\bar{z}) = -iS^{-1}(z,\bar{z})\bar{\partial}S(z,\bar{z}), \quad H(z,\bar{z}) = S^{-1}(z,\bar{z})\tilde{H}(z,\bar{z}), \tag{9}$$

where S takes values in $GL(N, \mathbf{C})$ and it is in the fundamental representation of $U(N)$ while the gauge singlet $\tilde{H}$ is an $N \times N$ complex matrix. There is an equivalence relation $(S, \tilde{H}) \sim (V(z)S, V(z)\tilde{H})$, where $V(z)$ is a holomorphic $GL(N, \mathbf{C})$ matrix with respect to z. The gauge group $U(N)$ and the flavor symmetry act as follows

$$S(z,\bar{z}) \to U_{\rm G}S(z,\bar{z}), \quad H_0(z) \to H_0(z)U_{\rm F}. \tag{10}$$

An important gauge invariant quantity is now constructed as

$$\Omega(z,\bar{z}) \equiv S(z,\bar{z})S(z,\bar{z})^\dagger. \tag{11}$$

With respect to the gauge invariant objects, the equations of motion are

$$4\bar{\partial}^2\left(\Omega\partial\Omega^{-1}\right) - \tilde{H}\bar{\partial}\left(\tilde{H}^\dagger\Omega^{-1}\right) + \bar{\partial}\tilde{H}\tilde{H}^\dagger\Omega^{-1} = 0, \tag{12}$$

$$\Omega\partial\left(\Omega^{-1}\bar{\partial}\tilde{H}\right) + \bar{\partial}\left(\Omega\partial\left(\Omega^{-1}\tilde{H}\right)\right) + \frac{\lambda^2}{4}\left(\Omega - \tilde{H}\tilde{H}^\dagger\right)\Omega^{-1}\tilde{H} = 0. \tag{13}$$

These equations must be solved with the boundary conditions for k vortices $\det \tilde{H} \to z^k$, $\Omega \to \tilde{H}\tilde{H}^\dagger$ as $z \to \infty$.

2.4. *BPS Limit*

For the later convenience, let us see the BPS limit $\lambda \to 1$. It can be done by just taking a holomorphic function $\tilde{H}$ with respect to z as

$$\tilde{H} = H_0(z). \tag{14}$$

Then the equations (12) and (13) reduce to the single matrix equation

$$\bar{\partial}\left(\Omega\partial\Omega^{-1}\right)+\frac{1}{4}\left(\mathbf{1}-H_0H_0^{\dagger}\Omega^{-1}\right)=0. \tag{15}$$

This is the master equation for the BPS non-Abelian vortex and the holomorphic matrix $H_0(z)$ is called the moduli matrix [4,13]. All the complex parameters contained in the moduli matrix are moduli of the BPS vortices. For example, the position of the vortices can be read from the moduli matrix as zeros of its determinant $\det H_0(z_i)=0$. Furthermore, the number of vortices (the units of magnetic flux of the configuration) corresponds to the degree of $\det H_0(z)$ as a polynomial with respect to z. The classification of the moduli matrix for the BPS vortices is given in Ref.[4,13].

Consider a $U(2)$ gauge theory. The minimal vortex is generated by

$$H_0^{(1,0)}=\begin{pmatrix} z-z_0 & 0 \\ -b' & 1 \end{pmatrix}, \qquad H_0^{(0,1)}=\begin{pmatrix} 1 & -b \\ 0 & z-z_0 \end{pmatrix}, \tag{16}$$

where z_0 corresponds to the position of the vortex and b and b' are the internal orientation. One can extract the orientation as the null eigenvector of $H_0(z)$ at the vortex position $z=z_0$ as

$$\vec{\phi}^{\,(1,0)}=\begin{pmatrix} 1 \\ b' \end{pmatrix} \quad \sim \quad \vec{\phi}^{\,(0,1)}=\begin{pmatrix} b \\ 1 \end{pmatrix}. \tag{17}$$

Here "$\sim$" stands for an identification up to complex non zero factors: $\vec{\phi}\sim\lambda\vec{\phi}$, $\lambda\in\mathbf{C}^*$, thus we identified $\mathbf{C}P^1$ [4,13]. We call two non-Abelian vortices with equal orientational vectors *parallel*, while orthogonal orientational vectors *anti-parallel*.

Two arbitrary vortices (the center of mass is fixed to be zero and the overall orientaion is fixed) is given by

$$H_{0\,\mathrm{red}}^{(1,1)}\equiv\begin{pmatrix} z-z_0 & -\eta \\ 0 & z+z_0 \end{pmatrix}. \tag{18}$$

The orientational vectors are then of the form

$$\vec{\phi}_1^{(1,1)}\Big|_{z=z_0}=\begin{pmatrix} 1 \\ 0 \end{pmatrix}, \qquad \vec{\phi}_2^{(1,1)}\Big|_{z=-z_0}=\begin{pmatrix} \eta \\ -2z_0 \end{pmatrix}. \tag{19}$$

3. Vortex interaction in the fine-tuned model

3.1. (k_1, k_2) *coincident vortices*

The minimal winding solution in the non-Abelian gauge theory is a mere embedding of the ANO solution into the non-Abelian theory. This embedding is also useful for another simple non-BPS configurations. Let us start with the moduli matrix for a configuration of k coincident vortices. The axial symmetry allows a reasonable ansatz for Ω and $\tilde{H}$

$$\Omega^{(0,1)} = \begin{pmatrix} 1 & 0 \\ 0 & w(r) \end{pmatrix}, \qquad \tilde{H}^{(0,1)} = \begin{pmatrix} 1 & 0 \\ 0 & f(r)z^k \end{pmatrix}. \tag{20}$$

We call this "$(0,k)$-vortex". When $k \geq 2$, it is possible that the ansatz (20) does not give the true solution (minimum of the energy) of the equations of motion (12) and (13). This is because there could be repulsive forces between the vortices. With ansatz (20) we fix the positions of all the vortices at the origin by hand. The master equation (15) is nevertheless still useful to investigate the interactions between two vortices. The results are listed in Table 1. For $\lambda = 1$, the masses are identical to integer values, up to 10^{-5}

λ	$k=1$	$k=2$
0.8	0.91231	1.77407
0.9	0.95737	1.88936
1	1.00000	2.00000
1.1	1.04053	2.10655
1.2	1.07922	2.20944

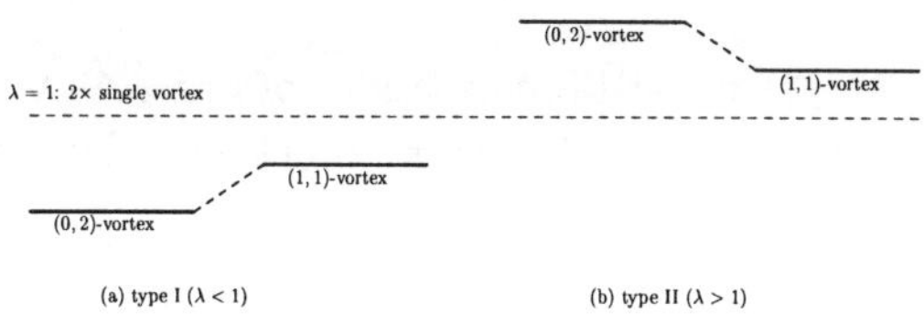

Fig. 1. Spectrum of the $(0,2)$ and $(1,1)$ coincident vortices.

order, which are nothing but the winding numbers of the vortices.

There is another type of composite configuration which can easily be analyzed numerically

$$\Omega^{(1,1)} = \begin{pmatrix} w_1(r) & 0 \\ 0 & w_2(r) \end{pmatrix}, \qquad \tilde{H}^{(1,1)} = \begin{pmatrix} f_1(r)z^{k_1} & 0 \\ 0 & f_2(r)z^{k_2} \end{pmatrix}. \tag{21}$$

This ansatz corresponds to a configuration with k_1 composite vortices which wind in the first diagonal $U(1)$ subgroup of $U(2)$ and with k_2 coincident vortices that wind the second diagonal $U(1)$ subgroup. We refer to these as a "(k_1,k_2)-vortex". The mass of a (k_1,k_2)-vortex is thus the sum of the mass of the $(k_1,0)$-vortex and that of the $(0,k_2)$-vortex.

We call the non-Abelian vortices in the fine-tuned model for $\lambda < 1$ type I, while they will be called type II for $\lambda > 1$. From Fig. 1, we can see that in the type I case, the $(0,2)$-vortex is energetically preferred to the $(1,1)$-vortex, while in type II case the $(1,1)$-vortex is preferred. If the two vortices are separated sufficiently, regardless of their orientations, the mass of two well separated vortices is twice that of the single vortex. This mass is equal to the mass of the $(1,1)$-vortex.

3.2. *Effective potential for coincident vortices*

The dynamics of BPS solitons can be investigated by the so-called moduli approximation [14]. The effective action is a massless non-linear sigma model whose target space is the moduli space. If the coupling constant λ is close to the BPS limit $\lambda = 1$, we can still use the moduli approximation, to investigate dynamics of the non-BPS non-Abelian vortices by adding a potential of order $|1-\lambda^2| \ll 1$. To this end, we write the Lagrangian

$$\tilde{\mathcal{L}} = \tilde{\mathcal{L}}_{\rm BPS} + \frac{(\lambda^2-1)}{4}\left(\mathbf{1}_N - HH^\dagger\right)^2. \tag{22}$$

We get non-BPS corrections of order $O(\lambda^2-1)$ by putting BPS solutions into Eq. (22). The energy functional thus takes the following form

$$\mathcal{E} = 2 + (\lambda^2-1)\mathcal{V}, \quad \mathcal{V} = \frac{1}{8\pi}\int dx^1 dx^2 \ {\rm Tr}\left(\mathbf{1} - |H_{\rm BPS}(\varphi_i)|^2\right)^2 \tag{23}$$

where $H_{\rm BPS}(\varphi_i)$ stands for the BPS solution. We have defined a reduced effective potential $\mathcal{V}$ which is independent of λ. The first term corresponds to the mass of two BPS vortices and the second term is the deviation from the BPS solutions which is nothing but the effective potential we want.

To have the effective potential on the moduli space of coincident vortices, it suffices to consider only the matrix (18) and turning off the relative distance z_0. In order to evaluate it, we need to solve the BPS equations with an intermediate value of η. Because of the axial symmetry and the boundary condition at infinity $\Omega \to H_0(z)H_0^\dagger(\bar{z})$, we can make an ansatz

$$\Omega^{(1,1)} = \begin{pmatrix} w_1(r) & -\eta e^{-i\theta} w_2(r) \\ -\eta e^{i\theta} w_2(r) & w_3(r) \end{pmatrix}. \tag{24}$$

The advantage of the moduli matrix formalism is that only three functions $w_i(r)$ are needed and the formalism itself is gauge invariant. The effective potential can be obtained by plugging numerical solutions into Eq. (23). The result is shown in Fig. 2.

The type II effective potential has the same qualitative behavior as showed in the figure. It has a minimum at $|\eta| = 0$. This matches the previous result that the $(1,1)$-vortex is energetically preferred to the $(2,0)$-vortex. The type I effective potential can be obtained just by flipping the overall sign of that of the type II case. Then the effective potential always takes

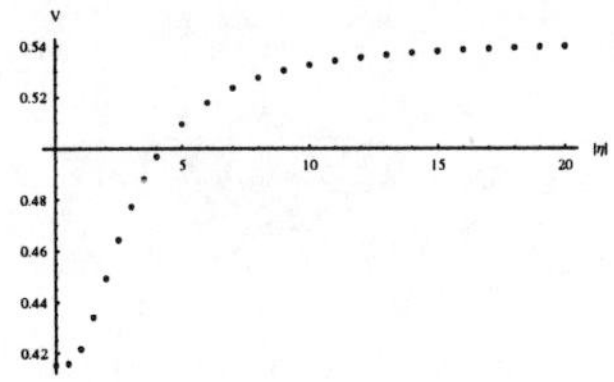

Fig. 2. Numerical plots of the effective reduced potential $\mathcal{V}(|\eta|)$.

a negative value, which is consistent with the fact that the masses of the type I vortices are less than those of the BPS vortices. Contrary to the type II case, the type I potential has a minimum at $|\eta| = \infty$, so that the $(2,0)$-vortex is preferred to the $(1,1)$ vortex.

3.3. *Interaction at generic vortex separation*

Next we go on investigating the interactions of non-Abelian vortices in the $U(2)$ gauge group at generic distances. We will again use the moduli space approximation. The generic configurations are described by the moduli matrices in Eq. (18). By putting the two vortices on the real axis, we can reduce z_0 to a real parameter d. So $2d$ is the relative distance and η the relative orientation. Now let us study the effective potential as function of η and d. As before, we first need the numerical solution to the BPS master equation. Despite the great complexity by broken axial symmetry, the moduli matrix formalism is a powerful tool and the relaxation method is very effective to solve the problem. Once we get the numerical solution, the effective potential is obtained by plugging them into Eq. (23), see Fig. 3. For the type II, it has the same shape, up to a small positive factor $(\lambda^2 - 1)$. The potential forms a hill whose top is at $(d, |\eta|) = (0, \infty)$. It clearly shows that two vortices feel repulsive forces, in both the real and internal space, for every distance and relative orientation. The minima of the potential has a flat direction along the d-axis where the orientations are anti-parallel $(\eta = 0)$ and along the η axis at infinite distance $(d = \infty)$. Therefore the anti-parallel vortices do not interact. In the type I case $(\lambda < 1)$ the effective potential is upside-down of that of the type II case. There is unique minimum of the potential at $(d, |\eta|) = (0, \infty)$. This means that attractive force works not only for the distance in real space but also among the internal orientations.

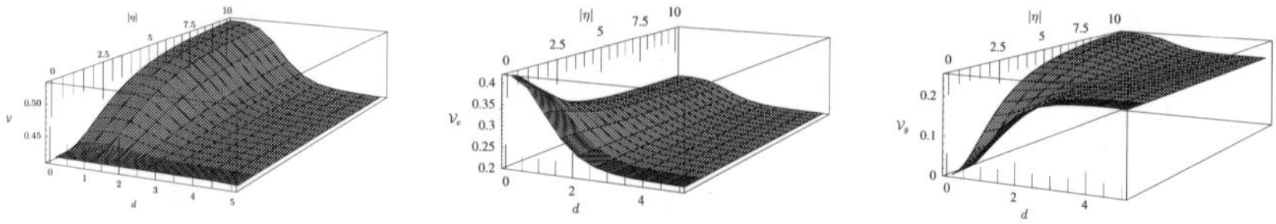

Fig. 3. The left panel shows the effective potential $\mathcal{V}(\eta, d)$. The Abelian potential $\mathcal{V}_e$ (in the middle) and the non-Abelian potential $\mathcal{V}_g$ (in the right) for $\gamma = 1$.

4. Vortices with generic couplings

In this section we study the general model defined in Eqs. (3) and (4). We have three effective couplings $\gamma = g/e, \lambda_e, \lambda_g$ after the rescaling (5). The masses of particles are rescaled as

$$M_{U(1)} = 1, \quad M_{SU(N)} = \gamma, \quad M_{\rm s} = \lambda_e, \quad M_{\rm ad} = \gamma\lambda_g. \tag{25}$$

In order to find the effective potential on the moduli space as before, we need to clarify BPS configurations. The moduli matrix in (14) is still valid, while the master equation (15) gets a modification

$$4\bar{\partial}\left(\Omega\partial\Omega^{-1}\right) = \Omega_0\Omega^{-1} - \mathbf{1}_N + (\gamma^2 - 1)\left(\Omega_0\Omega^{-1} - \frac{\mathrm{Tr}\left(\Omega_0\Omega^{-1}\right)}{N}\mathbf{1}_N\right) \tag{26}$$

where $\Omega = SS^\dagger$ is the same as before and $\Omega_0 \equiv H_0 H_0^\dagger$. It turns out that the effective potential consists of the Abelian and the non-Abelian potentials

$$\mathcal{V}_e(\eta, d; \gamma) = \int d\tilde{x}^2\ \mathrm{Tr}(F_{12}^0 T^0)^2, \quad \mathcal{V}_g(\eta, d; \gamma) = \int d\tilde{x}^2\ \mathrm{Tr}(\hat{F}_{12})^2. \tag{27}$$

The true potential is a linear combination of them

$$V(\eta, d; \gamma, \lambda_e, \lambda_g) = (\lambda_e^2 - 1)\mathcal{V}_e(\eta, d; \gamma) + \frac{\lambda_g^2 - 1}{\gamma^2}\mathcal{V}_g(\eta, d; \gamma). \tag{28}$$

4.1. *Equal gauge coupling* $\gamma = 1$ *revisited*

The effective potential with $\gamma = 1$ and $\lambda = \lambda_g = \lambda_e$ in the left panel of Fig. 3 should now be decomposed in the two potentials, see the middle and the right panels in Fig. 3. In the case of $\lambda_e^2 - 1 > 0$ and $\lambda_g^2 - 1 > 0$, the effective potential will have the same qualitative behavior as the reduced potentials in the Figs. 3. The figures show how $\mathcal{V}_e$ and $\mathcal{V}_g$ behave very differently. In particular, the Abelian potential is always repulsive, both in the real and internal space. The non-Abelian potential is on the contrary sensitive to the orientations. Fig. 3 shows that it is repulsive for parallel vortices while it is attractive for anti-parallel ones. When the two scalar couplings are equal, $\lambda_e^2 = \lambda_g^2$, the two potentials exactly cancel for anti-parallel vortices.

Of course, the true effective potential depends on λ_e and λ_g through the combination in Eq. (28). This indicates the interaction between non-Abelian vortices is quite rich in comparison with that of the ANO vortices.

4.2. *Different gauge couplings: $\gamma \neq 1$*

We now consider interactions between non-Abelian vortices with different gauge coupling $e \neq g$ ($\gamma \neq 1$). In Figs. 4 and 5 we show two numerical examples for the reduced effective potentials $\mathcal{V}_e$, $\mathcal{V}_g$ given in Eq. (27). These

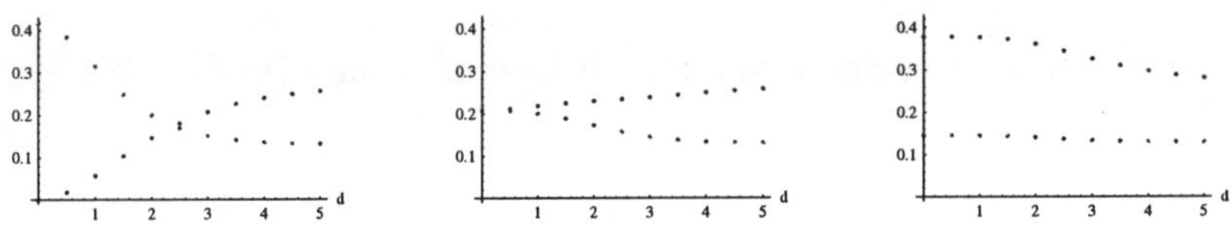

anti-parallel ($\eta = 0$) intermediate ($\eta = 4$) parallel ($\eta = \infty$)

Fig. 4. Upper line at $d = 5$ refers to $\mathcal{V}_g$; lower line at $d = 5$ refers to $\mathcal{V}_e$.

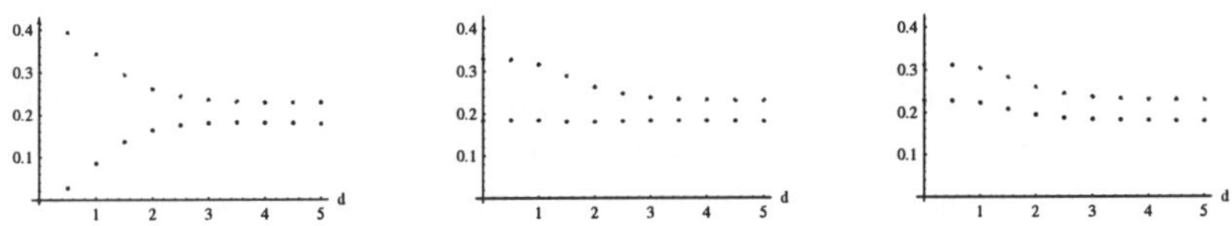

anti-parallel ($\eta = 0$) intermediate ($\eta = 4$) parallel ($\eta = \infty$)

Fig. 5. Upper line refers to $\mathcal{V}_e$; lower line refers to $\mathcal{V}_g$.

show that the qualitative features of $\mathcal{V}_e$ and $\mathcal{V}_g$ are basically the same as what is discussed in the equal gauge coupling case ($\gamma = 1$). The true effective potential in Eq. (28) depends on the three parameters γ, λ_e and λ_g. We can have potentials which develop a global minimum at some finite non-zero distance, see Fig. 6. The figure shows the presence of a minimum around $d \sim 2$. This kind of behavior has not been found for the ANO type I/II vortices and the possibility of bound vortices really results from the non-Abelian symmetry.

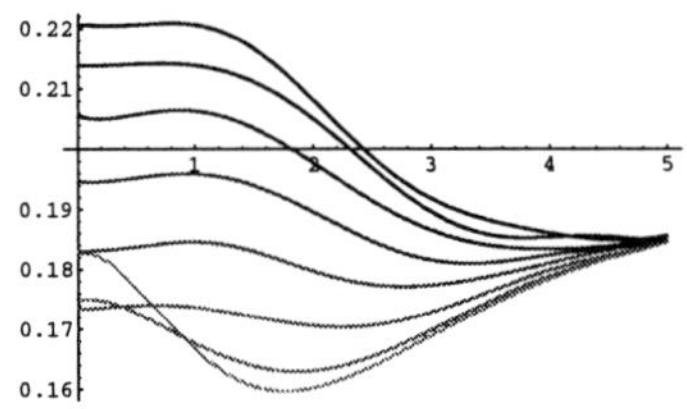

Fig. 6. The effective potential vs. separation d. $\gamma = 1/2$, $\lambda_e = 1.2$, $\lambda_g = 1.06$: From $\eta = 0$ (bottom) to $\eta = 7$ (top) with $d = 0 \sim 5$ for each η.

5. Interaction at large vortex separation

5.1. *Vortices in fine-tuned models $e = g$ and $\lambda_e = \lambda_g$*

We study asymptotic forces between vortices at large separation, following Ref. [10]. We need to find asymptotic behaviors around $(1,0)$-vortex

$$H_0(z)^{(1,0)} = \begin{pmatrix} z & 0 \\ 0 & 1 \end{pmatrix}, \qquad \vec{\phi}_1^{(1,0)} = \begin{pmatrix} 1 \\ 0 \end{pmatrix}. \tag{29}$$

We are lead to the well-known asymptotic behavior of the ANO vortex

$$H_{[1,1]} = \left(1 + \frac{q}{2\pi} K_0(\lambda r)\right) e^{i\theta}, \quad \bar{W}_{[1,1]} = -\frac{i}{2}\left(\frac{1}{r} - \frac{m}{2\pi} K_1(r)\right) e^{i\theta}, \tag{30}$$

where $K_1 \equiv -K_0'$ and we have defined $H_{[1,1]}$ and $\bar{W}_{[1,1]}$ as $[1,1]$ elements of H and $\bar{W}$ in Eq. (9) with the $k = 1$ ansatz (20).

Next we treat the vortices as point particles in a linear field theory coupled with a scalar source ρ and a vector current j_μ. To linearize the Yang-Mills-Higgs Lagrangian, we choose a gauge such that the Higgs fields is given by Hermitian matrix $H = \mathbf{1}_2 + \frac{1}{2} h^i \sigma_i$, $W_\mu = \frac{1}{2} w_\mu^i \sigma_i$ with $\sigma = (\mathbf{1}_2, \vec{\sigma})$. with all h^a, w_μ^a real. Then the quadratic part of the Lagrangian is

$$\mathcal{L}_{\text{free}}^{(2)} = \sum_{a=0}^{3} \left[-\frac{1}{4} f_{\mu\nu}^a f^{a\mu\nu} + \frac{1}{2} w_\mu^a w^{a\mu} + \frac{1}{2} \partial_\mu h^a \partial^\mu h^a - \frac{\lambda^2}{2} (h^a)^2 \right] \tag{31}$$

with $f_{\mu\nu}^a \equiv \partial_\mu w_\nu^a - \partial_\nu w_\mu^a$. We also take into account the external source terms to realize the point vortex

$$\mathcal{L}_{\text{source}} = \sum_{a=0}^{3} \left[\rho^a h^a - j_\mu^a w^{a\mu} \right]. \tag{32}$$

The scalar and the vector sources should be determined so that the asymptotic behavior of the fields in Eq. (30) are replicated. The solution to the equation of motion is

$$h^0 = h^3 = \frac{q}{2\pi} K_0(\lambda r), \quad \rho^0 = \rho^3 = q\delta(r),$$
$$\mathbf{w}^0 = \mathbf{w}^3 = -\frac{m}{2\pi} \hat{\mathbf{k}} \times \nabla K_0(r), \quad \mathbf{j}^0 = \mathbf{j}^3 = -m\hat{\mathbf{k}} \times \nabla \delta(r) \tag{33}$$

where $\hat{\mathbf{k}}$ is a spatial fictitious unit vector along the vortex world-volume. The vortex configuration with general orientation is also treated easily, since the origin of the orientation is the Nambu-Goldstone mode associated with the broken $SU(2)$ color-flavor symmetry $H_0 \to H_0(z)^{(1,0)} U_{\rm F}$, $\vec{\phi}_2 = U_{\rm F}^\dagger \vec{\phi}_1^{(1,0)}$. The interaction between a vortex at $\mathbf{x} = \mathbf{x}_1$ with the orientation $\vec{\phi}_1$ and another vortex at $\mathbf{x} = \mathbf{x}_2$ with the orientation $\vec{\phi}_2$ is given through the source term and is summarized as

$$V_{\rm int} = -\frac{\left|\vec{\phi}_1^\dagger \vec{\phi}_2\right|^2}{\left|\vec{\phi}_1\right|^2 \left|\vec{\phi}_2\right|^2} \left(\frac{q^2}{2\pi} K_0(\lambda r) - \frac{m^2}{2\pi} K_0(r)\right), \tag{34}$$

where $r \equiv |\mathbf{x}_1 - \mathbf{x}_2| \gg 1$. When two vortices have parallel orientations, this potential becomes that of two ANO vortices [10]. On the other hand, the potential vanishes when their orientations are anti-parallel. This agrees with the numerical results found in the previous sections. In the BPS limit $\lambda = 1$ $(q = m)$, the interaction becomes precisely zero.

5.2. *Vortices with general couplings*

It is quite straightforward to generalize the results of the previous section to the case of generic couplings. We find the total potential $V_{\rm int}$

$$V_{\rm int} = \frac{1}{2}\left(-\frac{(q^0)^2}{2\pi} K_0(\lambda_e r) + \frac{(m^0)^2}{2\pi} K_0(r)\right)$$
$$+ \left(\frac{\left|\vec{\phi}_1^\dagger \vec{\phi}_2\right|^2}{\left|\vec{\phi}_1\right|^2 \left|\vec{\phi}_2\right|^2} - \frac{1}{2}\right)\left(-\frac{(q^3)^2}{2\pi} K_0(\lambda_g \gamma r) + \frac{(m^3)^2}{2\pi} K_0(\gamma r)\right). \tag{35}$$

At large distance, the interactions between vortices are dominated by the particles with the lowest mass $M_{\rm low}$. There are four possible regimes $V_{\rm int} =$

$$\begin{cases} -\frac{(q^0)^2}{4\pi}\sqrt{\frac{\pi}{2\lambda_e r}}e^{-\lambda_e r} & \text{for } M_{\rm low} = M_{\rm s}, \quad \text{Type I} \\ -\left(\frac{|\vec{\phi}_1^\dagger \vec{\phi}_2|^2}{|\vec{\phi}_1|^2|\vec{\phi}_2|^2} - \frac{1}{2}\right)\frac{(q^3)^2}{2\pi}\sqrt{\frac{\pi}{2\lambda_g \gamma r}}e^{-\lambda_g \gamma r} & \text{for } M_{\rm low} = M_{\rm ad}, \quad \text{Type I}^* \\ \frac{(m^0)^2}{4\pi}\sqrt{\frac{\pi}{2r}}e^{-r} & \text{for } M_{\rm low} = M_{U(1)}, \quad \text{Type II} \\ \left(\frac{|\vec{\phi}_1^\dagger \vec{\phi}_2|^2}{|\vec{\phi}_1|^2|\vec{\phi}_2|^2} - \frac{1}{2}\right)\frac{(m^3)^2}{2\pi}\sqrt{\frac{\pi}{2\gamma r}}e^{-\gamma r} & \text{for } M_{\rm low} = M_{SU(2)}, \quad \text{Type II}^* \end{cases} \tag{36}$$

because of $K_0(\lambda r) \sim \sqrt{\pi/2\lambda r}e^{-\lambda r}$. This generalizes the type I/II classification of Abelian superconductors. We have found two new categories, called type I* and type II*, in which the force can be attractive or repulsive depending on the relative orientation. In the type I* case the forces between parallel vortices are attractive while anti-parallel vortices repel each other. The type II* vortices feel opposite forces of the type I*. The result in Eq. (36) is easily extended to the general case of $U(1) \times SU(N)$. This can be done by just thinking of the orientation vectors $\vec{\phi}$ as taking values in $\mathbf{C}P^{N-1}$.

It may be interesting to compare these results with the recently studied asymptotic interactions between non-BPS non-Abelian global vortices [15].

Acknowledgments

The author thanks the organizers of the conference "Continuous Advances in QCD 2008" for warm hospitality. The work of the author is supported by the Research Fellowships of the Japan Society for the Promotion of Science for Research Abroad.

References

1. A. Hanany and D. Tong, JHEP **0307** (2003) 037
2. R. Auzzi, S. Bolognesi, J. Evslin, K. Konishi and A. Yung, Nucl. Phys. B **673** (2003) 187
3. D. Tong, arXiv:hep-th/0509216; K. Konishi, arXiv:hep-th/0702102; M. Shifman and A. Yung, arXiv:hep-th/0703267.
4. M. Eto, Y. Isozumi, M. Nitta, K. Ohashi and N. Sakai, J. Phys. A **39** (2006) R315
5. A. A. Abrikosov, Sov. Phys. JETP **5** (1957) 1174 [Zh. Eksp. Teor. Fiz. **32** (1957) 1442].
6. H. B. Nielsen and P. Olesen, Nucl. Phys. **B61** (1973) 45.
7. S. Gustafson and I. M. Sigal, Commun. Math. Phys. **212**, 257 (2000).
8. L. Jacobs and C. Rebbi, Phys. Rev. B **19**, 4486 (1979); K. J. M. Moriarty, E. Myers and C. Rebbi, Phys. Lett. B **207**, 411 (1988).
9. L. M. A. Bettencourt and R. J. Rivers, Phys. Rev. D **51**, 1842 (1995)
10. J. M. Speight, Phys. Rev. D **55**, 3830 (1997)
11. J. Heo and T. Vachaspati, Phys. Rev. D **58**, 065011 (1998); R. Auzzi, M. Eto and W. Vinci, JHEP **0711**, 090 (2007)
12. R. Auzzi, M. Eto and W. Vinci, JHEP **0802**, 100 (2008)
13. M. Eto, Y. Isozumi, M. Nitta, K. Ohashi and N. Sakai, Phys. Rev. Lett. **96** (2006) 161601
14. N. S. Manton, Phys. Lett. B **110**, 54 (1982).
15. M. Nitta and N. Shiiki, Phys. Lett. B **658**, 143 (2008); E. Nakano, M. Nitta and T. Matsuura, arXiv:0708.4092 [hep-ph]; arXiv:0708.4096 [hep-ph].

VORTICES WHICH DO NOT ABELIANIZE DYNAMICALLY: SEMI-CLASSICAL ORIGIN OF NON-ABELIAN MONOPOLES

K. KONISHI

Department of Physics, "E. Fermi",
University of Pisa,
Largo Pontecorvo, 3, 56127, Pisa, Italy
** E-mail: konishi@df.unipi.it*
www.df.unipi.it/ konishi/konishi.html

After briefly reviewing the problems associated with non-Abelian monopoles, we turn our attention to the development in our understanding of non-Abelian *vortices* in the last several years. In the $U(N)$ model with $N_f = N$ flavors in which they were first found, the fluctuations of the orientational modes along the vortex length and in time become strongly coupled at long distances. They effectively reduce to Abelian ANO vortices. We discuss then a very recent work on non-Abelian vortices with $CP^{n-1} \times CP^{r-1}$ orientational moduli, which, unlike the ones so far extensively studied, do not dynamically Abelianize completely. The surviving vortex orientational moduli, fluctuating along the vortex length and in time, gets absorbed by the monopoles at the ends, turning into the dual gauge degrees of freedom for the latter.

Keywords: Vortex, monopole, quantum solitons, duality, confinement

1. Non-Abelian monopoles

Non-Abelian monopoles have been introduced as a natural generalization of the 't Hooft-Polyakov monopoles;[1] they arise in systems with partial gauge symmetry breaking,

$$G \xrightarrow{v_1} H \tag{1}$$

where H is a non-Abelian gauge group. The regular monopoles arising in such a system are characterized by the charges β such that

$$F_{ij} = \epsilon_{ijk} B_k = \epsilon_{ijk} \frac{r_k}{r^3} (\beta \cdot \mathbf{H}), \tag{2}$$

in an appropriate gauge, where **H** are the diagonal generators of H in the Cartan subalgebra. A straightforward generalization of the Dirac's quantization condition leads to

$$2\,\beta \cdot \alpha \in \mathbf{Z} \tag{3}$$

where α are the root vectors of H. The constant vectors β (with the number of components equal to the rank of the group H) label possible monopoles. It is easy to see that the solution of Eq. (3) is that β is any of the *weight vectors* of a group whose nonzero roots are given by

$$\alpha^* = \alpha/\alpha \cdot \alpha. \tag{4}$$

The group generated by Eq. (4) is known as the (GNOW) dual of H, let us call $\tilde{H}$. One is thus led to a set of semi-classical *degenerate* monopoles, with multiplicity equal to that of a representation of $\tilde{H}$; this has led to the so-called GNOW conjecture, i.e., that they form a multiplet of the group $\tilde{H}$, dual of H.[2] For simply-laced groups (with the same length of all nonzero roots) such as $SU(N)$, $SO(2N)$, the dual of H is basically the same group, except that the allowed representations tell us that

$$U(N) \leftrightarrow U(N); \qquad SO(2N) \leftrightarrow SO(2N), \tag{5}$$

while

$$SU(N) \leftrightarrow SU(N)/\mathbf{Z}_N; \qquad SO(2N+1) \leftrightarrow USp(2N). \tag{6}$$

There are however well-known difficulties with such an interpretation. The first concerns the topological obstruction discussed in:[3,4] in the presence of the classical monopole background, it is not possible to define a globally well-defined set of generators isomorphic to H. As a consequence, no "colored dyons" exist. In a simplest case with the breaking

$$SU(3) \stackrel{\langle\phi_1\rangle\neq 0}{\longrightarrow} SU(2) \times U(1), \tag{7}$$

this means that

$$\textit{no monopoles with charges} \quad (\underline{2}, 1^*) \quad \textit{exist}, \tag{8}$$

where the asterisk indicates a dual, magnetic charge.

The second can be regarded as an infinitesimal version of the same difficulty: certain bosonic zero modes around the monopole solution, corresponding to H gauge transformations, are non-normalizable (behaving as $r^{-1/2}$ asymptotically). Thus the standard procedure of quantization leading to H multiplets of monopoles does not work.[4]

Both of these difficulties concern the transformation properties of the monopoles under the subgroup H, while the relevant question should be how they transform under the dual group, $\tilde{H}$. As field transformation groups, H and $\tilde{H}$ are relatively nonlocal, the latter should look like a nonlocal transformation group in the original, electric description.

2. Light non-Abelian monopoles

In spite of these apparent difficulties, light non-Abelian monopoles do appear regularly in the low-energy effective action of a wide class of $\mathcal{N} = 2$ gauge theories with matter hypermultiplets.[5–7] It is however important to bare in mind that non-Abelian dual gauge groups (and associated monopoles) occur only in models with massless flavors in the underlying theory. The renormalization-group effects explain this fact. For instance, in the softly broken, $\mathcal{N} = 2$ supersymmetric $SU(N)$ ($N \geq 3$) gauge theory with N_f quarks, the low-energy, magnetic group $SU(r) \times U(1)^{N-r}$ appears only for

$$r \leq N_f/2 \,.$$

The reason is that the monopoles, via the Jackiw-Rebbi mechanism, form a degenerate flavor multiplet quantum mechanically. (Indeed the light monopoles appear in the above theory as a fundamental multiplet of the flavor $SU(N_f)$ symmetry group.) Their quantum effects attenuate the dual gauge interactions so that the $SU(r)$ dual gauge group is now infrared-free (the sign flip with respect to the underlying theory[8]). This is how these objects can appear in the infrared as a recognizable degrees of freedom. When this is not possible (e.g., pure $\mathcal{N} = 2$ gauge theories) the would-be non-Abelian monopoles interact too strongly and form (baryon-like) composites, which are the Abelian monopoles.[7]

The vacua $r = N_f/2$ constitute interesting, limiting class of theories: they are infrared fixed-point theories (SCFT).[9]

Now there must be ways to *understand* these massless non-Abelian monopoles, in spite of the above-mentioned difficulties, in terms of more familiar, semiclassical language. Below we shall show that this is indeed possible. We shall study the monopoles in terms of *vortices*, by putting the low-energy H gauge system in Higgs phase. A systematic study of non-Abelian *vortices* started only recently, but are much better understood than the non-Abelian *monopoles*. The monopoles and vortices are closely related to each other, through the homotopy map and by symmetry.[8,10,11] The moduli and non-Abelian transformation properties among the monopoles follow from those of the low-energy vortices which confine them.

3. Non-Abelian vortices

The non-Abelian vortices have been found several years ago,[12,13] in the context of $U(N)$ gauge theory with N_f flavors, $N_f \geq N$.[a] Even though such a model can be considered in its own right, we shall follow here the original approach[13] where the $U(N)$ model arises as the low-energy approximation after a partial symmetry breaking,

$$SU(N+1) \stackrel{v_1 \neq 0}{\longrightarrow} \frac{SU(N) \times U(1)}{\mathbf{Z}_N} \sim U(N) \,. \tag{9}$$

This approach allows us to connect the vortex properties to those of the monopoles appearing at the ends, as neither the vortices nor monopoles (arising from the symmetry breaking Eq. (9)) are truly stable.[8,10,11] The point is that at a much smaller mass scale, $v_2 \ll v_1$, the squark fields condense, and break the low-energy $U(N)$ gauge symmetry completely.

At scales much lower than v_1 but still neglecting the smaller squark VEV v_2 the theory reduces to an $SU(N) \times U(1)$ gauge theory[13] with N_f light quarks $q_i, \tilde{q}^i$ (the first N components of the original quark multiplets $Q_i, \tilde{Q}^i$). In the most frequently studied case, $N_f = N$, the light squark fields can be expressed as $N \times N$ color-flavor mixed matrix. The adjoint scalars are fixed to its VEV of the form,

$$\langle \Phi \rangle = \mathrm{diag}\,(m, m, \ldots, m, -N\, m),$$

with small fluctuations around it,

$$\Phi = \langle \Phi \rangle (1 + \langle \Phi \rangle^{-1}\, \tilde{\Phi}), \qquad |\tilde{\Phi}| \ll m. \tag{10}$$

In the consideration of the vortices of the low-energy theory, Φ will be in fact replaced by the constant VEV. The presence of the small terms Eq. (10), however, makes the low-energy vortices not strictly BPS (and this is important in the consideration of their stability).

The quark fields are replaced, consistently with the vanishing of the D-term potential, as

$$\tilde{q} \equiv q^{\dagger}, \qquad q \to \frac{1}{\sqrt{2}}\, q, \tag{11}$$

[a]Recently the construction has been generalized to any gauge group of the form, $G = G' \times U(1)$.[14]

where the second replacement brings back the kinetic term to the standard form.

We further replace the singlet coupling constant and the $U(1)$ gauge field appropriately: the net effect is

$$\mathcal{L} = \frac{1}{4g_N^2}(F^a_{\mu\nu})^2 + \frac{1}{4e^2}(\tilde{F}_{\mu\nu})^2 + |\mathcal{D}_\mu q|^2 - \frac{e^2}{2}\,|\,q^\dagger q - c\,\mathbf{1}\,|^2 - \frac{1}{2}\,g_N^2\,|\,q^\dagger\,t^a q\,|^2, \tag{12}$$

where $c = N(N+1)\sqrt{2}\,\mu\,m$. Neglecting the small terms left implicit, this is basically the $U(N)$ model, studied extensively.[10–12,15–19]

The transformation property of the vortices can be determined from the moduli matrix.[16] Indeed, the system possesses BPS saturated vortices described by the linearized equations

$$(\mathcal{D}_1 + i\mathcal{D}_2)\; q = 0, \tag{13}$$

$$F_{12}^{(0)} + \frac{e^2}{2}\left(c\,\mathbf{1}_N - q\,q^\dagger\right) = 0; \qquad F_{12}^{(a)} + \frac{g_N^2}{2}\,q_i^\dagger\,t^a\,q_i = 0. \tag{14}$$

The matter equation can be solved[15,16] ($z = x^1 + ix^2$) by setting

$$q = S^{-1}(z,\bar{z})\,H_0(z), \quad A_1 + i\,A_2 = -2\,i\,S^{-1}(z,\bar{z})\,\bar{\partial}_z S(z,\bar{z}), \tag{15}$$

where S is an $N \times N$ invertible matrix, and H_0 is the moduli matrix, holomorphic in z. S satisfies a simple second-order differential equation, which can be solved numerically.

The individual vortex solution breaks the color-flavor symmetry as

$$SU(N)_{C+F} \to SU(N-1) \times U(1), \tag{16}$$

leading to the moduli space of the minimum vortices which is

$$\mathcal{M} \simeq \mathbf{C}P^{N-1} = \frac{SU(N)}{SU(N-1) \times U(1)}. \tag{17}$$

The vortex represented by the moduli matrix (we consider here the vortices of minimal winding, $k = 1$)

$$H_0(z) \simeq \begin{pmatrix} 1 & 0 & 0 & -a_1 \\ & & & \\ 0 & \ddots & 0 & \vdots \\ 0 & 0 & 1 & -a_{N-1} \\ 0 & \dots & 0 & z \end{pmatrix}, \tag{18}$$

can be shown explicitly[11] to transform according to the fundamental representation of $SU(N)$.

4. Dynamical Abelianization

The fluctuations of internal, CP^{N-1} moduli in the above system are described in terms of a (2, 2) supersymmetric two-dimensional CP^{N-1} sigma model,[13,17,18] and the known results in this model assure that in the long-distance limit the system loses its orientation, and in the infrared there appear N vortex "vacua" (analogue of the Witten's index). Kinks appear connecting the different (vortex) vacua, which can be interpreted as the Abelian monopoles whose spectrum match nicely that found in the 4D theory.[12,17–19] It means that the "non-Abelian" vortex found in the $U(N)$ model dynamically Abelianize, and this is not the type of system we are seeking.

As the concept of dynamical Abelianization is quite central to our discussion, and as this point is somewhat subtle, let us make a pause of reflection. Dynamical Abelianzation, as normally understood, concerns the *gauge* symmetry. It means by definition that a non-Abelian gauge symmetry of a given theory reduces at low energies by quantum effects to an Abelian (*dual or not*) gauge theory. Related concepts are dynamical Higgs mechanism, or tumbling.[23] Example of the theories in which this is known to occur are the pure $\mathcal{N} = 2$ supersymmetric Yang-Mills theories[24,25] which reduce to Abelian gauge theories at low energies, and the $SU(2)$ $\mathcal{N} = 2$ theories with $N_f = 1, 2, 3$ matter hypermultiplets.[25] But as has been emphasized repeatedly and in Introduction above, $\mathcal{N} = 2$ supersymmetric $SU(N)$ QCD (with $N \geq 3$) *with quark multiplets*, do not Abelianize in general.[5,7,8] Whether or not the standard QCD with light quarks Abelianizes is not known. The 't Hooft-Mandelstam scenario implies a sort of dynamical Abelianization, as it assumes the Abelian $U(1)^2$ monopoles to be the dominant degrees of freedom at some relevant scales, but this has not been proven.

As the vortex orientation fluctuation modes are intimately connected to the way *dual* gauge symmetry emerges at low-energies ([8,11]), it is perfectly reasonable to use the same terminology for the vortex modes.

Nevertheless, one could *define* the concept of *non-Abelian or Abelian vortices*, independently of the usual meaning attributed to it in relation to a gauge symmetry. A vortex is non-Abelian, if it carries a non-trivial, internal non-Abelian moduli, which can fluctuate along its length and in time. We exclude from this consideration other vortex moduli associated with their (transverse) positions, shapes or sizes (in the case of higher-winding[26–28] or semi-local vortices[29,30]). Otherwise, a vortex is Abelian. The standard ANO vortex is Abelian, as it possesses no-continuous moduli. The vortices found in the context of $U(N)$ models[12,13] *are* indeed non-Abelian in this sense.

But just as a non-Abelian gauge group may or may not Abelianize depending on dynamics, a non-Abelian vortex may or may not dynamically Abelianize. In the very papers in which these vortices have been discovered[12,13] and in those which followed,[18] it was shown that they dynamically reduced to Abelian, ANO like vortices at long distances. The orientational moduli fluctuate strongly and at long distances they effectively lose their orientation. A recent observation[20] nicely exhibits this aspect through the Lüscher term of the string tension.

In a very recent paper[21] it was shown that this fate is not unavoidable. Semi-classical non-Abelian vortices which remain so at low-energies do exist; they can be found in appropariate vacua, selected by a careful tuning of the bare quark masses. This is quite similar to the situation in $\mathcal{N} = 1$ supersymmetric QCD, where a vacuum with a prescribed chiral symmetry breaking pattern can be selected out of the degenerate set of vacua by appropriately tuning the bare quark mass ratios, before sending them to zero. The symmetry breaking pattern in those theories is aligned with the bare quark masses, as is well-known.[22]

The construction of vortices which do not Abelianize completely closes the gap in matching the results in the $4D$ theories at fully quantum regimes (where all bare mass parameters are small) and those in semi-classical regimes where the vortices can be reliably studied. In other words the result of this work allows us to identify the semi-classical origin of the quantum non-Abelian monopoles found in.[5,7]

5. Non-Abelian vortices which do not dynamically reduce to ANO vortices

The model on which we shall base our consideration is the softly broken $\mathcal{N} = 2$ supersymmetric QCD with $SU(N)$ and $N_f = N$ flavors of quark multiplets,

$$\mathcal{L} = \frac{1}{8\pi}\mathrm{Im}\,\tau_{cl}\left[\int d^4\theta\,\mathrm{Tr}\,(\Phi^\dagger e^V \Phi e^{-V}) + \int d^2\theta\,\frac{1}{2}\mathrm{Tr}\,(WW)\right]$$
$$+\mathcal{L}^{(\mathrm{quarks})} + \int d^2\theta\,\mu\,\mathrm{Tr}\,\Phi^2;$$

where the quark Lagrangian is

$$\mathcal{L}^{(\mathrm{quarks})}$$
$$=\sum_i\left[\int d^4\theta\,(Q_i^\dagger\, e^V\, Q_i + \tilde{Q}_i\, e^{-V}\, \tilde{Q}_i^\dagger) + \int d^2\theta\,(\sqrt{2}\,\tilde{Q}_i\,\Phi\, Q^i + m_i\,\tilde{Q}_i\, Q^i)\right].$$

where $\tau_{cl} \equiv \theta_0/\pi + 8\pi i/g_0^2$ contains the coupling constant and the theta parameter, μ is the adjoint scalar mass, breaking softly $\mathcal{N}=2$ supersymmetry to $\mathcal{N}=1$. This is the same class of theories as that of Section 3, but this time we tune the bare quark masses as

$$m_1 = \ldots = m_n = m^{(1)}; \quad m_{n+1} = m_{n+2} = \ldots = m_{n+r} = m^{(2)} ,$$

$$N = n + r ; \qquad n\, m^{(1)} + r\, m^{(2)} = 0 , \tag{19}$$

or

$$m^{(1)} = \frac{r\, m_0}{\sqrt{r^2+n^2}}, \quad m^{(2)} = -\frac{n\, m_0}{\sqrt{r^2+n^2}}, \tag{20}$$

and their magnitude is taken as

$$|m_0| \gg |\mu| \gg \Lambda . \tag{21}$$

The adjoint scalar VEV can be taken to be

$$\langle \Phi \rangle = -\frac{1}{\sqrt{2}} \begin{pmatrix} m^{(1)}\, \mathbb{1}_{n\times n} & \mathbf{0} \\ \mathbf{0} & m^{(2)}\, \mathbb{1}_{r\times r} \end{pmatrix} \tag{22}$$

Below the mass scale $v_1 \sim |m_i|$ the system thus reduces to a gauge theory with gauge group

$$G = \frac{SU(n) \times SU(r) \times U(1)}{\mathbb{Z}_K}, \quad K = \mathrm{LCM}\{n, r\} \tag{23}$$

where K is the least common multiple of n and r. The higher n color components of the first n flavors (with the bare mass $m^{(1)}$) remain massless, as well as the lower r color components of the last r flavors (with the bare mass $m^{(2)}$): they will be denoted as $q^{(1)}$ and $q^{(2)}$, respectively. Our model then is:

$$\mathcal{L} = -\frac{1}{4g_0^2} F_{\mu\nu}^{0\,2} - \frac{1}{4g_n^2} F_{\mu\nu}^{n\,2} - \frac{1}{4g_r^2} F_{\mu\nu}^{r\,2} + \frac{1}{g_0^2} |\mathcal{D}_\mu \Phi^{(0)}|^2 + \frac{1}{g_n^2} |\mathcal{D}_\mu \Phi^{(n)}|^2 +$$
$$+ \frac{1}{g_r^2} |\mathcal{D}_\mu \Phi^{(r)}|^2 + \left|\mathcal{D}_\mu q^{(1)}\right|^2 + \left|\mathcal{D}_\mu \bar{\tilde{q}}^{(1)}\right|^2 + \left|\mathcal{D}_\mu q^{(2)}\right|^2 + \left|\mathcal{D}_\mu \bar{\tilde{q}}^{(2)}\right|^2 - V_D - V_F,$$

plus fermionic terms, where V_D and V_F are the D-term and F-term potentials. The light squark fields are the components

$$Q(x) = \begin{pmatrix} q^{(1)}(x)_{n\times n} & 0 \\ 0 & q^{(2)}(x)_{r\times r} \end{pmatrix}, \quad \tilde{Q}(x) = \begin{pmatrix} \tilde{q}^{(1)}(x)_{n\times n} & 0 \\ 0 & \tilde{q}^{(2)}(x)_{r\times r} \end{pmatrix}, \tag{24}$$

if written in a color-flavor mixed matrix notation. We keep

$$\tilde{q}^{(1)} = (q^{(1)})^\dagger, \qquad q^{(2)} = -(\tilde{q}^{(2)})^\dagger ; \tag{25}$$

the redefinition

$$q^{(1)} \to \frac{1}{\sqrt{2}}\, q^{(1)}, \quad \tilde{q}^{(2)} \to \frac{1}{\sqrt{2}}\, \tilde{q}^{(2)} \tag{26}$$

brings the kinetic terms for these fields back to the original form.

The VEVs of the adjoint scalars are given by $\langle \Phi^{(0)} \rangle = -m_0, \quad \langle \Phi^{(a)} \rangle = \langle \Phi^{(b)} \rangle = 0$, while the squark VEVs are given by

$$\langle Q \rangle = \begin{pmatrix} v^{(1)}\, \mathbb{1}_{n\times n} & 0 \\ 0 & -v^{(2)\,*}\, \mathbb{1}_{r\times r} \end{pmatrix}, \quad \langle \tilde{Q} \rangle = \begin{pmatrix} v^{(1)\,*}\, \mathbb{1}_{n\times n} & 0 \\ 0 & v^{(2)}\, \mathbb{1}_{r\times r} \end{pmatrix}, \tag{27}$$

with

$$|v^{(1)}|^2 + |v^{(2)}|^2 = \sqrt{n + r/n\,r}\, \mu\, m_0 \ . \tag{28}$$

There is a continuous vacuum degeneracy; we take

$$v^{(1)} \neq 0; \qquad v^{(2)} \neq 0 \ ,$$

in the following.

The vacuum breaks the gauge group G completely, leaving at the same time a color-flavor diagonal symmetry

$$[SU(n) \times SU(r) \times U(1)]_{C+F} \tag{29}$$

unbroken. The full global symmetry, including the overall global $U(1)$ is given by $U(n) \times U(r)$. The minimal vortex in this system

has e.g. the form,

$$q^{(1)} = \begin{pmatrix} e^{i\phi}\, f_1(\rho) & 0 \\ 0 & f_2(\rho)\, \mathbb{1}_{(n-1)\times(n-1)} \end{pmatrix}, \quad \tilde{q}^{(2)} = \begin{pmatrix} e^{i\phi}\, g_1(\rho) & 0 \\ 0 & g_2(\rho)\, \mathbb{1}_{(r-1)\times(r-1)} \end{pmatrix}, \tag{30}$$

where ρ and ϕ stand for the polar coordinates in the plane perpendicular to the vortex axis, $f_{1,2}, g_{1,2}$ are profile functions. The behavior of numerically integrated vortex profile functions $f_{1,2}, g_{1,2}$ is illustrated in Fig. 1.

We note here only that the necessary boundary conditions on the squark profile functions have the form,

$$f_1(\infty) = f_2(\infty) = v^{(1)}, \qquad g_1(\infty) = g_2(\infty) = v^{(2)},$$

while at the vortex core,

$$f_1(0) = 0, \quad g_1(0) = 0, \qquad f_2(0) \neq 0, \quad g_2(0) \neq 0, \tag{31}$$

The most important fact about these minimum vortices is that one of the $q^{(1)}$ *and* one of the $\tilde{q}^{(2)}$ fields must necessarily wind at infinity, simultaneously. As the individual vortex breaks the (global) symmetry of the

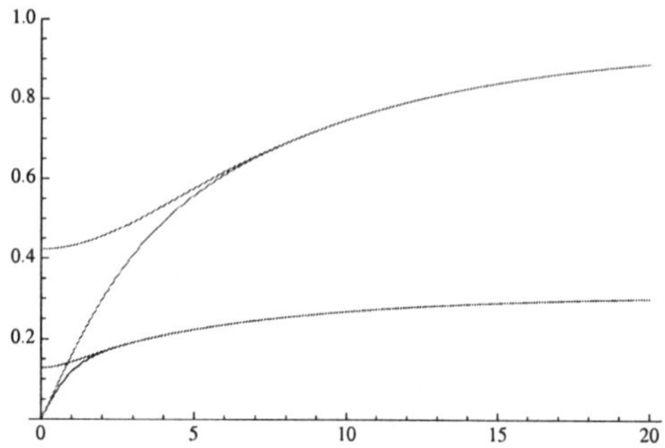

Fig. 1. Numerical result for the profile functions $f_{1,2}, g_{1,2}$ as functions of the radius ρ, for $SU(3) \times SU(2) \times U(1)$ theory. The coupling constants and the ratio of the VEVs are taken to be $g_0 = 0.1$, $g_3 = 10$, $g_2 = 1$, $|v_2|/|v_1| = 3$.

vacuum as

$$[SU(n) \times SU(r) \times U(1)]_{C+F} \to SU(n-1) \times SU(r-1) \times U(1)^3, \quad (32)$$

the vortex acquires Nambu-Goldstone modes parametrizing

$$CP^{n-1} \times CP^{r-1} \ : \quad (33)$$

they transform under the exact color-flavor symmetry $SU(n) \times SU(r)$ as the bi-fundamental representation, $(\underline{n}, \underline{r})$. Allowing the vortex orientation to fluctuate along the vortex length and in time, we get a $CP^{n-1} \times CP^{r-1}$ two-dimensional sigma model as an effective Lagrangian describing them. The details have been worked out in[12,18] and need not be repeated here.

The main idea is this. Let us assume without losing generality that $n > r$ (excluding the special case of $r = n$). As has been shown in[12,18] the coupling constant of the CP^{n-1} sigma models grows precisely as the coupling constant of the 4D $SU(n)$ gauge theory. At the point the CP^{n-1} vortex moduli fluctuations become strong and the dynamical scale Λ gets generated, with vortex kinks (Abelian monopoles) acquiring mass of the order of Λ, the vortex still carries the unbroken $SU(r)$ fluctuation modes (CP^{r-1}), as the $SU(r)$ interactions are still weak. Such a vortex will carry one of the $U(1)$ flux arising from the dynamical breaking of $SU(n) \times U(1) \to U(1)^n$, as well as an $SU(r)$ flux. As these vortices end at a massive monopole (arising from the high-energy gauge symmetry breaking, Eq. (22)), the latter necessarily carries a non-Abelian continuous moduli, whose points transform as in the fundamental representation of $SU(r)$. This can be interpreted as the (electric description of) dual gauge $SU(r)$ system observed in the infrared limit of the 4D SQCD.[5,7]

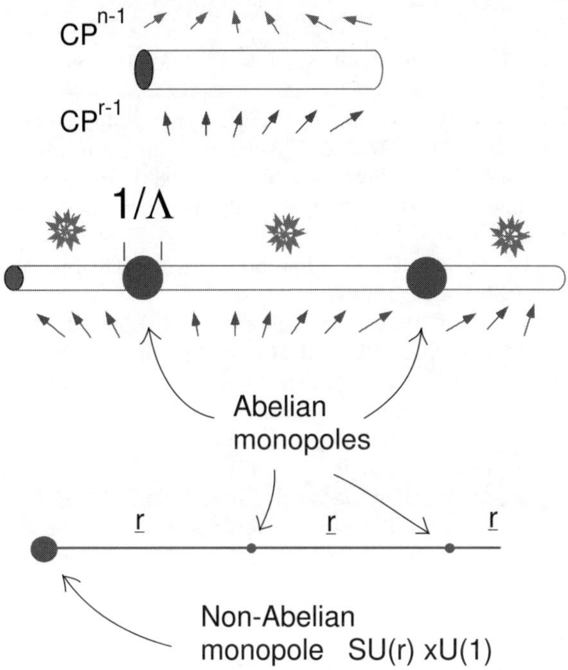

Fig. 2. The vortex with $CP^{n-1} \times CP^{r-1}$ orientational modes.

Thus vortices with non-Abelian moduli, which do not dynamically Abelianize completely, can be constructed in a natural way. Semi-classically, they are simply vortices carrying the $SU(n) \times SU(r) \times U(1)$ color-flavor flux. More precisely, they carry the Nambu-Goldstone modes Eq. (33) resulting from the partial breaking of the $SU(n) \times SU(r) \times U(1)$ global symmetry by the vortex. For $n > r$, CP^{n-1} field fluctuations propagating along the vortex length become strongly coupled in the infrared, the $SU(n) \times U(1)$ part dynamically Abelianizes; the vortex however still carries weakly-fluctuating $SU(r)$ flux modulations. In our theory where $SU(n) \times SU(r) \times U(1)$ model emerges as the low-energy approximation of an underlying $SU(N)$ theory, such a vortex is not stable. When the vortex ends at a monopole, its CP^{r-1} orientational modes are turned into the dual $SU(r)$ color modulations of the monopole.

References

1. G. 't Hooft, Nucl. Phys. B **79**, 817 (1974), A.M. Polyakov, JETP Lett. **20**, 194 (1974).
2. P. Goddard, J. Nuyts, D. Olive, Nucl. Phys. B **125**, 1 (1977); F.A. Bais, Phys. Rev. D **18**, 1206 (1978); E.J. Weinberg, Nucl. Phys. B **167**, 500 (1980).
3. A. Abouelsaood, Nucl. Phys. B **226**, 309 (1983); P. Nelson, A. Manohar, Phys. Rev. Lett. **50**, 943 (1983); A. Balachandran et. al., Phys. Rev. Lett. **50**, 1553 (1983).
4. N. Dorey, C. Fraser, T.J. Hollowood, M.A.C. Kneipp, hep-th/9512116; Phys.Lett. B **383**, 422 (1996).
5. P. C. Argyres, M. R. Plesser, N. Seiberg, Nucl. Phys. B **471**, 159 (1996).
6. A. Hanany, Y. Oz, Nucl. Phys. B **466**, 85 (1996).
7. G. Carlino, K. Konishi, H. Murayama, Nucl. Phys. B **590**, 37 (2000).
8. K. Konishi, Proc. of Continuous Advances in QCD 2004, Minneapolis, World Scientific (2004) 403; "The Magnetic Monopoles Seventy-Five Years Later", Lect. Notes in Phys. (Springer) 737: 471 (2008), hep-th/0702102.
9. P. C. Argyres, M. R. Douglas, Nucl. Phys. B **448**, 93 (1995).
10. R. Auzzi, S. Bolognesi, J. Evslin, K. Konishi, Nucl. Phys. B **686**, 119 (2004).
11. M. Eto, et.al., Nucl.Phys. **B780** 161-187, 2007, hep-th/0611313.
12. A. Hanany, D. Tong, JHEP **0307**, 037 (2003); JHEP **0404**, 066 (2004).
13. R. Auzzi, S. Bolognesi, J. Evslin, K. Konishi, A. Yung, Nucl. Phys. B **673**, 187 (2003).
14. M. Eto, T. Fujimori, S. B. Gudnason, K. Konish, M. Nitta, K. Ohashi, W. Vinci, [arXiv:0802.1020 [hep-th]].
15. M. Eto, Y. Isozumi, M. Nitta, K. Ohashi, N. Sakai, Phys. Rev. Lett. **96**, 161601 (2006).
16. M. Eto, et. al., J. Phys. A **39**, 315 (2006).
17. D. Tong, TASI lectures on solitons, hep-th/0509216.
18. M. Shifman and A. Yung, Phys. Rev. D **70**, 045004 (2004).
19. A. Gorsky, M. Shifman, A. Yung, Phys. Rev. D **71**, 045010 (2005).
20. M. Shifman and A. Yung, Phys.Rev. **D77**: 066008 (2008) arXiv:0712.3512 [hep-th].
21. D. Dorigoni, K. Konishi, K. Ohashi, arXiv:0801.3284 [hep-th].
22. K. Konishi, Phys.Lett. **B135**: 439 (1984).
23. S. Raby, S. Dimopoulos and L. Susskind, Nucl.Phys. **B169**:373 (1980).
24. P. C. Argyres, A. F. Faraggi, Phys. Rev. Lett **74**, 3931 (1995); A. Klemm, W. Lerche, S. Theisen, S. Yankielowicz, Phys. Lett. B **344**, 169 (1995); A. Hanany, Y. Oz, Nucl. Phys. B **452**, 283 (1995).
25. N. Seiberg, E. Witten, Nucl. Phys. B **426**, 19 (1994); Erratum *ibid.* B **430**, 485 (1994); Nucl. Phys. B **431**, 484 (1994).
26. K. Hashimoto and D. Tong, JCAP **0509**, 004 (2005) hep-th/0506022.
27. R. Auzzi, M. Shifman, A. Yung, Phys.Rev. **D73**:105012,2006, Erratum-ibid.D76:109901,2007. hep-th/0511150.
28. M. Eto, et.al., Phys. Rev. D **74**, 065021 (2006); hep-th/0607070.
29. M. Shifman, A. Yung, Phys. Rev. D **73**, 125012 (2006).
30. M. Eto, et. al., Phys.Rev. **D76** 105002 (2007), arXiv:0704.2218 [hep-th].

A GENERALIZED CONSTRUCTION FOR LUMPS AND NON-ABELIAN VORTICES

W. VINCI*

Department of Physics E.Fermi, University of Pisa,
INFN, Sezione di Pisa,
Largo Pontecorvo, 3,Ed. C, 56127, Pisa, Italy.
**E-mail: walter.vinci@pi.infn.it*

We construct the general vortex solution in a fully-Higgsed, color-flavor locked vacuum of a non-Abelian gauge theory, where the gauge group is taken to be the product of an arbitrary simple group and $U(1)$, with a Fayet-Iliopoulos term. The strict correspondence between vortices and lumps in the associated NLσM which arise in the limit of strong coupling is pointed out. The construction of the vortex moduli space is derived here as a consequence of this correspondence.

Keywords: Non-Abelian vortex; lump; moduli space; rational maps.

1. Introduction

Vortices play important roles in various areas of physics from condensed matter physics to particle physics[2] and cosmology.[3] Most importantly, vortices could play a crucial role in the confinement mechanism of QCD.[4]

On the other hand, non-linear sigma models (NLσM), and in particular the $\mathbb{C}P^1$ sigma model,[5] have been intensively studied because of their similarities with QCD, like asymptotic freedom, and non-perturbatively generated mass gaps.

Recently, there has been a significant progress in the understanding of non-Abelian vortices in the color-flavor locked vacuum of $U(N)$ gauge theories.[6,7] Unlike Abelian vortices,[8] they carry orientation moduli in the internal space, in addition to the usual position moduli. The most general Bogomol'nyi-Prasad-Sommerfield (BPS) vortex solutions and their moduli spaces have been found[9] and many other interesting features have been extensively explored.[10] Most studies so far have been restricted to the gauge group $U(N)$, with a few but notable exceptions[11,12] mainly devoted to the investigation of GNO duality[13] from the vortex side.

This contribution is based on Ref. 1, where we presented a simple framework to construct the most general non-Abelian BPS vortex solution in theories with an arbitrary gauge group of type $G = G' \times U(1)$. Here we will derive the same construction emphasizing the deep connections between vortices and lumps. In fact, our prescription corresponds to a "generalized" rational map construction[5] that can be used to construct both vortices and lump solutions in a wide class of gauge theories and non-linear sigma models.

We take G' to be a simple Lie group, but the method can be easily generalized to non-simple groups. The cases of classical groups $G' = SU, SO, USp$ will be worked out in more detail.

2. The Model and BPS Vortex Equations

Let us start with presenting the class of models in which we will construct vortices. It is a non-Abelian gauge theory with gauge group $G = G' \times U(1)$, whose Lagrangian is given by

$$\begin{aligned} \mathcal{L} = & -\frac{1}{4e^2} F^0_{\mu\nu} F^{0\mu\nu} - \frac{1}{4g^2} F^a_{\mu\nu} F^{a\mu\nu} + (\mathcal{D}_\mu H_A)^\dagger \mathcal{D}^\mu H_A \\ & - \frac{e^2}{2} \left| H_A^\dagger t^0 H_A - \frac{v^2}{\sqrt{2N}} \right|^2 - \frac{g^2}{2} |H_A^\dagger t^a H_A|^2 , \end{aligned} \tag{1}$$

where $\mathcal{D}_\mu H = (\partial_\mu + i W_\mu) H$, e and g are the gauge coupling constants for $U(1)$ and G', respectively, and A is the flavor index. Apart from the gauge bosons $W_\mu = W^0_\mu t^0 + W^a_\mu t^a$ the matter content of the model consists of N flavors of Higgs scalar fields in the fundamental representation, with a common $U(1)$ charge, written as a color-flavor mixed $N \times N$ matrix H. t^0 and t^a denote the generators of $U(1)$ and G' and are normalized as

$$t^0 = \frac{\mathbf{1}_N}{\sqrt{2N}} \quad \mathrm{Tr}(t^a t^b) = \frac{\delta^{ab}}{2} . \tag{2}$$

The flavor symmetry of the model is $SU(N)_{\mathrm{F}}$. Though our discussion concerns mainly the bosonic system (1), the model is really to be considered as the (truncated) bosonic sector of the corresponding $\mathcal{N} = 2$ supersymmetric gauge theory, which explains the particular form of the potential, ensuring at the same time its stability against radiative corrections. Another consequence of this is that we have several useful descriptions of the moduli space of vacua, which will be important in the upcoming discussion:[20]

$$\mathcal{M}_G = \{H \,|\, \text{D-term conditions}\}\,/G \tag{3a}$$
$$= \{H\}\,//G^{\mathbb{C}} \tag{3b}$$
$$= \left\{I_G^{ij},\ \text{holomorphic}\, G\text{-invariants}\right\}/\{\text{algebraic relations}\} \tag{3c}$$

Furthermore, $\mathcal{N}=2$ supersymmetry implies the existence of BPS saturated vortex solutions which are supported by the non-trivial first homotopy group $\pi_1(G = G' \times U(1)) = \mathbb{Z}$. The standard Bogomol'nyi completion for static, x^3-independent, configurations

$$T = \int d^2x \left[\frac{1}{2e^2}\left|F_{12}^0 - e^2\left(H_A^\dagger t^0 H_A - \frac{v^2}{\sqrt{2N}}\right)\right|^2 + \right.$$
$$\left. + 4\left|\mathcal{D}_{\bar{z}}H\right|^2 + \frac{1}{2g^2}\left|F_{12}^a - g^2 H_A^\dagger t^a H_A\right|^2 - \frac{v^2}{\sqrt{2N}}F_{12}^0\right]$$
$$\geq\ -\frac{v^2}{\sqrt{2N}}\int d^2x\, F_{12}^0\,, \tag{4}$$

gives the BPS equations for vortices.

We will focus our attention on the classical Lie groups $SU(N)$, $SO(2M)$ and $USp(2M)$. For $G' = SO(2M), USp(2M)$ their group elements are embedded into $SU(N)$ $(N = 2M)$ by constraints of the form, $U^T J U = J$, where J is the rank-2 invariant tensor

$$J = \begin{pmatrix} \mathbf{0}_M & \mathbf{1}_M \\ \epsilon\mathbf{1}_M & \mathbf{0}_M \end{pmatrix}, \tag{5}$$

where $\epsilon = +1$ for $SO(2M)$, while $\epsilon = -1$ for $USp(2M)$. Specializing (4) to these cases we get the BPS vortex equations

$$\mathcal{D}_{\bar{z}}H = 0, \tag{6a}$$
$$\frac{\sqrt{2N}}{e^2}F_{12}^0 - \left(\mathrm{Tr}(HH^\dagger) - v^2\right) = 0, \tag{6b}$$
$$\frac{4}{g^2}F_{12}^a t^a - \left(HH^\dagger - J^\dagger (HH^\dagger)^T J\right) = 0, \tag{6c}$$

where the complex coordinate $z \equiv x^1 + ix^2$ has been introduced. Eq. (6c) reads for $G' = SU(N)$ instead:

$$\frac{2}{g^2}F_{12}^a t^a - \left[HH^\dagger - \frac{\mathbf{1}_N}{N}\mathrm{Tr}(HH^\dagger)\right] = 0\,. \tag{7}$$

3. The Strong Coupling Limit and the BPS Lump Equations

Another class of theories we are interested in, is the set of non-linear sigma models which can be obtained from the theories (3) by taking the strong coupling limit: $e, g \to \infty$. In a fully Higgsed vacuum, all gauge fields are massive and can be integrated out, and the limit is reliable. The resulting theory is an $\mathcal{N} = 2$ NLσM in 3+1 dimensions whose target space is the vacuum manifold of the gauge theory:

$$\mathcal{L}_{\mathrm{NL}\sigma\mathrm{M}} = g_{i,\bar{j}} \partial_\mu \phi_i \partial_\mu \bar{\phi}_j, \quad g_{i,\bar{j}} = \partial_{\phi_i} \partial_{\bar{\phi}_j} K(\phi_i, \bar{\phi}_j),$$
$$\mathcal{M}_{\mathrm{NL}\sigma\mathrm{M}} = \mathcal{M}_G. \tag{8}$$

The fields ϕ_i are a conveniently chosen set of independent coordinates on the target space while K is the Kähler potential of the gauge theory.

This class of models also admits BPS, x^3-independent, stringy soliton called lumps. This kind of solitons are supported by a non-trivial homotopy group $\pi_2(\mathcal{M}_{\mathrm{NL}\sigma\mathrm{M}})$. It is indeed very easy to find the equations for lumps without going into the explicit computation of the Kähler potential. In fact, taking the strong coupling limit in (1) without explicitly intagrating out the gauge fields, we obtain a description of the models (8) in terms of the fields of the original gauge theory (H, W_μ). Lump equations are simply given by the strong coupling limit of the vortex equations (6).

$$\mathcal{D}_{\bar{z}} H = 0, \tag{9a}$$
$$\mathrm{Tr}(HH^\dagger) - v^2 = 0, \tag{9b}$$
$$HH^\dagger - J^\dagger (HH^\dagger)^T J = 0. \tag{9c}$$

Notice that Eq. (9b) and Eq. (9c) are nothing but the D-term vacuum equations, whose solutions define $\mathcal{M}_{\mathrm{NL}\sigma\mathrm{M}}$.

A general property of the models (8) is the classical scale invariance. Lumps develop a moduli space of degenerate solutions with arbitrary size and consequently there appear singular solutions with zero size, called small-lump singularities. Another way to understand these singularities is to notice that the NLσMs are well defined only on a fully Higgsed vacuum, while the theories (1) have a moduli space of vacua that usually includes many branches with different physical properties: Higgs and Coulomb branches, mixed Higgs-Coulomb phases in which the gauge symmetry is only partially broken. Lump solutions can be understood as maps from the $\mathbb{C}$-plane orthogonal to the x^3-axis, into the target space $\mathcal{M}_{\mathrm{NL}\sigma\mathrm{M}}$ and we expect singularities when this map hit a point which corresponds to a phase of the original gauge theory with an unbroken gauge symmetry. In fact, in this points, the appearance of massless particles gives rise to singularities in the Kähler potential of the gauge theory, and thus on the target manifold of the NLσM. For example, small-lump singularities are usually related to conical singularities in the target space[a].

Before giving the recipes for the construction of lumps, we will comment, in the next section, on the correspondence between vortices in the gauge theories (1) and lumps in the related NLσM (8). A consequence of this correspondence is that we get, as promised, the moduli space of vortices.

4. The Vortex-Lump Correspondence

The deep connection between vortices and lumps was first noticed by Hindmarsh.[22] In the extended Abelian Higgs model "semilocal" vortices[23,24] appear, which are generalizations of the usual Abrikosov-Nielsen-Olesen (ANO) vortices: they possess zero modes related to an arbitrary size parameter. Semilocal vortices "interpolate" between ANO vortices and $\mathbb{C}P^n$ lumps, in the sense that zero-size semilocal vortices are ANO ("local") vortices, while semilocal vortices with increasing size become identical to lumps.

We consider this relationship from another, but equivalent, point of view:[17] taking the strong coupling limit $e, g \to \infty$ we map the gauge theory (1) into the sigma model (8); at the same time, we map any vortex appearing in the first theory into a lump of the latter:

$$\begin{aligned} G' \times U(1) \text{ Gauge theory} &\longrightarrow \text{NL}\sigma\text{M on } \mathcal{M}_G \\ \text{Vortices} &\longrightarrow \text{Lumps}. \end{aligned} \tag{10}$$

This correspondence can also be better understood if one consider more quantitative relations.

BPS correspondence. Vortices and lumps are BPS saturated objects. Their masses do not depend on the gauge couplings, being proportional to the central charge of the supersymmetry algebra: $T_{vor} = M_{lum}$

Topological correspondence. In a generic case in which the vacuum manifold is simply connected[b], the following relations hold: $\pi_2(\mathcal{M}_G) = \pi_2(\mathcal{M}_{vac\,man}/G) = \pi_1(G)$, which define a correspondence between the topological charges of vortices and lumps.

[a] $\mathcal{M}_{\text{NL}\sigma\text{M}}$ can contain also different kind of singularities, like curvature singularities. They give rise to different kinds of singular lumps.[19]

[b] This property of the vacuum manifold is the necessary condition for the existence of semilocal vortices.[24]

It is important, however, to notice that some vortex configurations will be mapped into singular lumps. For example, ANO vortices, with a typical size $\sim 1/gv$, shrink to singular spikes of energy, and are mapped into small-lump singularities. Conversely, small-lump singularities are regularized into ANO vortices when finite gauge couplings are restored. The discussion above leads to the following relation among the moduli spaces of vortices and lumps:[14c]

$$\mathcal{M}_{vor} = \mathcal{M}_{lum} \oplus \{\text{singular lumps}\} \tag{11}$$

5. The Generalized Rational Map Construction for Lumps

After the general discussion of the previous Section, we are now ready to explicitly construct the moduli space of lumps arising from (8).

We choose the fully Higgsed, color-flavor locked vacuum: $H_{\text{vev}} = \frac{v}{\sqrt{N}}\mathbf{1}_N$. The $G' \times U(1) \times SU(N)_{\text{F}}$ invariance of the theory is broken to the global color-flavor diagonal $G'_{\text{C+F}}$. The first step toward the solutions of Eq. (9) is to switch from the description (3a) for the moduli space of vacua to that of (3b). We can thus forget the D-term vacuum equations (9b) and (9c), at the prize of lifting the gauge symmetry G to its complexification $G^{\mathbb{C}}$. Eq. (9a) has now a very intuitive meaning. It can be thought as a "covariant holomorphicity" condition. This means that H is holomorphic, up to a complexified gauge transformation:

$$H = S^{-1}H_0(z) = S_e^{-1}S'^{-1}H_0(z) \; , \tag{12}$$

where $S \in G^{\mathbb{C}}$, while $H_0(z)$ is a matrix whose elements are holomorphic in z. We denote $H_0(z)$ the moduli matrix,[16] as it encodes all moduli parameters[c]. Eq. (12) does not fix completely $G^{\mathbb{C}}$. $H_0(z)$ is in fact only defined up to holomorphic gauge transformations $V(z)$:

$$H_0 \sim V(z)H_0 = V_e V'(z)H_0 \; , \quad V' \in G'^{\mathbb{C}}, \quad V_e \in \mathbb{C}^* \; . \tag{13}$$

The next step is to consider the holomorphic invariants $I_G^{(i,j)}(H)$ made of H, which are invariant under $G^{\mathbb{C}}$, with (i,j) labeling them. This is equivalent to changing our description of $\mathcal{M}_G$ from Eq. (3b) to Eq. (3c). The key point here is that, thanks to Eq. (12), these invariant are also holomorphic functions of the coordinate z:

$$I_G^{(i,j)}(H) = I_G^{(i,j)}\left(S^{-1}H_0(z)\right) = I_G^{(i,j)}(H_0)(z). \tag{14}$$

[c]To reconstruct the solution in terms of the original Higgs fields H, one has still to solve Eq. (9) as algebraic equations for S.

We have thus identified a lump as a holomorphic map from the complex plane $\mathbb{C}$ to the set of holomorphic invariants $I_G^{(i,j)}$[d].

The last step is to impose the following boundary conditions to the holomorphic invariants:

$$I_G^{(i,j)}(H)\Big|_{|z|\to\infty} = I_{\rm vev}^{(i,j)} \,. \tag{15}$$

It is convenient to translate the above conditions in terms of the holomorphic invariants of G', $I_{G'}^j(H_0)$:

$$I_G^{(i,j)}(H_0) \equiv \frac{I_{G'}^i(H_0)}{I_{G'}^j(H_0)} \,, \tag{16}$$

where we must take ratios of invariants with the same $U(1)$ charge: $n_i = n_j$[e]. Eq. (16) defines the generalized rational map construction for lumps.

It is easy to check that Eq. (15) and Eq. (16) together imply the following:

$$I_{G'}^i(H_0)\Big|_{|z|\to\infty} = I_{\rm vev}^i \, z^{\nu n_i} \tag{17}$$

where ν is an integer common to all invariants. As $I_{G'}^i(H_0(z))$ are holomorphic, the above condition means that $I_{G'}^i(H_0(z))$ are *polynomials* in z. Furthermore, $\nu\, n_i$ must be a positive integer for all i:

$$\nu\, n_i \in \mathbb{Z}_+ \quad \to \quad \nu = k/n_0 \,, \quad k \in \mathbb{Z}_+ \,, \tag{18}$$

where (GCD = the greatest common divisor)

$$n_0 \equiv {\rm GCD}\{n_i \mid I_{\rm vev}^i \neq 0\} \,. \tag{19}$$

Note that a $U(1)$ gauge transformation $e^{2\pi i/n_0}$ leaves $I_{G'}^i(H)$ invariant:

$$I_{G'}^i(H') = e^{2\pi i n_i/n_0} I_{G'}^i(H) = I_{G'}^i(H) : \tag{20}$$

the phase rotation $e^{2\pi i/n_0} \in \mathbb{Z}_{n_0}$ changes no physics, and the true gauge group is thus $G = U(1) \times G'/\mathbb{Z}_{n_0}$, where $\mathbb{Z}_{n_0}$ is the center of G'. A simple homotopy argument tells us that $1/n_0$ is the $U(1)$ winding for the minimal, $k = 1$, topologically non-trivial configuration. Finally, for a given k the following important relation holds

$$I_{G'}^i(H_0) = I_{\rm vev}^i z^{kn_i/n_0} + \mathcal{O}(z^{kn_i/n_0 - 1}) \,, \tag{21}$$

---

[d] All the points at the infinity of z are in fact identified, so that the map is in fact defined on $\mathbb{C}P^1$.

[e] The $U(1)$ charge n_i is simply defined as: $I_{G'}^i(\alpha H_0) = \alpha^{n_i} I_{G'}^i(H_0)$, for any complex number α.

which implies nontrivial constraints on $H_0(z)$. The set of all inequivalent $H_0(z)$ satisfying Eq. (21) identify a lump solution, and thus defines the full moduli space.

6. Moduli Space of Vortices

The lump construction explained in the last Section is only partially complete, because one has to identify all singular lump solutions, and this is not easy in general. This problem disappears when we lift the construction for lump to that for vortices in the related gauge theory, according to the discussion of Sec. 4. As explained there, singular lumps are regularized into vortices as we turn on finite gauge couplings, and Eq. (21) now give the full and complete description of the moduli space of non-Abelian vortices.

To reconstruct the original Higgs fields $H = S^{-1}H_0$ we now have to solve the BPS differential equations Eq. (6b) and Eq. (6c) (or Eq. (7)) in terms of S, while from Eq. (6a) we find the gauge fields:

$$W_1 + iW_2 = -2i\, S^{-1}\bar{\partial}S. \tag{22}$$

The tension of the BPS vortices can be written as

$$T = -\frac{v^2}{\sqrt{2N}}\int d^2x\; F^0_{12} = 2v^2\int d^2x\; \bar{\partial}\partial\,\log(S_e S_e^\dagger)\;. \tag{23}$$

The asymptotic behavior $S_e \sim |z|^\nu$ then determines the tension

$$T = 2\pi v^2 \nu\;, \tag{24}$$

thus ν has to be identified with the $U(1)$ winding number of the vortex configuration.

Let us now apply the general discussion above to concrete examples. For $G' = SU(N)$, with N flavors, there only exists one invariant

$$I_{SU} = \det(H)\;, \tag{25}$$

with charge N. Thus the minimal winding is equal to $1/N$ and the condition for k vortices is given by:

$$A_{N-1} : \det H_0(z) = z^k + \mathcal{O}(z^{k-1}), \quad \nu = k/N\;. \tag{26}$$

For $G' = SO(2M), USp(2M)$, there are $N(N\pm 1)/2$ invariants

$$(I_{SO,USp})^r{}_s = (H^{\mathrm{T}}JH)^r{}_s, \quad 1 \le r \le s \le N\;, \tag{27}$$

in addition to (25). The constraints are:

$$C_M, D_M : H_0^T(z)JH_0(z) = z^k J + \mathcal{O}(z^{k-1})\;, \quad \nu = k/2\;. \tag{28}$$

Thus, vortices in the $SO(2M)$ and $Usp(2M)$ theories are quantized in half integers of the $U(1)$ winding.[12]

Explicitly, the minimal vortices in $SU(N)$ and $SO(2M)$ or $USp(2M)$ theories are given respectively by the moduli matrices:

$$H_0 = \begin{pmatrix} z-a & 0 \\ \mathbf{b} & \mathbf{1}_{N-1} \end{pmatrix}, \begin{pmatrix} z\mathbf{1}_M - \mathbf{A} & \mathbf{C}_{S/A} \\ \mathbf{B}_{A/S} & \mathbf{1}_M \end{pmatrix}. \tag{29}$$

The moduli parameters are all complex. For $SU(N)$, a is just a number; $\mathbf{b}$ is a column vector. For $SO(2M)$ or $USp(2M)$, the matrix $\mathbf{C}_{S/A}$ for instance is symmetric or antisymmetric, respectively. And vice versa for $\mathbf{B}$. Moduli matrices for $SO(2M+1)$ as well as those for $k=2$ vortices in SU, SO, USp theories, will soon be given explicitly in Ref. 19.

The index theorem gives the complex dimension of the moduli space

$$\dim_{\mathbb{C}}(\mathcal{M}_{G',k}) = \frac{k\,N^2}{n_0}\,. \tag{30}$$

This was obtained in Ref. 6 for $SU(N)$; a proof in other cases will be reported elsewhere.[19] In all cases studied we have checked that the dimension of the moduli space inferred from the moduli matrices agrees with the one given in Eq. (30).

Except for the $SU(N)$ case, our model has a non-trivial Higgs branch (flat directions). The color-flavor locked vacuum $H_{\rm vev} \propto \mathbf{1}_N$ is just one of the possible (albeit the most symmetric) choices for the vacuum; our discussion can readily be generalized to a generic vacuum on the Higgs branch. This fact, however, implies that our non-Abelian vortices have semilocal moduli even for $N_f = N$. In contrast to the Abelian or $SU(N)$ cases, moreover, they exhibit new, interesting phenomena such as "fractional" vortices.[19]

The generalization to exceptional groups can be done if the invariant tensors of each group are known. Indeed, this is the case, and Table 1 lists them all.

Table 1. The dimension of the fundamental representation (R), the rank of the other invariants[21] and the minimal tension $\nu = 1/n_0$ i.e. the center $\mathbb{Z}_{n_0}$ of G'. The determinant of the $R \times R$ matrix gives one invariant with charge dim R.

G'	A_{N-1}	B_M	C_M, D_M	E_6	E_7	E_8	F_4	G_2
R	N	$2M+1$	$2M$	27	56	248	26	7
rank inv	–	2	2	3	2, 4	2, 3, 8	2, 3	2, 3
n_0	N	1	2	3	2	1	1	1

7. Local (ANO-like) Vortices

For various considerations, we are interested in knowing which of the moduli parameters describe the local vortices, the ANO-type vortices with exponential tails. The moduli space of fundamental local vortices, for example, is completely generated by the global symmetries of the vacuum (see below) and it can eventually survive non-BPS deformations that preserve these symmetries. Furthermore, recently it was found that local vortices correspond to the subset of solutions with the maximum number of normalizable moduli.[17]

Local vortices correspond, as we mentioned, to small lump singularities in the lump construction. These configurations are obtained when the rational map construction is degenerate, e.g., when all the invariants $I^i_{G'}$ share a common zero:

$$I^i_{G'} = (z - z_0)^{n_i/n_0}\, I'^i_{G'}. \tag{31}$$

The rational map in Eq. (16), does not feel any of these common zeros, which can be interpreted as insertions of singular spikes of energy, or local vortices. Configurations with only local vortices are obtained imposing a complete degeneration of the rational map:

$$I^i_{G'}(H_{0,\mathrm{local}}) = \left[\prod_{\ell=1}^{k}(z - z_{0\ell})\right]^{n_i/n_0} I^i_{\mathrm{vev}}\ . \tag{32}$$

For $G' = SO(2M), USp(2M)$ with $I_{SO,USp}$ of Eq. (27) we find that the condition for vortices to be of local type is

$$H^T_{0,\mathrm{local}}(z) J H_{0,\mathrm{local}}(z) = \prod_{\ell=1}^{k}(z - z_{0\ell})\, J\ . \tag{33}$$

Let us now discuss a few concrete examples. The general solution for the minimal vortex (29) for $G' = \{SU(N), SO(2M), USp(2M)\}$ is reduced to a local vortex if we restrict it to be of the form:

$$H_{0,\mathrm{local}} = \begin{pmatrix} z-a & 0 \\ \mathbf{b} & \mathbf{1}_{N-1} \end{pmatrix}, \begin{pmatrix} (z-a)\mathbf{1}_M & 0 \\ \mathbf{B}_{A/S} & \mathbf{1}_M \end{pmatrix}. \tag{34}$$

The vortex position is given by a, while $\mathbf{b}$ for $SU(N)$ and $\mathbf{B}_{A/S}$ for $SO(2M)$ or $USp(2M)$ encode the Nambu-Goldstone modes associated with the breaking of the color-flavor symmetry by the vortex $G'_{\mathrm{C+F}} \to H_{G'}$. The moduli spaces are direct products of a complex number and the Hermitian symmetric spaces

$$\mathcal{M}^{\mathrm{local}}_{G',k=1} \simeq \mathbb{C} \times G'_{\mathrm{C+F}}/H_{G'}\ , \tag{35}$$

$H_{SU(N)} = SU(N-1) \times U(1)$ while $H_{SO(2M),USp(2M)} = U(M)$. The results for $SU(N)$ and $SO(2M)$ are well-known.[6,7,12] The matrices (34) describe just one patch of the moduli space. In order to define the manifold globally we need a sufficient number of patches. The number of patches is N for $G' = SU(N)$ and 2^M for $G' = SO(2M), USp(2M)$. The transition functions correspond to the V-equivalence relations.[9] In the case of $G' = SO(2M)$, the patches are given by permutation of the i-th and the $(M+i)$-th columns in (34). We find that no regular transition functions connect the odd and even permutations (patches), hence the moduli space consists of two disconnected copies of $SO(2M)/U(M)$.[12] The complex dimensions of the moduli spaces are $\dim_{\mathbb{C}} \mathcal{M}^{\rm local}_{SO(2M),k=1} = \frac{1}{2}M(M-1)+1$ and $\dim_{\mathbb{C}} \mathcal{M}^{\rm local}_{USp(2M),k=1} = \frac{1}{2}M(M+1)+1$.

8. Conclusion

We have given all the necessary tools to construct vortex solutions in a wide class of non-Abelian gauge theories. In fact, our method may be potentially extended to any gauge theory with any matter content. The only requirement is $\mathcal{N}=2$ supersymmetry. The same method gives also a generalization of the rational map construction for lump, which can be applied to any NLσM whose target space is the vacuum manifold of an $\mathcal{N}=2$ gauge theory. Our method can also be extended to other BPS solitons such as domain walls, monopoles and instantons, and hopefully opens powerful new windows for their investigation.

Acknowledgments:

The author thanks the Organizers of the conference for the warm hospitality.

References

1. M. Eto, T. Fujimori, S. B. Gudnason, K. Konishi, M. Nitta, K. Ohashi and W. Vinci, "Constructing Non-Abelian Vortices with Arbitrary Gauge Groups," arXiv:0802.1020 [hep-th].
2. A. Achucarro and T. Vachaspati, Phys. Rept. **327**, 347 (2000) [arXiv:hep-ph/9904229]; R. Jeannerot, J. Rocher and M. Sakellariadou, Phys. Rev. D **68**, 103514 (2003) [arXiv:hep-ph/0308134].
3. A. Vilenkin and E. P. S. Shellard, Cosmic Strings and Other Topological Defects, Cambridge Univ. Press (1994); M. B. Hindmarsh and T. W. B. Kibble, Rept. Prog. Phys. **58**, 477 (1995).

4. G. 't Hooft, Nucl. Phys. B **190**, 455 (1981); S. Mandelstam, Phys. Lett. B **53**, 476 (1975).
5. A. M. Polyakov and A. A. Belavin, JETP Lett. **22**, 245 (1975) [Pisma Zh. Eksp. Teor. Fiz. **22**, 503 (1975)].
6. A. Hanany and D. Tong, JHEP **0307**, 037 (2003) [arXiv:hep-th/0306150].
7. R. Auzzi, S. Bolognesi, J. Evslin, K. Konishi and A. Yung, Nucl. Phys. B **673**, 187 (2003) [arXiv:hep-th/0307287].
8. A. A. Abrikosov, Sov. Phys. JETP **5** (1957) 1174 [Zh. Eksp. Teor. Fiz. **32** (1957) 1442]; H. B. Nielsen and P. Olesen, Nucl. Phys. B **61** (1973) 45.
9. M. Eto, Y. Isozumi, M. Nitta, K. Ohashi and N. Sakai, Phys. Rev. Lett. **96**, 161601 (2006) [arXiv:hep-th/0511088]; M. Eto, K. Konishi, G. Marmorini, M. Nitta, K. Ohashi, W. Vinci and N. Yokoi, Phys. Rev. D **74**, 065021 (2006) [arXiv:hep-th/0607070].
10. M. Eto, K. Hashimoto, G. Marmorini, M. Nitta, K. Ohashi and W. Vinci, Phys. Rev. Lett. **98**, 091602 (2007) [arXiv:hep-th/0609214]; M. Eto, Y. Isozumi, M. Nitta, K. Ohashi and N. Sakai, J. Phys. A **39**, R315 (2006) [arXiv:hep-th/0602170]; M. Eto, T. Fujimori, M. Nitta, K. Ohashi, K. Ohta and N. Sakai, Nucl. Phys. B **788**, 120 (2008) [arXiv:hep-th/0703197].
11. D. Dorigoni, K. Konishi and K. Ohashi, arXiv:0801.3284 [hep-th] (2008).
12. L. Ferretti, S. B. Gudnason and K. Konishi, Nucl. Phys. B **789**, 84 (2008)
13. P. Goddard, J. Nuyts and D. I. Olive, Nucl. Phys. B **125**, 1 (1977).
14. M. Shifman and A. Yung, Phys. Rev. D **70**, 045004 (2004) [arXiv:hep-th/0403149]; A. Hanany and D. Tong, JHEP **0404**, 066 (2004) [arXiv:hep-th/0403158]. M. Eto, Y. Isozumi, M. Nitta, K. Ohashi and N. Sakai, Phys. Rev. D **72**, 025011 (2005) [arXiv:hep-th/0412048].
15. D. Tong, arXiv:hep-th/0509216; M. Shifman and A. Yung, arXiv:hep-th/0703267.
16. Y. Isozumi, M. Nitta, K. Ohashi and N. Sakai, Phys. Rev. D **71**, 065018 (2005) [arXiv:hep-th/0405129]; Phys. Rev. Lett. **93**, 161601 (2004) [arXiv:hep-th/0404198]; Phys. Rev. D **70**, 125014 (2004) [arXiv:hep-th/0405194].
17. M. Shifman and A. Yung, Phys. Rev. D **73**, 125012 (2006) [arXiv:hep-th/0603134]; M. Eto, J. Evslin, K. Konishi, G. Marmorini, M. Nitta, K. Ohashi, W. Vinci and N. Yokoi, Phys. Rev. D **76**, 105002 (2007) [arXiv:0704.2218 [hep-th]].
18. M. Eto, L. Ferretti, K. Konishi, G. Marmorini, M. Nitta, K. Ohashi, W. Vinci and N. Yokoi, Nucl. Phys. B **780**, 161 (2007) [arXiv:hep-th/0611313].
19. M. Eto, T. Fujimori, S. B. Gudnason, K. Konishi, M. Nitta, K. Ohashi and W. Vinci, in preparation.
20. M. A. Luty and W. Taylor, Phys. Rev. D **53**, 3399 (1996).
21. P. Cvitanovic, Phys. Rev. D **14**, 1536 (1976).
22. M. Hindmarsh, Phys. Rev. Lett. **68**, 1263 (1992).
23. T. Vachaspati and A. Achucarro, Phys. Rev. D **44**, 3067 (1991);
24. J. Preskill, Phys. Rev. D **46**, 4218 (1992) [arXiv:hep-ph/9206216].

SECTION 6.

DYNAMICS IN SUPERSYMMETRIC THEORIES

CUSP ANOMALOUS DIMENSION IN PLANAR MAXIMALLY SUPERSYMMETRIC YANG–MILLS THEORY

B. BASSO

Laboratoire de Physique Théorique, Université de Paris XI,
91405 Orsay Cédex, France
E-mail: basso@th.u-psud.fr

The conjectured all-loop integrability of the planar maximally supersymmetric $\mathcal{N} = 4$ Yang-Mills theory offers an efficient way to compute the cusp anomalous dimension at any value of the 't Hooft coupling constant through its relation with the logarithmic scaling of anomalous dimensions of high-spin Wilson operators. In this talk, I report on calculation of the cusp anomalous dimension in this theory at strong coupling.

Keywords: Cusp Anomalous Dimension; Integrability; AdS/CFT.

1. Introduction

The cusp anomalous dimension is an ubiquitous observable in four-dimensional Yang-Mills theories. It controls asymptotic behavior of various gauge invariant quantities (logarithmic scaling of anomalous dimensions of high-spin Wilson operators, infared singularities of on-shell scattering amplitudes, double-logarithmic asymptotics of form-factors, etc.) and its determination as a function of the coupling constant is an important physical problem. At present, the cusp anomalous dimension is known at weak coupling in a generic Yang-Mills theory at first few orders of perturbative expansion. In the case of the planar maximally supersymmetric $\mathcal{N} = 4$ Yang-Mills theory (SYM), the conjectured all-loop integrability offers an efficient way to compute the cusp anomalous dimension at any value of the 't Hooft coupling constant through its relation with the logarithmic scaling of anomalous dimensions of high-spin Wilson operators. In this talk, I report on calculation of the cusp anomalous dimension in planar $\mathcal{N} = 4$ SYM theory at strong coupling.

2. Cusp Anomalous Dimension

2.1. *Definition*

Wilson loops are important observables in any gauge theory. They were introduced a long ago[1] as convenient collective variables in description of long-distance dynamics of gauge fields and in formulation of the criterium of quark confinement in QCD. The vacuum expectation value of the Wilson loop is defined as

$$W(\mathrm{C}) = \langle 0|\mathrm{Tr}\big[P \exp \imath g_{\mathrm{YM}} \int_{\mathrm{C}} A_\mu(x) dx^\mu\big]|0\rangle, \tag{1}$$

where the circulation of the gauge field $A_\mu(x)$ is evaluated over a closed contour C in Minkowski space-time and g_{YM} is the coupling constant of the gauge theory. Here, the path ordering P and the trace over the color indices insures the gauge invariance of $W(\mathrm{C})$.

The Wilson loops form the starting point of the reformulation of any gauge theory in terms of gauge invariant degrees of freedom.[2] They also play a distinguished role in an attempt to unravel a gauge/string duality.[3] In this framework, the *cusp anomaly* emerges as an unfortunate property of a Wilson loop with a cusp on the integration contour C (see Fig. 1). Namely, the expectation values of cusped Wilson loops develop extra ultraviolet (UV) divergences which are not related to the conventional ones of the underlying Lagrangian. The cusp divergences can be however eliminated by a multiplicative renormalization of the Wilson loop.[2] The dependence of the cusped Wilson loop on the renormalization scale μ is governed by the renormalization group equation

$$\mu \frac{d}{d\mu} W(\mathrm{C}) = -\Gamma_{\mathrm{cusp}}(g_{\mathrm{YM}}, \vartheta)\, W(\mathrm{C}), \tag{2}$$

where $\Gamma_{\mathrm{cusp}}(g_{\mathrm{YM}}, \vartheta)$ is a function of the coupling constant g_{YM} and of the cusp angle ϑ (see Fig. 1).

The function $\Gamma_{\mathrm{cusp}}(g_{\mathrm{YM}}, \vartheta)$ is known up to two loops in pure Yang-Mills theory[2,4] and it displays a complicated dependence on the cusp angle ϑ. In Minkowsky space-time, depending on the orientation of the tangent vectors at the cusp point, the angle ϑ can be either trigonometric, or hyperbolic. In particular, in the latter case, when the two tangent vectors are almost light-like, the cusp angle can take arbitrary large values. In this limit of large hyperbolic angle, $\Gamma_{\mathrm{cusp}}(g_{\mathrm{YM}}, \vartheta)$ scales linearly at any order in perturbation theory[4]

$$\Gamma_{\mathrm{cusp}}(g_{\mathrm{YM}}, \vartheta) = \Gamma_{\mathrm{cusp}}(g_{\mathrm{YM}})\, \vartheta + O(\vartheta^0), \qquad |\vartheta| \gg 1, \tag{3}$$

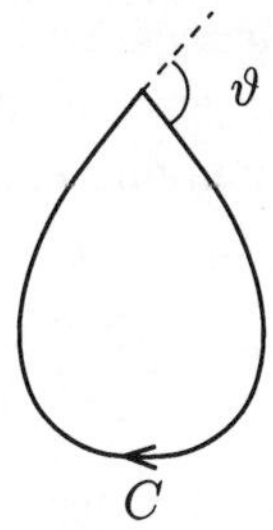

Fig. 1. Contour C with a cusp angle ϑ.

and its leading asymptotic behavior is controlled by a function of the coupling constant only, $\Gamma_{\rm cusp}(g_{\rm YM})$, to which we shall refer following Ref. 4 as the *cusp anomalous dimension*.

2.2. *Weak Coupling Expansion from Gauge Theory*

The cusp anomalous dimension $\Gamma_{\rm cusp}(g_{\rm YM})$ has been computed in a generic (super) Yang-Mills theory up to two loops.[2,4,6] In $\mathcal{N}=4$ SYM theory, its planar expansion is known up to three loops analytically[7] and to four loops numerically,[8]

$$\Gamma_{\rm cusp}(g) = 4g^2 - \frac{4}{3}\pi^2 g^4 + \frac{44}{45}\pi^4 g^6 + \mathcal{O}\left(g^8\right), \tag{4}$$

where $g^2 = g_{\rm YM}^2 N_c/16\pi^2$ is the 't Hooft coupling constant, kept fixed as $N_c \to \infty$. Its weak coupling expansion is believed to satisfy the maximal transcendentality principle[9] that constraints the contribution at order g^{2n} in perturbation theory to involve special numbers with assigned degree of transcendentality equal to $2n-2$ (π^2 at two loops, π^4 at three loops, π^6 and/or $\zeta(3)^2$ at four loops, for example). Under this assumption, the numerical four-loop correction to the cusp anomalous dimension is consistent with the analytical expression[8]

$$-8\left(\frac{73}{630}\pi^6 + 4\zeta(3)^2\right). \tag{5}$$

2.3. *Strong Coupling Expansion from String Theory*

At strong coupling, the cusp anomalous dimension in the planar $\mathcal{N}=4$ SYM theory can be computed using the *AdS/CFT correspondence*.[10] In this approach, the cusp anomalous dimension controls the logarithmic scaling of the semiclassical energy of a folded string rotating in AdS_3 with a large angular momentum. The classical and one-loop stringy corrections to the energy of the rotating string were found in Refs. 11 and 12, respectively. They lead to the strong coupling expansion of the cusp anomalous dimension

$$\Gamma_{\text{cusp}}(g) = 2g - \frac{3\ln 2}{2\pi} + \mathcal{O}\left(g^{-2}\right). \tag{6}$$

Equations (4) and (6) define the cusp anomalous dimension at weak and strong coupling, respectively. A natural question arises whether it is possible to determine an exact expression for $\Gamma_{\text{cusp}}(g)$ which interpolates between the known weak and strong coupling results? As we will explain in the next sections, the answer is affirmative and it relies on the conjectured all-loop *integrability* of the planar $\mathcal{N} = 4$ SYM theory.

3. Renormalization of Wilson Operators and Integrability

As was already mentioned, the cusp anomalous dimension $\Gamma_{\text{cusp}}(g_{\text{YM}})$ controls leading asymptotic behavior of many physical observables in gauge theories and it also plays an important role in QCD. For our purposes, we will use its relation to the logarithmic scaling of anomalous dimensions of high-spin Wilson operators.[5,6]

3.1. *Wilson Operators and Anomalous Dimensions*

Let us consider the single-trace Wilson operators in $\mathcal{N} = 4$ SYM theory

$$O_{\mathbf{n}}(0) = \text{Tr}\left[D_+^{n_1}\mathcal{Z}(0)...D_+^{n_L}\mathcal{Z}(0)\right], \tag{7}$$

built from L scalar complex fields $\mathcal{Z}(x = 0)$ and $N = n_1 + ... + n_L$ lightcone covariant derivatives D_+. The operators $O_{\mathbf{n}}(0)$ carry a Lorentz spin N and a twist and a R-charge both equal to L.

Since the scalar fields in Eq. (7) are evaluated at the same point $x = 0$, the Wilson operators $O_{\mathbf{n}}(0)$ develop extra UV divergences and mix under renormalization with all other operators in $\mathcal{N} = 4$ SYM having the same quantum numbers. A distinguished property of the operators $O_{\mathbf{n}}(0)$ is that they form a closed sector under renormalization in the planar limit. Equivalently, under a change of the renormalization scale μ they only mix with themselves and satisfy the Callan-Symanzik equation

$$\mu\frac{\partial}{\partial\mu}O_{\mathbf{n}}(0) = \left(\mathbb{H}\cdot O\right)_{\mathbf{n}}(0). \tag{8}$$

Here, the mixing matrix $\mathbb{H}$ defines the action of the dilatation operator of the planar $\mathcal{N} = 4$ SYM theory on the operators $O_{\mathbf{n}}(0)$. Its diagonalization provides the spectrum of anomalous dimensions $\delta_{N,L}(g)$ of the Wilson operators with twist-L and spin-N.

The analysis of the spectrum of anomalous dimensions simplifies in the limit of large Lorentz spin, $N \to \infty$ with L kept fixed, and reveals an interesting relation with the cusp anomalous dimension.[5,6] Namely, for a given twist L, the anomalous dimensions $\delta_{N,L}(g)$ are bounded in the large$-N$ limit by two logarithmic branches,

$$2\Gamma_{\text{cusp}}(g) \ln N \leqslant \delta_{N,L}(g) \leqslant L\Gamma_{\text{cusp}}(g) \ln N, \tag{9}$$

to any order in perturbation theory. In particular, the twist-L minimal anomalous dimensions have universal (twist-independent) scaling

$$\delta_{N,L}^{\text{min}}(g) \sim 2\Gamma_{\text{cusp}}(g) \ln N, \qquad \text{when} \qquad N \gg 1. \tag{10}$$

3.2. *Integrable Spin Chain and Bethe Ansatz*

The diagonalization of the dilatation operator in Eq. (8) relies on its remarkable property of being completely integrable in the planar limit. This property holds at one loop both in planar QCD for a restricted class of operators[13] and in planar $\mathcal{N} = 4$ SYM theory[14,15] where it extends to all operators.[16]

The precise statement is that the planar one-loop dilatation operator is identical to the Hamiltonian of a closed Heisenberg $\text{XXX}_{s=-1/2}$ spin chain, with a length equal to the twist L. This Heisenberg spin chain turns out to be a completely integrable model that can be solved by means of the algebraic Bethe ansatz.[17] The Wilson operator $O_{\mathbf{n}}(0)$ with no covariant derivatives, $n_1 = \ldots = n_L = 0$, is identified with the vacuum of the spin chain, and an operator with Lorentz spin N is interpreted as a state of the spin chain with N excitations (magnons) over the vacuum. An eigenstate of the Heisenberg spin chain is characterized by a set of Bethe roots $\{u_k, k = 1, ..., N\}$ satisfying the Bethe ansatz equations

$$\left(\frac{u_k + \imath/2}{u_k - \imath/2}\right)^L = \prod_{j \neq k} \frac{u_k - u_j - \imath}{u_k - u_j + \imath}, \qquad \text{with} \qquad k, j = 1, ..., N, \tag{11}$$

wich play the role of quantization conditions. Moreover, because of the cyclicity of the trace in the definition of the Wilson operators in Eq. (7), only the states of the spin chain invariant under a cylic permutation are of physical interest. This condition is implemented as the additional constraint on the Bethe roots

$$\prod_{k=1}^{N} \frac{u_k + \imath/2}{u_k - \imath/2} = 1. \tag{12}$$

Once the solutions to Eqs. (11) and (12) are found, the algebraic Bethe ansatz leads to the following expression for the spectrum of one-loop anomalous dimensions of the Wilson operators, Eq. (7),

$$\delta_{N,L}(g) = 2g^2 \sum_{k=1}^{N} \frac{1}{u_k^2 + 1/4} + \mathcal{O}\left(g^4\right). \tag{13}$$

3.3. *Large Spin Limit*

In general, the Bethe equations, Eqs. (11) and (12), are too complicated to be solved exactly. An exception is the case of twist $L = 2$ for which an explicit solution is known.[18] In that case, there is only one distinct anomalous dimension for each Lorentz spin N given by

$$\delta_{N,L=2}(g) = 8g^2 (\psi(N+1) - \psi(1)) + \mathcal{O}\left(g^4\right), \tag{14}$$

where $\psi(x)$ is the logarithmic derivative of the Euler gamma function $\psi(x) = \Gamma'(x)/\Gamma(x)$. For large spin N we find that, in agreement with Eq. (10), $\delta_{N,L=2}(g)$ scales logarithmically

$$\delta_{N,L=2}(g) \sim 8g^2 \ln N + \mathcal{O}\left(g^4\right), \qquad N \gg 1. \tag{15}$$

Moreover, for $L = 2$, Eq. (10) states that the coefficient in front of $\ln N$ should be twice the cusp anomalous dimension. Then, an immediate consequence of Eq. (15) is that the one-loop expression for the cusp anomalous dimension is

$$\Gamma_{\text{cusp}}(g) = 4g^2 + \mathcal{O}\left(g^4\right), \tag{16}$$

in agreement with Eq. (4).

From the spin chain point of view, the logarithmic scaling follows from an accumulation of Bethe roots at the origin.[18] To analyse this effect it is convenient to introduce a distribution density of Bethe roots, defined by

$$\rho_0(u) = \frac{1}{N} \sum_{k=1}^{N} \delta\left(u - u_k\right). \tag{17}$$

In the large-N limit (large number of excitations), the distribution density associated to the minimal anomalous dimensions admits a continuous interpolation which is explicitely known[18]

$$\rho_0(u) = \frac{1}{N\pi} \ln \frac{1+\sqrt{1-4u^2/N^2}}{1-\sqrt{1-4u^2/N^2}}, \qquad \text{with} \qquad u \in [-N/2, N/2]. \quad (18)$$

This distribution is manifestly independent on the twist L and it is singular at the origin, $\rho_0(u) \sim \ln u$. It is this singular behavior of the distribution density of Bethe roots that is responsible for the logarithmic scaling of the anomalous dimensions

$$\delta_{N,L}(g) = 2g^2 N \int_{-N/2}^{N/2} du \, \frac{\rho_0(u)}{u^2+1/4} + \mathcal{O}\left(g^4\right) \sim 8g^2 \ln N + \mathcal{O}\left(g^4\right). \quad (19)$$

4. All-Loop Integrability and the BES Equation

The dilatation operator of planar $\mathcal{N} = 4$ SYM theory is believed to be integrable to all loops. This expectation is supported by the AdS/CFT correspondence since the dual string theory turns out to be integrable at the classical level.[19] The investigation of high-loop integrability in the gauge theory[20] and the comparison with the string integrable structures[21–23] have led to an all-loop proposal.

4.1. *All-loop asymptotic Bethe ansatz*

The conjectured Bethe ansatz equations generalizing Eq. (11) to all loops are[24]

$$\left(\frac{x_k^+}{x_k^-}\right)^L = \prod_{j\neq k}^{N} \frac{x_k^- - x_j^+}{x_k^+ - x_j^-} \frac{1-g^2/x_k^+ x_j^-}{1-g^2/x_k^- x_j^+} \exp\left(2\imath\theta(u_k, u_j)\right), \quad (20)$$

where $x_k^\pm = x^\pm(u_k)$ are defined by $u \pm \imath/2 = x^\pm(u) + g^2/x^\pm(u)$. They involve a nontrivial dressing phase $\theta(u_k, u_j)$ whose explicit expression, constrained by the crossing symmetry,[25] was proposed in Refs. 22,26,27. Equation (20) is supplemented by the cyclicity condition for physical states

$$\prod_{k=1}^{N} \frac{x^+(u_k)}{x^-(u_k)} = 1, \quad (21)$$

and the all-loop anomalous dimension is given by

$$\delta_{N,L}(g) = 2g^2 \sum_{k=1}^{N} \left(\frac{\imath}{x^+(u_k)} - \frac{\imath}{x^-(u_k)}\right). \quad (22)$$

These equations are only asymptotic—they describe the spectrum of anomalous dimensions corresponding to the Wilson operators in Eq. (7) carrying an infinitely large twist L. When applied illegally at finite twist, they are expected to reproduce the gauge theory answer at least up to $\mathcal{O}\left(g^{2L-2}\right)$. Nevertheless, the cusp anomalous dimension being extracted from the twist-independent logarithmic scaling of the minimal anomalous dimensions, its determination from Eqs. (20), (21), (22) is achieved to all loops.

4.2. *Large Spin Limit*

Similarly to the one-loop analysis, the minimal anomalous dimensions in the large-spin limit can be described by a continuous distribution density of roots $\rho(u) = \rho(-u)$ that admits the following decomposition[27,28]

$$\rho(u) = \rho_0(u) - 8g^2 \frac{\ln N}{N} \sigma(u) + o\left(\frac{\ln N}{N}\right). \tag{23}$$

Here $\rho_0(u)$ is the singular leading-order distribution density, Eq. (18), whereas $\sigma(u)$ is a regular contribution, named density of fluctuations, that absorbs the higher-loop corrections.[27,28] The latter is independent on N and is defined on the real line in the strict large-spin limit, $N \to \infty$. It is also independent on the twist L enforcing the universality of the logarithmic scaling. Plugging Eq. (23) into Eq. (22) leads to the desired large-spin logaritmic scaling of the minimal anomalous dimensions

$$\delta_{N,L}(g) = 2g^2 N \int_{-N/2}^{N/2} \left(\frac{\imath}{x^+(u)} - \frac{\imath}{x^-(u)}\right) \rho(u) \sim 2\Gamma_{\text{cusp}}(g) \ln N, \tag{24}$$

and expresses $\Gamma_{\text{cusp}}(g)$ in terms of the density of fluctuations $\sigma(u)$.[27,28]

4.3. *BES Equation*

The density of fluctuations $\sigma(u)$ is completely fixed by the asymptotic Bethe ansatz and can be defined as the solution to an integral equation that encodes its all-loop dependence. This equation was proposed by Beisert–Eden–Staudacher (BES) in Ref. 27. It generalizes an ealier attempt done in Ref. 28 by incorporating the contribution of the dressing phase, whose effects are crucial to correctly reproduce both the gauge theory four-loop result[8,27] and the string theory strong coupling expansion.[29–32]

In terms of the 'Fourier transform' of the density of fluctuations

$$\hat{\sigma}(t) = e^{-\frac{|t|}{2}} \int_{-\infty}^{\infty} du\, e^{\imath t u} \sigma(u), \tag{25}$$

the BES equation reads

$$\hat{\sigma}(t) = \frac{t}{e^t - 1}\left[K(2gt, 0) - 4g^2 \int_0^\infty dt'\, K(2gt, 2gt')\hat{\sigma}(t')\right]. \quad (26)$$

It involves a complicated kernel, $K(t,t')$, wich receives correction from the dressing phase. Its explicit form can be found in Refs. 27,28. The all-loop cusp anomalous dimension is obtained from the solution to the BES equation through the relation $\Gamma_{\text{cusp}}(g) = 8g^2\hat{\sigma}(0)$.

4.3.1. *Solution at Weak Coupling*

The solution to the BES equation at weak coupling is obtained by iteration of Eq. (26)

$$\hat{\sigma}(t) = \frac{t}{e^t - 1}\Bigg[K(2gt, 0) \\ - 4g^2 \int_0^\infty dt'\, K(2gt, 2gt')\frac{t'}{e^{t'} - 1}K(2gt', 0) + ...\Bigg]. \quad (27)$$

It leads to the following weak coupling expansion for the cusp anomalous dimension[27]

$$\Gamma_{\text{cusp}}(g) = 4g^2 - \frac{4}{3}\pi^2 g^2 + \frac{44}{45}\pi^4 g^6 - 8\left(\frac{73}{630}\pi^6 + 4\zeta_3^2\right)g^8 \\ +32\left(\frac{887}{14175}\pi^8 + \frac{4}{3}\pi^2\zeta_3^2 + 40\zeta_3\zeta_5\right)g^{10} + \mathcal{O}\left(g^{12}\right), \quad (28)$$

in agreement with the four-loop gauge theory results, Eqs. (4) and (5). Equation (28) fulfills the maximal transcendentality principle to all loops[30] and its first few coefficients indicate that it has a finite radius of convergence.[27]

4.3.2. *Solution at Strong Coupling*

At strong coupling, one cannot rely on the iteration procedure, Eq. (27), to solve the BES equation and the construction of the solution is more involved.[29–31,33,34] At intermediate coupling, the BES equation was investigated numerically in Ref 29. The obtained solution for $\Gamma_{\text{cusp}}(g)$ was found to be a smooth function of g that matches for $g > 1$ the string theory prediction, Eq. (6), with high accuracy. This agreement was precised by the exact solution to the BES equation, constructed in the strong coupling limit in Ref. 30,31 and to an arbitrary order in a $1/g$-expansion in Ref. 33,34.

5. Cusp Anomalous Dimension at Strong Coupling

The recursive procedure proposed in Refs. 33,34 solves the BES equation for $g \gg 1$ and determines the strong coupling expansion of the cusp anomalous dimension order by order in $1/g$. The first few terms of this expansion are[33]

$$\Gamma_{\text{cusp}}(g+c_1) = 2g\Big[1 - c_2\, g^{-2} - c_3\, g^{-3} - \left(c_4 + 2c_2^2\right) g^{-4} \\ - \left(c_5 + 23 c_2 c_3\right) g^{-5} + \mathcal{O}\left(g^{-6}\right)\Big], \qquad (29)$$

with coefficients given by

$$c_1 = \frac{3\ln 2}{4\pi}, \qquad c_2 = \frac{\mathrm{K}}{16\pi^2}, \qquad c_3 = \frac{27}{2^{11}\pi^3}\zeta(3), \\ c_4 = \frac{21}{2^{10}\pi^4}\beta(4), \qquad c_5 = \frac{43065}{2^{21}\pi^5}\zeta(5), \qquad c_6 = \frac{1605}{2^{15}\pi^6}\beta(6). \qquad (30)$$

Here $\zeta(x)$ denotes the Riemann zeta function, $\beta(x) = \sum_{n\geqslant 0}(-1)^n(2n+1)^{-x}$ the Dirichlet beta function and $\mathrm{K} = \beta(2)$ the Catalan's constant. To simplify the expression of the result, we used in Eq. (29) the empirical observation that the c_1-dependent terms appearing in the $1/g$-expansion of $\Gamma_{\text{cusp}}(g)$ can be eliminated by a shift of the coupling constant $g \to g + c_1$.

A distinguished feature of the expansion in Eq. (29) is that the coefficients in front of $1/(\pi g)^n$ are given by a linear combination of the product of $\zeta(2p+1)$ and $\beta(2r)$ such that the sum of their arguments equals n. This property is reminiscent of the maximal transcendentality principle that shapes the weak coupling expansion of $\Gamma_{\text{cusp}}(g)$. However, in contrast to the *convergency* of its weak coupling expansion, the strong coupling expansion of $\Gamma_{\text{cusp}}(g)$ was found to be merely *asymptotic* and not Borel-summable.[33]

The result for $c_2 = \mathrm{K}/16\pi^2$ coincides exactly with the recent two-loop stringy correction computed in Ref. 35, providing a striking confirmation of the AdS/CFT correspondence. Another contact with the string theory follows from the non Borel-summability of the series in Eq. (29) that indicates that $\Gamma_{\text{cusp}}(g)$ receives a nonpertubative correction at strong coupling. This contribution was estimated from the large orders behavior of the $1/g$-expansion of $\Gamma_{\text{cusp}}(g)$ to be $\sim g^{1/2}e^{-2\pi g}$,[33] in a perfect agreement with the findings of Ref. 36 where this nonperturbative correction was identified in the string theory.[a]

[a]In Ref. 36, the scale $m \sim g^{1/4}e^{-\pi g}$ is the mass gap of the O(6) sigma model that is embedded into the world-sheet action of the string theory with a bare coupling constant $e_0^2 \sim 1/g$. The cusp anomalous dimension has the meaning of a density of energy and is expected to receive a nonperturbative correction $\sim m^2$.

6. Conclusion

The cusp anomalous dimension is an ubiquitous observable in any gauge theory and its dependence on the coupling constant is of physical interest. In the planar limit, the conjectured all-loop integrability of the $\mathcal{N} = 4$ super Yang-Mills theory allows to predict the cusp anomalous dimension, at any value of the 't Hooft coupling, as a solution to an integral equation. This prediction extrapolates the known four-loop result of the gauge theory to all loops and implements a resummation of the pertubative expansion. Moreover, when expanded at strong coupling, the solution precisely matches the string theory answer up to two loops. Altogether this gauge/string interpolation provides a remarkable confirmation of the AdS/CFT correspondence.

Acknowledgments

I would like to thank the organizers of CAQCD 2008 for hospitality. I am indebted to Gregory Korchemsky for valuable discussions. This work was supported in part by the French Agence Nationale de la Recherche under grant ANR-06-BLAN-0142-02.

References

1. K. G. Wilson, *Phys. Rev. D* **10**, 2445 (1974).
2. A. M. Polyakov, *Nucl. Phys. B* **164**, 171 (1980).
3. A. M. Polyakov, *Nucl. Phys. B* **486**, 23 (1997); A. M. Polyakov, *Int. J. Mod. Phys. A* **14**, 645 (1999).
4. G. P. Korchemsky and A. V. Radyushkin, *Nucl. Phys. B* **283**, 342 (1987).
5. G. P. Korchemsky, *Mod. Phys. Lett. A* **4**, 1257 (1989); G. P. Korchemsky and G. Marchesini, *Nucl. Phys. B* **406**, 225 (1993).
6. A. V. Belitsky, A. S. Gorsky and G. P. Korchemsky, *Nucl. Phys. B* **667**, 3 (2003); A. V. Belitsky, A. S. Gorsky and G. P. Korchemsky, *Nucl. Phys. B* **748**, 24 (2006).
7. A. V. Kotikov, L. N. Lipatov, A. I. Onishchenko and V. N. Velizhanin, *Phys. Lett. B* **595**, 521 (2004); [Erratum-*ibid. B* **632**, 754 (2006)]; Z. Bern, L. J. Dixon and V. A. Smirnov, *Phys. Rev. D* **72**, 085001 (2005).
8. Z. Bern, M. Czakon, L. J. Dixon, D. A. Kosower and V. A. Smirnov, *Phys. Rev. D* **75**, 085010 (2007); F. Cachazo, M. Spradlin and A. Volovich, *Phys. Rev. D* **75**, 105011 (2007).
9. A. V. Kotikov and L. N. Lipatov, *Nucl. Phys. B* **661**, 19 (2003); [Erratum-*ibid. B* **685**, 405 (2004)]; A. V. Kotikov, L. N. Lipatov and V. N. Velizhanin, *Phys. Lett. B* **557**, 114 (2003).
10. J. M. Maldacena, *Adv. Theor. Math. Phys.* **2**, 231 (1998); S. S. Gubser, I. R. Klebanov and A. M. Polyakov, *Phys. Lett. B* **428**, 105 (1998); E. Witten, *Adv. Theor. Math. Phys.* **2**, 253 (1998).

11. S. S. Gubser, I. R. Klebanov and A. M. Polyakov, *Nucl. Phys. B* **636**, 99 (2002).
12. S. Frolov and A. A. Tseytlin, *JHEP* **0206**, 007 (2002).
13. V. M. Braun, S. E. Derkachov and A. N. Manashov, *Phys. Rev. Lett.* **81**, 2020 (1998); V. M. Braun, S. E. Derkachov, G. P. Korchemsky and A. N. Manashov, *Nucl. Phys. B* **553**, 355 (1999); A. V. Belitsky, *Phys. Lett. B* **453**, 59 (1999); A. V. Belitsky, *Nucl. Phys. B* **574**, 407 (2000).
14. L. N. Lipatov, *Prepared for ICTP Conference on Perspectives in Hadronic Physics, Trieste, Italy, 12-16 May 1997.*
15. J. A. Minahan and K. Zarembo, *JHEP* **0303**, 013 (2003).
16. N. Beisert and M. Staudacher, *Nucl. Phys. B* **670**, 439 (2003); N. Beisert, *Nucl. Phys. B* **676**, 3 (2004).
17. L. D. Faddeev, arXiv:hep-th/9605187.
18. G. P. Korchemsky, *Nucl. Phys. B* **462**, 333 (1996).
19. I. Bena, J. Polchinski and R. Roiban, *Phys. Rev. D* **69**, 046002 (2004).
20. N. Beisert, C. Kristjansen and M. Staudacher, *Nucl. Phys. B* **664**, 131 (2003); D. Serban and M. Staudacher, *JHEP* **0406**, 001 (2004); N. Beisert, V. Dippel and M. Staudacher, *JHEP* **0407**, 075 (2004).
21. V. A. Kazakov, A. Marshakov, J. A. Minahan and K. Zarembo, *JHEP* **0405**, 024 (2004); N. Beisert, V. A. Kazakov, K. Sakai and K. Zarembo, *Commun. Math. Phys.* **263**, 659 (2006).
22. G. Arutyunov, S. Frolov and M. Staudacher, *JHEP* **0410**, 016 (2004).
23. M. Staudacher, *JHEP* **0505**, 054 (2005).
24. N. Beisert and M. Staudacher, *Nucl. Phys. B* **727**, 1 (2005); N. Beisert, arXiv:hep-th/0511082.
25. R. A. Janik, *Phys. Rev. D* **73**, 086006 (2006); G. Arutyunov and S. Frolov, *Phys. Lett. B* **639**, 378 (2006).
26. N. Beisert and A. A. Tseytlin, *Phys. Lett. B* **629**, 102 (2005); R. Hernandez and E. Lopez, *JHEP* **0607**, 004 (2006); N. Beisert, R. Hernandez and E. Lopez, *JHEP* **0611**, 070 (2006).
27. N. Beisert, B. Eden and M. Staudacher, *J. Stat. Mech.* **0701**, P021 (2007).
28. B. Eden and M. Staudacher, *J. Stat. Mech.* **0611**, P014 (2006).
29. M. K. Benna, S. Benvenuti, I. R. Klebanov and A. Scardicchio, *Phys. Rev. Lett.* **98**, 131603 (2007).
30. A. V. Kotikov and L. N. Lipatov, *Nucl. Phys. B* **769**, 217 (2007).
31. L. F. Alday, G. Arutyunov, M. K. Benna, B. Eden and I. R. Klebanov, *JHEP* **0704**, 082 (2007); I. Kostov, D. Serban and D. Volin, *Nucl. Phys. B* **789**, 413 (2008); M. Beccaria, G. F. De Angelis and V. Forini, *JHEP* **0704**, 066 (2007).
32. P. Y. Casteill and C. Kristjansen, *Nucl. Phys. B* **785**, 1 (2007); A. V. Belitsky, *Phys. Lett. B* **659**, 732 (2008); N. Gromov, arXiv:0805.4615 [hep-th].
33. B. Basso, G. P. Korchemsky and J. Kotanski, *Phys. Rev. Lett.* **100**, 091601 (2008).
34. I. Kostov, D. Serban and D. Volin, arXiv:0801.2542 [hep-th].
35. R. Roiban and A. A. Tseytlin, *JHEP* **0711**, 016 (2007).
36. L. F. Alday and J. M. Maldacena, *JHEP* **0711**, 019 (2007).

$SO(2M)$ AND $USp(2M)$ (HYPER)KÄHLER QUOTIENTS AND LUMPS

SVEN BJARKE GUDNASON
Department of Physics, University of Pisa,
INFN, Sezione di Pisa,
Largo Pontecorvo, 3, Ed. C, 56127 Pisa, Italy
E-mail: gudnason@df.unipi.it

The properties of supersymmetric gauge theories in the Higgs phase at low energies can appropriately be studied by means of a non-linear σ model, which has the target space being Kähler for $\mathcal{N} = 1$ supersymmetric models and hyperKähler for $\mathcal{N} = 2$ models. By construction of the Kähler and hyperKähler quotients for the gauge theories with SO and USp gauge symmetries, we obtain the explicit metrics on their respective manifolds. Furthermore, we study the lumps in the non-linear σ models and their effective description, using the Kähler quotients.

Keywords: Low energy effective theories, non-linear σ model lumps, (hyper)Kähler quotients

1. Introduction

The target space of the $\mathcal{N} = 2$ supersymmetric non-linear sigma model (NLσM), with eight supercharges, must be hyperKähler.[2] By using this fact, the notion of the hyperKähler quotient was first found in physics[3,4] and was later formulated mathematically.[5] A $U(1)$ hyperKähler quotient[3] recovers the Calabi metric on the cotangent bundle over the projective space, $T^\star\mathbb{C}P^{N-1}$, while its $U(N)$ generalization leads to the cotangent bundle over the complex Grassmann manifold, $T^\star G_{M,N}$.[4] The hyperKähler quotient offers a powerful tool to construct hyperKähler manifolds such as instanton moduli spaces,[6] gravitational instantons[7] and monopole moduli spaces.[8]

The Higgs branch of $\mathcal{N} = 2$ supersymmetric QCD is hyperKähler. The low energy effective theory on the Higgs branch is described by an $\mathcal{N} = 2$ NLσM on the hyperKähler manifold.[9–11] In the cases of an $SU(N)$ or a $U(N)$ gauge theory with hypermultiplets charged commonly under $U(1)$, the explicit metrics on the Higgs branch and their Kähler potentials are known explicitly. The latter is nothing but the Lindström-Roček metric.[4,12] A $U(1) \times U(1)$ gauge theory for instance gives the space: $T^{\star}F_n$ with F_n being the Hirzebruch surface.[13]

This contribution has two main concerns, the first is the construction of the metric and Kähler potential on the Higgs branch of $\mathcal{N} = 2$ supersymmetric gauge theories with $SO(2N)$, $USp(2N)$, $U(1) \times SO(2N)$, $U(1) \times USp(2N)$ gauge groups.[1]

The second motive is concerned with σ model lumps, or σ model instantons. A lump solution was first found in the $O(3)$ sigma model, or the $\mathbb{C}P^1$ model.[14] It was then generalized to the $\mathbb{C}P^n$ model and the Grassmann model. Gauge theories coupled to several Higgs fields often admit semi-local vortex-strings.[15] In the strong gauge coupling limit, the gauge theories reduce to NLσMs and in this limit, semi-local strings reduce to lump-strings. In the gauge theories at finite couplings, the large distance behavior of semi-local strings is well approximated by lump solutions. It was demonstrated in Ref. 16 that non-Abelian semi-local strings[17,18] in a $U(N)$ gauge theory reduce to the Grassmann lumps at large distance.

This work has been done in collaboration with M. Eto, T. Fujimori, K. Konishi, T. Nagashima, M. Nitta, K. Ohashi and W. Vinci. Many details are left in the Ref. 1.

2. Obtaining the Low Energy Effective Theory

Obtaining the low-energy effective theory of supersymmetric gauge theories on the Higgs branch has been well studied in the literature.

To obtain the target space, we can do one of the following[19]

(i) Fix the gauge to the Wess-Zumino gauge and find the potential zeroes ($D = 0$, $F = 0$) and then mod out the remaining gauge group.
(ii) Take the infinite gauge coupling limit immediately and then mod out the full complexified gauge group.
(iii) Construct all gauge invariants and find all relations between them. This set constitutes the target space.

In the next Section we will construct the metrics following the method (ii) and rewrite them according to method (iii) for the gauge theories with $SO(N_{\rm c})$ and $USp(2M_{\rm c})$ gauge groups. Similarly, the result can be transformed onto the form of method (i).

3. The $SO(N_c)$ and $USp(2M_c)$ Kähler Quotients

The Kähler potential for an $SO(N_c)$ or a $USp(2M_c)$ gauge theory is given by

$$K_{SO,USp} = \mathrm{Tr}\left[QQ^\dagger e^{-V'}\right] , \tag{1}$$

where V' takes a value in the $\mathfrak{so}(N_c)$ or $\mathfrak{usp}(2M_c)$ algebra and hence satisfies

$$V'^{\mathrm{T}}J + JV' = 0 \quad \leftrightarrow \quad e^{-V'^{\mathrm{T}}}Je^{-V'} = J . \tag{2}$$

Note that, this condition implies that $\det(e^{-V'}) = 1$.[a] Here the matrix J is the invariant tensor of the SO or USp group, $g^{\mathrm{T}}Jg = J$ with $g \in SO(N_c), USp(2M_c)$, satisfying

$$J^{\mathrm{T}} = \epsilon J , \quad J^\dagger J = \mathbf{1}_{N_c} , \qquad \epsilon = \begin{cases} +1 \text{ for } & SO(N_c) , \\ -1 \text{ for } & USp(N_c = 2M_c) . \end{cases}$$

Conversely, a matrix J satisfying the above equations defines a representation of the SO and USp groups. We will (mainly) use the convention (in the even case)

$$J \equiv \begin{pmatrix} \mathbf{0}_{M_c} & \mathbf{1}_{M_c} \\ \epsilon\mathbf{1}_{M_c} & \mathbf{0}_{M_c} \end{pmatrix} .$$

First we will discuss the breaking pattern of the gauge symmetry and the flat directions of the vacuum. For this we will consider both the gauge and the global symmetries. The vacuum expectation value of Q^{SO}_{wz} in the case of $SO(N_c)$ can be put on diagonal form after fixing both the local and the global symmetry[21]

$$\begin{aligned} Q^{SO}_{\mathrm{wz}} &= \left(A_{N_c \times N_c}, \mathbf{0}_{N_c \times (N_f - N_c)}\right) , \\ \text{with} \quad A_{N_c \times N_c} &= \mathrm{diag}(a_1, a_2, \cdots, a_{N_c}) , \quad J = \mathbf{1}_{N_c} , \end{aligned}$$

where all the parameters a_i are taken to be real and positive, which indeed parametrize flat directions of the Higgs branch. In generic points on the vacuum manifold with non-degenerate a_i, there is no global symmetry in the vacuum and the flavor symmetry is $U(N_f)$ apart from $U(N_f - N_c)$ which freely acts on the vacuum configuration. At a generic point, the vacuum manifold can be written as

$$\mathcal{M}^{\mathrm{generic,K}}_{SO(N_c)} \simeq \mathbb{R}^{N_c}_{\geq 0} \times \frac{U(N_f)}{U(N_f - N_c)} . \tag{3}$$

[a] In the $SO(N_c)$ cases, we remove an integral region with $\det e^{-V} = -1$ in the functional integral of V.

The flat directions $\mathbb{R}^{N_c}_{\geq 0}$ have no origin due to symmetry breaking, and their coordinates $\{a_i\}$ are quasi Nambu-Goldstone (NG) modes.[20] When two parameters coincide, $a_i = a_j, (i \neq j)$, a color-flavor locking symmetry $SO(2)$ emerges. In such degenerate points on the manifold, the above quotient space attached to $\mathbb{R}^{N_c}_{\geq 0}$ shrinks to a space with a smaller dimension. The existence of quasi-NG modes are strongly related to the emergence of an unbroken phase (classically). When any two of the a_i's vanish, an $SO(2)$ subgroup of the gauge symmetry remains unbroken. Thus, in the Higgs phase with completely broken gauge symmetry, the rank of $Q_{\rm wz}$ has to be greater than $N_c - 2$. We will only consider this latter case here, namely the models with $N_f \geq N_c - 1$.

For the $USp(2M_c)$ case,[9,22] it is known that the flat directions are parametrized by

$$Q_{\rm wz}^{USp} = \left(A_{M_c \times M_c}, \mathbf{0}_{M_c \times (M_f - M_c)}\right) \otimes \mathbf{1}_2 \ ,$$

where the number of flavors is even $N_f = 2M_f$. Even in generic points with non-degenerate $\{a_i\}$, the color-flavor symmetry $USp(2)^{M_c} \simeq SU(2)^{M_c}$ remains unbroken in the vacuum. Therefore, the vacuum manifold can, at generic points, be written as

$$\mathcal{M}^{\rm generic,K}_{USp(2M_c)} \simeq \mathbb{R}^{M_c}_{\geq 0} \times \frac{U(N_f)}{SU(2)^{M_c} \times U(N_f - 2M_c)} \ , \tag{4}$$

except for sub-manifolds where the quotient space shrinks. In this case the completely broken gauge symmetry needs $M_f \geq M_c$.

The D-flatness conditions in the Wess-Zumino gauge (i.e. method (i)) are

$$D^A = {\rm Tr_f}\left(Q_{\rm wz}^\dagger T^A Q_{\rm wz}\right) = 0 \ ,$$

with T_A being the generators in the Lie algebra $\mathfrak{so}$ or $\mathfrak{usp}$. However, these conditions are rather difficult to solve.[b] Without taking the Wess-Zumino gauge, we can eliminate the superfield V' directly within the superfield formalism by using the following trick; let us consider V' taking a value in a larger algebra, namely $\mathfrak{u}(N_c)$ and then introduce an $N_c \times N_c$ matrix of Lagrange multipliers[c] λ to restrict V' to take a value in the $\mathfrak{so}(N_c)$ or the

[b]To our knowledge the D-flatness conditions have not been solved in the case of an SO or a USp, $\mathcal{N} = 1$ supersymmetric gauge theory.

[c]Hermiticity of λ is defined such that $\lambda e^{-V'^{\rm T}} J$ is a vector superfield, that is, $\lambda^\dagger = e^{V'^{\rm T}} J \lambda e^{-V'^{\rm T}} J$.

$\mathfrak{usp}(N_c = 2M_c)$ sub-algebra which leaves us with

$$K_{SO,USp} = \mathrm{Tr}\left[QQ^\dagger e^{-V'} + \lambda\left(e^{-V'^{\mathrm{T}}} J e^{-V'} - J\right)\right] , \tag{5}$$

where Q are N_f chiral superfields as earlier and V' is a vector superfield of $U(N_c)$. The added term breaks the complexified gauge transformations down to $SO(N_c), USp(2M_c)$ and the equations of motion for λ gives the constraint (2) which reduces the Kähler potential (5) back to Eq. (1). The equation of motion for V' takes the form

$$QQ^\dagger e^{-V'} + (\lambda + \epsilon\lambda^{\mathrm{T}}) J = 0 ,$$

where we have used (2). λ can be eliminated by combining the equations of motion with their transposed ones. The resultant equation contains the square of the manifest Hermitian and positive definite matrix X that traces to the Kähler potential (1) that we set out to find:

$$X^2 = \left(Q^{\mathrm{T}} J \sqrt{QQ^\dagger}\right)^\dagger \left(Q^{\mathrm{T}} J \sqrt{QQ^\dagger}\right) , \quad X \equiv \sqrt{QQ^\dagger} e^{-V'} \sqrt{QQ^\dagger} .$$

We can uniquely obtain V' from X if and only if $\mathrm{rank}\, M > N_c - 2$, where M are the holomorphic invariants for the SO, USp gauge theories, that is, the vacuum is in the full Higgs phase. See Appendix B in Ref. 1 for a uniqueness proof, in the case of $\mathrm{rank}\, M = N_c - 1$. It is possible to switch to Q_{wz} from Q by the complexified gauge transformation $Q_{\mathrm{wz}} = u'^{-1} Q$ with $u'u'^\dagger = e^{V'}$. Without using an explicit solution for V', we obtain the Kähler potential of the NLσM (according to method (ii))

$$K_{SO,USp} = \mathrm{Tr}\, X = \mathrm{Tr}\sqrt{\left(Q^{\mathrm{T}} J \sqrt{QQ^\dagger}\right)^\dagger \left(Q^{\mathrm{T}} J \sqrt{QQ^\dagger}\right)} . \tag{6}$$

Now we can naturally switch to an expression according to method (iii) for this NLσM. With the help of $\mathrm{Tr}_c \sqrt{AA^\dagger} = \mathrm{Tr}_f \sqrt{A^\dagger A}$, one can rewrite the Kähler potential (6) as follows

$$K_{SO,USp} = \mathrm{Tr}_f \sqrt{MM^\dagger} , \qquad M^{\mathrm{T}} = \epsilon M , \tag{7}$$

where M is nothing but the holomorphic invariants of the gauge symmetry

$$M \equiv Q^{\mathrm{T}} J Q , \qquad B^{\langle A \rangle} \equiv \det Q^{\langle A \rangle} .$$

The first one is the "mesonic" invariant while the second is the "baryonic" one which appears for $N_f \geq N_c$ and $\langle A \rangle \equiv \langle A_1, \ldots, A_{N_c} \rangle$ and $Q^{\langle A \rangle}$ is the N_c-by-N_c minor matrix, i.e. $\left(Q^{\langle A \rangle}\right)_{i,j} \equiv Q_{i,A_j}$ with $A_j \in [1, N_f]$. The two kinds of invariants should be subject to constraints in order to correctly express the NLσM. There are relations between the mesons and the baryons:

$$\begin{aligned} SO(N_c) : \det(J)\, B^{\langle A\rangle} B^{\langle B\rangle} &= \det M^{\langle A\rangle\langle B\rangle} \\ USp(2M_c) : \qquad \mathrm{Pf}(J)\, B^{\langle A\rangle} &= \mathrm{Pf}\, M^{\langle A\rangle\langle A\rangle}. \end{aligned} \tag{8}$$

The Plücker relation among the baryonic invariants $B^{\langle A\rangle}$ is derived from the above relation. Actually, from the invariants M and $B^{\langle A\rangle}$ with the constraints we can reconstruct Q modulo the complexified gauge symmetry as follows. By using an algorithm similar to the Cholesky decomposition of an Hermitian matrix, we show in Ref. 1 that

> An arbitrary $n \times n$ (anti-)symmetric complex matrix X can always be decomposed as $X = p^{\mathrm{T}} J p$ with p being a rank$(X) \times n$ matrix. (9)

In the USp case, with a decomposition of the meson M, we can completely reconstruct Q modulo $USp(2M_c)^{\mathbb{C}}$ transformations. This fact corresponds to that there are no independent baryons $B^{\langle A\rangle}$ in a $USp(2M_c)$ theory and only the meson fields describe the full Higgs phase

$$\mathcal{M}_{USp} = \left\{ M \mid M \in N_f \times N_f \text{ matrix}, \quad M^{\mathrm{T}} = -M, \quad \operatorname{rank} M = 2M_c \right\} .$$

On the contrary, in the $SO(N_c)$ case, a decomposition of M gives Q modulo $O(N_c)^{\mathbb{C}}$ and one finds two candidates for Q since $\mathbb{Z}_2 \simeq O^{\mathbb{C}}/SO^{\mathbb{C}}$ which is fixed by the sign of the baryons.[d] Therefore we have to take the degrees of freedom of the baryons into account to consider the full Higgs phase

$$\begin{aligned} \mathcal{M}_{SO} = \Big\{ M, B^{\langle A\rangle} \mid M &: \text{symmetric } N_f \times N_f, \\ &\text{Relation (8)}, \quad N_c - 1 \leq \operatorname{rank} M \leq N_c \Big\} . \end{aligned}$$

For large N_c, it is a hard task to obtain an explicit metric from the formula (7), since we need to calculate the eigenvalues of $MM^\dagger$. In Ref. 1 we calculate explicitly an expansion of the Kähler potential around its vacuum value from which we are able to obtain the metric and curvature.

4. The $U(1) \times SO(N_c)$ and $U(1) \times USp(2M_c)$ Kähler Quotients

Next, we would like to consider a Kähler quotient by gauging an overall $U(1)$ phase in addition to the $SO(N_c)$ or $USp(2M_c)$ gauge symmetry. We

[d] In the case of rank $M = N_c - 1$, $g \in \mathbb{Z}_2$ acts trivially on Q as $g\,Q = Q$, although all the baryons vanish.

turn on the FI D-term associated with the additional $U(1)$ gauge group. The Kähler potential can be written as

$$K_{U(1)\times(SO,USp)} = \mathrm{Tr}\left[QQ^\dagger e^{-V'}e^{-V_e} + \lambda\left(e^{-V'^{\mathrm{T}}}Je^{-V'} - J\right)\right] + \xi V_e\ ,$$

where V_e is the vector multiplet of the additional $U(1)$ gauge field. We have already solved the $SO(N_\mathrm{c})$ and $USp(2M_\mathrm{c})$ part in the previous section, so the Kähler potential can be rewritten as

$$K_{U(1)\times(SO,USp)} = \mathrm{Tr}\left[\sqrt{MM^\dagger}\right]e^{-V_e} + \xi V_e\ .$$

The equation of motion for V_e can be solved by $V_e = \log\left[\mathrm{Tr}\left(\sqrt{MM^\dagger}\right)/\xi\right]$. Plugging this into the Kähler potential, we obtain

$$K_{U(1)\times(SO,USp)} = \xi\log\left[\mathrm{Tr}\left(\sqrt{MM^\dagger}\right)\right]\ , \qquad M \equiv Q^{\mathrm{T}}JQ\ . \tag{10}$$

5. The $SO(N_\mathrm{c})$, $USp(2M_\mathrm{c})$ HyperKähler Quotients

Our next task is lifting up the $SO(N_\mathrm{c})$ and $USp(N_\mathrm{c} = 2M_\mathrm{c})$ Kähler quotients to hyperKähler quotients. In order to construct the $SO(N_\mathrm{c}), USp(2M_\mathrm{c})$ hyperKähler quotients we need to consider $\mathcal{N}=2$ hypermultiplets. Hence, we consider an $\mathcal{N}=2$ extension of the $\mathcal{N}=1$ Kähler potential (5), together with the superpotential

$$\tilde{K}_{SO,USp} = \mathrm{Tr}\left[QQ^\dagger e^{-V'} + \tilde{Q}^\dagger\tilde{Q}e^{V'} + \lambda\left(e^{-V'^{\mathrm{T}}}Je^{-V'} - J\right)\right]\ , \tag{11}$$

$$W = \mathrm{Tr}\left[Q\tilde{Q}\Sigma' + \chi\left(\Sigma'^{\mathrm{T}}J + J\Sigma'\right)\right]\ , \tag{12}$$

where (V', Σ') denote the $SO(N_\mathrm{c})$ and $USp(2M_\mathrm{c})$ vector multiplets, $(Q, \tilde{Q})$ are N_f hypermultiplets in the fundamental representation of $SO(N_\mathrm{c})$ or $USp(2M_\mathrm{c})$, and (λ, χ) are the Lagrange multipliers which are $N_\mathrm{c}\times N_\mathrm{c}$ matrix valued superfields. We can rewrite the Kähler potential (11) as follows

$$\tilde{K}_{SO,USp} = \mathrm{Tr}\left[\mathcal{Q}\mathcal{Q}^\dagger e^{-V'}\right]\ , \qquad \text{with} \qquad \mathcal{Q} \equiv \left(Q,\ J\tilde{Q}^{\mathrm{T}}\right)\ ,$$

where we have used that $e^{V'^{\mathrm{T}}} = J^{\mathrm{T}}e^{-V'}J$. This trick relates the superfields in the anti-fundamental representation with those of the fundamental representation via the algebra. This Kähler potential is nothing but the $\mathcal{N}=1$ Kähler potential of $SO(N_\mathrm{c})$ and $USp(2M_\mathrm{c})$ with $\mathcal{Q}$ being a set of $2N_\mathrm{f}$ chiral superfields. We can straightforwardly borrow the result of Sec. 3 and hence the Kähler potential reads

$$\tilde{K}_{SO,USp} = \mathrm{Tr}\left[\sqrt{\mathcal{M}\mathcal{M}^\dagger}\right]\ , \qquad \mathcal{M} \equiv \mathcal{Q}^{\mathrm{T}}J\mathcal{Q}\ . \tag{13}$$

The constraint coming from the superpotential (12) is

$$\mathcal{Q}\tilde{J}\mathcal{Q}^{\mathrm{T}} = 0 \ , \quad \text{with } \tilde{J} \equiv \begin{pmatrix} \mathbf{0}_{N_\mathrm{f}} & \mathbf{1}_{N_\mathrm{f}} \\ -\epsilon\mathbf{1}_{N_\mathrm{f}} & \mathbf{0}_{N_\mathrm{f}} \end{pmatrix} .$$

Therefore, we again find the constraints for the meson field $\mathcal{M}$

$$\mathcal{M}^{\mathrm{T}} = \epsilon\mathcal{M} \ , \quad \mathcal{M}\tilde{J}\mathcal{M} = 0 \ , \quad N_\mathrm{c} - 2 < \operatorname{rank}\mathcal{M} \leq N_\mathrm{c} \ .$$

As is well-known, the $SO(N_\mathrm{c})$ case has a $USp(2N_\mathrm{f})$ flavor symmetry while the $USp(2M_\mathrm{c})$ case has an $SO(2N_\mathrm{f})$ flavor symmetry.

Like in the case of the Kähler manifolds, the vacuum manifolds in the hyperKähler case can be written down for a generic point, which for the $SO(N_\mathrm{c})$ case contains the space of Eq. (3)

$$\mathcal{M}^{\text{generic,HK}}_{SO(N_\mathrm{c})} \simeq \mathbb{R}^{N_\mathrm{c}}_{\geq 0} \times \frac{USp(2N_\mathrm{f})}{USp(2N_\mathrm{f} - 2N_\mathrm{c})} \underset{\substack{\text{Kähler}\\ \text{submfd.}}}{\supset} \mathcal{M}^{\text{generic,K}}_{SO(N_\mathrm{c})} \ .$$

Similarly, in a generic point on the vacuum manifold of the $USp(2M_\mathrm{c})$ hyperKähler case, we can write

$$\mathcal{M}^{\text{generic,HK}}_{USp(2M_\mathrm{c})} \simeq \mathbb{R}^{M_\mathrm{c}}_{\geq 0} \times \frac{SO(2N_\mathrm{f})}{SO(2N_\mathrm{f} - 4M_\mathrm{c}) \times SU(2)^{M_\mathrm{c}}} \underset{\substack{\text{Kähler}\\ \text{submfd.}}}{\supset} \mathcal{M}^{\text{generic,K}}_{USp(2M_\mathrm{c})} \ ,$$

where Eq. (4) is a special Lagrangian sub-manifold of the hyperKähler manifold and analogously for the SO case.

Let us make a comment on the hyperKähler quotient of the $U(1) \times SO(N_\mathrm{c})$ and $U(1) \times USp(2M_\mathrm{c})$ theories. We succeeded in constructing the hyperKähler quotient of $SO(N_\mathrm{c})$ and $USp(2M_\mathrm{c})$ thanks to the fact that $J\tilde{Q}^{\mathrm{T}}$ is in the anti-fundamental representation, which is the same representation as Q. Although, we want to make use of the same strategy for $U(1) \times SO(N_\mathrm{c})$ and $U(1) \times USp(2M_\mathrm{c})$ as before, $J\tilde{Q}^{\mathrm{T}}$ still has charge -1 with respect to the the $U(1)$ gauge symmetry while Q has $U(1)$ charge $+1$. Therefore, it is not an easy task to construct the $U(1) \times SO(N_\mathrm{c})$ and $U(1) \times USp(2M_\mathrm{c})$ hyperKähler quotients.

6. 1/2 BPS states: NLσM Lumps

In this section, we will study the σ model lumps which are 1/2 BPS states. Lumps are stringy topological defects in $d = 1 + 3$ dimensional spacetime and are supported by the non-trivial homotopy group $\pi_2(\mathcal{M})$ associated with a holomorphic mapping from the 2 dimensional spatial plane $z = x_1 + ix_2$ to the target space of the NLσMs.[14]

Let us concentrate on $U(1)\times G'$ Kähler quotient models as NLσMs. In these cases, (inhomogeneous) complex coordinates on the Kähler manifold $\{\phi^\alpha\}$, which are the lowest scalar components of the chiral superfields, are given by some set of holomorphic G' invariants; I^i modulo $U(1)^{\mathbf{C}}$, $\phi^\alpha \in \{I^i\}/\!\!/U(1)^{\mathbf{C}}$. Lump solutions can be obtained by just imposing ϕ^α to be a holomorphic function with respect to z

$$\phi^\alpha(t,z,\bar{z},x^3) \to \phi^\alpha(z;\varphi^i)\,, \tag{14}$$

where φ^i denote complex constants. The mass of the lumps can be obtained by plugging the solution back into the Lagrangian

$$E = 2\int_{\mathbf{C}} K_{\alpha\bar{\beta}}(\phi,\bar{\phi})\;\partial\phi^\alpha\bar{\partial}\bar{\phi}^{\bar{\beta}}\bigg|_{\phi\to\phi(z)}\,.$$

We would like to stress that all the parameters φ^i are nothing but the moduli parameters of the 1/2 BPS lumps.

We assume that the boundary i.e. $z=\infty$ is mapped to a point $\phi^\alpha = \phi^\alpha_{\rm vev}$ on the vacuum manifold in a lump solution. Since the functions $\phi^\alpha(z)$ should be single valued, $\phi^\alpha(z)$ can be expressed with a finite number of poles as

$$\phi^\alpha(z) = \phi^\alpha_{\rm vev} + \sum_{i=1}^{k}\frac{\phi^\alpha_i}{z-z_i} + \mathcal{O}(z^{-2})\,.$$

Strictly speaking, we have to change patches of the manifold at the poles in order to describe the solutions correctly. To pick up such global information of lumps thoroughly, it is convenient to use the holomorphic invariants I^i as homogeneous coordinates. By fixing some components of I^i to be constants, we can construct the invariants I^i in terms of $\phi^\alpha(z)$ and find that I^i also be holomorphic functions $I^i(z) = I^i_{\rm vev} + \mathcal{O}(z^{-1})$. We can redefine the functions $I^i(z)$ by using $U(1)^{\mathbf{C}}$ transformations $I^i(z) \simeq I'^i(z) = (z^\nu)^{n_i} I^i(z)$, such that all the invariants $I^i(z)$ are polynomials

$$I^i(z) = I^i_{\rm vev} z^{n_i\nu} + \mathcal{O}(z^{n_i\nu-1})\,,$$

where n_i is the $U(1)$ charge of the holomorphic invariant I^i and $n_i\,\nu \in \mathbb{Z}_{>0}$. These polynomials are basic tools to study lump solutions and their moduli, and $\phi^\alpha(z)$ can be written as ratios of these polynomials, which are known as rational maps in the Abelian case. Here we assume that the invariants $I^i(z)$ have no common zeroes, in order to fix $U(1)^{\mathbf{C}}$, $I^i(z) \simeq (z-a)I^i(z)$. If a common zero accidentally emerges by varying the moduli parameters, the behavior of lumps cannot be defined from the view point of the NLσM, since a common zero corresponds to the Coulomb phase for $U(1)$ in the original gauge theory. This can also be understood as the emergence of a local vortex (see Ref. 1 for details).

Using the so-called *moduli matrix*, which describes different BPS solitons in supersymmetric gauge theories,[23] we can indeed identify the lowest component ϕ^α with the moduli matrix. The key observation is that the gauge symmetry G in the supersymmetric theory is naturally complexified: $G^{\mathbb{C}}$. Hence, the moduli matrix naturally appears in the superfield formulation, while if we fix $G^{\mathbb{C}}$ in the Wess-Zumino gauge, the scalar field $Q_{\rm wz}$ appears as the usual bosonic component in the Lagrangian.

7. Effective Action of Lumps

Now we have a great advantage, thanks to the above superfield formulation of the NLσMs. A supersymmetric low energy effective theory of the 1/2 BPS lumps is immediately obtained merely by plugging the 1/2 BPS solution (14) into the Kähler potential which we have obtained in the previous section after promoting the moduli parameters φ^i to fields on the lump world-volume

$$\phi^\alpha(t,z,\bar{z},x^3) \to \phi^\alpha(z;\varphi^i(t,x^3)) \ .$$

The resulting (effective) expression for the Kähler potential is

$$\mathcal{K}_{\rm lump} = \int_{\mathbb{C}} K\left(\phi(z,\varphi^i(t,x^3)),\ \phi^\dagger(\bar{z},\bar{\varphi}^i(t,x^3))\right) \ .$$

8. Lumps in $U(1)\times SO(N_{\rm c})$ Kähler Quotients

Let us start with a very simple example of a theory with the gauge group $U(1)\times SO(2)$ and only two flavors $N_{\rm f}=2$. The target space is[1] $\mathbb{C}P^1_+\times\mathbb{C}P^1_-$. Lump solutions are classified by a pair of integers (k_+,k_-) and

$$\pi_2\left(\mathcal{M}^{U(1)\times SO(2)}_{N_{\rm f}=2}\right) = \mathbb{Z}_+\otimes\mathbb{Z}_- \ ,$$

where $\mathbb{Z}_\pm$ denote integers. A solution with (k_+,k_-) lumps is given by

$$Q(z) = \begin{pmatrix} Q_{+1}(z) & Q_{+2}(z) \\ Q_{-1}(z) & Q_{-2}(z)\end{pmatrix} = \begin{pmatrix}\vec{Q}_+ \\ \vec{Q}_-\end{pmatrix} \ ,$$

where $Q_{\pm i}(z)$ are holomorphic functions of z of degree $k_\pm$, respectively. The energy reads

$$E = 2\int_{\mathbb{C}} \partial\bar{\partial} K_{U(1)\times SO(2)} = \pi\xi(k_+ + k_-) \equiv \pi\xi k \ ,$$

$$K_{U(1)\times SO(2)} = \frac{\xi}{2}\log|\vec{Q}_+|^2 + \frac{\xi}{2}\log|\vec{Q}_-|^2 \ .$$

Interestingly, the tension of the minimal lump $(k_+, k_-) = (1,0), (0,1)$ is half of $2\pi\xi$ which is that of the minimal lump in the usual $\mathbb{C}P^1$ model.

We would now like to consider lump configurations in slightly more complicated models by considering general $SO(2M_c)$ Kähler quotients, where we set $M_c \geq 2$, $N_f = 2M_c$ and $M_{vev} = J$. However, we should take into account the following constraint on the holomorphic invariants of the $SO(2M_c)$ group for k lump configurations[24]

$$M_{SO(2M_c)} = Q^{\mathrm{T}}(z)JQ(z) = Jz^k + \mathcal{O}(z^{k-1}) \,. \tag{15}$$

As an example for $k = 1$, we take

$$Q_{k=1} = \begin{pmatrix} z\mathbf{1}_{M_c} - \mathbf{A} & \mathbf{C} \\ 0 & \mathbf{1}_{M_c} \end{pmatrix} , \qquad \begin{cases} \mathbf{A} = \mathrm{diag}(z_1, z_2, \cdots, z_{M_c}) \,, \\ \mathbf{C} = \mathrm{diag}(c_1, c_2, \cdots, c_{M_c}) \,. \end{cases}$$

This choice of diagonal matrices allows us to treat the invariants as if they where simply invariants of M different $SO(2)$ subgroups. Note that non-zero parameters c_i keep the rank $M \geq 2M_c$ even at $z = z_i$. Thus, the Kähler potential in Eq. (10) becomes

$$K = \xi \log \left(2 \sum_{i=1}^{M} \sqrt{|z - z_i|^2 + |c_i|^2} \right) . \tag{16}$$

The corresponding energy density is obtained by calculating $\mathcal{E} = 2\partial\bar{\partial}K$ with this potential. If we take some c_i to vanish, the energy density becomes singular at $z = z_i$

$$E = 2\xi\partial\bar{\partial}\log\left[\sqrt{|z - z_i|^2} + \cdots\right] \sim \text{const.} \times \frac{1}{|z - z_i|} + \mathcal{O}\left((z - z_i)^0\right) ,$$

This is due to the occurrence of a curvature singularity in the manifold when $\mathrm{rank}\, M = 2M_c - 2$. The trace part of $\mathbf{C}$ determines the overall size of the configuration and the trace part of $\mathbf{A}$ the center of mass, where only the latter turns out to be a normalizable mode.[1]

9. Conclusion

We have studied the NLσM lumps in gauge theories with $SO(N_c)$, $USp(2M_c)$, $U(1) \times SO(N_c)$ and $U(1) \times USp(2M_c)$ gauge groups and obtained Kähler metrics and for the two first cases also the hyperKähler metrics. Furthermore, we have constructed NLσM lumps in these models.

Acknowledgments

The author thanks the Organizers for warm hospitality.

References

1. M. Eto, T. Fujimori, S. B. Gudnason, M. Nitta, K. Ohashi and W. Vinci, *in preparation*
2. L. Alvarez-Gaume and D. Z. Freedman, Commun. Math. Phys. **80**, 443 (1981).
3. T. L. Curtright and D. Z. Freedman, Phys. Lett. B **90**, 71 (1980) [Erratum-ibid. B **91**, 487 (1980)]; L. Alvarez-Gaume and D. Z. Freedman, Phys. Lett. B **94**, 171 (1980); M. Rocek and P. K. Townsend, Phys. Lett. B **96**, 72 (1980).
4. U. Lindström and M. Roček, Nucl. Phys. B **222**, 285 (1983).
5. N. J. Hitchin, A. Karlhede, U. Lindström and M. Roček, Commun. Math. Phys. **108**, 535 (1987).
6. M. F. Atiyah, N. J. Hitchin, V. G. Drinfeld and Yu. I. Manin, Phys. Lett. A **65**, 185 (1978).
7. P. B. Kronheimer, J. Diff. Geom. **29**, 665 (1989).
8. G. W. Gibbons, R. Goto and P. Rychenkova, Commun. Math. Phys. **186**, 585 (1997) [arXiv:hep-th/9608085].
9. P. C. Argyres, M. Ronen Plesser and A. D. Shapere, Nucl. Phys. B **483**, 172 (1997) [arXiv:hep-th/9608129].
10. N. Seiberg and E. Witten, Nucl. Phys. B **431**, 484 (1994) [arXiv:hep-th/9408099].
11. P. C. Argyres, M. R. Plesser and N. Seiberg, Nucl. Phys. B **471**, 159 (1996) [arXiv:hep-th/9603042].
12. I. Antoniadis and B. Pioline, Int. J. Mod. Phys. A **12**, 4907 (1997) [arXiv:hep-th/9607058].
13. M. Eto, Y. Isozumi, M. Nitta, K. Ohashi, K. Ohta, N. Sakai and Y. Tachikawa, Phys. Rev. D **71**, 105009 (2005) [arXiv:hep-th/0503033].
14. A. M. Polyakov and A. A. Belavin, JETP Lett. **22**, 245 (1975) [Pisma Zh. Eksp. Teor. Fiz. **22**, 503 (1975)].
15. T. Vachaspati and A. Achucarro, Phys. Rev. D **44**, 3067 (1991); A. Achucarro and T. Vachaspati, Phys. Rept. **327**, 347 (2000) [arXiv:hep-ph/9904229].
16. M. Eto *et al.*, Phys. Rev. D **76**, 105002 (2007) [arXiv:0704.2218 [hep-th]].
17. M. Shifman and A. Yung, Phys. Rev. D **73**, 125012 (2006) [arXiv:hep-th/0603134].
18. A. Hanany and D. Tong, JHEP **0307**, 037 (2003) [arXiv:hep-th/0306150].
19. M. A. Luty and W. Taylor, Phys. Rev. D **53**, 3399 (1996) [arXiv:hep-th/9506098].
20. M. Bando, T. Kuramoto, T. Maskawa and S. Uehara, Phys. Lett. B **138**, 94 (1984).
21. K. A. Intriligator and N. Seiberg, Nucl. Phys. B **444**, 125 (1995) [arXiv:hep-th/9503179].
22. K. A. Intriligator and P. Pouliot, Phys. Lett. B **353**, 471 (1995) [arXiv:hep-th/9505006].
23. M. Eto, Y. Isozumi, M. Nitta, K. Ohashi and N. Sakai, J. Phys. A **39**, R315 (2006) [arXiv:hep-th/0602170].
24. M. Eto, T. Fujimori, S. B. Gudnason, K. Konishi, M. Nitta, K. Ohashi and W. Vinci, arXiv:0802.1020 [hep-th].

SECTION 7.

OTHER DEVELOPMENTS

Gluinos Condensing at the CCNI: 4096 CPUs Weigh In*

Joel Giedt

giedtj@rpi.edu
Department of Physics, Applied Physics and Astronomy,
Rensselaer Polytechnic Institute,
110 8th Street, Troy NY 12065 USA

Richard Brower

brower@bu.edu
Physics Department, Boston University
590 Commonwealth Avenue, Boston MA 02215

Simon Catterall

smc@syracuse.edu
Department of Physics, Syracuse University, Syracuse, NY 13244 USA

George T. Fleming

George.Fleming@yale.edu
Department of Physics, Sloane Laboratory, Yale University,
New Haven, Connecticut 06520, USA

Pavlos Vranas

vranas@llnl.gov
Physical Sciences Directorate, Lawrence Livermore National Laboratory,
7000 East Ave., Livermore, CA 94550

We report preliminary results of lattice super-Yang-Mills computations using domain wall fermions, performed at an actual rate of 1000 Gflop/s, over the course of six months, using two BlueGene/L racks at Rensselaer's CCNI supercomputing center. This has allowed us to compute the gluino condensate and string tension over a wide range of lattice parameters, setting the stage for continuum, chiral extrapolations.

*Talk given by Joel Giedt at "Continuous Advances in QCD 2008," Fine Theoretical Physics Institute, University of Minnesota, Minneapolis, MN, May 15-18, 2008.

In this talk, I present preliminary results obtained in collaboration with coauthors of a forthcoming paper.[1] We have used domain wall fermions (DWF)[2,3] to study nonperturbative aspects of pure $\mathcal{N} = 1$ super-Yang-Mills (SYM).[4] The strong dynamics of supersymmetric gauge theories underlie most models of spontaneous supersymmetry (SUSY) breaking, and the development of a first-principles tool is needed. Though at finite lattice spacing SUSY is violated, it is automatically recovered[5] in the continuum limit with a massless gluino.[a] When the DWF formulation is employed, this "chiral limit" is achieved without the need for a computationally expensive, nonperturbatively determined fine-tuning of the bare gluino mass;[6,7] in the limit of infinite domain wall separation $L_s \to \infty$, DWF realize the lattice chiral symmetry[8] associated with Ginsparg-Wilson fermions,[9] which protects against additive mass renormalization. These nice features of the DWF approach are to be contrasted with the Wilson fermion formulation, which was pursued for several years by the DESY-Münster-Roma collaboration.[10–16]

Lattice studies can provide details that other approaches cannot, such as "snapshots" of the gauge field configurations that are dominating the gluino condensate. For instance, in the work of Fleming et al.,[17] the only DWF simulation of SYM to date, it was suggested that spikes in the gluino condensate may correspond to configurations with fractional topological charge, as would be expected from Ref.[18] Further pursuit of this conjecture, consisting of lattice studies of monopoles and topological charge, is on our agenda. Also, we will compute the low-lying spectrum of composite states, consisting of strongly bound gluons and gluinos. Apart from the inconclusive results of DESY-Münster-Roma, this aspect of SYM is completely unknown from continuum methods, and ideally suited to the lattice approach.

At early stages in such studies, understanding numerical behavior of important quantities such as the gluino condensate will teach us a lot about the lattice formulation that we currently do not know. For instance, it is important to set "benchmarks" regarding compute time, and lattice artifacts such as discretization and finite size effects. The first place that we will examine this is in the gluino condensate, which is believed to be known exactly by continuum methods,[18–22] and is therefore ideal for calibrating the lattice methods. Understanding of the lattice theory and simulation

[a] Here we use the terms "gluon" and "gluino" by way of analogy. It should be kept in mind that the strongly coupled gauge theory would be an extension to the gauge group of the Standard Model.

performance is already emerging from our preliminary results, as we now briefly discuss.

Domain wall fermion simulations require world-class computing resources, such as are are available to Giedt at Rensselaer; namely, the Computational Center for Nanotechnology Innovations (CCNI), one of the world's most powerful university-based supercomputing centers, and a top 25 supercomputing center of any kind in the world. We are presently the third heaviest user of this facility, and have been generating lattice configurations and measurements continuously at a sustained actual rate of 1000 Gflop/s since the end of January 2008. For comparison the DESY-Münster-Roma collaboration performed their computations at a rate of 10 Gflop/s for a cumulative time of one year. Thus our study represents a hundred-fold improvement over what has been done previously, just in terms of raw computation power.

As an example of our results, we have obtained the bare gluino condensate from dynamical domain wall fermion simulations for a variety of bare gauge couplings g, parameterized in terms of $\beta = 4/g^2$, as is conventional in SU(2) lattice gauge theory. The results for a $16^3 \times 32$ lattice (i.e., the number of sites in spatial and temporal directions) with domain wall separation $L_s = 16$ sites are displayed in Fig. 1. We note that such data for the condensate versus β has never been obtained before; it is important because the continuum limit corresponds to $\beta \to \infty$ (with the physical size of the lattice held fixed). A fit to the data obviously yields a vanishing condensate at a finite gauge coupling $\beta \sim 2.7$. This just reflects the fact that as β increases the lattice spacing shrinks, and thus so does the physical size of the lattice in its entirety. In a small enough "box" confinement will disappear and the condensate "melts." Thus we already gain an important benchmark: to go much beyond $\beta = 2.5$ will require larger lattices, and in fact one should carefully measure systematic errors due to finite size effects at $\beta \approx 2.5$. This is consistent with what is already known from the so-called "quenched" theory, which has no gluinos.

Fig. 2 shows the gluino condensate for decreasing values of the residual mass $m_{\rm res}$, which is a measure of explicit chiral symmetry breaking due to finite L_s.[23] As expected, larger L_s values have the smallest $m_{\rm res}$, and a nonzero gluino condensate appears to occur in the $m_{\rm res} \to \infty$ limit. Also as expected, smaller values of $m_{\rm res}$ occur for the weaker coupling $\beta = 2.4$.

Finally, we have looked at Creutz ratios,[24]

$$\chi(R,R) = -\ln \frac{W(R,R)W(R-1,R-1)}{W(R,R-1)^2} \sim \sigma a^2, \qquad (1)$$

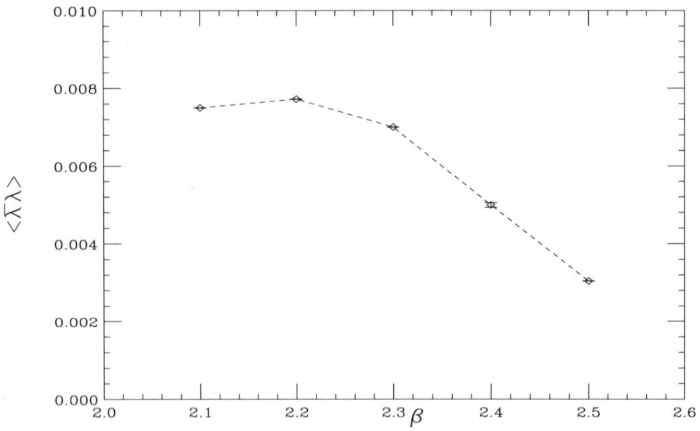

Fig. 1. The gluino condensate versus β for a $16^3 \times 32$ lattice with domain wall separation $L_s = 16$ (dashed line drawn to guide the eye). It can be estimated from the figure that for the $16^3 \times 32$ lattice studied here, the system will deconfine at $\beta \sim 2.7$, as a result of finite size effects.

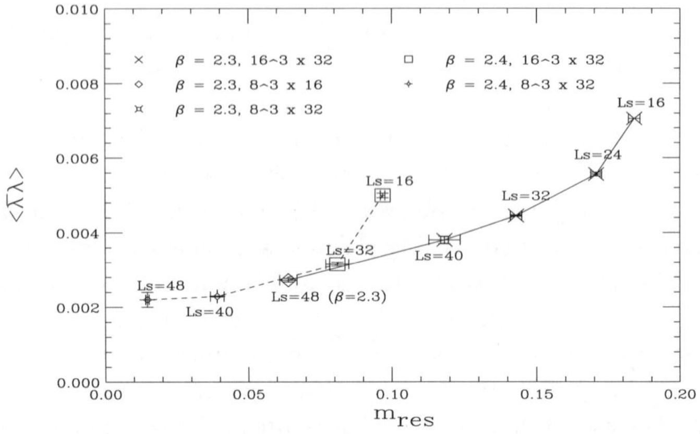

Fig. 2. Our simulation results for the gluino condensate versus m_{res}, where the latter is a measure of explicit chiral symmetry breaking due to finite domain wall separation L_s. The solid line corresponds to $\beta = 2.3$ whereas the dashed line is for $\beta = 2.4$.

where $W(R, R')$ is an $R \times R'$ Wilson loop, in order to extract the string tension σ in lattice units, as well as to delineate the scaling regime where the continuum limit may be extracted. In the process we obtain an estimate of the lattice spacing in units of the string tension. Results for the $16^3 \times 32$, $L_s = 16$ lattice are shown in Fig. 3. Although the errors are somewhat large,

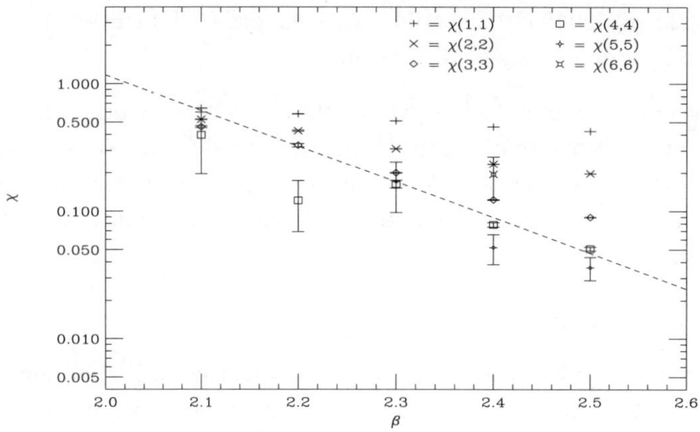

Fig. 3. Creutz ratios for the $16^3 \times 32$ lattice with $L_s = 16$. The dashed line indicates the 2-loop SUSY prediction for the dependence $\chi \sim \sigma a^2$ where σ is the string tension and $a = a(\beta)$ is the lattice spacing.

scaling is clearly setting in at around $\beta = 2.3$. To see this one notes that the larger Creutz ratios appear to coalesce on an envelope, corresponding to the distance scale at which an area law begins to take hold in the Wilson loops.

We emphasize that all of the results mentioned above are groundbreaking as far as lattice SYM is concerned. Several important goals related to the lattice SYM project have already been achieved, laying the groundwork for the more extensive studies that will follow:

- Developed parallel simulation code for SYM by modification of the current version of the Columbia Physics System (CPS) QCD package.
- This extends DOE funded code (CPS, part of USQCD's SciDAC program) to "beyond the Standard Model" physics, which is a realization of one of the USQCD Collaboration objectives.[25]
- We have reproduced the results of[17] as a check on our code.
- Our software runs successfully on IBM's Bluegene (BG) architecture, taking full advantage of BG specific communications utilities.
- Developed Landau gauge-fixing and Fourier space propagator code for adjoint fermion representations, essential for nonperturbative renormalization of the condensate, in the RI/MOM scheme.[26,27]
- Established timing and statistical uncertainty benchmarks. For example, we have found that for small single-node volumes ($16^3 \times$

$32 \times L_s/2048$ CPU's $= 64 \times L_s$ sites per CPU) the efficiency of the BG parallel code is 10 percent.

In the course of our studies, we intend to investigate other domain wall fermion formulations (e.g., "gap"[28] and "Mobius"[29]) as ways to approach the chiral limit more quickly. Also, we will implement recent optimizations of fermion matrix inverters (to increase efficiency) and improved actions (to reduce systematic errors).

Acknowledgements

We benefited from a copy of the code that was used in,[17] and employed it as the basis for our modifications to the current version of the Columbia Physics System. At various points JG benefited from technical assistance provided by Chulwoo Jung (Brookhaven National Lab) and Adam Todorski (SCOREC and CCNI at Rensselaer). The computational efforts on this project mainly utilized the Computational Center for Nanotechnology Innovations (CCNI), and JG expresses his appreciation for continuous access to that facility. The project also, at times, utilized the SUR BlueGene/L at Rensselaer, which is supported by NSF grant 0420703 entitled "MRI: Acquisition of Infrastructure for Research in Grid Computing and Multiscale Systems Computation" and a gift by the IBM Corporation of a BlueGene/L computer. JG acknowledges support from Rensselaer faculty development funds.

References

1. J. Giedt, R. Brower, S. Catterall, G. T. Fleming, P. Vranas, in progress.
2. D. B. Kaplan, "A Method for simulating chiral fermions on the lattice," Phys. Lett. B **288** (1992) 342 [arXiv:hep-lat/9206013].
3. Y. Shamir, "Chiral fermions from lattice boundaries," Nucl. Phys. B **406** (1993) 90 [arXiv:hep-lat/9303005].
4. S. Ferrara and B. Zumino, "Supergauge Invariant Yang-Mills Theories," Nucl. Phys. B **79** (1974) 413.
5. G. Curci and G. Veneziano, "Supersymmetry And The Lattice: A Reconciliation?," Nucl. Phys. B **292** (1987) 555.
6. H. Neuberger, "Vector like gauge theories with almost massless fermions on the lattice," Phys. Rev. D **57** (1998) 5417 [arXiv:hep-lat/9710089].
7. D. B. Kaplan and M. Schmaltz, "Supersymmetric Yang-Mills theories from domain wall fermions," Chin. J. Phys. **38** (2000) 543 [arXiv:hep-lat/0002030].
8. M. Luscher, "Exact chiral symmetry on the lattice and the Ginsparg-Wilson relation," Phys. Lett. B **428** (1998) 342 [arXiv:hep-lat/9802011].
9. P. H. Ginsparg and K. G. Wilson, "A Remnant Of Chiral Symmetry On The Lattice," Phys. Rev. D **25** (1982) 2649.

10. I. Campos, R. Kirchner, I. Montvay, J. Westphalen, A. Feo, S. Luckmann, G. Münster, K. Spanderan [DESY-Munster Collaboration], "Monte Carlo simulation of SU(2) Yang-Mills theory with light gluinos," Eur. Phys. J. C **11** (1999) 507 [arXiv:hep-lat/9903014].
11. F. Farchioni, C. Gebert, R. Kirchner, I. Montvay, A. Feo, G. Münster, T. Galla, A. Vladikas [DESY-Munster-Roma Collaboration], "The supersymmetric Ward identities on the lattice," Eur. Phys. J. C **23** (2002) 719 [arXiv:hep-lat/0111008].
12. I. Montvay, "Supersymmetric Yang-Mills theory on the lattice," Int. J. Mod. Phys. A **17** (2002) 2377 [arXiv:hep-lat/0112007].
13. R. Peetz, F. Farchioni, C. Gebert and G. Munster, "Spectrum of SU(2) SUSY Yang-Mills theory with a light gluino," Nucl. Phys. Proc. Suppl. **119** (2003) 912 [arXiv:hep-lat/0209065].
14. F. Farchioni, G. Muenster and R. Peetz, "The volume source technique for flavor singlets: A second look," Eur. Phys. J. C **38** (2004) 329 [arXiv:hep-lat/0404004].
15. F. Farchioni and R. Peetz, "The low-lying mass spectrum of the N = 1 SU(2) SUSY Yang-Mills theory with Wilson fermions," Eur. Phys. J. C **39** (2005) 87 [arXiv:hep-lat/0407036].
16. R. Peetz, "Spectrum of $\mathcal{N} = 1$ Super Yang Mills Theory on the Lattice with a Light Gluino," doctoral dissertation, University of Münster, Germany, 2003, available at: `http://deposit.ddb.de/cgi-bin/dokserv?idn=97018249X`.
17. G. T. Fleming, J. B. Kogut and P. M. Vranas, "Super Yang-Mills on the lattice with domain wall fermions," Phys. Rev. D **64** (2001) 034510 [arXiv:hep-lat/0008009].
18. N. M. Davies, T. J. Hollowood, V. V. Khoze and M. P. Mattis, "Gluino condensate and magnetic monopoles in supersymmetric gluodynamics," Nucl. Phys. B **559** (1999) 123 [arXiv:hep-th/9905015].
19. I. Affleck, M. Dine and N. Seiberg, "Dynamical Supersymmetry Breaking In Supersymmetric QCD," Nucl. Phys. B **241** (1984) 493.
20. I. Affleck, M. Dine and N. Seiberg, "Dynamical Supersymmetry Breaking In Four-Dimensions And Its Phenomenological Implications," Nucl. Phys. B **256** (1985) 557.
21. V. A. Novikov, M. A. Shifman, A. I. Vainshtein and V. I. Zakharov, "Supersymmetric instanton calculus: Gauge theories with matter," Nucl. Phys. B **260** (1985) 157 [Yad. Fiz. **42** (1985) 1499].
22. F. Cachazo, M. R. Douglas, N. Seiberg and E. Witten, "Chiral rings and anomalies in supersymmetric gauge theory," JHEP **0212** (2002) 071 [arXiv:hep-th/0211170].
23. T. Blum *et al.*, "Quenched lattice QCD with domain wall fermions and the chiral limit," Phys. Rev. D **69** (2004) 074502 [arXiv:hep-lat/0007038].
24. M. Creutz, "Asymptotic Freedom Scales," Phys. Rev. Lett. **45** (1980) 313.
25. R. Brower, N. Christ, M. Creutz, P. Mackenzie, J. Negele, C. Rebbi, D. Richards, S. Sharpe, R. Sugar [Lattice QCD Executive Committee, USQCD Collaboration], "Lattice Gauge Theory for Physics Beyond the Standard Model," whitepaper available at:

http://www.usqcd.org/documents/bsm.pdf.
26. G. Martinelli, C. Pittori, C. T. Sachrajda, M. Testa and A. Vladikas, "A General Method For Nonperturbative Renormalization Of Lattice Operators," Nucl. Phys. B **445** (1995) 81 [arXiv:hep-lat/9411010].
27. T. Blum *et al.*, "Non-perturbative renormalisation of domain wall fermions: Quark bilinears," Phys. Rev. D **66** (2002) 014504 [arXiv:hep-lat/0102005].
28. P. M. Vranas, "Gap domain wall fermions," Phys. Rev. D **74** (2006) 034512 [arXiv:hep-lat/0606014].
29. R. C. Brower, H. Neff and K. Orginos, "Moebius fermions: Improved domain wall chiral fermions," Nucl. Phys. Proc. Suppl. **140** (2005) 686 [arXiv:hep-lat/0409118].

Baryon Regge Trajectories and the $1/N_c$ Expansion

J. L. GOITY*

Department of Physics, Hampton University,
Hampton, VA 23668, USA
and Jefferson Lab,
Newport News, VA 23606, USA

N. MATAGNE**

Institut für Theoretische Physik, Universität Giessen,
D-35392 Giessen, Germany

E-mail: *goity@jlab.org, **Nicolas.Matagne@theo.physik.uni-giessen.de

Baryon Regge trajectories are discussed in the light of the $1/N_c$ expansion. The approximate dynamical symmetry $SU(6) \times O(3)$ is used to identify the spin-flavor singlet component of baryon masses. By fitting to the known baryon spectrum, this component is shown to produce distinct Regge trajectories for the SU(6) **56**- and **70**-plets which are remarkably linear.

Keywords: $1/N_c$ expansion; baryons; Regge trajectories

1. Introduction

One of the best established and remarkable properties of meson and baryon spectra is the ordering of states into approximately linear trajectories in the Chew-Frautschi plot, also known as Regge trajectories. They are the most clear manifestation of the confining dynamics of QCD, supporting the picture of confinement as the result of string-like behavior of the interaction between color charges at long distances. In the light meson sector, the plot M^2 vs J shows very clearly an approximately linear law; for instance, non-strange mesons with masses up to about 2.5 GeV fall into several nearly linear trajectories: π-trajectory, ρ-ω-trajectory, f-trajectory, etc., with slopes in the range 1.1 to 1.4 GeV^2. One interesting aspect is the possibility of assigning an "orbital" quantum number ℓ to each meson, as one would do in a quark model when spin-orbit interactions are neglected. Plotting instead M^2 vs ℓ leads to a set of trajectories with similar slopes, and with intercepts that are similar (with exception of the π-trajectory, where the Goldstone Boson nature of the pion forces a nearly vanishing intercept). One can proceed further and fit a single linear trajectory M^2 vs ℓ, which shows that all states are along this trajectory within a rather narrow spread, the only clear discrepancy resulting from the hyperfine splitting at $\ell = 0$ between the pion and ρ meson. There is, therefore, an important phenomenological manifestation of an approximate $O(3)$ orbital symmetry in the light mesons, which is also well known to be important in baryons, as we further discuss. In passing, we should mention that there are no established linear Regge trajectories in heavy mesons or heavy baryons. There is, however, an interesting observation one can make for D-mesons based on one assumption: the $D^*(2640)$ exists (only one group has claimed observation of this state[1]) and has $\ell = 2$. In that case, a linear relation can be found: $(M_{D^*(\ell)} - M_{\bar{D}})^2 = 0.21\,\ell\,\mathrm{GeV}^2$, where $M_{\bar{D}} = (M_D + 3M_{D^*})/4$ is the spin-flavor singlet component of the ground state D-mesons. This seems to be a plausible relation whose validity will be clarified with progress in establishing of the excited heavy meson spectrum. It may as well turn out to extend to heavy baryons.

For light baryons, the topic of this talk, Regge trajectories have been discussed for a long time, with a consensus that they are approximately linear and with similar slopes as those for mesons. The discussions are based on plotting trajectories for different types of baryons, say N, Δ, etc. Remarkably enough, baryon Regge trajectories in the framework presented in,[2,3] which makes use of approximate dynamical symmetries, had not been discussed in the past. The motivation is based on the expectation that Regge trajectories are primarily determined by quark-spin independent components of the strong interactions. Thus, if one can separate in a clear manner the quark-spin independent component of baryon masses, a more pristine version of Regge trajectories should result for that mass component [a]. In reference[3] and in this talk, we show how to identify the quark-spin and flavor independent component of baryon masses in the framework of the $1/N_c$ expansion, and show that indeed a much clearer picture for the trajectories results.

The key to the implementation of the $1/N_c$ expansion in baryons is the emergent *contracted spin-flavor* dynamical symmetry in the large N_c limit

[a]We should emphasize that associating the terminology "Regge trajectory" with a piece of the baryon masses, in this case the spin-flavor singlet piece, and plotting it against ℓ instead of the baryon's J is not rigorous terminology from the point of view of Regge theory.

of QCD.[4,5] This dynamical symmetry results from the requirement of unitarity of elastic pion-nucleon scattering at large N_c. The generators of the contracted group are: T^a, S^i and X^{ia}, where the first are the flavor generators, the second the spin generators and the last correspond to the axial currents at zero momentum scaled by a factor $1/N_c$. In the large N_c limit the generators X^{ia} become semiclassical, and this leads to the mentioned contracted symmetry.[4,5] Those generators generate a contracted $SU(2N_f)$ group, where N_f is the number of light flavors. In general the irreducible representations of the contracted symmetry are infinite dimensional. There is an alternative and equally rigorous way of implementing the consistency requirements by using instead the compact Lie group $SU(2N_f)$ at finite N_c (whose generators are T^a, S^i and $G^{ia} = N_c X^{ia}$). The latter has the advantage that one works with finite dimensional representations and the implementation of the expansion is more straightforward.

On a less fundamental footing, and similarly to the case of mesons, one can phenomenologically identify an $O(3)$ approximate symmetry; excited baryons can be accommodated into multiplets of this orbital $O(3)$. Its origin is dynamical, without a known fundamental explanation from QCD, and there is no assurance that it will hold at large N_c, although this seems plausible. It is then natural to have a classification of baryons in multiplets of $SU(6) \times O(3)$, as it was realized back in the 1960's. This provides a framework for implementing the $1/N_c$ expansion, which in the particular case of baryon masses has been studied in various works.[5,6,8]

2. Baryon masses

The effective mass operators for baryons are constructed as linear combinations of composite operators built with the generators of $SU(6) \times O(3)$. This is possible if one neglects the possibility of configuration mixings, i.e., mixings of states in different representations of $SU(6) \times O(3)$.[7] For most phenomenological purposes this seems to be a reasonable approximation, which the conclusions of this work will further reaffirm. The $1/N_c$ power counting is determined by the $n-$bodyness of the effective operator: a suppression factor $1/N_c^{n-1}$ accompanies every n-body operator. In addition, matrix elements have N_c dependencies: so called coherent operators, such as the axial currents G^{ia} present $\mathcal{O}(N_c)$ matrix elements, while incoherent ones have $\mathcal{O}(1)$ or even $\mathcal{O}(1/N_c)$ matrix elements.

All baryons discussed here will be accommodated into the symmetric (**S**) or the mixed symmetric representations (**MS**) of $SU(6)$ which for $N_c = 3$ correspond to the **56**- and the **70**-plets respectively. For the excited baryons we distinguish one box in the Young tableux of $SU(6)$, which in a quark model picture one would identify with an excited quark. Coupling this box to the totally symmetric Young tableux with $N_c - 1$ boxes (known as the core) one can build the representations of interest.

The ground state (GS) baryons belong into the $[\mathbf{56}, 0^+]$ multiplet. The mass formula is the old and well known Gürsey-Radicati mass formula, which is the most general one to $\mathcal{O}(1/N_c)$ and $\mathcal{O}(\epsilon)$, where ϵ measures the breaking of flavor $SU(3)$:[5]

$$M_{GS} = c_1 \, N_c + \frac{1}{N_c} \, c_{\mathrm{HF}} \, (\vec{S}^2 - \frac{3}{4} N_c) - c_S \, \mathcal{S} \tag{1}$$

The first term is the leading $\mathcal{O}(N_c)$ spin-flavor symmetric term, the second is the hyperfine term defined in such a way that in a quark model picture corresponds to the form $1/N_c \sum_{i \neq j} \vec{s}_i \cdot \vec{s}_j$. Different definitions of this term can be absorbed into changes $\mathcal{O}(1/N_c)$ to the first term. Finally, the third term gives the breaking of $SU(3)$ through the strangeness operator $\mathcal{S}$.

For excited baryons, the number of terms in the mass formulas proliferate. For instance, for the $[\mathbf{56}, \ell > 0]$ multiplets there are in addition to the $\mathcal{O}(N_c)$ term, two $SU(3)$ symmetric $\mathcal{O}(1/N_c)$ terms (one of them a spin-orbit type term and the other the hyperfine type term), and three $SU(3)$ breaking terms, and for the $[\mathbf{70}, \ell > 0]$ multiplets there are in addition to the $\mathcal{O}(N_c)$ term, four $\mathcal{O}(N_c^0)$ terms (three of them involving the coupling of the $O(3)$ generators, and thus called spin-orbit type), seven $SU(3)$ symmetric $\mathcal{O}(1/N_c)$ terms and three $SU(3)$ breaking terms. The various analyses of baryon masses[6] have established several important facts: i) The breaking of spin-flavor symmetry is dominated by the hyperfine type interaction, which is $\mathcal{O}(1/N_c)$ in all multiplets. ii) The $\mathcal{O}(N_c^0)$ breaking of the $SU(6) \times O(3)$ symmetry (as it happens in the **MS** multiplets) is dynamically suppressed (this is the reason why there is approximate $O(3)$ symmetry to start with), and of smaller magnitude than the breaking due to the hyperfine interaction. In the strict large N_c limit one would eventually find a contracted spin-flavor symmetry and an approximate $O(3)$ symmetry as it was discussed in.[9,10] iii) $SU(3)$ breaking is to a good first approximation given by the strangeness operator. Based on these observations, and for the present purpose, it is sufficient to keep only the three types of operators, namely leading spin-flavor singlet, hyperfine and strangeness. In our original work,[3] we included an $SU(3)$ breaking hyperfine term, which improved the fits, but which is of little effect for the point of our discussion. Disregarding that term gives small numerical differences between the results quoted here and in.[3]

For excited baryons one can, therefore, write down the mass formula:

$$M^* = c_1\, N_c + \frac{1}{N_c}\, c_{\rm HF}\, (\vec{S_c}^{\,2} - \frac{3}{4}(N_c - 1)) - c_S\, \mathcal{S} \tag{2}$$

where $\vec{S}_c$ indicates the spin operator restricted to acting on the core of the baryon. Notice that these mass formulas are not dependent on a quark model picture of baryons. Rather, the only assumptions that are being made concern the assignment of states to specific multiplets, which is typically what is done in effective theories. The only tuning in Eqn.(2) is in the choice of the hyperfine term: first, it has been shown in the analysis of the **70**-plet baryon masses that the $\vec{S_c}^{\,2}$ hyperfine is the only relevant one (the other one being $\vec{s}\cdot\vec{S}_c$ ($\vec{s}$ is the excited quark spin), which turns out to be almost entirely suppressed); it is natural that the same should occur for the **56**-plets, although in this case one cannot separate the two mentioned hyperfine terms explicitly because they are linearly dependent operators; second, we subtract the $\frac{3}{4}\,(N_c-1)$ in analogy with what was done for the GS baryons. All this implies some $\mathcal{O}(1/N_c)$ arbitrariness for the leading terms, or equivalently $\mathcal{O}(1/N_c^2)$ corrections to c_1. It is important to emphasize that with these very simple mass formulas the main features of the multiplets considered in here are well described. The finer features of the spectrum naturally requires the inclusion of further mass operators in the $1/N_c$ expanded basis.[6]

Explicit evaluation gives the following results for the excited baryon masses:

$$\begin{aligned}
M'_{\mathbf{S}}(S) &= N_c\, c_1 + \frac{N_c-2}{N_c^2} c_{\rm HF} \left(S(S+1) - \frac{3}{4}N_c\right),\\
M'_{\mathbf{MS}}(S{=}I) &= N_c\, c_1 + \frac{c_{\rm HF}}{N_c}\left(\frac{N_c+2}{N_c}S(S+1) - \frac{3}{4}N_c + \frac{1}{2}\right),\\
M'_{\mathbf{MS}}(S{=}I{-}1) &= N_c\, c_1 + \frac{c_{\rm HF}}{N_c}\left(S(S+2) - \frac{3}{4}(N_c-2)\right),\\
M'_{\mathbf{MS}}(S{=}I{+}1) &= N_c\, c_1 + \frac{c_{\rm HF}}{N_c}\left(S^2 - \frac{3}{4}N_c + \frac{1}{2}\right),
\end{aligned} \tag{3}$$

which for $N_c = 3$ read:

$$\begin{aligned}
&N_{\rm GS} = 3\,c_1 - \frac{1}{2}\,c_{\rm HF},\ \Delta_{\rm GS} = 3\,c_1 + \frac{1}{2}\,c_{\rm HF},\\
&N_{\bf 56} = 3\,c_1 - \frac{1}{6}\,c_{\rm HF},\ \Delta_{\bf 56} = 3\,c_1 + \frac{1}{6}\,c_{\rm HF},\\
&N_{\bf 70}\left(S=\frac{1}{2}\right) = 3\,c_1 - \frac{1}{6}\,c_{\rm HF},\\
&N_{\bf 70}\left(S=\frac{3}{2}\right) = \Delta_{\bf 70}\left(S=\frac{1}{2}\right) = 3\,c_1 + \frac{1}{6}\,c_{\rm HF},
\end{aligned} \tag{4}$$

where we denote $N \equiv M_N$, etc. Note that for the MS states we need to specify the total quark spin S. The case of strange baryons is obvious; one needs to add the SU(3) singlet Λ states in the **70**-plets, whose mass formula becomes:

$$\Lambda^1_{\bf 70} = 3\,c_1 - \frac{1}{2}\,c_{\rm HF} + c_S \tag{5}$$

The effective coefficients c_1, $c_{\rm HF}$ and c_S are obtained for each multiplet by fitting to the corresponding masses (we use here the Breit-Wigner masses as provided by the Particle Data Group (PDG)[1]). The coefficients are $\mathcal{O}(N_c^0)$ and change by amounts $\mathcal{O}(1/N_c)$ as one goes from multiplet to multiplet. The following multiplets are considered for the analysis: i) the GS $[\mathbf{56}, 0^+]$ where all states are empirically established, ii) the $[\mathbf{56}, 2^+]$ where all non-strange states are established, but only a few strangeness -1 states are known, namely two Λ and two Σ states (one in an octet and one in a decuplet), iii) the $[\mathbf{56}, 4^+]$ are sparsely known, with only one N, three Δ and one Λ states known, several of them with only one star status, iv) two states, which we assign to the $[\mathbf{56}, 6^+]$, and which represent a very loose input in the fit, v) the $[\mathbf{70}, 1^-]$ which is one of the best known excited multiplets, where some of the Σ states and only one state with strangeness smaller than -1 ($\Xi(1823)$) are known, vi) the $[\mathbf{70}, 2^+]$ where three N, one Δ and two Λ states are known, vii) the $[\mathbf{70}, 3^-]$ with three N, one Δ and one Λ states known, and finally viii) the $[\mathbf{70}, 5^-]$ where similarly to the case of the $[\mathbf{56}, 6^+]$, the assignment is loose given that only one state is known. Clearly, the most critical aspect of the discussion is the assignment of states into the multiplets. This issue has been studied at length, and for the states in the second resonance region it seems to be clearly understood. In addition to the masses, the analyses of strong and electromagnetic transitions have been used for that purpose.

In the table the results of the fits using the mass formulas in Eqns. (4) are shown. We also include a fit where the masses of the non-strange states have

Table 1. The coefficients c_1, $c_{\rm HF}$ and c_S from fitting to the PDG masses.

Multiplet	c_1 [MeV]	$c_{\rm HF}$ [MeV]	c_S [MeV]
$[\mathbf{56}, 0^+]$	363 ± 0.4 361	269 ± 2	156 ± 1
$[\mathbf{56}, 2^+]$	603 ± 5 589	671 ± 53	126 ± 18
$[\mathbf{56}, 4^+]$	770 ± 20 361	398 ± 372	110 ± 94
$[\mathbf{56}, 6^+]$	954 ± 40 954	342 ± 720	–
$[\mathbf{70}, 1^-]$	529 ± 5 545	443 ± 19	148 ± 13
$[\mathbf{70}, 2^+]$	640 ± 16 671	~ 400 (input)	120 ± 86
$[\mathbf{70}, 3^-]$	731 ± 17 756	249 ± 315	30 ± 159
$[\mathbf{70}, 5^-]$	900 ± 20 (estimated) 900	–	–

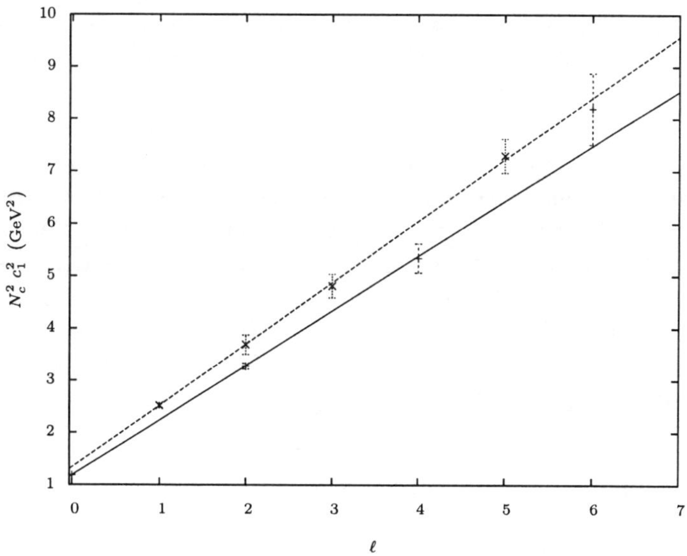

Fig. 1. Plot $(N_c\, c_1)^2$ vs ℓ for the **56**-plets (+) and the **70**-plets (×). The linear fits to the respective trajectories are shown.

been fitted using only the leading term, i.e., the coefficients c_1. Clearly, the results for c_1 in the latter fit are close to the ones in the more accurate fit. In the figure, we plot the square of the spin-flavor singlet component of the baryon masses $(N_c\, c_1)^2$ vs ℓ. It shows very distinctly two Regge trajectories, which are clearly determined up to $\ell = 5$, and distinguish between the **56**- and **70**-plets. As indicated earlier, the assignments to specific multiplets of the $\ell = 5, 6$ inputs are debatable and therefore these inputs should be taken with a grain of salt. The purpose to include them here is to argue that they seem to consistently fit into the linear trend established by the lower ℓ inputs.

The linear fits to the trajectories give:

$$\begin{aligned}(3\, c_1([\mathbf{56}, \ell]))^2 &= ((1.186 \pm 0.003) + (1.05 \pm 0.03)\, \ell)\ \mathrm{GeV}^2, \\ (3\, c_1([\mathbf{70}, \ell]))^2 &= ((1.34 \pm 0.02) + (1.18 \pm 0.02)\, \ell)\ \mathrm{GeV}^2. \end{aligned} \tag{6}$$

For a given multiplet, the coefficient c_1 has a dependence on ℓ which is $\mathcal{O}(1/N_c)$. Therefore, the slopes of the fitted trajectories are $\mathcal{O}(N_c)$. For comparison with the slopes between baryons and mesons it would be more reasonable to fit $N_c\, c_1^2$ instead. Thus, it seems in some sense fortuitous that in the real world, meson and baryon trajectories have similar slopes. In the large N_c limit a plot $N_c\, c_1$ vs ℓ would obviously be the appropriate one.

One important point to be emphasized is that for a given ℓ, the splitting between the spin-flavor singlet pieces of the masses, namely $N_c \times (c_1(\mathbf{70}) - c_1(\mathbf{56}))$, is a quantity $\mathcal{O}(N_c^0)$. To ascertain this, one can simply consider a quark picture of the large N_c baryon in the Hartree approximation, and show that the exchange terms in the matrix elements of the spin-flavor independent piece of the quark-quark interaction give rise to mass contributions $\mathcal{O}(N_c^0)$ for **56**-plets and $\mathcal{O}(1/N_c)$ for **70**-plets. Therefore, one should expect on these grounds the two distinct trajectories we have found, and they should persist in a large N_c world. For the splitting between **56**- and **70**-plet, the following linear relation gives a very accurate approximation:

$$(c_1([\mathbf{56}, \ell]) - c_1([\mathbf{70}, \ell]))^2 = (5.2 + 4.2\, \ell) \times 10^{-4}\, \mathrm{GeV}^2. \tag{7}$$

This corresponds to a mass splitting that increases with ℓ, going from 70 MeV at the $\ell = 0$ intersect to 170 MeV at $\ell = 6$. The hyperfine effects are in general larger than the splitting between trajectories, and for this reason the two distinct trajectories cannot be distinguished when one plots using physical masses. This is again a QCD dynamical issue in which an effect leading in $1/N_c$ is smaller than a sub-leading one.

The validity of the $1/N_c$ expansion should translate into a dependence of $c_{\rm HF}$ and c_S on ℓ of $\mathcal{O}(1/N_c)$. One observes, however, a rather important dependence of $c_{\rm HF}$ (modulo the large error bars). This could be explained by the fact that the hyperfine interaction is more sensitive to the effective size of the core than the other terms in the mass formulas, and for $N_c = 3$ this effect becomes important. In particular, in the quark-diquark picture of baryons, this sensitivity in the hyperfine effect indicates a reduction in the size of the diquark for $\ell > 0$. Concerning c_S, the quark model[2] gives very similar results, except for the $[\mathbf{70}, 3^-]$ multiplet, which is larger than in the present fit (note however the large error bar). The quark model calculation indeed points to a more or less constant value of c_S for all multiplets, which is to be expected.

As a summary, we draw some conclusions from our analysis.

i) The QCD string-like confinement dynamics giving rise to approximately linear Regge trajectories clearly resides in the $SU(6) \times O(3)$ singlet component of the baryon masses.

ii) The clear separation of the two trajectories corresponding to **56** and **70**-plets strongly suggest that configuration mixings are relatively small, thus emphasizing the correctness classification of multiplets of observed excited baryons into multiplets of spin-flavor. This is also seen to be correct in the analyses of strong and electromagnetic transitions.

iii) Unlike mesons, the widths of excited baryons are $\mathcal{O}(N_c^0)$, and one may therefore expect some consequences for the behavior of Regge trajectories for highly excited states. Intuitively, in mesons the string breakup is suppressed in large N_c limit, and thus it is possible that linear trajectories extend out to highly excited states. In baryons, because the decays are not suppressed in large N_c, the string breakup should as well be unsuppressed, and therefore one would expect that the linear character of the trajectories should end at some finite excitation.

iv) Finally, in this meeting the framework of AdS/QCD was addressed in several talks. In this framework the spectrum of excited hadrons depends on the model used to introduce the QCD scale, which corresponds, in the models discussed in those talks, to modifying the AdS geometry in the 5-th dimension by introducing an IR cut-off in different ways. Regge trajectories impose, therefore, a selection criterion for these models. In this context, the results of this work provide useful constraints for AdS/QCD models of baryons.

Acknowledgements

One of us, JLG, thanks the organizers of CAQCD-08 for the opportunity to attend this very interesting meeting and present this work. This work was supported by DOE (USA) through contract DE-AC05-84ER40150, and by the NSF (USA) grant # PHY-0555559 (JLG).

References

1. Particle Data Group, W.-M. Yao et al., *J. Phys. G* **33**, 1 (2006).
2. C. Semay, F. Buisseret, N. Matagne and F. Stancu, *Phys. Rev. D* **75**, 096001 (2007).
3. J. L. Goity and N. Matagne, *Phys. Lett. B* **655**, 223 (2007).
4. J. L. Gervais and B. Sakita, *Phys. Rev. Lett.* **52**, 87 (1984); *Phys. Rev. D* **30**, 1795 (1984).
5. R. Dashen and A. V. Manohar, *Phys. Lett. B* **315**, 425 (1993); ibid, 438 (1993).
6. C. E. Carlson, C. D. Carone, J. L. Goity and R. F. Lebed, *Phys. Lett. B* **438**, 327 (1998), and *Phys. Rev. D* **59**, 114008 (1999).
   C. L. Schat, J. L. Goity and N. N. Scoccola, *Phys. Rev. Lett.* **88**, 102002 (2002).
   J. L. Goity, C. L. Schat and N. N. Scoccola, *Phys. Rev. D* **66**, 114014 (2002).
   N. Matagne and Fl. Stancu, *Phys. Lett. B* **631**, 7 (2005), *Phys. Rev. D* **71**, 014010 (2005), and *Phys. Rev. D* **74**, 034014 (2006).
7. J. L. Goity, *Phys. Atom. Nucl.* **68**, 624 (2005), and *Large N_c QCD 2004*, World Scientific, 211 (2005) [arXiv:hep-ph/0504101].
8. J. L. Goity, *Phys. Lett. B* **414**, 140 (1997).
   D. Pirjol and T. M. Yan, *Phys. Rev. D* **57** 1449 (1998); ibid. **D57**, 5434 (1998).
9. D. Pirjol and C. L. Schat, *Phys. Rev. D* **67**, 096009 (2003).
10. T. D. Cohen and R. F. Lebed, *Phys. Rev. Lett.* **91**, 012001 (2003).

Infrared Behavior of the Fermion Propagator in Unquenched QED_3 with Finite Threshold Effects

Yuichi Hoshino

Kushiro National College of Technology, Otanoshike Nishi 2-32-1, Kushiro City, Hokkaido 084, Japan
E-mail: hoshino@ippan.kushiro-ct.ac.jp

Infrared behavior of the fermion propagator in QED3 is analysed by dispersion method with dressed photon spectral function. Propagator has no one-partcle states. Finite threshold effects of fermion pair cause oscillation of the propagator.However this does not affect the order parameter. Screening effect is large for small N and the N dependence of the order parameter is not significant.

1. Introduction

To study the infrared behavior of the propagator in QED_{2+1},we evaluated the spectral function which is known as the Bloch-Nordsieck approximation in four dimension.In quenched case linear and logarithmic infrared divergences appear in the order e^2 spectral function F.Exponentiation of F yields full propagator with all order of infrared divergences.In four dimension anomalous dimension modifies the short distance singularity at least for weak coupling,which leads cut structure near the mass shell.For fixed infrared cut-off we expand the function F and find a mass shift,its log correction and anomalous dimension of wave function.To avoid the linear infrared divergences we include effects of massive fermion loops to the photon spectral function.The resutls have no infrared divergence but the effects of finite threshold seems to make oscillation of the propagator.If we set the anomalous dimension to be unity short distance singularity disappears,chiral symmetry breaks dynamically, and the spectral function has $-\delta'(p^2/m^2-1)$ like singularity which shows the absence of one particle state and vanishment of Z_2^{-1}.

2. Spectral representation of the propagator

2.1. *Fermion*

In this section we show how to evalute the fermion propagator non pertubatively by the spectral represntation which preserves unitarity and analyticity.[1–3] Assuming parity conservation we adopt 4-component spinor. The spectral function of the fermion in (2+1) dimension is defined

$$\langle 0|T(\psi(x)\overline{\psi}(y)|0\rangle = i\int \frac{d^3p}{(2\pi)^3}e^{-ip\cdot(x-y)}\int_0^\infty ds\frac{\gamma\cdot p\rho_1(s)+\rho_2(s)}{p^2-s+i\epsilon}, \quad (1)$$

$$\rho(p) = -\frac{1}{\pi}\,\mathrm{Im}\,S_F(p) = \gamma\cdot p\rho_1(p)+\rho_2(p)$$
$$= (2\pi)^2\sum_n \delta^{(3)}(p-p_n)\,\langle 0|\psi(x)|n\rangle\,\langle n|\overline{\psi}(0)|0\rangle\,. \quad (2)$$

In the quenched approximation the state $|n>$ stands for a fermion and arbitrary numbers of photons,

$$|n>=|r;k_1,...,k_n>, r^2=m^2. \quad (3)$$

In deriving the matrix element $\langle 0|\psi(x)|n\rangle$ we must take into occount the soft photon emission vertex which is written in the textbook for the scattering of charged particle by external electromagnetic fied or collision of charged particles.Based on low-energy theorem the most singular contribution for the matrix element $T_n = \langle\Omega|\psi|r;k_1,....,k_n\rangle$ is known as the soft photons attached to external line.Ward-Takahashi-identity

$$T_n = \epsilon^n_\mu T^\mu_n,$$
$$k^n_\mu T^\mu_n(r;k_1,..kn) = eT_{n-1}(r;k_1,..k_{n-1}), k^{n2}_\mu \neq 0 \quad (4)$$

is proved to be satisfied with the use of LSZ reduction formula.[2]We have an approximate solution of (4)

$$T_n|_{k_n^2=0} = e\frac{\gamma\cdot\epsilon}{\gamma\cdot(r+k_n)-m}T_{n-1}. \quad (5)$$

From this relation the n-photon matrix element is replaced by the products of T_1

$$T_nT_n^+\gamma_0 \to \prod_{j=1}^n T_1(k_j)T_1^+(k_j)\gamma_0. \quad (6)$$

By one-photon matrix element

$$T_1 = -ie\frac{\gamma\cdot(r+k)+m}{(r+k)^2-m^2+i\epsilon}\gamma_\mu\epsilon^\mu_\lambda(k)U_S(r), \quad (7)$$

and the function

$$F = \int\frac{d^3k}{i(2\pi)^2}\exp(ik\cdot x)\theta(k_0)\delta(k^2)\sum_{\lambda,S}T_1\overline{T_1}, \quad (8)$$

the infinite sum $\sum_{n=0} T_n T_n^+ \gamma_0/n!$ leads an $\exp(F)$.In this way we obtain a dressed fermion propagator with soft photon

$$S_F(x) = -(\frac{i\gamma\cdot\partial}{m}+1)\int \frac{md^2r}{(2\pi)^2\sqrt{r^2+m^2}}\exp(ir\cdot x)\exp(F), \qquad (9)$$

where F is known as model independent

$$F = -\frac{e^2}{2}(\frac{\gamma\cdot r+m}{m})\int \frac{d^3k}{i(2\pi)^2}\exp(ik\cdot x)\theta(k^0)\delta(k^2)[\frac{m^2}{(r\cdot k)^2}+\frac{1}{(r\cdot k)}+\frac{d-1}{k^2}], \qquad (10)$$

where we used covaiant d gauge photon propagator and $\delta(k^2)$ is read as the imaginary part of the free photon propagator at on shell.In our approximation two kinds of spectral function satisfy $\rho_2 = m\rho_1$.

2.2. *Photon*

For unquenched case we use the dressed photon with massive fermion loop with N flavours.Spectral functions for dressed photon are given by vacuum polarization.[4,5]

$$\begin{aligned}\Pi_{\mu\nu}(k) &\equiv ie^2\int \overline{d^3p}Tr(\gamma_\mu\frac{1}{\gamma\cdot p-m}\gamma_\nu\frac{1}{\gamma\cdot(p-k)-m})\\ &= -e^2\frac{T_{\mu\nu}}{8\pi}[(\sqrt{k^2}+\frac{4m^2}{\sqrt{k^2}})\ln(\frac{2m+\sqrt{k^2}}{2m-\sqrt{k^2}})-4m], \qquad (11)\\ T_{\mu\nu} &= -(g_{\mu\nu}-\frac{k_\mu k_\nu}{k^2}), \overline{d^3p} = \frac{d^3p}{(2\pi)^3},\end{aligned}$$

$$D^{-1}_{\mu\nu}(k) \equiv -T_{\mu\nu}(k^2+dk_\mu k_\nu)+\Pi_{\mu\nu}(k). \qquad (12)$$

Polarization function $\Pi(k)$ is

$$\begin{aligned}\Pi(k) &= -\frac{e^2}{8\pi}[(\sqrt{-k^2}+\frac{4m^2}{\sqrt{-k^2}})\ln(\frac{2m+\sqrt{-k^2}}{2m-\sqrt{-k^2}})-4m],\\ &= -\frac{e^2}{8}i\sqrt{-k^2}(-k^2>0, m=0). \qquad (13)\end{aligned}$$

Fermion mass is assumed to be generated dynamically.In quenched case it is shown that m is proportional to e^2.[2] For massless case or high-energy limit we have for number of N fermion flavour

$$\rho^D_\gamma(k) = \frac{1}{\pi}\operatorname{Im} D_F(k) = \frac{c\sqrt{k^2}}{k^2(k^2+c^2)}, c = \frac{e^2N}{8}, \qquad (14)$$

$$D_{\mu\nu}(k) = -T_{\mu\nu}\int_0^\infty \frac{\rho_\gamma(\mu^2)d\mu^2}{k^2-\mu^2+i\epsilon} - d\frac{k_\mu k_\nu}{(k^2-i\epsilon)^2}. \tag{15}$$

If we include finite threshold effects of massive fermion pair we have

$$\rho_\gamma^F(k) = \frac{1}{\pi}\operatorname{Im} D_F(k) = \frac{\operatorname{Im}\Pi(k)}{\pi(-k^2+\operatorname{Re}\Pi(k))^2+(\operatorname{Im}\Pi(k))^2}\theta(k^2-4m^2). \tag{16}$$

3. Analysis in position space

To evaluate the function F it is helpful to use the exponential cut-off(infrared cut-off).[2]Using the bare photon propagator with bare mass μ ;$D_F^{(0)}(x)_+ = \exp(-\mu|x|)/8\pi i|x|$.The function F is written in the following form

$$F = ie^2m^2\int_0^\infty \alpha d\alpha D_F(x+\alpha r) - e^2\int_0^\infty d\alpha D_F(x+\alpha r) - i(d-1)e^2\frac{\partial}{\partial\mu^2}D_F(x,\mu^2). \tag{17}$$

The above formulea are derived by the parameter tric

$$\lim_{\epsilon\to 0}\int_0^\infty d\alpha\exp(-\alpha(\epsilon - ik\cdot r)) = \frac{i}{k\cdot r},$$
$$\lim_{\epsilon\to 0}\int_0^\infty \alpha d\alpha\exp(-\alpha(\epsilon - ik\cdot r)) = -\frac{1}{(k\cdot r)^2}. \tag{18}$$

In this case we have

$$F = \frac{e^2}{8\pi}\Big[\frac{\exp(-\mu|x|)-\mu|x|E_1(\mu|x|)}{\mu} - \frac{E_1(\mu|x|)}{m} + \frac{(d-1)\exp(-\mu\,|x|))}{2\mu}\Big], |x| = \sqrt{x^2} \tag{19}$$

where μ is a bare photon mass.Short distance behaiviour of F has the following form

$$F \sim \frac{e^2(1+d)}{16\pi\mu} + \frac{e^2}{8\pi m}(\gamma + (1+m|x|)\ln(\mu\,|x|)) - \frac{(d+1-2\gamma)e^2|x|}{16\pi}, (\mu|x| \ll 1). \tag{20}$$

Long distance behaviour is given by the asymptotic expansion of $E_1(\mu|x|)$

$$E_1(z) \sim \frac{\exp(-z)}{z}\{1 - \frac{1}{z} + \frac{1\cdot 2}{z^2} - \frac{1\cdot 2\cdot 3}{z^3} + ...\}, (|\arg z| < \frac{3}{2}\pi), \tag{21}$$

$$F \sim -\frac{e^2}{8\pi}\Big[\frac{\exp(-\mu|x|)}{\mu^2|x|} + \frac{\exp(-\mu|x|)}{m\mu|x|} + \frac{(d-1)\exp(-\mu|x|)}{2\mu}\Big], (\mu|x| \gg 1). \tag{22}$$

where γ is an Euler constant.In (20) linear term in $|x|$ is understood as the finite mass shift from the form of the propagator in position space (9,25) and $|x|\ln(\mu\,|x|)$ term is position dependent mass

$$m = m_0 + \frac{e^2}{16\pi}(1+d-2\gamma) + \frac{e^2}{8\pi m}\mu, \tag{23}$$

$$m(x) = m - \frac{e^2}{8\pi}\ln(\mu\,|x|), \tag{24}$$

which has mass changing effects at short distance and it will be discussed in section 4.These mass terms has different gauge dependence from that obtained by self-energy in the Dyson-Schwinger equation.Here we notice that the propagator in poition space can be written

$$S_F(x) = -(i\gamma\cdot\partial + m)\frac{1}{4\pi|x|}\begin{pmatrix}\exp(-m|x|)A(\mu|x|)^{D+C|x|}(\mu|x| \ll 1)\\ \exp(-m|x|)(\mu|x| \gg 1)\end{pmatrix},$$

$$= i\gamma\cdot\partial S_V(x) + S_S(x), \tag{25}$$

where

$$A = \exp(\frac{e^2(1+d)}{16\pi\mu} + \frac{e^2\gamma}{8\pi m}), C = \frac{e^2}{8\pi}, D = \frac{e^2}{8\pi m}.$$

For the finitenes of the value $S_F(0)$ we imply $D = 1$.In this case the physical mass equals to $m = e^2/8\pi$ and fermion may be confined for finite μ.Here we apply the spectral function of photon to evaluate the unquenched fermion proagator.We simply integrate the function $F(x,\mu)$ for quenched case which is given in (19),where μ is a photon mass.Spectral function of photon with massless fermion loop is given in (13) and we have

$$\rho_\gamma^D(\mu) = \frac{c}{\pi\mu(\mu^2+c^2)}, Z_3^{-1} = \int_0^\infty 2\rho_\gamma^D(\mu)\mu d\mu = 1. \tag{26}$$

In this case the spectral function of the fermion is given

$$\widetilde{\rho}(x) = \int_0^\infty 2\rho_\gamma^D(\mu)\exp(F(x,\mu))d\mu. \tag{27}$$

In this way the short distance fermion propagator with N flavours is modified in the Landau gauge $d = 0$

$$S_F(x) = -(i\gamma\cdot\partial + m)\frac{\exp(-m\,|x|)}{4\pi\,|x|}\widetilde{\rho}(x). \tag{28}$$

For the case of N fermion flavour we assume the physical mass m as $m = c/N\pi$.In the whole region of $|x|$ we evaluate $\widetilde{\rho}(x)$ numerically including finite threshold effect for the photon spectral function with massive fermion loop

$$\widetilde{\rho}(x) = \int_{4m^2}^\infty \exp(F(s,x))\rho_\gamma^F(s)ds. \tag{29}$$

The renormalization constant $Z_3^{-1}$is given

$$Z_3^{-1} = \int_{4m^2}^{\infty} \rho_\gamma^F(s) ds. \tag{30}$$

We set the mass $m = c/N\pi$ and see the N dependence of Z_3^{-1} in unit of $e^2 : c = 1/8$ in Fig.1.

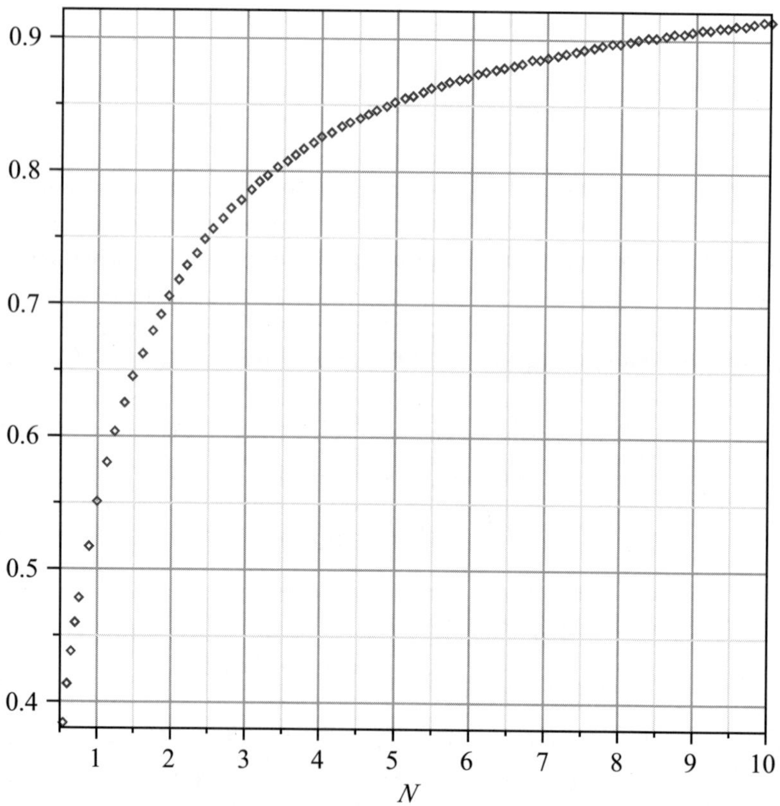

Fig.1 Z_3^{-1} for $N = 1/2..10, c = 1/8$.

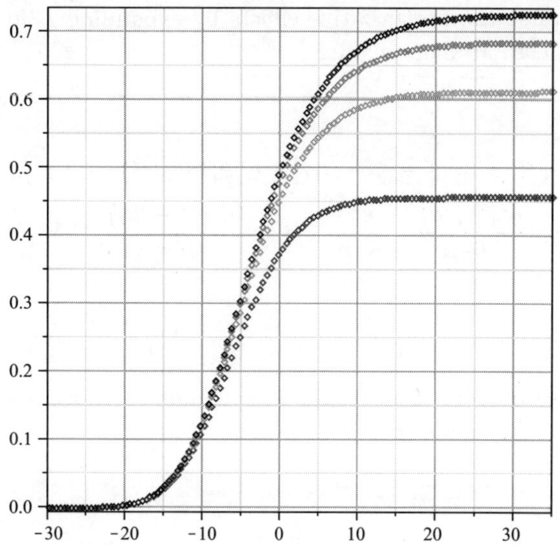

Fig.2.$\widetilde{\rho}(x)$ for $N = 1(bottom), 2, 3, 4(top)$ in unit of e^2.

In Fig.2 we see the screenig effect leads infrared finite spectral function with dynamical fermion mass $m = c/N\pi$ from $N = 1$ to 4 in the Landau gauge $d = 0, c = 1/8$.In comparison with quenched case with finite cut-off μ these are reduced from unity by screening effects at large distance.

4. Analysises in momentum space

4.1. *structure in Euclid space*

Now we turn to the fermion propagator in momentum space.The momentum space propagator at short distance is given by Fourier transform

$$S_F(p) = -\int d^3x \exp(-ip\cdot x)(i\gamma\cdot\partial + m)\frac{\exp(-m\,|x|}{4\pi|x|}\exp(F(x)), \quad (31)$$

where

$$\exp(F(x)) = A\exp(D\gamma)(\mu\,|x|)^{D+C|x|}, D = \frac{c}{N\pi m}, C = \frac{c}{N}. \quad (32)$$

First we examine the perturbative effects by expanding F in powers of e^2 for quenched case at short distace

$$\exp(F) = 1 + F(x) = 1 + A - B|x| + C|x|\ln(\mu|x|) + D\ln(\mu|x|),$$
$$A = \exp(\frac{e^2(1+d)}{16\pi\mu} + \frac{e^2\gamma}{8\pi m}), B = \frac{(d+1-2\gamma)}{16\pi}, C = \frac{e^2}{8\pi}, D = \frac{e^2}{8\pi m}. \tag{33}$$

Terms proportional to B, C represents dynamical mass generation correctly.On the other hand it is known that

$$\int d^3x \exp(-ip\cdot x)\frac{\exp(-m\,|x|)}{4\pi\,|x|}(\mu\,|x|)^D = \frac{2m\mu}{(p^2+m^2)^2} \text{ for } D = 1.$$

for Euclidean momentum $p^2 \geq 0$.However above formule are restricted for small $|x|$,we may include the finite range effect and test the high ernergy behaviour for small μ

$$\int_0^{1/\mu} \frac{\exp(-m|x|)}{|x|}(\mu|x|)\frac{\sin(p|x|)}{p|x|}x^2 dx$$
$$\to -\frac{\mu\exp(-m/\mu)\cos(p/\mu)}{p^2} + \frac{\mu\exp(-m/\mu)\sin(p/\mu)(m-\mu)}{p^3} + O(1/p^4). \tag{34}$$

Above formula shows the oscillation and the propagator does not dump as $1/p^4$.For large N this oscillation effects becomes small as $\mu \to 0$.Numerical solutions of the scalar part of the propagator $S_S(p)$ with various N are shown in Fig.3 for $c = 1, N = 1, 2, 3$.The boundary condition at $p = 0$ is not specfied in our approximation.For $c = 1$ zero momentum mass for $N = 1$ is the largest among them,which has been seen in the Dyson-Schwinger analysis.[8]

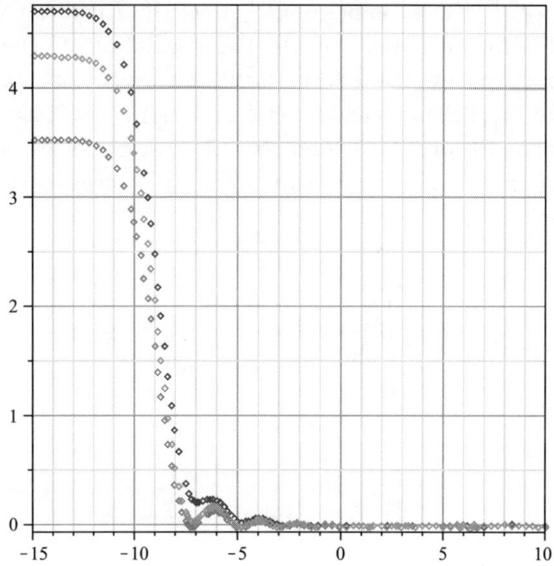

Figure.3 $S_S(p)$ for $c = 1$,
$N = 1(bottom), 2(middle), 3(top), p =$
$\exp(\pi/2 \sinh((m/5))$

4.2. *structure in Minkowski space*

Now we derive the propagator in Minkowski momentum region.To do this it is helpful in real time to derive the spectral function.[1]In Minkowski space we change the variable from $r = \sqrt{x^2}$ to iT.[6]We may define

$$\rho(s) = \frac{1}{\pi} \int_0^\infty \exp(-i(s-1)T) \operatorname{Im}(\exp(\widetilde{F}(iT)))dT, s = p^2/m^2 \tag{35}$$

where

$$\exp(\widetilde{F}(iT)) = \int_{4m^2}^\infty d\sigma \rho^F(\sigma) \exp(F(iT, \sqrt{\sigma})). \tag{36}$$

Taking the imaginary part of the function $\exp(F(\mu, iT))$ at short distance in the Landau gauge for quenched case for $D = 1$

$$\begin{aligned} &\operatorname{Im}\exp(F(\mu, iT)) \\ &= \exp(-\frac{\pi}{2}CT)\mu T \sin(\frac{\pi}{2} + CT\ln(\mu T) - C(1/2 - \gamma)T), \end{aligned} \tag{37}$$

provided $(i\mu T)^{iCT+D} = \exp(\frac{\pi}{2}(-CT + Di))(\mu T)^D(\cos(CT\ln(\mu T)) + i\sin(CT\ln(\mu T)))$.In this way we see the oscillation of the propagator in Minkowski space by the effects position dependent mass as $m = iCT\ln(iCT)$.In Fig.4 the spectral function $\rho(s)$ for $c = 1/8$ has sharp peaks for both side of $s = 1$ for $N = 2, 3$.For $N = 1$ these peaks are wide.At small T the function $\exp(F(iT))$ is approximate as $i\mu T$ suggests the singularity such as the $-\delta'(s-1)$.

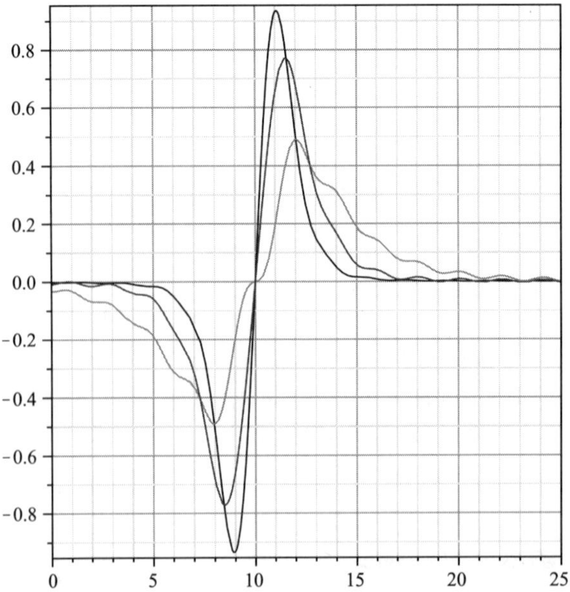

Fig.4 $\rho(s), N = 1(bottom), 2, 3(top),\ s = l/10.$

5. Renormalization constant and order parameter

In this section we consider the renormalization constant in our model.It is easy to evaluate the renormalization constant and bare mass which are defined by assuming multiplicative renormalization

$$\frac{Z_2^{-1}}{\gamma \cdot p - m_0} = S_F(p). \tag{38}$$

$$\begin{aligned} Z_2^{-1} &= \lim_{p\to\infty} \frac{1}{4} tr(\gamma \cdot p S_F(p)) \\ &= \lim_{p\to\infty} \int \frac{p^2\rho(s)ds}{p^2 - s + i\epsilon} = \int \rho(s)ds, \end{aligned} \tag{39}$$

where Z_2 is defined for one particle state in the intial and final state in a weak sense $\psi(x)_{t\to+\infty,-\infty} \to \sqrt{Z_2}\psi(x)_{out,in}$

$$Z_2(2\pi)^{-2}\theta(p_0)\delta(p^2-m^2) \equiv |\langle p|\psi(0)|0\rangle|^2_{p^2=m^2}. \tag{40}$$

However pole part is absent in our approximation and Z_2 vanishes for $D > 0$ case.For $D = 1$ we have

$$\begin{aligned} Z_2^{-1} &= \int_0^\infty ds \int_{-\infty}^\infty \frac{dt}{2\pi} \exp(-i(s-1)t) \operatorname{Im}(\exp(F(it))) \\ &= \int_{-\infty}^\infty dt \delta_+(t) \operatorname{Im}(\exp(F(it))) = 0, \end{aligned} \tag{41}$$

provided $\operatorname{Im}(\exp(F(0))) = 0$.Finally we notice the analytic solution of scalar part of linear Schwinger-Dyson equation in quenched Landau gauge[7]

$$S_S(p) = \frac{m^3}{(p^2+m^2)^2}, m = \frac{e^2}{4\pi}. \tag{42}$$

If we continue to Minkowski space,we have the imaginary part of the propagator

$$\Im \frac{m^3}{(p^2-m^2+i\epsilon)^2} = \frac{\pi\delta(p^2-m^2)}{p^2-m^2} = -\pi\delta'(p^2-m^2). \tag{43}$$

Order parameter for each flavour $\langle\overline{\psi}\psi\rangle$ is given

$$\langle\overline{\psi}\psi\rangle = -trS_F(x). \tag{44}$$

We obtain the values of $\langle\overline{\psi}\psi\rangle$ are $6.8(3.4, 2.3)\times10^{-3}e^4$ for $N = 1(2,3)$.These are relatively large in comparison with $1.2(0.13, 0.0002) \times 10^{-3}e^4$for $N = 1(2,3)$ in the CP vertex with massless loop correction.[8]If we take into account Z_3^{-1} for $N = 1, 2$ case, our numerical solution is reduced by 1/3 and may be equivalent with massless fermion loop case where $Z_3^{-1} = 1$.

6. Summary

We evaluated the fermion propagator in three dimensional QED with dressed photon by the dispersion method.In the evaluation of lowest order matrix element for fermion spectral function we obtain finite mass shift,wave function renormalization and gauge invariant position dependent mass with bare photon mass μ as an infrared cut-off.To remove linear infrared divergences we include finite threshold effects of massive fermion loop to the photon spectral function.We set the anomalous dimension $D = c/N\pi m$ to be unity which is consistent with perurabative analysis of the mass at high energy which is proportional to$1/p^2$.In this case the order parameter $\langle\overline{\psi}\psi\rangle$ becomes finite.In our approximation vacumm expectation value and infrared mass are not sensitive to flavour number N which may due to the screening effects of Z_3^{-1} for small N.In Minkowski space the spectral fuction has a form such as $-\delta'(p^2/m^2 - 1)$ which leads to the vanishment of Z_2^{-1} at least for weak coupling.In our analysis chiral symmetry breaks dynamically by anomalous dimension.Our results suggests the importance of anomalous dimension to understand confinement.

7. Aknowledgement

The author would like to than Professor Robert Delbourgo for critiques of linear divergence.

References

1. R.Jackiw and L.Soloviev, "Low-Energy-Theorem Approach to Single-Particle Singularity in the Presence of Massless Bosons," Phys. Rev. 137.3 (1968) 1485
2. Y.Hoshino, "Mass singulaity and confining property in QED_3," JHEP 0409:048 (2004), where I have a wrong sign of $|x|\ln(\mu|x|)$ term.
3. Y.Hoshino, "Determination of the propagator in QED_3 by spectral function," Nucl.Phys. A790 (2007) 613c-618c.
4. A.B.Waites,R.Delbourgo, "Non perturbative Behavior in Three-Dimensional QED," Int. J. Mod. Phys. **A7** (1992) 6857.
5. C.Itzykson,J.B.Zuber, "Quantum field theory," McGRAW-HILL.
6. J.Schwinger, "Particle Sources and Fields," vol.I,Pereseus Books Publishing, L,L,C(1970).
7. M.Koopman, "Dynamical Symmetry Breaking in QED_3," Ph.D thesis. Groningen University (1990).
8. C.S.Fisher,R.Alkofer,T.Dahm,P.Maris, "Dynamical Chiral symmetry Breaking in QED_3," Phys.Rev. D70 073007 (2004).

Gauge Fields in Accelerated Frames

Frieder Lenz

Institute for Theoretical Physics III, University of Erlangen-Nürnberg
91058 Erlangen, Germany
* *E-mail:flenz@theorie3.physik.uni-erlangen.de*

Quantized fields in accelerated frames (Rindler spaces) with emphasis on gauge fields are investigated. Important properties of the dynamics in Rindler spaces are shown to follow from the scale invariance of the corresponding Hamiltonians. Origin and consequences of this extraordinary property of Hamiltonians in Rindler spaces are elucidated. Characteristics of the Unruh radiation, the appearance of a photon condensate and the interaction energy of vector and scalar static charges are discussed and implications for Yang-Mills theories and QCD in Rindler spaces are indicated.

1. Introduction

In this contribution I will report on an investigation of gauge fields in static space times which has been carried out in collaboration with K. Yazaki and K. Ohta.[1] Our interest in this subject was triggered by the Ads/CFT correspondence which opened the possibility to formulate effective theories of e.g. QCD in terms of quantum fields in gravitational backgrounds. The studies I report on are intended to improve and extend our understanding of the peculiar properties of the dynamics of quantized fields in static space-times.

The following discussion will focus on results concerning quantized gauge fields in Rindler spaces or equivalently quantized gauge fields as seen by a uniformly accelerated observer. The metric of Rindler spaces incorporates the physics of the relativistic generalization of a homogeneous gravitational field and can therefore be used locally as an approximation to more complicated spaces such as the space close to the horizon of a Schwarzschild black hole. Since the discovery of the so called Unruh effect,[2–4] i.e. the appearance of a heat bath as a result of the acceleration and its relation to Hawking radiation about 30 years ago, conceptual questions concerning the presence or absence of radiation have played an important role (for a review cf.[5]). Applications of such studies have addressed the motion of particles in accelerators (cf. [6]), acceleration induced decays of particles[7,8] or, very recently, the possibility of thermalization in relativistic heavy ion collisions.[9] While most of the results have been obtained for scalar fields no systematic investigations of the properties of gauge fields in Rindler spaces have been carried out.

2. Kinematics

A uniformly accelerated observer in $d + 1$ dimensional Minkowski space moves along the hyperbola[10]

$$x^2 - t^2 = \frac{1}{a^2}, \quad \mathbf{x}_\perp = 0. \tag{1}$$

The acceleration is denoted by a and the d-1 coordinates transverse to the motion by $\mathbf{x}_\perp$. The initial conditions

$$x(0) = \frac{1}{a}, \quad \frac{dx}{dt}\Big|_{t=0} = 0,$$

have been chosen. To describe quantum fields as seen by the accelerated observer we transform into his rest frame and consider the coordinate transformation

$$t, x, x_\perp \to \tau, \xi, x_\perp : \quad t(\tau,\xi) = \frac{1}{a}\, e^{a\xi} \sinh a\tau, \quad x(\tau,\xi) = \frac{1}{a}\, e^{a\xi} \cosh a\tau, \tag{2}$$

with the inverse

$$x^2 - t^2 = \frac{1}{a^2}\, e^{2a\xi}, \quad \frac{t}{x} = \tanh a\tau.$$

By construction, $\xi = 0$ corresponds to the hyperbolic motion (1) or, more generally, a particle at rest in the observers system at $\xi = \xi_0$ =const. corresponds to the uniformly accelerated motion in Minkowski space with acceleration $a \exp\{-a\xi_0\}$. Trajectories of uniformly accelerated particles for different values of ξ_0 are shown in Fig. 1 together with the lines τ =const.. We note that the mapping (2) is not one-to-one. The coordinates $-\infty < \tau, \xi < \infty$ cover only one quarter of the Minkowski space, the right "Rindler wedge" R_+ (region I)

$$R_\pm = \{x^\mu \big| \, |t| \leq \pm x\}. \tag{3}$$

Upon reversion of the sign of x in Eq. (2) the left Rindler wedge R_- (region III) is covered by the corresponding parametrization. As illustrated in the Figure, the light cone $x = t$ corresponding to $\xi = -\tau = -\infty$ is an event

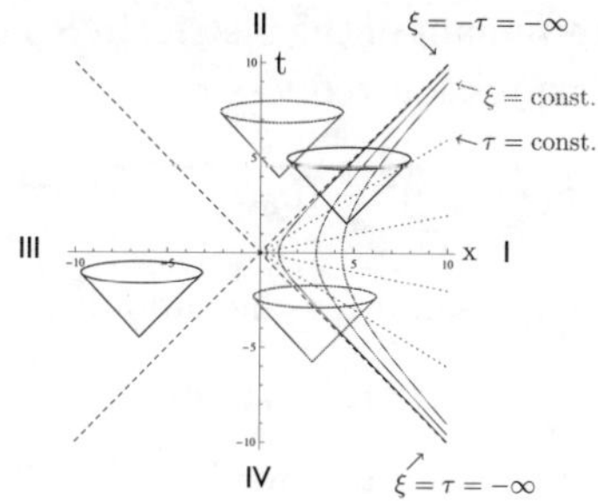

Fig. 1. Kinematics of uniform acceleration

horizon. The light cones indicate that in region I signals can be transmitted to region II but not received from it. Signals received from region IV appear to have originated from the horizon $\xi = \tau = -\infty$. The space-time defined by the coordinate transformation (2) is called Rindler space and its metric is given by

$$ds^2 = e^{2a\xi}(d\tau^2 - d\xi^2) - d\mathbf{x}_\perp^2 \,. \tag{4}$$

Observations in the accelerated frame can be interpreted equivalently as observations of a stationary observer in a gravitational background. Indeed the Rindler metric accounts for the gravitational field in the near horizon limit of a Schwarzschild black hole (cf. [11]) with the black hole horizon at $x = t = 0$ and the acceleration given by the inverse of the Schwarzschild radius. In the non-relativistic limit, $a\tau \ll 1$, the Rindler metric yields a linear potential

$$U \sim \frac{c^2}{2}\big(g_{00} - 1\big) = \frac{c^2}{2}\big(e^{2a\xi} - 1\big) \approx ac^2\xi\,,$$

generating the parabolic motion of test particles around $t = 0$ (cf. Fig. 1).

3. Scalar fields in Rindler spaces

In preparation for the discussion of gauge fields I will introduce some of the relevant concepts in the context of scalar fields in Rindler spaces. The action of a non-interacting scalar with self-interaction $V(\phi)$ in curved space-time is given by

$$S = \int \sqrt{|g|} d\tau\, d\xi\, d^{d-1}x_\perp \big\{\frac{1}{2}g^{\mu\nu}\partial_\mu\phi\partial_\nu\phi - V(\phi)\big\}\,,$$

with g denoting the determinant of the metric. In Rindler space, and keeping only a mass term this expression reduces to

$$S = \frac{1}{2}\int d\tau\, d\xi\, d^{d-1}x_\perp \{(\partial_\tau\phi)^2 - (\partial_\xi\phi)^2 - (m^2\phi^2 + (\partial_\perp\phi)^2)\, e^{2a\xi}\}\,. \quad (5)$$

Since $\sqrt{|g|}g^{00} = 1$, the expression for the canonical momentum is standard

$$\pi(x) = \partial_\tau\phi(x)\,,$$

and so is the equal time commutator for ϕ and π. The fields are expanded in terms of the normal modes of the associated wave equation

$$\phi(\tau,\xi,\mathbf{x}_\perp) \sim \int \frac{d\omega}{\sqrt{2\omega}} d^{d-1}k_\perp (a(\omega,\mathbf{k}_\perp)e^{-i\omega\tau + i\mathbf{k}_\perp\mathbf{x}_\perp} + \text{h.c.})\, K_{i\frac{\omega}{a}}\left(m_\perp\, e^{a\xi}\right), \quad (6)$$

with

$$m_\perp^2 = (m^2 + \mathbf{k}_\perp^2)/a^2\,, \quad (7)$$

and the normalization chosen such that the commutator of creation and annihilation operators $a^{(\dagger)}(\omega,\mathbf{k}_\perp)$ is standard. The non-trivial ξ-dependence, given by the McDonald functions $K_{i\frac{\omega}{a}}\left(m_\perp\, e^{a\xi}\right)$ satisfying the equation

$$\Big(-\frac{d^2}{d\xi^2} + a^2 m_\perp^2\, e^{2a\xi} - \omega^2\Big)K_{i\frac{\omega}{a}}\left(m_\perp\, e^{a\xi}\right) = 0 \quad (8)$$

is illustrated in Fig. 2.

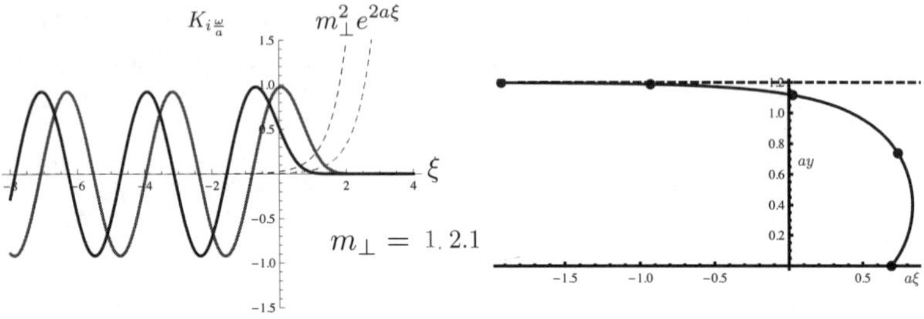

Fig. 2. Left: Eigenfunctions of the wave equation (8) for $\omega/a = 2$ and two values of the transverse mass (7) and corresponding potentials of the wave-equation (8). Right: Trajectory of a particle moving with constant speed in Minkowski as seen by a uniformly accelerated observer (solid line) and asymptotic value of the transverse position (dashed line). The points indicate the values $a\tau = 0, 1, 2, 3, 4$.

The repulsive exponential potential prevents propagation of the wave for positive ξ. This repulsion accounts for the fact that a particle moving with arbitrary constant speed in Minkowski space is seen by the accelerated observer to approach $\xi = -\infty$ and the speed of light for large times τ. This is shown in Fig. 2 for a particle starting at $\tau = 0$ at $ax_0 = 2$ with velocity $v^x = 0.5$, $v^y = 0.3$. At $a\tau = 0.55$, the x-component of the the velocity of accelerated observer and particle coincide and the ξ-component of the velocity of the particle vanishes. In the accelerated frame, the transverse velocity of the particle vanishes exponentially for large times ($\sim \exp\{-2a\tau\}$) as a result of the forever increasing time dilation induced by the acceleration. Around $\tau = 3$, the particle has essentially reached its asymptotic transverse position $a\tilde{y}^y_\infty = ax_0 v^y/(1-v^x)$ in the accelerated frame.

3.1. *Scale invariant Hamiltonian*

The Hamiltonian of the scalar field in Rindler space is given by

$$H = \int d^{d-1}k_\perp \int_0^\infty d\omega\, \omega a^\dagger(\omega, \mathbf{k}_\perp) a(\omega, \mathbf{k}_\perp). \tag{9}$$

Unlike in Minkowski space, the energy ω of Rindler particles is independent of the transverse momentum $\mathbf{k}_\perp$ and mass though the solutions of the wave equation (8) depend on these quantities as is illustrated in Fig. 2 for two eigenfunctions with the same energy. The kinematical origin of the degeneracy of the energy eigenvalues with respect to the transverse momenta is the time dilation in the accelerated frame leading to asymptotically vanishing transverse velocities. The degeneracy can also be interpreted as a consequence of a symmetry. For a massless scalar field, the invariance of the action (5) under the scale transformation

$$\tau' = \tau\,, \quad e^{a\xi'} = e^{a\xi_0} e^{a\xi}\,, \quad \mathbf{x}'_\perp = e^{a\xi_0}\mathbf{x}_\perp\,, \tag{10}$$

$$\phi(\tau, \xi, \mathbf{x}_\perp) \longrightarrow e^{(d-1)a\xi_0/2}\,\phi(\tau, \xi', \mathbf{x}'_\perp)\,,$$

is manifest. The generator of this transformation

$$Q = \int d\xi d^{d-1}x_\perp \pi(\xi, \mathbf{x}_\perp) \left\{\partial_\xi + a\left(\frac{d-1}{2} + \mathbf{x}_\perp \boldsymbol{\partial}_\perp\right)\right\}\phi$$

commutes with the Hamiltonian but not with the transverse component of the momentum operator

$$[H_{m=0}, Q] = 0, \quad [P_i, H_{m=0}] = 0\,, \quad [P_i, Q] = -iaP_i\,, \quad i \neq 1\,. \tag{11}$$

Together with the rotational symmetry in the transverse space these commutators imply the degeneracy of the Hamiltonian (9). In the presence of a mass term, the generator of the symmetry transformations reads

$$Q=\int d\xi d^{d-1}x_\perp \pi(\xi,\mathbf{x}_\perp)\Big\{\partial_\xi + a\Big(\frac{d-1}{2}+\mathbf{x}_\perp\boldsymbol{\partial}_\perp\Big)\phi + a\,m^2\Big(\frac{d-3}{2}+\mathbf{x}_\perp\boldsymbol{\partial}_\perp\Big)$$

$$\int d^{d-1}y_\perp g(|\mathbf{x}_\perp-\mathbf{y}_\perp|)\phi(\tau,\xi,\mathbf{y}_\perp)\Big\} \quad \text{with} \quad \Delta_\perp\, g(\mathbf{x}_\perp) = -\delta(\mathbf{x}_\perp)\,.$$

It is remarkable that the dimensionful Hamiltonian is invariant under scale transformation and that the introduction of a mass term does not destroy this invariance.

3.2. *Unruh heat bath*

Here I will discuss the dynamics of scalar fields in Rindler space in terms of the dynamics in Minkowski space observed in a uniformly accelerated frame. The starting point for establishing the relation between the scalar field theory in inertial and accelerated frames is the identity of the fields in accelerated and inertial frame (Rindler wedge)

$$\phi(\tau,\xi,\mathbf{x}_\perp) = \tilde{\phi}(t,\mathbf{x})\Big|_{t,\mathbf{x}=t,\mathbf{x}(\tau,\xi)}\,.$$

Projection of this equation onto the Rindler space normal modes (6) yields the following relation (Bogoliubov transformation) between the creation and annihilation operators in the two frames

$$a(\Omega,\mathbf{k}_\perp) = \frac{1}{\sqrt{a\sinh\pi\frac{\Omega}{a}}}\int_{-\infty}^{\infty}\frac{dk}{\sqrt{4\pi\omega_k}}e^{i\frac{\Omega}{a}\beta_\mathbf{k}}\left[e^{\frac{\pi\Omega}{2a}}\tilde{a}(k,\mathbf{k}_\perp) + e^{-\frac{\pi\Omega}{2a}}\tilde{a}^\dagger(k,-\mathbf{k}_\perp)\right]. \tag{12}$$

Observations in the accelerated frame are performed in the Minkowski ($|0_M\rangle$) rather than the Rindler space vacuum (barring a local cooling in the observers rocket). A fundamental quantity is the number of particles measured in the accelerated frame which with the help of (12) is found to be

$$\langle 0_M|a^\dagger(\Omega,\mathbf{k}_\perp)a(\Omega',\mathbf{k}'_\perp)|0_M\rangle = \frac{1}{e^{2\pi\frac{\Omega}{a}}-1}\delta(\Omega-\Omega')\delta(\mathbf{k}_\perp-\mathbf{k}'_\perp)\,. \tag{13}$$

In the accelerated frame, a thermal distribution of (Rindler) particles is observed. For a detailed discussion of the "Unruh effect" cf.[5,12]. I will discuss the Unruh radiation in the context of the electromagnetic field.

4. Gauge fields in Rindler spaces

Starting point of the canonical quantization of the electromagnetic field is the Rindler space action in Weyl gauge $A_0 = 0$

$$S[A] = \frac{1}{2}\int d\tau d\xi d^{d-1}x_\perp \Big\{ e^{-2a\xi}(\partial_0 A_1)^2 + \sum_{I=2}^{d} \Big[(\partial_0 A_I)^2 - (\partial_1 A_I - \partial_I A_1)^2\Big] - e^{2a\xi} \sum_{J>I=2}^{d} (\partial_I A_J - \partial_J A_I)^2 \Big\}. \quad (14)$$

In Weyl gauge, the quantization of gauge fields in Rindler space is standard

$$\Pi_i = \partial_0 A^i, \quad [\Pi^i(\tau,\xi,\mathbf{x}_\perp), A_j(\tau,\xi',\mathbf{x}'_\perp)] = \frac{1}{i}\delta_{ij}\delta(\xi-\xi')\delta(\mathbf{x}_\perp - \mathbf{x}'_\perp).$$

The Gauß law is implemented as a constraint on the space of physical states

$$\partial_i \Pi^i |\psi\rangle = 0,$$

and the Hamiltonian reads

$$H = \frac{1}{2}\int d\xi d^{d-1}x_\perp \Big\{ e^{2a\xi}\Pi^{1\,2} + \sum_{I=2}^{d} \Pi^{I\,2} + \sum_{I=2}^{d}(F_{1I}^2 + e^{2a\xi}\sum_{J>I}^{d} F_{IJ}^2)\Big\}. \quad (15)$$

Following the method developed in section 3.1 one easily derives the degeneracy of the energy eigenstates without explicit construction of the normal modes of the system. Under the scale transformations (cf. Eq. (10))

$$\begin{aligned} A_1(\tau,\xi,\mathbf{x}_\perp) &\to e^{(d-3)a\xi_0/2} A_1(\tau,\xi',\mathbf{x}'_\perp), \\ A_I(\tau,\xi,\mathbf{x}_\perp) &\to e^{(d-1)a\xi_0/2} A_I(\tau,\xi',\mathbf{x}'_\perp), \end{aligned} \quad (16)$$

the action (14) remains invariant and the generator of the symmetry transformations is given by

$$Q_A = \int d\xi dx_\perp \Big\{ \sum_{i=1}^{d} \Pi^i \Big[\partial_\xi + a(\frac{d-1}{2} + \mathbf{x}_\perp \boldsymbol{\partial}_\perp)\Big] A_i - a\,\Pi^1 A_1 \Big\}. \quad (17)$$

The generator Q_A, the Hamiltonian H_A and the transverse momentum operator of the electromagnetic field satisfy the operator algebra (11) with the same consequences for the spectrum as for scalar fields.

4.1. *QED and QCD in Rindler space time*

The transformation property of the covariant derivative

$$\begin{aligned} &\partial_1 - ieA_1(\tau,\xi,\mathbf{x}_\perp) \to \partial'_1 - i\hat{e}(\xi_0)\, A_1(\tau,\xi',\mathbf{x}'_\perp), \quad \hat{e}(\xi_0) = e^{(d-3)a\xi_0/2} e, \\ &\partial_I - ieA_I(\tau,\xi,\mathbf{x}_\perp) \to e^{a\xi_0}\big(\partial'_I - i\hat{e}(\xi_0)\, A_I(\tau,\xi',\mathbf{x}'_\perp)\big), \end{aligned} \quad (18)$$

is essential for the symmetry analysis of interacting gauge theories. The Rindler space action of the electromagnetic field coupled to massless fermions (cf.[13,14])

$$S[A,\psi,e] = S[A] + \int d\tau d\xi d^{d-1}x_\perp$$
$$\bar\psi i\Big\{\gamma^0(\partial_0 - ieA_0) + \gamma^1(\partial_1 + \frac{a}{2} - ieA_1) + e^{a\xi}\gamma^I(\partial_I - ieA_I)\Big\}\psi\,,$$

transforms under the combined transformation of gauge (16) and fermion fields

$$\psi(\tau,\xi,\mathbf{x}_\perp) \to e^{(d-1)a\xi_0/2}\psi(\tau,\xi',\mathbf{x}'_\perp) \tag{19}$$

as

$$S[A,\psi,e] \to S[A,\psi,\hat e(\xi_0)]\,,$$

with generator (cf. (17))

$$Q = Q_A - i\int d\xi dx_\perp \psi^\dagger\Big[\partial_\xi + a\big(\frac{d-1}{2} + \mathbf{x}_\perp\boldsymbol{\partial}_\perp\big)\Big]\psi$$

and with the following change in the coupling constant

$$\hat e(\xi_0) = e^{(d-3)a\xi_0/2}e\,. \tag{20}$$

This result applies also to Yang-Mills theories coupled to massless quarks. The transformation of gauge fields (16), of covariant derivatives (18) and of fermion fields (19) have to be applied to each color component and the coupling constant e in (20) replaced by the Yang Mills coupling constant g_{YM} with the result

$$S[A,\psi,g_{YM}] \to S[A,\psi,\hat g_{YM}(\xi_0)]\,.$$

In 3+1 dimensional Rindler space time (d=3) the combined transformation leaves the (tree level) Hamiltonians invariant

$$[H_{QED},Q] = 0\,,\quad [H_{QCD},Q] = 0\,,$$

implying degeneracies in the spectra.

4.2. *Electromagnetic fields in accelerated frames*

In order to study detailed properties of electromagnetic fields in Rindler space, the Gauß law constraint has to be resolved and the normal modes have to be constructed.[1] The resolution of the Gauß law constraint is most efficiently achieved by decomposition of the gauge fields in transverse and longitudinal fields

$$\hat{A}^i = A^i + \partial^i \frac{1}{\Delta}\, \partial_j A^j\,, \quad \Delta = -\partial_i \partial^i = \partial_\xi e^{-2a\xi} \partial_\xi + \boldsymbol{\partial}_\perp^2\,. \tag{21}$$

The normal mode decomposition of the transverse field operators $\hat{A}^i$ is carried out in terms of the d-1 annihilation and creation operators $a_i^{(\dagger)}(\omega, \mathbf{k}_\perp)$. The normal modes are given by the McDonald functions (8) and powers of their argument. In order to establish the relation between measurements in inertial and accelerated frames we again have to identify appropriate field operators in the two systems (cf. Eq. (3.2)). Here the transverse gauge field $\hat{A}_i(x)$ in the accelerated frame $x = \{\tau, \xi, \mathbf{x}_\perp\}$ has to be identified with the transformed inertial frame (transverse) gauge field $\hat{\tilde{A}}_i(\tilde{x})$ in the Rindler wedge $\tilde{x} = \{t, x^1, \mathbf{x}_\perp\} = \tilde{x}(\tau, \xi, \mathbf{x}_\perp)$

$$\hat{A}_i(\tau, \xi, \mathbf{x}_\perp) = \frac{\partial \tilde{x}^k}{\partial x^i} \hat{\tilde{A}}_k(\tilde{x}(\tau, \xi, \mathbf{x}_\perp)) - \frac{\partial}{\partial x^i} \int^\tau d\tau' \, \frac{\partial \tilde{x}^k}{\partial \tau'} \hat{\tilde{A}}_k(\tilde{x}(\tau', \xi, \mathbf{x}_\perp))\,. \tag{22}$$

The first term on the right hand side accounts for the coordinate transformation of a vector field, the second represents the necessary gauge transformation to the Weyl gauge. Projection on the normal modes of the components of the transverse gauge fields yields as above (12) the expression of the Rindler space creation and annihilation operators in terms of the corresponding Minkowski space operators

$$\begin{aligned} a_i(\Omega, \mathbf{k}_\perp) = {} & \frac{1}{\sqrt{a \sinh \pi \frac{\Omega}{a}}} \int_{-\infty}^{\infty} \frac{dk}{\sqrt{4\pi\omega_k}} e^{i\frac{\Omega}{a}\beta_k} \sum_{j=1}^{d-1} \Big[e^{\frac{\pi\Omega}{2a}} R_{ij}(k, \mathbf{k}_\perp) \tilde{a}_j(k, \mathbf{k}_\perp) \\ & + e^{-\frac{\pi\Omega}{2a}} R_{ij}(k, -\mathbf{k}_\perp) \tilde{a}_j^\dagger(k, -\mathbf{k}_\perp) \Big]\,, \end{aligned} \tag{23}$$

where the matrix R describes the mixing of the components of the gauge fields under the coordinate and gauge transformations of Eq. (22).

4.3. *The electromagnetic Unruh heat bath*

The relation (23) of the creation and annihilation operators in inertial and accelerated frames implies the following expression for the number of Rindler photons in the Minkowski vacuum (cf. Eq. (13))

$$\langle 0_M | a_i^\dagger(\omega, \mathbf{k}_\perp) a_j(\omega', \mathbf{k}'_\perp) | 0_M \rangle = \frac{1}{e^{2\pi\frac{\omega}{a}} - 1} \delta_{ij}\, \delta(\omega - \omega') \delta(\mathbf{k}_\perp - \mathbf{k}'_\perp)\,.$$

Although of the same structure as the corresponding expression for the number of Minkowski space photons at finite temperature, the different dispersion law of Rindler photons $\frac{\partial\omega}{\partial \mathbf{k}_\perp} = 0$ gives rise to significant changes in Unruh as compared to blackbody radiation. I illustrate this by a discussion of the energy density $\mathcal{H}$ (cf. Eq. (15)) of the Unruh radiation. As in finite temperature field theory, divergences in $\mathcal{H}$ are avoided by normal ordering with respect to the (Rindler) ground state and the following result is obtained

$$\langle 0_M| : \mathcal{H}(\xi, \mathbf{x}_\perp) :_R |0_M\rangle = N_d a^{d+1} e^{(1-d)a\xi} \int_0^\infty \frac{1}{a} d\omega\, \sigma\Big(\frac{\omega}{a}\Big) , \tag{24}$$

$$\sigma(\kappa) = \frac{\kappa^2 + \frac{1}{4}(d-1)^2}{e^{2\pi\kappa} + (-1)^d} \begin{cases} \kappa\, P^o_{\frac{d-3}{2}}\left(\kappa^2\right) \\ P^e_{\frac{d-2}{2}}\left(\kappa^2\right) \end{cases},$$

where $P^o_{\frac{d-3}{2}}$ and $P^e_{\frac{d-2}{2}}$ denote everywhere non-zero polynomials for odd and even space dimensions respectively with the degree given by the index and with asymptotics

$$\kappa\, P^o_{\frac{d-3}{2}}(\kappa^2), \quad P^e_{\frac{d-2}{2}}(\kappa^2) \xrightarrow[\kappa\to\infty]{} (\kappa)^{d-2} .$$

After redefining the energy density with respect to the measure $\sqrt{|g|}d\xi dx_\perp$ instead of the flat measure (15), $\mathcal{H}'(\xi, \mathbf{x}_\perp) = e^{-2a\xi}\mathcal{H}(\xi, \mathbf{x}_\perp)$, Eq. (24) is rewritten as

$$\langle 0_M| : \mathcal{H}'(\xi, \mathbf{x}_\perp) :_R |0_M\rangle \sim \left(a e^{-a\xi}\right)^{d+1} \sim T_\xi^{d+1} ,$$

with T_ξ satisfying Tolman's law of relativistic thermodynamics[15]

$$T_\xi \sqrt{g_{00}} = \text{const} .$$

The integrand in Eq. (24) suggests significant differences in the frequency distributions of Unruh and blackbody radiation Fig. 3. The two integrands exhibit the same high frequency behavior ($\sim \omega^d e^{-\omega}$); at threshold the integrand of the Unruh radiation approaches a constant while that of the blackbody radiation vanishes like $\sim \omega^{d-1}$.

The appearance of a photon condensate is another consequence of the different dispersion laws of Rindler and Minkowski photons. Unlike in Minkowski space at finite temperature, in Rindler space the photon condensate is in general different from zero

$$\langle 0_M| : \mathcal{H}_E - \mathcal{H}_B :_R |0_M\rangle = N_d d^d\, e^{(1-d)a\xi} \frac{(d-1)(3-d)}{2}$$
$$\cdot \int_0^\infty d\omega \frac{1}{e^{2\pi\frac{\omega}{a}} + (-1)^d} \begin{cases} \frac{\omega}{a}\, P^o_{\frac{d-3}{2}}\left(\frac{\omega^2}{a^2}\right) \\ P^e_{\frac{d-2}{2}}\left(\frac{\omega^2}{a^2}\right) \end{cases} .$$

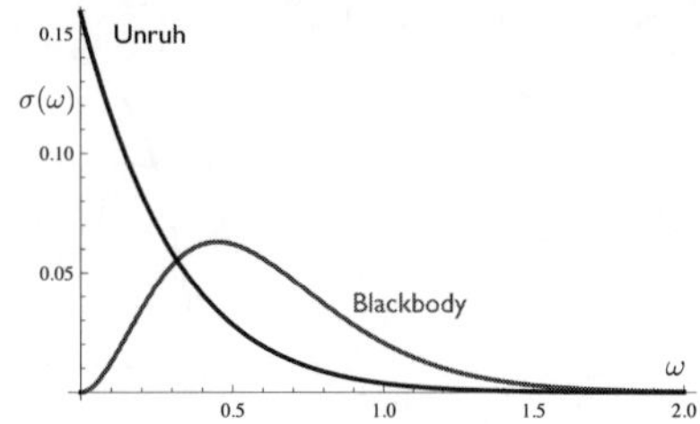

Fig. 3. The integrand of Eq. (24) in comparison with the corresponding integrand of the energy density in the blackbody radiation in 3 space dimensions as a function of ω in units of the acceleration (a) and the temperature ($T/2\pi$) respectively

It vanishes for $d = 3$ and is dominated for higher dimensions by the magnetic field contribution.

4.4. *Interaction energy of static charges in accelerated frames*

In the last application I will discuss the interaction energy of two uniformly accelerated static charges measured by a co-accelerated observer. The Coulomb energy of oppositely charged pointlike sources in Rindler space is given by

$$V_C = e^2 D(\xi_1, \mathbf{x}_\perp, \xi_2, \mathbf{0}_\perp)\,, \quad D(\mathbf{x}, \mathbf{x}') = \langle \mathbf{x}|\frac{1}{\Delta}|\mathbf{x}'\rangle\,,$$

with the Laplace operator Δ defined in (21). The static propagator can be evaluated analytically in terms of Legendre functions of the second kind

$$\frac{V_C}{a} = -\frac{e^{-i\frac{\pi}{2}(d-2)}}{2^{3/2}\pi^{d/2}}\,\tilde{e}^2\,(u^2-1)^{-(d-2)/4}\,Q_{1/2}^{\frac{d}{2}-1}(u)\,. \tag{25}$$

The quantity $u(\xi_1, \mathbf{x}_{\perp\,1}, \xi_2, \mathbf{x}_{\perp\,2})$ is given by the geodesic distance d_g in units determined by the average of the coordinates ξ_1, ξ_2 of the 2 charges

$$u = 1+\frac{1}{2}s^2\,, \quad s^2 = \frac{d_g^2}{a^{-2}e^{a(\xi_1+\xi_2)}}\,, \quad d_g^2 = a^{-2}(e^{a\xi_1}-e^{a\xi_2})^2+(\mathbf{x}_{\perp\,1}^2-\mathbf{x}_{\perp\,2}^2)^2\,.$$

The left side of Fig. 4 shows the geodesic connecting the two charges. The strength of the interaction $\tilde{e}^2$ also "runs" with the average ξ coordinate. Its definition

$$\tilde{e}^2 = \frac{e^2}{(a^{-2}e^{a(\xi_1+\xi_2)})^{(d-3)/2}} = \Big(\frac{a}{\sqrt{2}}\Big)^{d-3}\hat{e}^2\Big(-\frac{1}{2}(\xi_1+\xi_2)\Big)\,,$$

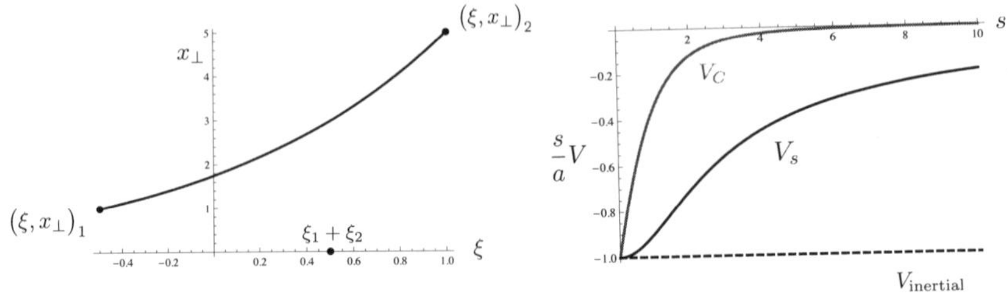

Fig. 4. Left: The geodesic connecting two charges at equal time. Right: Interaction energy of static vector (V_C) and scalar (V_s) charges multiplied with their distance in comparison to the same quantities in the inertial system

is in accordance with the definition (20) of $\hat{e}$ determined from the scaling properties of the action of the Maxwell field. The running of $\tilde{e}$ with the ξ coordinates of the sources is reminiscent of the Ads/CFT duality where the scale is set by the coordinate transverse to the 4-dimensional Minkowski like space.[16] The interaction energy displays power law behavior for large and small distances

$$\frac{V_C}{a} \xrightarrow[s\to 0]{} -\frac{\tilde{e}^2}{s^{d-2}}, \quad \frac{V_C}{a} \xrightarrow[s\to\infty]{} -\frac{\tilde{e}^2}{s^{d+1}},$$

agrees at small distances with the inertial frame result and is suppressed at large distances by three powers. In three space dimensions (as for any odd d), the interaction energy (25) can be expressed by elementary functions

$$\frac{V_C}{a} = \frac{e^2}{4\pi}\Big(1 - \frac{1+\frac{1}{2}s^2}{\sqrt{s^2+\frac{1}{4}s^4}}\Big). \tag{26}$$

In Fig. 4 this interaction energy in an accelerated frame is compared with the inertial frame result and with the interaction energy V_s of sources coupled to a massless scalar field. Common to the three interaction energies is the familiar "$1/r$" behavior at small distances while at large distances scalar and vector interaction energies differ from each other and both from the inertial frame result. In Rindler space, the scalar propagator is the inverse of (cf.(21))

$$\Delta_s = -e^{2a\xi}\partial^i\partial_i = \partial_\xi^2 + e^{2a\xi}\boldsymbol{\partial}_\perp^2,$$

and, independent of the dimension, the following ratio of interaction energies is obtained

$$\frac{V_s}{V_C} = e^{-a(\xi_1+\xi_2)}\Big(1 + \frac{1}{2}s^2 + s\sqrt{1+\frac{1}{4}s^2}\Big).$$

5. Conclusions

I conclude with a few remarks, concerning Yang-Mills fields and QCD as seen by an accelerated observer. Due to its kinematical origin, the characteristic Rindler space degeneracy must show up also in the spectrum of interacting theories. Whatever the elementary excitations, their energy measured by an accelerated observer will neither depend on the transverse momentum, nor on the mass of e.g. glueballs or mesons. Furthermore the analogy with the finite temperature Yang-Mills theory or QCD leads one to expect that either, as a function of the acceleration, these systems exhibit a phase transition or, because of the high degeneracy, do not exhibit at all a confined or a chirally broken phase. In line with such an expectation is the observed weakening of the electrostatic interaction of charges at large distances.

Acknowledgment

I thank K. Yazaki for discussions and for a careful reading of the manuscript.

References

1. F. Lenz, K. Ohta and K. Yazaki, Canonical quantization of gauge fields in static space-times with applications to Rindler spaces **hep-th/0803.2001**, to appear in *Phys. Rev. D*
2. S. A. Fulling, *Phys. Rev. D* **7**, 2850, (1973)
3. P. C. W. Davies, *J. Phys. A* **8**, 609, (1975)
4. W. G. Unruh, *Phys. Rev. D* **14**, 870, (1976)
5. L. C. B. Crispino, A. Higuchi and G. E. A. Matsas, *Rev. Mod. Phys.* **80**, 787, (2008), **gr-qc/0710.5373**
6. J. S. Bell and J. M. Leinaas, *Nucl. Phys. B* **284**, 488, (1987)
7. R. Müller, *Phys. Rev. D* **56**, 953, (1997)
8. D. A. T. Vanzella and G. E. A. Matsas *Phys. Rev. D* **63**, 014010, (2001), **hep-ph/0002010**
9. D. Kharzeev, K. Tuchin, *Nucl. Phys. A* **753**, 316, (2005), **hep-ph/0501234**
10. W. Rindler, Relativity, Oxford University Press 2001
11. L. Susskind and J. Lindesay, Black Holes, Information and the String Theory Revolution, World Scientific, 2005
12. W. G. Unruh and R. M. Wald *Phys. Rev. D* **29**, 1047, (1984)
13. D. R. Brill and J. A. Wheeler, *Rev. Mod. Phys.* **29**, 465, (1957), and Errata *Rev. Mod. Phys.* **33**, 623, (1961)
14. M. Nakahara, Geometry, Topology and Physics, Adam Hilger 1990
15. R. C. Tolman, "Relativity, Thermodynamics and Cosmology", Dover 1987
16. J. Polchinski, M. J. Strassler, *Phys. Rev. Lett.*, **88**, 031601, (2002), **hep-th/0109174**

QCD AT COMPLEX COUPLING, LARGE ORDER IN PERTURBATION THEORY AND THE GLUON CONDENSATE

Y. MEURICE

Department of Physics and Astronomy, University of Iowa,
Iowa City, IA 52242, USA
[*] *E-mail: yannick-meurice@uiowa.edu*

We discuss the relationship between the large order behavior of the perturbative series for the average plaquette in pure gauge theory and singularities in the complex coupling plane. We discuss simple extrapolations of the large order for this series. We point out that when these extrapolated series are subtracted from the Monte Carlo data, one obtains (naive) estimates of the gluon condensate that are significantly larger than values commonly used in the continuum for phenomelogical purpose. We present numerical results concerning the zeros of the partition function in the complex coupling plane (Fisher's zeros). We report recent attempts to solve this problem using the density of states. We show that weak and strong coupling expansions for the density of states agree surprisingly well for values of the action relevant for the crossover regime.

Keywords: Quantum Chromodynamics, large order in pertubation theory, gluon condensate.

1. Motivations

Perturbation theory has played a major role in the establishment of the standard model of electro-weak and strong interactions. However, it is well known[1] that perturbative series have a zero radius of convergence and that at some critical order, adding more terms does not improve the accuracy of the result.

There exists a connection between large field configurations and the large order of perturbative series[2] that can be illustrated with the simple integral

$$\int_{-\infty}^{+\infty} d\phi e^{-\frac{1}{2}\phi^2-\lambda\phi^4} \neq \sum_{0}^{\infty} \frac{(-\lambda)^l}{l!} \int_{-\infty}^{+\infty} d\phi e^{-\frac{1}{2}\phi^2}\phi^{4l} \tag{1}$$

The truncation of the exponential at order l is justified if the argument is much smaller than l. However, the peak of the integrand is located at $\phi = \sqrt{4l}$. For this value of ϕ, the argument of the exponential is $\lambda 16 l^2$, which for l large enough will be larger than l. In other words, the peak of the integrand of the r.h.s. moves too fast when the order increases. On the other hand, if we introduce a field cutoff, the peak moves outside of the integration range and

$$\int_{-\phi_{max}}^{+\phi_{max}} d\phi e^{-\frac{1}{2}\phi^2-\lambda\phi^4} = \sum_0^\infty \frac{(-\lambda)^l}{l!} \int_{-\phi_{max}}^{+\phi_{max}} d\phi e^{-\frac{1}{2}\phi^2} \phi^{4l} \tag{2}$$

This example suggests that one should use perturbation theory to treat small quantum fluctuations and semi-classical methods to treat the large field configurations.

For QCD, an important challenge is to describe the large distance behavior of the theory in terms of degrees of freedom which are weakly coupled at short distance. This question can be addressed in the framework of the lattice formulation. We consider the simplest case of Wilson's action which is the sum over the plaquettes in the fundamental $SU(N)$ representation:

$$S = \sum_p (1 - (1/N) ReTr(U_p)) \ . \tag{3}$$

With the usual notation $\beta = 2N/g^2$, the partition function reads

$$Z = \prod_l \int dU_l e^{-\beta S} \ . \tag{4}$$

For $N = 3$, this theory has no phase transition when one goes from small coupling (large β) to large coupling (small β). Recently, convincing argument have been given[3,4] in favor of the smoothness of the RG flows between the two corresponding fixed points. Consequently, there does not seem to exist any fundamental obstruction to match the two regimes. One general question that we would like to address is if it is possible to construct a modified weak coupling expansion that could bridge the gap to the strong coupling regime.

Lattice gauge theories with a *compact* group have a build-in large field cutoff: the group elements associated with the links are integrated with dU_l, the compact Haar measure. For $SU(2)$ and $SU(3)$, the action density has an upper bound which is saturated when the group element is a nontrivial element of the center. It is remarkable that this formulation has a UV regularization *and* a large field regularization that preserves gauge invariance. Does the presence of large field cutoff means that perturbative series are convergent? Not necessarily, because in constructing perturbative series, one decompactifies[5] the gauge field integration which at low order amounts to neglect exponentially small tails of integration, but modifies the large order drastically.

In continuous field theories, it is expected[6–9] that the large order of perturbative series can be calculated from classical solutions at small negative λ for scalar theories or small negative g^2 for gauge theories. For lattice gauge theory with compact groups, the theory is mathematically well defined at negative g^2 ,i. e., negative β, but some Wilson loops oscillate when their size change and the average plaquette has a discontinuity when g^2 changes sign.[10] For small negative g^2, i. e., large negative β, the behavior of expectation values is dominated by the large field configurations. This statement can be made more precise by introducing the spectral decomposition

$$Z = \int_0^{S_{max}} dS \; n(S) e^{-\beta S} \; , \tag{5}$$

with $n(S)$ the so-called density of states. It is clear that for large negative β, what matters is the behavior of $n(S)$ near S_{max}.

We can make the discussion more concrete by considering the case of the single plaquette model[11] with $SU(2)$ gauge group. In this case, $S_{max} = 2$ and we have

$$n_{1pl.}(S) = \frac{2}{\pi}\sqrt{S(2-S)} \; . \tag{6}$$

The weak coupling expansion of Z is determined by the behavior of near $S = 2$. When we expand $\sqrt{2-S}$ about $S = 0$, the large order of this series is determined by the cut at $S = 2$. After integration (from 0 to $+\infty$) over S, the series with a finite radius of convergence becomes asymptotic (inverse Borel transform). Alternatively, we can maintain the finite range of integration and construct a converging weak coupling expansion but with coupling dependent coefficients as done by Hadamard a century ago in his study of Bessel functions. It would be interesting to know if the features of the one plaquette model persist in the infinite volume limit of lattice gauge theory.

In these proceedings, we discuss the large order behavior of the weak coupling expansion of the plaquette (Sec. 2) and the possibility of defining its non-perturbative part (the gluon condensate? Sec. 3) . More details can be found in Ref.[12] In Sec. 4, we present numerical results concerning the zeros of the partition function in the complex coupling plane[13,14] (Fisher's zeros). This problem could be solved using the density of states. Preliminary numerical results concerning the density of states are provided in Sec.

5 where we also show that weak and strong coupling expansions for the density of states agree surprisingly well for values of the actions relevant for the crossover regime. After the conference, we wrote a more detailed preprint[15] concerning this question.

2. Perturbative series in lattice gauge theory

In this section, we denote $\mathcal{N}_p \equiv L^D D(D-1)/2$ the number of plaquettes and the average plaquette:

$$P(\beta) = (1/\mathcal{N}_p)\left\langle \sum_p (1-(1/N)ReTr(U_p)) \right\rangle . \tag{7}$$

The weak coupling expansion of this quantity

$$P(\beta) \simeq \sum_{m=1} b_m \beta^{-m} + \ldots . \tag{8}$$

has been calculated up to order ten[16] and 16.[17] Series analysis[18,19] suggest a singularity $P \propto (1/5.74 - 1/\beta)^{1.08}$ and consequently a finite radius of convergence. This is not expected since the plaquette changes discontinuously[10] at $\beta \to \pm\infty$. This behavior is also not seen in the 2d derivative of P and would require massless glueballs.

A simple alternative is that the critical point in the fundamental-adjoint plane[20] has mean field exponents and in particular $\alpha = 0$. On the $\beta_{adj.} = 0$ line, we assume an approximate logarithmic behavior (mean field)

$$-\partial P/\partial\beta \propto \ln((1/\beta_m - 1/\beta)^2 + \Gamma^2) . \tag{9}$$

Γ cannot be too small (absence of singularities) or too large (this would create visible oscllations in the perturbative coefficients). From these constraints, we get the approximate bounds[12] $0.001 < \Gamma < 0.01$.

Integrating, we get the approximate form

$$\sum_{k=0} b_k \beta^{-k} \simeq C(\mathrm{Li}_2(\beta^{-1}/(\beta_m^{-1} + i\Gamma)) + \mathrm{h.c} , \tag{10}$$

with

$$\mathrm{Li}_2(x) = \sum_{k=0} x^k/k^2 . \tag{11}$$

We fixed $\Gamma = 0.003$ and obtained $C = 0.0654$ and β_m=5.787 using of b_9 and b_{10}. The low order coefficients depend very little on Γ (when $\Gamma < 0.01$), larger series are needed to get a better estimate of Γ for $SU(3)$. It interesting to notice that we get very good predictions of the values

Table 1. Predicted values of b_m with the dilog model.

order	predicted	numerical[16]	rel. error
1	0.7567	2	-0.62
2	1.094	1.2208	-0.10
3	2.811	2.961	-0.05
4	9.138	9.417	-0.03
5	33.79	34.39	-0.017
6	135.5	136.8	-0.009
7	575.1	577.4	-0.004
8	2541	2545	-0.0016
9	exact	11590	
10	exact	54160	

of $b_8, b_7, \ldots$! We believe that these regularities should have a Feynman diagram interpretation.

Another possible model[21,22] is based on IR renormalons

$$\sum_{k=0} b_k \bar{\beta}^{-k} \simeq K \int_{t_1}^{t_2} dt e^{-\bar{\beta}t} \, (1 - t\, 33/16\pi^2)^{-1-204/121} \tag{12}$$

$$\bar{\beta} = \beta(1 + d_1/\beta + \ldots) \tag{13}$$

$t_1 = 0$ corresponds to the UV cutoff while $t_2 = 16\pi^2/33$ corresponds to the Landau pole. For $t_2 = \infty$ we get a perturbative series with a factorial growth, in contrast with the previous model which had a power growth. Unfortunately a clear distinction between the two types of large order behavior requires numerical calculations at order larger than 20.

3. The Gluon Condensate

Using the two large order extrapolations described in the previous section, one can see good evidence[12] for

$$P(\beta) - P_{pert.}(\beta) \simeq B(a/r_0)^4 \,, \tag{14}$$

with $a(\beta)$ defined with the so-called force scale,[23,24] and P_{pert} appropriately truncated for the second model. B is sensitive to resummation. $B \simeq 0.7$ with the bare series[12] and 0.4 with the tadpole improved series.[17]

It is tempting but potentially dangerous to try to relate B to the numerical value of the gluon condensate.[25] If we identify[26] for $N = 3$

$$P(\beta) - P_{pert.}(\beta) \simeq a^4 \frac{\pi^2}{36} < \frac{\alpha}{\pi} GG >, \tag{15}$$

we obtain for $r_0 = 0.5\ fm$ and $B = 0.4$ that $< \frac{\alpha}{\pi} GG > \simeq 0.035\ GeV^4$ which is about 3 times the original estimate.[25] It is not clear to us that there is a precise correspondence between the continuum and the lattice definitions. Also, different values for the continuum value have been proposed. A negative value,[27] subsequently criticized[28] can even be found. For these reasons, some authors[29] prefer to use a value 0 with error bars when estimating α_s at some large scale. If we use the correspondence between the lattice and the continuum discussed above at face value with $B = 0.4 - 0.7$, these error bars should be multiplied by a factor 3 to 6.

4. Zeros of the Partition Function

The existence of complex singularities near the real axis can be tested by studying the complex zeros of the partition function.[30] The basic technique is the reweighting[31,32] of action distributions at given β_0

$$Z(\beta_0 + \Delta\beta) = Z(\beta_0) < \exp(-\Delta\beta S) >_{\beta_0} \ . \tag{16}$$

As mentioned in the introduction, $Z(\beta)$ is the Laplace transform of the density of states $n(S)$. For $SU(2)$ with even numbers of sites in each direction[10] $Z(-\beta) = e^{2\beta\mathcal{N}_p} Z(\beta)$ and $n(S) = n(2\mathcal{N}_p - S)$. In the crossover, we have attempted local parametrizations:[13,14]

$$n(S) \propto e^{-(a_1 S + a_2 S^2 + a_3 S^4 + a_4 S^4)} \tag{17}$$

For $SU(3)$ on a 4^4 lattice, stronger deviations from the Gaussian behavior are observed than for $SU(2)$. This is illustrated in Fig. 4 that has a larger scale than its $SU(2)$ counterpart. The histogram were made with 50,000 configurations prepared for a study of the third and fourth moments.[19] As the volume increases, these features tend to disappear in the statistical noise. These local model give results that can be compared to MC reweighting in the region where the errors on the zero level curves for the real and imaginary parts are not too large.[13] An important consistency test is to show that approximately the same zeros are obtained for different β_0. This is illustrated in Fig. 2. We plan to pursue this study using the more global information provided by the density of states.

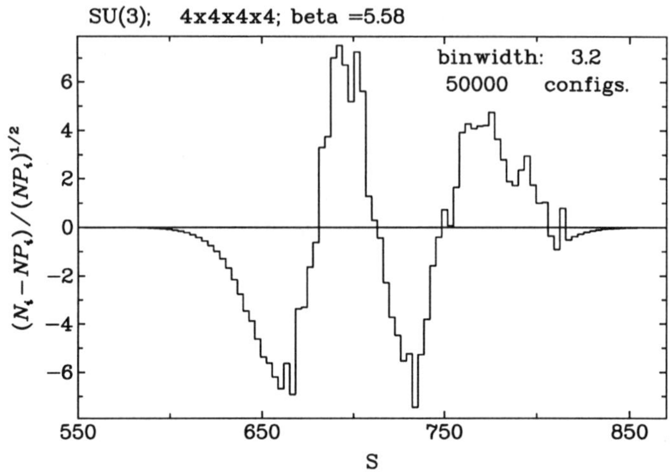

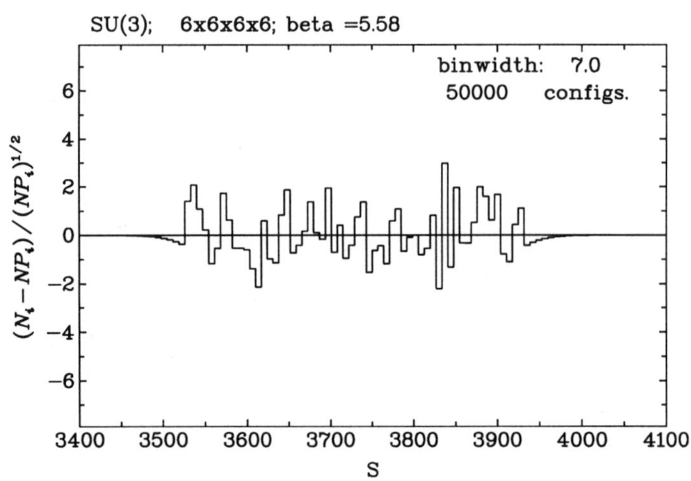

Fig. 1. The relative difference with a Gaussian distribution for 50,000 values of S in an histogram with 100 bins for a $SU(3)$ pure gauge theory on 4^4 and 6^4 lattices at $\beta = 5.58$

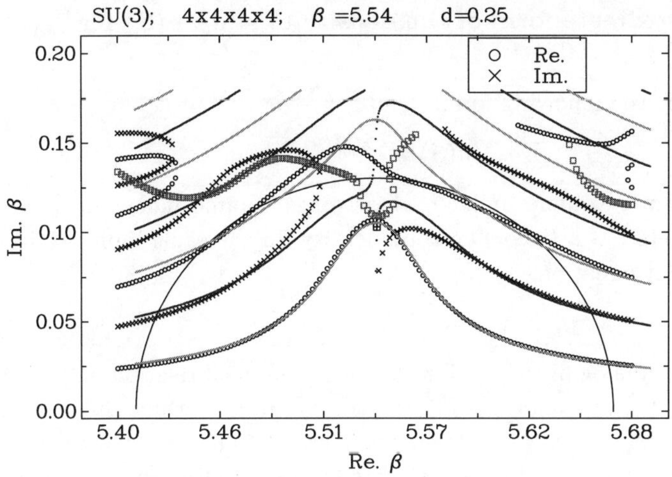

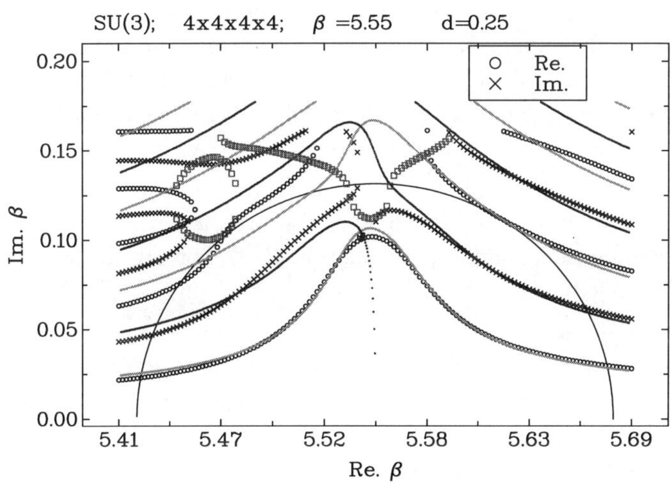

Fig. 2. Zeros of the real (circles) and imaginary (crosses) for $SU(3)$ on a 4^4 lattice using reweighting of MC data at $\beta_0 = 5.54$ and 5.55. The small dots are the values for the real (green) and imaginary (blue) parts obtained from the 4 parameter model. The crossed box at ((5.541,0.103) above and (5.542,0.102) below) have been obtained with a perturbative method. Red boxes: boundary of the MC confidence region. The solid line is the circle of confidence of the Gaussian approximation. The locations of the complex zeros are consistent.

5. Approximate form of the density of state $n(S)$ for $SU(2)$

We assume the following form:

$$n(S) = e^{\mathcal{N}_p f(S/\mathcal{N}_p)} . \tag{18}$$

In the infinite volume limit, $f(x)$ becomes volume independent and can be interpreted as a (color) entropy density. In the same limit, we have the saddle point equation

$$f'(x) = \beta . \tag{19}$$

This is the analog of the familiar thermodynamical relation $dS/dE = 1/T$. Knowing $f(x)$ amounts to solve the theory (in a thermodynamical sense). For $SU(2)$, $f(x) = f(2-x)$ (symmetric about 1) and we don't need to calculate f for $x > 1$. We have constructed weak coupling and strong coupling expansions for $SU(2)$ and compared with numerical data on a 4^4 lattice.

For the weak coupling, we use the large beta expansion near $x = 0$, $x =< S/\mathcal{N}_p >= \frac{3}{4}\frac{1}{\beta} + 0.156\frac{1}{\beta^2} + \ldots$. We assume $f(x) \simeq A\, ln(x) + B + Cx + \ldots$, plug the expansion in the saddle point equation and solve for A and other unknowns. We obtained $f(x) = \frac{3}{4}ln(x) + 0.208x + 0.0804x^2 + \ldots$. For the strong coupling, we use the low beta expansion[33] to solve near $x = 1$ for $SU(2)$, and obtained $x - 1 =< S/\mathcal{N}_p - 1 >= -\frac{\beta}{4} + \frac{\beta^3}{96} - \frac{7\beta^5}{1536} + \frac{31\beta^7}{23040} + \cdots + \frac{1826017873\beta^{15}}{68491306598400}$. With periodic boundary conditions, the coefficients have no volume dependence for graphs with trivial topology (volume dependence should appear at order β^{2L}). Solving the saddle point equation for an expansion about 1, we get $f(x) = -2(x-1)^2 - \frac{2}{3}(x-1)^4 + \cdots + \frac{163150033(x-1)^{16}}{255150} + \ldots$.

The numerical construction of $n(S)$ by patching was done by A. Denbleyker and is illustrated in Fig. 3. Series expansions of $f(x)$ are compared with the numerical data in Fig. 4. Note the good agreement in the intermediate region. After the conference, we wrote a more detailed preprint[15] where details can be found.

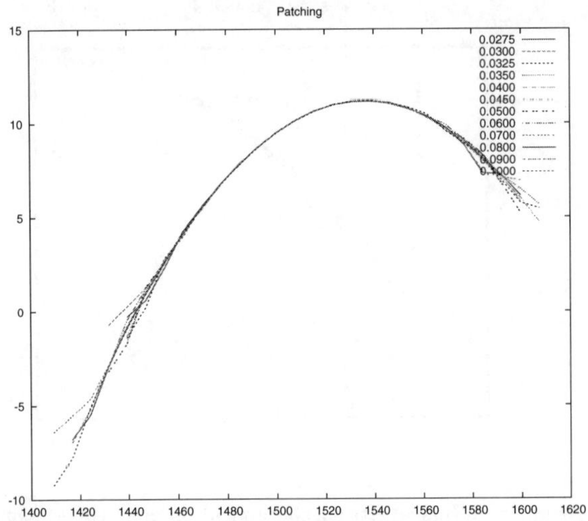

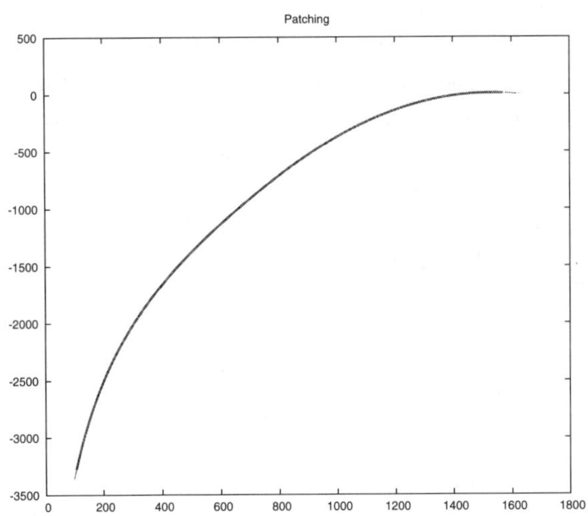

Fig. 3. $n(S) \propto P_{\beta_0}(S)e^{\beta_0 S}$ for different β_0 (patching). Collection of overlapping data (A. Denbleyker).

6. Conclusions

The density of state show a nice overlapping of the strong and weak coupling expansions. We plan to use the density of states to study the Fisher's zeros.

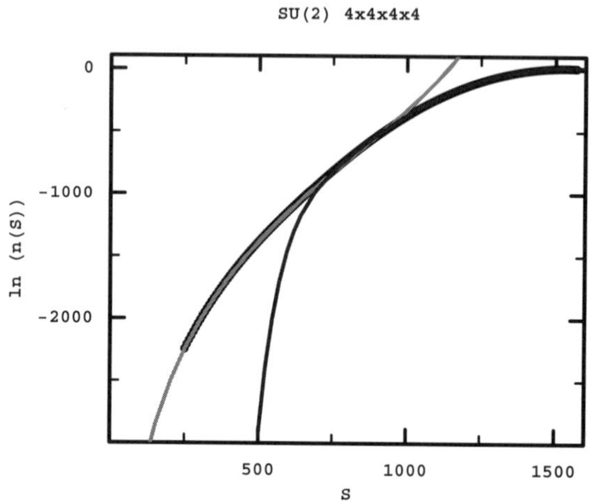

Fig. 4. $ln(n(S))$ numerical (black), strong coupling at order 16 (blue) and weak coupling with dilog model at order 10 (green) for $SU(2)$ on a 4^4 lattice.

We need numerical confirmation of guesses made for the weak coupling expansion for $SU(2)$ where $Im\beta_F/Re\beta_F$ is 5 times larger than for $SU(3)$. We need better understanding of the lattice and the continuum definitions of the gluon condensate. We need a better understanding of the large order behavior of QCD series in terms of the behavior at small complex coupling (a picture analog to metastability at $\lambda < 0$ for the anharmonic oscillator[6–9]).

This research was supported in part by the Department of Energy under Contract No. FG02-91ER40664.

References

1. F. Dyson, *Phys. Rev.* **85**, p. 631 (1952).
2. Y. Meurice, *Phys. Rev. Lett.* **88**, p. 141601 (2002).
3. E. T. Tomboulis, arXiv 0707.2179 (2007).
4. E. T. Tomboulis, *PoS* **LATTICE2007**, p. 336 (2007).
5. U. M. Heller and F. Karsch, *Nucl. Phys.* **B251**, p. 254 (1985).
6. C. Bender and T. T. Wu, *Phys. Rev.* **184**, p. 1231 (1969).
7. G. Parisi, *Phys. Lett.* **68B**, p. 361 (1977).
8. E. Brezin, J. L. Guillou and J. Zinn-Justin, *Phys. Rev. D* **15**, p. 1544 (1977).
9. J. C. LeGuillou and J. Zinn-Justin, *Large-Order Behavior of Perturbation Theory* (North Holland, Amsterdam, 1990).
10. L. Li and Y. Meurice, *Phys. Rev. D* **71**, p. 016008 (2005).
11. L. Li and Y. Meurice, *Phys. Rev.* **D71**, p. 054509 (2005).
12. Y. Meurice, *Phys. Rev.* **D74**, p. 096005 (2006).
13. A. Denbleyker, D. Du, Y. Meurice and A. Velytsky, *Phys. Rev.* **D76**, p. 116002 (2007).
14. A. Denbleyker, D. Du, Y. Meurice and A. Velytsky, *PoS* **LAT2007**, p. 269 (2007).

15. A. Denbleyker, D. Du, Y. Liu, Y. Meurice and A. Velytsky, arXiv 0807.0185 (2008) (*Phys. Rev. D* (submitted)).
16. F. Di Renzo and L. Scorzato, *JHEP* **10**, p. 038 (2001).
17. P. E. L. Rakow, *PoS* **LAT2005**, p. 284 (2006).
18. R. Horsley, P. E. L. Rakow and G. Schierholz, *Nucl. Phys. Proc. Suppl.* **106**, 870 (2002).
19. L. Li and Y. Meurice, *Phys. Rev.* **D73**, p. 036006 (2006).
20. G. Bhanot and M. Creutz, *Phys. Rev.* **D24**, p. 3212 (1981).
21. A. H. Mueller Talk given at Workshop on QCD: 20 Years Later, Aachen, Germany, 9-13 Jun 1992.
22. F. Di Renzo, E. Onofri and G. Marchesini, *Nucl. Phys.* **B457**, 202 (1995).
23. M. Guagnelli, R. Sommer and H. Wittig, *Nucl. Phys.* **B535**, 389 (1998).
24. S. Necco and R. Sommer, *Nucl. Phys.* **B622**, 328 (2002).
25. M. A. Shifman, A. I. Vainshtein and V. I. Zakharov, *Nucl. Phys.* **B147**, 385 (1979).
26. A. Di Giacomo and G. C. Rossi, *Phys. Lett.* **B100**, p. 481 (1981).
27. M. Davier, S. Descotes-Genon, A. Hocker, B. Malaescu and Z. Zhang, arXiv 0803.0979 (2008).
28. K. Maltman and T. Yavin, arXiv 0807.0650 (2008).
29. C. T. H. Davies *et al.*, arXiv 0807.1687 (2008).
30. M. Fisher, *in Lectures in Theoretical Physics Vol. VIIC* (University of Colorado Press, Boulder, Colorado, 1965), Boulder, Colorado.
31. M. Falcioni, E. Marinari, M. L. Paciello, G. Parisi and B. Taglienti, *Phys. Lett.* **B108**, 331 (1982).
32. N. A. Alves, B. A. Berg and S. Sanielevici, *Phys. Rev. Lett.* **64**, 3107 (1990).
33. R. Balian, J. M. Drouffe and C. Itzykson, *Phys. Rev.* **D11**, p. 2104 (1975).

511 KeV Line and Other Diffuse Emissions as a Trace of the Dark Matter

Ariel R. Zhitnitsky

Department of Physics and Astronomy, University of British Columbia, Vancouver, Canada

In this talk I argue that dark matter in the form of dense matter/antimatter nuggets could provide a natural and unified explanation for several distinct bands of diffuse radiation from the core of the galaxy spanning over 13 orders of magnitude in frequency. It includes: SPI/INTEGRAL observations of 511 keV gamma rays; detection by the CHANDRA satellite of diffuse X-ray emission; COMPTEL observation of the flux of gamma rays in the 1-20 MeV range; finally, the so-called "WMAP- haze" which is a diffuse microwave excess observed from the core of our galaxy. If correct, our proposal identifies the nature of the dark matter, explains baryogenesis, and provides a means to directly probe the matter distribution in our Galaxy by analyzing several different types of diffuse emissions.

1. Introduction

Recent observations of the galactic centre have presented a number of puzzles for our current understanding of galactic structure and astrophysical processes. In particular a series of independent observations have detected an excess flux of photons across a broad range of energies. In particular, these observations include:

• SPI/INTEGRAL observations of the galactic centre have detected an excess of 511 keV gamma rays resulting from low momentum electron-positron annihilations. The observed intensity is a mystery. After accounting for known positron sources, only a small fraction of the emission may be explained[1 – 9].

• Detection by the CHANDRA satellite of diffuse X-ray emission from across the galactic bulge provides a puzzling picture: after subtracting known X-ray sources one finds a residual diffuse thermal X-ray emission consistent with a two-temperature plasma with the hot component close to $T \simeq 8\ keV$. According to[10] the hot component is very difficult to understand within the standard picture. Such a plasma would be too hot to be bound to the galactic center. The authors of ref.[10] also remark that the energy required to sustain a plasma of this temperature corresponds to the entire kinetic energy of one supernova every 3000 yr, which is unreasonably high.

• The flux of gamma rays in the 1-20 MeV range measured by COMPTEL represents yet another mystery. As discussed in[11] the best fit models for diffuse galactic γ rays fit the observed spectrum well for a very broad range of energies, 20 MeV- 100 GeV. It also gives a good representation of the latitude distribution of the emission from the plane to the poles, and of the longitudinal distribution. However, the model fails to explain the excess in the 1-20 MeV range observed by COMPTEL in the inner part of the galaxy ($l = 330^0 - 30^0, |b| = 0^0 - 5^0$), see below for details. As claimed in [11] some additional γ ray sources are required to explain this energy region.

• WMAP has detected an excess of GHz microwave radiation – dubbed the "WMAP- haze" – from the inner 20° core of our galaxy.[12–14]

• These data, when taken together, suggest the existence of an energy source beyond currently established astrophysical phenomenon. The main goal of this paper is to argue that these (seemingly unrelated) observations may be explained by a single mechanism.

2. Dark Matter as Dense Nuggets

A number of the outstanding cosmological mysteries—the natures of dark matter, baryogenesis, some unexplained sources of energy in different frequency bands, which includes, see above (511 keV line, X rays, 1-20 MeV continuum, microwave radiation) —may be explained by one and the same idea that dark matter consists of Compact Composite Objects (CCOs),[15–17] similar to Witten's strangelets.[18] The idea is that these form at the same QCD phase transition as conventional baryons (neutrons and protons), providing a natural explanation for the similar scales $\Omega_{DM} \approx 5\Omega_B$. Baryogenesis proceeds through a charge separation mechanism: both matter and antimatter nuggets form, but the natural CP violation of the so-called θ term in QCD[a] associated drives the formation of more antimatter nuggets

[a] If θ is non-zero, one must confront the so-called strong CP problem whe2reby some mechanism must be found to make the effective θ parameter extremely small today in accordance with measurements. One of the most natural resolutions is through a dynamical axion. Domain walls associated with this field (or ultimately, whatever mechanism resolves the strong CP problem) play an important role in forming these nuggets, and may play in important role in their ultimate stability. See[15,16] for details.

than matter nuggets, resulting in the left-over baryonic matter that forms visible matter today (see[16] for details). The idea that CP violation may be able to effectively drive charge separation may already have found experimental support through the Relativistic Heavy Ion Collider (RHIC) at Bookhaven.[19]

This mechanism requires no fundamental baryon asymmetry to explain the observed matter/antimatter asymmetry: $B_{\text{univ}} = B_{\text{DM}} + B_{\text{Visible}} - \bar{B}_{\text{DM}} = 0$ where $B_{\text{DM}}(\bar{B}_{\text{DM}})$ is the total (anti)baryon charge contained in the dark (anti)matter nuggets. From this and the observed relation $\Omega_{\text{DM}} \approx 5\Omega_{\text{B}}$ we have the approximate ratios $\bar{B}_{\text{DM}}:B_{\text{DM}}:B_{\text{Visible}} \simeq 3:2:1$.

Unlike conventional dark matter candidates, dark matter/antimatter nuggets will be strongly interacting, but macroscopically large objects. They do not contradict any of the many known observational constraints on dark matter or antimatter[17] for three reasons:

(1) They carry a huge (anti)baryon charge $|B| \approx 10^{20}$ - 10^{33}, so they have an extremely tiny number density. This explains why they have not been directly observed on earth. The local number density of dark matter particles with these masses is small enough that interactions with detectors are exceedingly rare and fall within all known detector and seismic constraints,[17] see also,[20][21] and references therein[b]
(2) They have nuclear densities, so their interaction cross-section is small $\sigma/M \approx 10^{-13}$ - 10^{-9} cm^2/g. This is well below typical astrophysical and cosmological limits which are on the order of $\sigma/M < 1$ cm^2/g. Dark-matter–dark-matter interactions between these nuggets are thus negligible.
(3) They have a large binding energy such that baryons in the nuggets are not available to participate in big bang nucleosynthesis (BBN) at $T \approx$ 1 MeV. In particular, we suspect that the baryons in these nuggets form a superfluid with a gap of the order $\Delta \approx 100$ MeVas this scale provides a natural explanation for the observed photon to baryon $n_B/n_\gamma \sim 10^{-10}$ which requires that formation of the nuggets stop at precisely $T_{\text{formation}} = 41$ MeV.[16,22] At temperatures below this, incident baryons

---

[b]It has been estimated in[21] that the abundance of the nuggets with mass ~ 10 kg (which corresponds to $B \sim 10^{28}$) to one ton (which corresponds to $B \sim 10^{30}$) must have at least an order of magnitude less than would saturate the dark matter in the solar neighborhood. Our remark here is that such a constraint comes from a very specific assumption on the nugget-mass distribution assumed in.[21] While our preferable range of B is somewhat smaller, in range of $B \sim (10^{25} - 10^{27})$ we still believe that $B \geq 10^{28}$ is not completely excluded by Apollo data.[21]

with energies below the gap would Andreev reflect rather than become incorporated into the nugget.

Thus, on large scales, the nuggets are sufficiently dilute that they behave as standard collisionless cold dark matter (CCDM). When the number densities of both dark and visible matter become sufficiently high, however, dark-antimatter–visible-matter collisions may release significant radiation and energy.

The main goal of this talk is to demonstrate that these observations suggest a common origin for a number of different emissions: the 511 keV radiation, excess of the 1-20 MeV diffuse γ radiation, diffuse keV X-ray emission and WMAP haze.

Assuming dark antimatter nuggets to be the common source, we extract some phenomenological parameters describing their properties. We show that dark antimatter is consistent with all of these data, and that it may fully explain the missing sources of emission without requiring the invention of new non-baryonic fields for dark matter. Finally, we argue that the spectrum of the microwave emission should be independent of position, and that there should be a spatial correlation between the various emissions and the distributions of dark and visible matter. These concrete predictions should allow our proposal to be verified or ruled out by observations and analysis in the near future. We postulate a few basic properties:

A.1. The antimatter nuggets provide a virtually unlimited source of positrons (e^+) such that impinging electrons (e^-) will readily annihilate at their surface through the formation of positronium.[23,24] About a quarter of the positronium annihilations release back-to-back 511 keV photons. If this reaction occurs on the surface, on average one of these photons will be absorbed by the nugget while the other will be released.

A.2. The nuggets provide a significant source of anti-baryonic matter such that impinging protons will annihilate. We assume that the proton annihilation rate is directly related to that of electrons through a suppression factor $f < 1$ as discussed in.[25] Proton annihilation events will release about $2m_p \approx 2$ GeV of energy per event and will occur close to the surface of the nugget creating a hot spot that will mainly radiate X-ray photons with keV energies. The rest of the surface will radiate photons with frequencies $\omega \leq$ eV, see next item.

A.3. As argued in[25] a considerable portion of the available energy from these proton annihilations will be thermalized inside the bulk of the nuggets. This energy will be eventually emitted as the photon emission in the broad frequency bands $\omega \leq$ eV such that the low energy tail of this emission is in microwave band and can be analyzed by precise WMAP measurements. While we predict that the very same mechanism leads to the diffuse emission in eV range as well (which corresponds to the visible and infrared bands), it is not likely to be observable due to much more intense sources in these energy bands.

This allows us to use the 511 keV observations to estimate the electron annihilation rate and the *Chandra* flux to estimate the proton annihilation rate. Normalization to the *Chandra* flux will allow us to estimate the diffuse WMAP haze intensity.

3. Proposal:

The primary feature of our proposal is that the antimatter nuggets provide a single annihilation target for both electrons and protons. As a result, all four: the 511 keV emission from electron annihilation, 1-20 MeV diffuse emission, the X-ray emission from proton annihilation concentrated in the hot spots (regions close to the surface of the nugget where annihilation took place) and microwave emission from proton annihilation from entire surface of the nuggets should originate from the same regions of space, with the local rates of annihilation proportional to the product of the number densities of visible and dark matter $n_V(\vec{r})n_D(\vec{r})$.

By comparing several different observational sources, one may remove the dependence on the dark and visible matter distributions because all observations integrate the radiation along the same line of sight from the earth to the core of the galaxy. We make the approximation that the relevant line-of-sight average is the same for all sets of data we consider. As a result, direct comparisons between the data provide non-trivial insights about the properties and nature of dark antimatter, independent of the matter distributions.

We should remark here that it is quite non-trivial observation that so different frequency bands have very similar (related) energetic budgets. It has very natural explanation within our framework when dark matter is presented in the form of macroscopical matter/antimatter nuggets. However, this "coincidense" of the observed excesses of the 511 keV line,[23,24] the diffuse X-ray emission,[25] the diffuse 1-20 MeV γ emission,[26] and diffuse microwave emissions[27] would be very difficult to explain by some other mechanisms which are not internally related. Observation of the morphological correlations between these very different emissions would be very strong evidence supporting our proposal. This is the main goal of the present talk: look for the correlations between so different emissions! In the next two sections we shall concentrate on two specific applications of our proposal: excess in 1-20 MeV emission[26] and WMAP haze.[27]

4. Proposal to Explain 1-20 MeV emission

As suggested previously in refs,[23 24] the 511 keV line can be naturally explained as a result of positronium formation when a non-relativistic electron (e^-) hits the antimatter nugget surrounded by positrons (e^+) with chemical potential μ. A certain fraction of galactic electrons incident on the CCO will annihilate directly $e^+e^- \to 2\gamma$ (rather than participate in resonance positronium formation) resulting in the creation of photons of energy greater than 511keV with a maximum energy up to $\simeq \mu$. The corresponding fraction of electrons obviously depends on lepton's chemical potential μ and on specific properties of the nugget's surface structure and the resulting distribution of positrons. It is not the goal of the present paper to calculate the corresponding fraction based on a specific model. Instead, we shall introduce this ratio as theoretically unknown phenomenological parameter χ which will be fixed to match the observational data. However, the spectrum is not sensitive to value of χ and is predicted by the proposal.

It will be demonstrated that radiation arising from this DM model can account for the broad spectrum across the 1-20 MeV range observed by COMPTEL. Anticipating this conclusion, we should mention here, that it is quite remarkable that the maximum energy where the excess has been observed by COMPTEL (~ 20 MeV) coincides with a typical estimation for μ in quark matter,[28 29] As we shall see below, the maximum photon energy within our mechanism exactly coincides with the lepton chemical potential in the nugget $\mu \sim 20$ MeV. We stress here that this energy scale arises naturally from the properties of quark matter and has not been introduced in order to fit with the COMPTEL observations. This is very robust prediction of our mechanism which is not sensitive to the specific details of the nugget's structure nor to DM and visible matter distributions.

Below we shall present the sketch of our calculations. We refer to the original paper[26] for details. We treat the positrons at the CCO surface as a non-interacting Fermi gas of chemical potential μ. The density of states in the momentum range p to $p + dp$ is then given by, $dn(p) = \frac{2d^3p}{(2\pi)^3}$ so that the rate of direct electron-positron annihilation resulting in a photon of momentum k and involving a positron of momentum p is given by,

$$\frac{dI(k,\mu)}{dkdt} = \int dn(p)v(p)\frac{d\sigma(p,k)}{dk} = \int \frac{2d^3(p)}{(2\pi)^3}\frac{p}{E}\frac{d\sigma(p,k)}{dk}, \tag{1}$$

where $\frac{d\sigma(p,k)}{dk}$ is the electron rest frame cross section for direct electron-positron annihilation resulting in a photon of momentum k. In this formula we assume that incoming electron has a velocity v_e well below that of a typical positron $(v(p))$ Within the nugget. Therefore, calculations are carried out in the rest frame of the incident electron, i.e. $v_e = 0$. By comparing the flux for the 511 keV line with the flux in the 1-20 MeV range, one may remove the dependence on the dark and visible matter distributions because, provided they have a common origin, the radiation for both fluxes should be integrated along almost the same line of sight from the earth to the core of the galaxy. As a result, direct comparisons between the data provides non-trivial insight into the properties of the nuggets, independent of the matter distributions. Therefore, we can avoid the corresponding uncertainties related to $\rho_{DM}(r), \quad \rho_B(r)$ as well as uncertainties related to typical sizes of the nuggets, their size-distribution etc by normalizing the spectrum of these 1-20 MeV photons using the well-measured intensity of the 511 keV line with an average flux observed to be $\frac{d\Phi}{d\Omega} \simeq 0.025$ photons $\mathrm{cm}^{-2}\mathrm{s}^{-1}\mathrm{sr}^{-1}$ coming from a circle of half angle 6^{02} from the inner part of the galaxy. This region strongly overlaps with the region of interests where COMPTEL data are available, $(l = 330^0 - 30^0, |b| = 0^0 - 5^0)$. In what follows we neglect any differences resulting from the slightly different lines of sight for measurements by INTEGRAL and COMPTEL.

In addition, we introduce the coefficient χ as the ratio of electrons which experience direct $e^+e^- \to 2\gamma$ annihilation in comparison with the number of electrons which experience resonance positronium formation. Following this procedure one arrives at an expression for the flux of 1-20 MeV photons from the bulge of the galaxy normalized to the 511 keV line as described above,

$$\frac{d\Phi(k)}{d\Omega dk} = 0.025 \cdot \frac{4\chi}{\mathrm{MeV} \cdot \mathrm{s} \cdot \mathrm{cm}^2 \cdot \mathrm{sr}} \cdot \left(\frac{2}{k^2\mu^2 \ln\left(\frac{\mu}{m}\right)} \right) \tag{2}$$

$$\times \left[k(k^2 + 2mk - 2m^2) \ln \left(\frac{(2k-m)(\mu+m-k)}{mk} \right) \right.$$

$$-\frac{3}{2}k^3 - (\mu + 5m)k^2 + (\frac{1}{2}\mu^2 + 3\mu m + \frac{9}{2}m^2)k - m^2(\mu + m)$$

$$\left. + \frac{k^2m^2}{\mu + m - k} + (8k^4 - 8mk^3 - \frac{5}{2}m^2k^2 + 4m^3k - m^4)\frac{k}{(2k-m)^2} \right]$$

where we have taken into account that the total number of positroniums formed is 4 times the number of positroniums in the $^1S_0$ state emitting 511 keV photons with the flux $\frac{d\Phi}{d\Omega} \simeq 0.025$ photons $\mathrm{cm}^{-2}\mathrm{s}^{-1}\mathrm{sr}^{-1}$. In eq.(2) mass, m, photon energy, k and chemical potential, μ are all measured in MeV units.

This normalization allows us to analyze the MeV spectra without a precise model of dark matter /visible matter distributions within the galactic bulge, assuming of course that both emissions (511 keV line and 1-20 MeV photons) come from the same source, the antimatter nuggets. The results of the calculations are presented on Figure 1. below along with gamma-ray spectrum of inner galaxy for optimized model.[30] Our mechanism can easily explain an excess measured by COMPTEL if the values of μ within the physically relevant range of tens of MeV and parameter $\chi \sim 0.1$.

5. Explaining the WMAP Haze

In our proposal, the matter annihilates on antimatter nuggets. The nuggets then radiate this energy over a wide range of frequencies. The model thus makes definite predictions relating these emissions: they should have similar morphologies, and the relative intensities should be related by an overall energy budget determined by the local annihilation rate. Four types of emission are from "hot spots" at the annihilation sites, and should be observable from the core of our galaxy:

B.1. Electron annihilations through positronium produce a well defined 511 keV emission[23,24] that is consistent with, and could possibly explain the puzzling diffuse 511 keV emission observed by SPI/INTEGRAL.

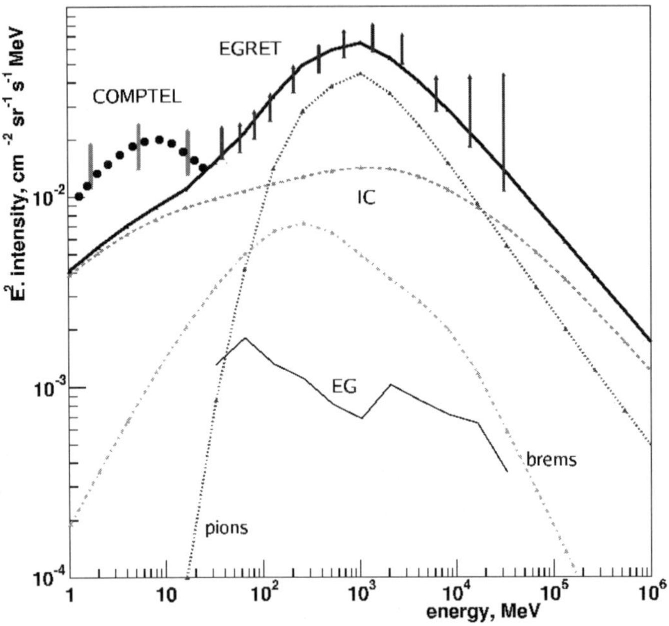

Fig. 1. Gamma- ray spectrum of inner galaxy for optimized model. Green vertical bars: COMPTEL data. Heavy solid line: total calculated flux for optimized model. Heavy black dots: Combination of calculated emission spectrum from electron-nugget annihilation and optimized model.

B.2. Direct electron annihilation can also produce emission in the 1-20 MeV band[26] which is consistent with, and could explain part of the diffuse gamma-ray emissions observed by COMPTEL.

B.3. Proton annihilation produces keV X-ray emission from a hot spot at the annihilation site[25] that is consistent with, and could possibly explain the puzzling diffuse X-ray emissions observed by *Chandra*.

B.4. Proton annihilation occasionally produces GeV photons[25] that are consistent with, and could partially account for the gamma-ray emissions observed by EGRET.

All of these emissions are "direct" in the sense that the timescale for the emission is much shorter than the time between successive annihilations. Thus, the intensity of these emissions depends only on the rate of annihilation events which is proportional to $n_{VM}(r)n_{DM}(r)$ – the product of the local visible and dark matter distributions at the annihilation site. The emitted spectrum is also independent of the local density. We emphasize that the model makes two the highly non-trivial predictions: 1) that the morphology of these emissions is very strongly correlated, and 2) that the spectral properties of these emissions are *independent of position.*

A comparison between observations of the "direct" emissions B.1 through B.4 along the same line-of-sight is possible because the local emission depends only on the local rate of annihilation $\phi(\vec{r}) \propto n_{VM}(\vec{r})$. The observed flux thus depends on the uncertain matter distribution through the same line-of-sight integral:

$$\Phi_{511,\text{X-ray,etc.}} \propto \int \mathrm{d}\Omega \mathrm{d}l \, n_{VM}(l) n_{DM}(l), \tag{3}$$

which cancels when comparing emissions from the same position in the sky.

There is an additional emission from the nuggets:

B.5. Energy not directly released through one of the mechanisms B.1-B.4 heats the nuggets, ultimately being thermally radiated.

As discussed in details in[27] the heat-capacity and energy budget ensure that the nuggets have a well-defined temperature scale of $T \sim 1eV$ in the core of the galaxy. The resulting thermal bremsstrahlung emission thus "averages" the annihilations over time, and the resulting emissivity and spectrum will depend on the temperature $T(n_{VM})$ which is a function of the local visible matter density. Finally, in principle, we may compare the *total* thermal emission with the "direct" emissions because thermal equilibrium relates the rate of total emission to the rate of annihilation, both of which are proportional to $n_{VM}(vectr)$. In practise, however, the thermal eV scale emission cannot be seen against the bright stellar background.

In,[25] the "direct" emissions B.1, B.3, and B.4 were compared, showing that our proposal is consistent with the current observations, and using the observations to constrain the properties of the nuggets. In particular, two parameters were introduced describe complicated properties of the nuggets: The parameter $f < 1$ was introduced to describe the suppression of the proton annihilation rate with respect to the electron annihilation rate, and the parameter $g < 1/2$ was introduced to describe the fraction of the proton annihilation energy that is directly released as X-rays. We emphasize that these parameters are not free, but they depend on detailed models of the nuggets and are beyond the reach of present day calculational techniques.

By hypothesizing that emissions from the nuggets completely explain the puzzling 511 keV (B.1) and diffuse X-ray (B.3) emissions, one obtains $fg \sim 6 \cdot 10^{-3}$, which is consistent with the theoretical estimates, and provides a non-trivial test of the theory. The thermal energy input into the antimatter nuggets is the complementary fraction, $1-g$ of the total proton annihilation energy not directly released as X-rays. Thus, if we use the observed X-ray flux $\Phi_{Chandra}$ to provide the energy normalization, then the total thermal emission will be approximately

$$\Phi_T \approx \frac{1-g}{g}\Phi_{Chandra}. \tag{4}$$

The total thermal emission Φ_T may then be used to estimate the observed microwave emission in a specified frequency band giving us the relationship,[27]

$$\frac{\mathrm{d}\Phi_{\text{WMAP}}}{\mathrm{d}\omega} \approx \frac{2}{T_{\text{eff}}}\frac{1-g}{g}\Phi_{Chandra}. \tag{5}$$

Observationally, *Chandra* observes a total flux[10]

$$\Phi_{Chandra} \approx 2\times 10^{-6}\frac{\text{erg}}{\text{cm}^2\cdot\text{s}\cdot\text{sr}} \tag{6a}$$

while the WMAP haze flux is[12–14]

$$\frac{\mathrm{d}\Phi_{\text{WMAP}}}{\mathrm{d}\omega} = (3\text{ - }6)\frac{\text{kJy}}{\text{sr}} \approx \frac{(3\text{ - }6)\times 10^{-20}\text{erg}}{\text{cm}^2\cdot\text{s}\cdot\text{sr}\cdot\text{Hz}}. \tag{6b}$$

Combining these, and converting 1 Hz $\approx 4\times 10^{-15}$ eV, we predict that the observed WMAP haze intensity will be saturated by thermal anti-nugget emission if the parameters which enter in our estimate satisfy the following constraint

$$\frac{\text{eV}}{T_{\text{eff}}}\cdot\frac{1-g}{g} \approx (2\text{ - }4). \tag{7}$$

Although this relationship is only approximate, it is quite amazing that it is satisfied (without any adjustment) if our previous estimate for $T_{\text{eff}} \sim 1$ eV is used. Thus, the non-trivial relationship (7) – which depends strongly on the observed intensity of the GHz WMAP haze – is automatically satisfied.

6. Conclusion

Our dark matter proposal not only explains many astrophysical and cosmological puzzles, but makes definite predictions about the correlations of the dark and visible matter distributions $n_V n_D$ with 511 keV, X-ray, gamma-ray in 1-20 MeV range and microwave emissions. In addition, it makes the definite prediction that the spectra of the emissions should be virtually independent of the local environment. Such correlations and spectral properties would be very difficult to account for with other dark matter candidates. Future observations may thus confirm or rule out this theory. If confirmed, it would provide a key for many cosmological and astrophysical secrets, and finally unlock nature of dark matter.

Acknowledgements

I am thankful to Michael Forbes and Kyle Lawson for the collaboration on the project. This work was supported in part by the National Science and Engineering Research Council of Canada.

References

1. Jurgen Knödlseder, V. Lonjou, Pierre Jean, M. Allain, P. Mandrou, J.-P. Roques, G.K. Skinner, G. Vedrenne, P. von Ballmoos, G. Weidenspointner, P. Caraveo, B. Cordier, V. Schönfelder, and B.J. Teegarden. Early SPI/INTEGRAL contraints on the morphology of the 511 keV line emission in the 4th galactic quadrant. *Astron. Astrophys.*, 411:L457–L460, 2003, arXiv:astro-ph/0309442.
2. Pierre Jean et al. Early SPI/INTEGRAL measurements of galactic 511 keV line emission from positron annihilation. *Astron. Astrophys.*, 407:L55, 2003, arXiv:astro-ph/0309484.
3. Jurgen Knödlseder, Pierre Jean, V. Lonjou, G. Weidenspointner, N. Guessoum, W. Gillard, G. Skinner, P. von Ballmoos, G. Vedrenne, J.-P. Roques, S. Schanne, B. Teegarden, V. Schönfelder, and C. Winkler. The all-sky distribution of 511-keV electron positron annihilation emission. *Astron. Astrophys.*, 441:513–532, 2005, arXiv:astro-ph/0506026.
4. Pierre Jean, Jurgen Knödlseder, W. Gillard, N. Guessoum, K. Ferriere, A. Marcowith, V. Lonjou, and J. P. Roques. Spectral analysis of the galactic e^+e^- annihilation emission. *Astron. Astrophys.*, 445:579–589, 2006, arXiv:astro-ph/0509298.
5. Hasan Yuksel. Positron annihilations at the galactic center: Generating more questions than answers. 2006, arXiv:astro-ph/0609139.
6. John F. Beacom and Hasan Yuksel. Stringent constraint on galactic positron production. *Phys. Rev. Lett.*, 97:071102, 2006, arXiv:astro-ph/0512411.
7. George Weidenspointner, C. R. Shrader, Jurgen Knödlseder, Pierre Jean, V. Lonjou, N. Guessoum, R. Diehl, W. Gillard, M.J. Harris, G.K. Skinner, P. von Ballmoos, G. Vedrenne, J.-P. Roques, S. Schanne, P. Sizun, B.J. Teegarden, V. Schoenfelder, and C. Winkler. The sky distribution of positronium annihilation continuum emission measured with SPI/INTEGRAL. *Astron. Astrophys.*, 450:1013–1021 2006, arXiv:astro-ph/0601673.
8. Celine Boehm, Dan Hooper, Joseph Silk, Michel Casse, and Jacques Paul. MeV dark matter: Has it been detected? *Phys. Rev. Lett.*, 92:101301, 2004, arXiv:astro-ph/0309686.
9. Dan Hooper et al. MeV dark matter in dwarf spheroidals: A smoking gun? *Phys. Rev. Lett.*, 93:161302, 2004, arXiv:astro-ph/0311150.
10. Michael P. Muno et al. Diffuse x-ray emission in a deep chandra image of the galactic center. *Astrophys. J.*, 613:326–342, 2004, arXiv:astro-ph/0402087.

11. A. W. Strong, I. V. Moskalenko and O. Reimer, Astrophys. J. **537**, 763 (2000) [Erratum-ibid. **541**, 1109 (2000)] [arXiv:astro-ph/9811296], A. W. Strong, I. V. Moskalenko and O. Reimer, Astrophys. J. **613**, 962 (2004) [arXiv:astro-ph/0406254], A. W. Strong, H. Bloemen, R. Diehl, W. Hermsen and V. Schoenfelder, Astrophys. Lett. Commun. **39**, 209 (1999) [arXiv:astro-ph/9811211].
12. D. P. Finkbeiner, Astrophys. J. **614**, 186 (2004) [arXiv:astro-ph/0311547].
13. D. P. Finkbeiner, G. I. Langston and A. H. Minter, Astrophys. J. **617**, 350 (2004) [arXiv:astro-ph/0408292].
14. D. P. Finkbeiner, arXiv:astro-ph/0409027.
15. Ariel R. Zhitnitsky. 'Nonbaryonic' dark matter as baryonic color superdonductor. *JCAP*, 0310:010, 2003, arXiv:hep-ph/0202161.
16. David H. Oaknin and Ariel Zhitnitsky. Baryon asymmetry, dark matter and quantum chromodynamics. *Phys. Rev. D*, 71:023519, 2005, arXiv:hep-ph/0309086.
17. Ariel Zhitnitsky. Cold Dark Matter as Compact Composite Objects. *Phys. Rev.*, D74:043515, 2006, arXiv:astro-ph/0603064.
18. Edward Witten. Cosmic separation of phases. *Phys. Rev. D*, 30:272–285, 1984.
19. D. Kharzeev and A. Zhitnitsky. Charge separation induced by P-odd bubbles in QCD matter. *Nucl. Phys.*, A797:67–79, 2007, arXiv:0706.1026 [hep-ph].
20. E. S. Abers, A. K. Bhatia, D. A. Dicus, W. W. Repko, D. C. Rosenbaum and V. L. Teplitz, arXiv:0712.4300 [astro-ph].
21. E. T. Herrin, D. C. Rosenbaum and V. L. Teplitz, Phys. Rev. D **73**, 043511 (2006) [arXiv:astro-ph/0505584].
22. Edward W. Kolb and Michael S. Turner. *The Early Universe.* Westview Press, 5500 Central Avenue, Boulder, Colorado, 80301, 1994.
23. David H. Oaknin and Ariel R. Zhitnitsky. 511-keV photons from color superconducting dark matter. *Phys. Rev. Lett.*, 94:101301, 2005, arXiv:hep-ph/0406146.
24. A. Zhitnitsky, "The Width of the 511 KeV Line from the Bulge of the Galaxy," Phys. Rev. D **76**, 103518 (2007) [arXiv:astro-ph/0607361].
25. M. M. Forbes and A. R. Zhitnitsky, "Diffuse X-rays: Directly Observing Dark Matter?", to appear in JCAP, January 2008. arXiv:astro-ph/0611506.
26. Kyle Lawson and Ariel R. Zhitnitsky. Diffuse cosmic gamma-rays at 1-20 MeV: A trace of the dark matter? 2007, arXiv:0704.3064 [astro-ph], to appear in JCAP, January 2008.
27. M. M. Forbes and A. R. Zhitnitsky, arXiv:0802.3830 [astro-ph].
28. C. Alcock, E. Farhi and A. Olinto, Astrophys. J. **310**, 261 (1986).
29. Jes Madsen. Color-flavor locked strangelets. *Phys. Rev. Lett.*, 87:172003, 2001, arXiv:hep-ph/0108036.
30. I. V. Moskalenko *et al*, [arXiv:0704.1328];
Andrew W. Strong, Igor V. Moskalenko, and Olaf Reimer. *Astrophys. J.*, 613:962–976, 2004, arXiv:astro-ph/0406254.

SECTION 8.

GLIMPSES OF THE CONFERENCE

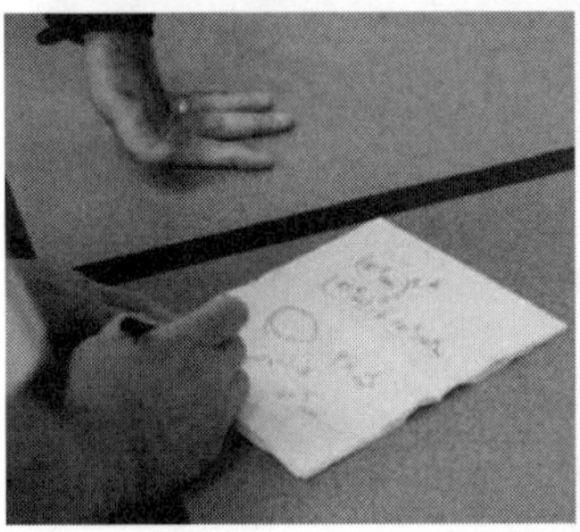

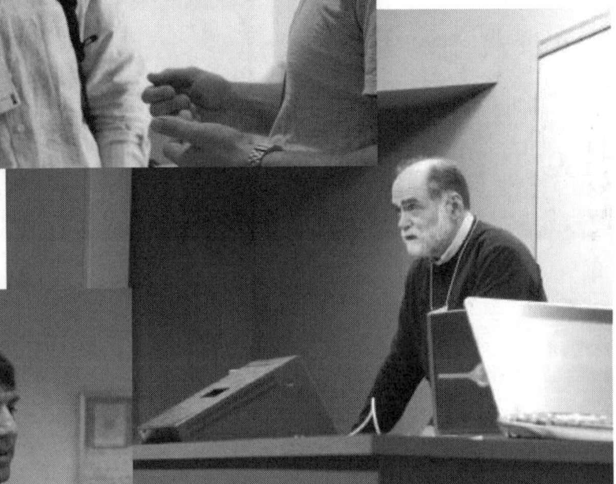

Mathematics of Molecular
and Cellular Biology

BEVERAGES

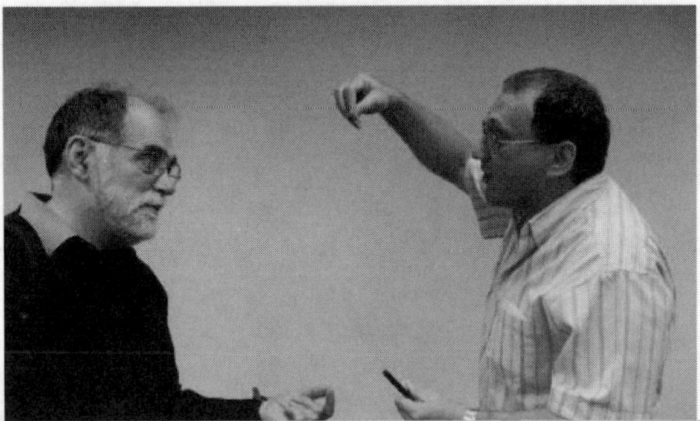

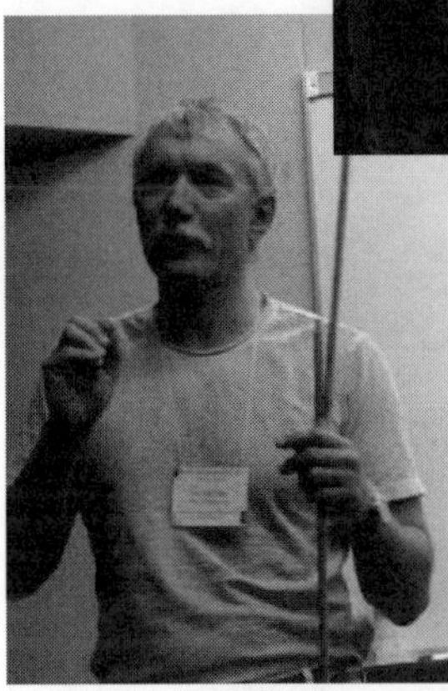

BEVERAGES

by Richard !

LIST OF THE PARTICIPANTS OF CAQCD-08

Roberto Auzzi, Swansea University
Ian Balitsky, ODU/Jefferson Lab
Benjamin Basso, Universite Paris Sud
Brian Batell, University of Minnesota
Zvi Bern, UCLA
Stefano Bolognesi, University of Minnesota
Eric Braaten, Ohio State University
Michal Czakon, Wuerzburg University
Andrzej Czarnecki, University of Alberta
Sergey Dubynskiy, University of Minnesota
Gerald Dunne, University of Connecticut
Minoru Eto, University of Pisa
Nick Evans, University of Southampton
Hilmar Forkel, University of Heidelberg
Gregory Gabadadze, New York University
Tony Gherghetta, University of Minnesota
Joel Giedt, Rensselaer Polytechnic Institute
Jose Goity , Hampton University/Jefferson Lab
Sven Bjarke Gudnason, INFN, Pisa
Richard Hill, Fermilab
Yuichi Hoshino, Kushiro National College of Technology
Tom Kelley, University of Minnesota
Ingo Kirsch, ETH Zurich
Kenichi Konishi, University of Pisa and INFN, Pisa
Johann Kuehn, University of Karlsruhe
Richard Lebed, Arizona State University
Frieder Lenz , Universitaet Erlangen
Alexander Lenz, Universitaet Regensburg
Juan Maldacena, Institute for Advanced Study
Larry McLerran, Brookhaven National Lab
Yannick Meurice, University of Iowa

Sven-Olaf Moch, DESY
Robert Myers, Perimeter Institute
Carlos Nunez, Swansea University
Keith Olive, University of Minnesota
Leopoldo Pando Zayas, University of Michigan
Andrey Parnachev, SUNY
Marco Peloso, University of Minnesota
Alexey Petrov, Wayne State University &
MCTP, University of Michigan
Jan Piclum, University of Alberta
Erich Poppitz, University of Toronto
Adam Ritz, University of Victoria
Serge Rudaz, University of Minnesota
Gregory Ruffa, University of Minnesota
Roger Rusack, University of Minnesota
Veronica Sanz , Boston University
Thomas Schaefer, North Carolina State University
Mikhail Shifman, University of Minnesota
Kim Splittorff, The Niels Bohr Institute
Matthias Steinhauser, University of Karlsruhe
George Sterman, SUNY Stony Brook
Shigeki Sugimoto, IPMU, University of Tokyo
Bayram Tekin, Middle East Technical University
Mithat Unsal, SLAC, Stanford University
Arkady Vainshtein, University of Minnesota
Diana Vaman, University of Virginia
Jacobus Verbaarschot, Stony Brook University
Walter Vinci, INFN, Pisa
Mikhail Voloshin, University of Minnesota
John Wittig, University of Cincinnati
Alberto Zaffaroni, Universita' di Milano-Bicocca
Ariel Zhitnitsky, University of British Columbia